**Nitride Semiconductor Devices:
Principles and Simulation**

Edited by
Joachim Piprek

1807–2007 Knowledge for Generations

Each generation has its unique needs and aspirations. When Charles Wiley first opened his small printing shop in lower Manhattan in 1807, it was a generation of boundless potential searching for an identity. And we were there, helping to define a new American literary tradition. Over half a century later, in the midst of the Second Industrial Revolution, it was a generation focused on building the future. Once again, we were there, supplying the critical scientific, technical, and engineering knowledge that helped frame the world. Throughout the 20th Century, and into the new millennium, nations began to reach out beyond their own borders and a new international community was born. Wiley was there, expanding its operations around the world to enable a global exchange of ideas, opinions, and know-how.

For 200 years, Wiley has been an integral part of each generation's journey, enabling the flow of information and understanding necessary to meet their needs and fulfill their aspirations. Today, bold new technologies are changing the way we live and learn. Wiley will be there, providing you the must-have knowledge you need to imagine new worlds, new possibilities, and new opportunities.

Generations come and go, but you can always count on Wiley to provide you the knowledge you need, when and where you need it!

William J. Pesce
President and Chief Executive Officer

Peter Booth Wiley
Chairman of the Board

Nitride Semiconductor Devices

Principles and Simulation

Edited by
Joachim Piprek

WILEY-VCH Verlag GmbH & Co. KGaA

The Editor
Joachim Piprek
NUSOD Institute, Newark, USA
e-mail: piprek@nusod.org

Cover illustration
Transient temperature change within a GaN-based vertical-cavity surface-emitting laser diode (FEM simulation by J. Piprek)

All books published by Wiley-VCH are carefully produced. Nevertheless, authors, editors, and publisher do not warrant the information contained in these books, including this book, to be free of errors. Readers are advised to keep in mind that statements, data, illustrations, procedural details or other items may inadvertently be inaccurate.

Library of Congress Card No.:
applied for

British Library Cataloguing-in-Publication Data
A catalogue record for this book is available from the British Library.

Bibliographic information published by the Deutsche Nationalbibliothek
The Deutsche Nationalbibliothek lists this publication in the Deutsche Nationalbibliografie; detailed bibliographic data are available in the Internet at <http://dnb.d-nb.de>.

© 2007 WILEY-VCH Verlag GmbH & Co. KGaA, Weinheim

All rights reserved (including those of translation into other languages). No part of this book may be reproduced in any form – by photoprinting, microfilm, or any other means – nor transmitted or translated into a machine language without written permission from the publishers. Registered names, trademarks, etc. used in this book, even when not specifically marked as such, are not to be considered unprotected by law.

Typesetting Uwe Krieg, Berlin
Printing Strauss GmbH, Mörlenbach
Binding Litges & Dopf GmbH, Heppenheim
Wiley Bicentennial Logo Richard J. Pacifico

Printed in the Federal Republic of Germany
Printed on acid-free paper

ISBN: 978-3-527-40667-8

Contents

Preface *XV*

List of Contributors *XVII*

Part 1 Material Properties *1*

1 Introduction *3*
Joachim Piprek
1.1 A Brief History *3*
1.2 Unique Material Properties *4*
1.3 Thermal Parameters *5*
   References *10*

2 Electron Bandstructure Parameters *13*
Igor Vurgaftman and Jerry R. Meyer
2.1 Introduction *13*
2.2 Band Structure Models *14*
2.3 Band Parameters *17*
2.3.1 GaN *19*
2.3.2 AlN *25*
2.3.3 InN *28*
2.3.4 AlGaN *30*
2.3.5 InGaN *32*
2.3.6 InAlN *34*
2.3.7 AlGaInN *34*
2.3.8 Band Offsets *35*
2.4 Conclusions *36*
   References *37*

Nitride Semiconductor Devices: Principles and Simulation. Joachim Piprek (Ed.)
Copyright © 2007 WILEY-VCH Verlag GmbH & Co. KGaA, Weinheim
ISBN: 978-3-527-40667-8

3 Spontaneous and Piezoelectric Polarization: Basic Theory vs. Practical Recipes 49
Fabio Bernardini
3.1 Why Spontaneous Polarization in III-V Nitrides? 49
3.2 Theoretical Prediction of Polarization Properties in AlN, GaN and InN 51
3.3 Piezoelectric and Pyroelectric Effects in III-V Nitrides Nanostructures 54
3.4 Polarization Properties in Ternary and Quaternary Alloys 58
3.5 Orientational Dependence of Polarization 64
References 67

4 Transport Parameters for Electrons and Holes 69
Enrico Bellotti and Francesco Bertazzi
4.1 Introduction 69
4.2 Numerical Simulation Model 70
4.2.1 Scattering in the Semi-Classical Boltzmann Equation 72
4.3 Analytical Models for the Transport Parameters 76
4.4 GaN Transport Parameters 79
4.4.1 Electron Transport Coefficients 79
4.4.2 Hole Transport Coefficients 81
4.5 AlN Transport Parameters 84
4.5.1 Electron Transport Coefficients 84
4.5.2 Hole Transport Coefficients 86
4.6 InN Transport Parameters 87
4.6.1 Electron Transport Coefficients 88
4.6.2 Hole Transport Coefficients 89
4.7 Conclusions 91
References 91

5 Optical Constants of Bulk Nitrides 95
Rüdiger Goldhahn, Carsten Buchheim, Pascal Schley, Andreas Theo Winzer, and Hans Wenzel
5.1 Introduction 95
5.2 Dielectric Function and Band Structure 95
5.2.1 Fundamental Relations 95
5.2.2 Valence Band Ordering, Optical Selection Rules and Anisotropy 97
5.3 Experimental Results 99
5.3.1 InN 99
5.3.2 GaN and AlN 102
5.3.3 AlGaN Alloys 105
5.3.4 In-rich InGaN and InAlN Alloys 107

5.4	Modeling of the Dielectric Function	*108*
5.4.1	Analytical Representation of the Dielectric Function	*109*
5.4.2	Calculation of the Dielectric Function for Alloys	*111*
5.4.3	Influence of Electric Fields on the Dielectric Function	*112*
	References	*114*

6 Intersubband Absorption in AlGaN/GaN Quantum Wells *117*
Sulakshana Gunna, Francesco Bertazzi, Roberto Paiella, and Enrico Bellotti

6.1	Introduction	*117*
6.2	Theoretical Model	*118*
6.2.1	Spontaneous and Piezoelectric Polarization	*122*
6.3	Numerical Implementation	*123*
6.3.1	Achieving Self-consistency: The Under-Relaxation Method	*127*
6.3.2	Predictor–Corrector Approach	*128*
6.4	Absorption Energy in AlGaN-GaN MQWs	*129*
6.4.1	Numerical Analysis of Periodic AlGaN-GaN MQWs	*130*
6.4.2	Numerical Analysis of Non-periodic AlGaN-GaN MQWs and Comparison with Experimental Results	*138*
6.5	Conclusions	*141*
	References	*142*

7 Interband Transitions in InGaN Quantum Wells *145*
Jörg Hader, Jerome V. Moloney, Angela Thränhardt, and Stephan W. Koch

7.1	Introduction	*145*
7.2	Theory	*146*
7.2.1	Bandstructure and Wavefunctions	*146*
7.2.2	Semiconductor Bloch Equations	*149*
7.2.3	Semiconductor Luminescence Equations	*151*
7.2.4	Auger Recombination Processes	*152*
7.3	Theory–Experiment Gain Comparison	*154*
7.4	Absorption/Gain	*156*
7.4.1	General Trends	*156*
7.4.2	Structural Dependence	*159*
7.5	Spontaneous Emission	*161*
7.6	Auger Recombinations	*164*
7.7	Internal Field Effects	*164*
7.8	Summary	*166*
	References	*167*

8		**Electronic and Optical Properties of GaN-based Quantum Wells with (10$\bar{1}$0) Crystal Orientation** *169*
		Seoung-Hwan Park and Shun-Lien Chuang
8.1		Introduction *169*
8.2		Theory *170*
8.2.1		Non-Markovian gain model with many-body effects *175*
8.3		Results and Discussion *177*
8.4		Summary *188*
		References *189*
9		**Carrier Scattering in Quantum-Dot Systems** *191*
		Frank Jahnke
9.1		Introduction *191*
9.2		Scattering Due to Carrier–Carrier Coulomb Interaction *193*
9.2.1		Formulation of the Problem and Previous Developments *193*
9.2.2		Kinetic Equation and Scattering Rates *195*
9.2.3		Results for Carrier–Carrier Scattering *198*
9.3		Scattering Due to Carrier–Phonon Interaction *200*
9.3.1		Perturbation Theory Versus Polaron Picture *200*
9.3.2		Polaron States and Kinetics *202*
9.3.3		Results for Carrier Scattering Due to LO-phonons *204*
9.4		Summary and Outlook *207*
		References *208*

Part 2   Devices *211*

10		**AlGaN/GaN High Electron Mobility Transistors** *213*
		Tomás Palacios and Umesh K. Mishra
10.1		Introduction *213*
10.2		Physics-based Simulations *216*
10.2.1		Basic Material Properties *217*
10.2.2		Polarization *218*
10.2.3		Surface States *222*
10.2.4		Electron Mobility *224*
10.2.5		Breakdown Voltage *227*
10.2.6		Energy Balance Models *228*
10.3		Conclusions *230*
		References *231*
11		**Intersubband Optical Switches for Optical Communications** *235*
		Nobuo Suzuki
11.1		Introduction *235*

11.2	Physics of ISBT in Nitride MQWs 236
11.2.1	Dipole Moment 236
11.2.2	Rate Equations 236
11.2.3	Absorption 238
11.2.4	Relaxation Time 239
11.2.5	Dephasing Time and Spectral Linewidth 240
11.3	Calculation of Absorption Spectra 242
11.3.1	Transition Wavelength and Built-In Field 242
11.3.2	Absorption Spectra 243
11.4	FDTD Simulator for GaN/AlGaN ISBT Switches 244
11.4.1	Model 245
11.4.2	Saturation of Absorption 246
11.4.3	Temporal Response 248
11.4.4	Future Applications 249
	References 251

12 Intersubband Electroabsorption Modulator   253
Petter Holmström

12.1	Introduction 253
12.2	Modulator Structure 256
12.2.1	Multiple-Quantum-Well Structure 256
12.2.2	Waveguide and Contacting 259
12.3	Model 261
12.3.1	Conduction Band Potential and Active Layer Biasing 261
12.3.2	Intersubband Transitions 263
12.3.3	Optical Mode and the Plasma Effect 265
12.4	Results 266
12.4.1	Electroabsorption 266
12.4.2	Chirp Parameter 269
12.4.3	Electrical Properties 270
12.4.4	Figure of Merit 271
12.4.5	Absorption Saturation 272
12.4.6	Thermal Properties and Current 273
12.4.7	Significance of the Linewidth 275
12.5	Summary 276
	References 276

13 Ultraviolet Light-Emitting Diodes   279
Yen-Kuang Kuo, Sheng-Horng Yen, and Jun-Rong Chen

13.1	Introduction 279
13.2	Device Structure 281
13.3	Physical Models and Parameters 282

	13.3.1	Band Structure *283*
	13.3.2	Polarization Effects *285*
	13.3.3	Carrier Transport Model *287*
	13.3.4	Thermal Model *288*
	13.3.5	Spontaneous Emission *288*
	13.3.6	Ray Tracing *290*
13.4		Comparison Between Simulated and Experimental Results *291*
13.5		Performance Optimization *293*
	13.5.1	Optimal Aluminum Composition in p-AlGaN Electron Blocking Layer *293*
	13.5.2	Optimal Number of Quantum Wells *294*
	13.5.3	Lattice-matched AlInGaN Electron Blocking Layer *296*
13.6		Conclusion *299*
		References *300*

14   Visible Light-Emitting Diodes *303*

Sergey Yu. Karpov

14.1   Introduction *303*
14.2   Simulation Approach and Materials Properties *304*
14.3   Device Analysis *309*
14.3.1 Band Diagrams, Carrier Concentrations, and Partial Currents *310*
14.3.2 Internal Quantum Efficiency and Carrier Leakage *312*
14.3.3 Emission Spectra *315*
14.3.4 Polarity Effects *317*
14.4   Novel LED Structures *320*
14.4.1 LED with Indium-free Active Region *320*
14.4.2 Hybrid ZnO/AlGaN LED *321*
14.5   Conclusion *323*
         References *324*

15   Simulation of LEDs with Phosphorescent Media for the Generation of White Light *327*

Norbert Linder, Dominik Eisert, Frank Jermann, and Dirk Berben

15.1   Introduction *327*
15.2   Requirements for a Conversion LED Model *328*
15.3   Color Metrics for Conversion LEDs *330*
15.4   Phosphor Model *332*
15.4.1 Phosphor Materials *332*
15.4.2 Luminescence and Absorption of Phosphor Particles *334*
15.4.3 Scattering of Phosphor Particles *335*
15.4.4 Determination of Material Parameters *341*
15.4.5 LED Ray Tracing Model *344*

15.5	Simulation Examples *346*
15.6	Conclusions *350*
	References *350*

16   Fundamental Characteristics of Edge-Emitting Lasers *353*
Gen-ichi Hatakoshi

16.1	Introduction *353*
16.2	Basic Equations for the Device Simulation *354*
16.2.1	Electrical and Optical Simulation *354*
16.2.2	Simulation Model for Thermal Analysis *357*
16.3	Simulation for Electrical Characteristics and Carrier Overflow Analysis *359*
16.4	Perpendicular Transverse Mode and Beam Quality Analysis *366*
16.5	Thermal Analysis *370*
16.6	Conclusions *378*
	References *378*

17   Resonant Internal Transverse-Mode Coupling in InGaN/GaN/AlGaN Lasers *381*
Gennady A. Smolyakov and Marek Osiński

17.1	Introduction *381*
17.2	Internal Mode Coupling and the Concept of "Ghost Modes" *382*
17.3	Device Structure and Material Parameters *384*
17.4	Calculation Technique *385*
17.5	Results of Calculations *386*
17.5.1	Resonant Conditions *386*
17.5.2	Spatial Characteristics of Laser Emission under the Resonant Internal Mode Coupling *391*
17.5.3	Spectral Effects of the Resonant Internal Mode Coupling *394*
17.5.4	Carrier-Induced Resonant Internal Mode Coupling *396*
17.6	Discussion and Conclusions *399*
	References *401*

18   Optical Properties of Edge-Emitting Lasers: Measurement and Simulation *405*
Ulrich T. Schwarz and Bernd Witzigmann

18.1	Introduction *405*
18.2	Waveguide Mode Stability *406*
18.3	Optical Waveguide Loss *412*
18.4	Mode Gain Analysis *417*
18.5	Conclusion *420*
	References *422*

19	**Electronic Properties of InGaN/GaN Vertical-Cavity Lasers** *423*	
	Joachim Piprek, Zhan-Ming Li, Robert Farrell, Steven P. DenBaars, and Shuji Nakamura	
19.1	Introduction to Vertical-Cavity Lasers *423*	
19.2	GaN-based VCSEL Structure *424*	
19.3	Theoretical Models and Material Parameters *425*	
19.3.1	Carrier Transport *426*	
19.3.2	Electron Band Structure *429*	
19.3.3	Built-In Polarization *432*	
19.3.4	Photon Generation in the Quantum Wells *434*	
19.3.5	Optical Mode *436*	
19.4	Simulation Results and Device Analysis *437*	
19.4.1	Current Confinement *438*	
19.4.2	Polarization Effects *438*	
19.4.3	Threshold Current *440*	
19.4.4	AlGaN Doping *442*	
19.4.5	AlGaN Composition *443*	
19.5	Summary *443*	
	References *443*	
20	**Optical Design of Vertical-Cavity Lasers** *447*	
	Włodzimierz Nakwaski, Tomasz Czyszanowski, and Robert P. Sarzała	
20.1	Introduction *447*	
20.2	The GaN VCSEL Structure *449*	
20.3	The Scalar Optical Approach *453*	
20.4	The Vectorial Optical Approach *454*	
20.5	The Self-consistent Calculation Algorithm *458*	
20.6	Simulation Results *460*	
20.7	Discussion and Conclusions *464*	
	References *465*	
21	**GaN Nanowire Lasers** *467*	
	Alexey V. Maslov and Cun-Zheng Ning	
21.1	Introduction *467*	
21.2	Nanowire Growth and Characterization *469*	
21.3	Nanowire Laser Principles *470*	
21.4	Anisotropy of Material Gain *471*	
21.5	Guided Modes *475*	
21.5.1	Guided Modes, Dispersions, and Mode Spacing *476*	
21.5.2	Reflection from Facets *479*	
21.5.3	Far-field Pattern *481*	
21.5.4	Confinement Factors for Anisotropic Nanowires *483*	

21.5.5	Spontaneous Emission Factors	*486*
21.6	Modal Gain and Threshold	*488*
21.7	Conclusion	*489*
	References	*490*

Index *493*

Preface

Gallium nitride and related semiconductor materials enable a wide range of novel devices, some of which already improve our everyday life. Examples are full-color video displays, solid-state lighting, and high-definition DVD players. Nanometer-scale nitride semiconductor structures are often at the heart of such applications, allowing for the generation of short-wavelength light ranging from green to blue to ultraviolet. GaN-based devices also enable innovations in high-speed and high-power electronics. Driven by vast consumer markets, research and development of nitride semiconductor devices has experienced tremendous growth worldwide. Recently, advanced software tools have emerged which facilitate the design and understanding of ever more sophisticated device structures. As these trends continue, physics-based nitride device modeling and simulation is expected to gain importance in the coming years.

This is the first book to be published on physical principles, mathematical models, and practical simulation of GaN-based semiconductor devices. It is intended for scientists and engineers who are interested in employing computer simulation for nitride device design and analysis. The book presents the joint effort of more than forty leading researchers. Its first part covers essential material parameters of nitride semiconductors which are crucial ingredients of device models. Some of these material properties are still under investigation, for instance thermal properties of quaternary compounds, which are briefly described in the introduction. Chapters 2 to 5 summarize electronic and optical properties of bulk nitrides, while Chapters 6 to 9 focus on nanometer-scale structures such as quantum wells and quantum dots. The second and main part of the book investigates a broad selection of state-of-the-art devices, from transistors to light-emitting diodes to laser diodes. Several novel device concepts are described in detail, such as optical switches and modulators, as well as vertical-cavity lasers and nanowire lasers. Each chapter provides a background in device theory, plus practical simulation results which offer deep insight into internal device physics. Some of the software packages are available to the public, on a commercial or noncommercial basis.

Nitride Semiconductor Devices: Principles and Simulation. Joachim Piprek (Ed.)
Copyright © 2007 WILEY-VCH Verlag GmbH & Co. KGaA, Weinheim
ISBN: 978-3-527-40667-8

I would like to sincerely thank all authors for their contributions to this book. Interested readers are encouraged to contact me or any author with questions or suggestions.

Newark, December 2006 *Joachim Piprek*

List of Contributors

Steven P. DenBaars Ch. 19
Materials Department
University of California
Santa Barbara, CA 93106
USA

denbaars@engineering.ucsb.edu

Enrico Bellotti Ch. 4, 6
Boston University
Electrical & Computer Engineering Dept.
8 Saint Mary's Street
Boston, MA 02215-2421
USA

bellotti@bu.edu

Dirk Berben Ch. 15
Osram GmbH
Hellabrunner Str. 1
81536 München
Germany

d.berben@osram.de

Fabio Bernardini Ch. 3
Dipartimento di Fisica
Università di Cagliari
Cittadella Universitaria
09042 Monserrato
Italy

fabio.bernardini@dsf.unica.it

Francesco Bertazzi Ch. 4, 6
Boston University
Electrical & Computer Engineering Dept.
8 Saint Mary's Street
Boston, MA 02215-2421
USA

bertazzi@bu.edu

and

Dipartimento di Elettronica
Politecnico di Torino
Corso Duca Degli Abruzzi 24
10129 Torino
Italy

francesco.bertazzi@polito.it

Carsten Buchheim Ch. 5
Technische Universität Ilmenau
Institut für Physik
PF 100565
98684 Ilmenau
Germany

carsten.buchheim@tu-ilmenau.de

Jun-Rong Chen Ch. 13
Institute of Photonics
National Changhua University of Education
Changhua 500
Taiwan

Shun-Lien Chuang Ch. 8
University of Illinois
Dept of ECE (EL 374B, MC-702)
1406 W. Green Street
Urbana, IL 61801
USA

s-chuang@uiuc.edu

Nitride Semiconductor Devices: Principles and Simulation. Joachim Piprek (Ed.)
Copyright © 2007 WILEY-VCH Verlag GmbH & Co. KGaA, Weinheim
ISBN: 978-3-527-40667-8

List of Contributors

Tomasz Czyszanowski   Ch. 20
Laboratory of Computer Physics
Institute of Physics
Technical University of Łódź
ul. Wólczańska 219
93-005 Łódź
Poland
czyszan@p.lodz.pl

Dominik Eisert   Ch. 15
Osram Opto Semiconductors GmbH
Leibnizstr. 4
93055 Regensburg
Germany
dominik.eisert@osram-os.com

Robert M. Farrell   Ch. 19
Materials Department
University of California
Santa Barbara, CA 93106
USA
rmf@engineering.ucsb.edu

Rüdiger Goldhahn   Ch. 5
Technische Universität Ilmenau
Institut für Physik
PF 100565
98684 Ilmenau
Germany
ruediger.goldhahn@tu-ilmenau.de

Sulakshana Gunna   Ch. 6
Boston University
Electrical & Computer Engineering Dept.
8 Saint Mary's Street
Boston, MA 02215-2421
USA
spasnoor@bu.edu

Jörg Hader   Ch. 7
Nonlinear Control Strategies Inc.
5669 N. Oracle Rd.
Suite 2201
Tucson, AZ 85704
USA
j.hader@nlcstr.com

and

Optical Sciences Center
University of Arizona
Tucson, AZ 85721
USA
jhader@acms.arizona.edu

Gen-ichi Hatakoshi   Ch. 16
Toshiba Research Consulting Corporation
1 Komukai Toshiba-cho
Saiwai-ku
Kawasaki 212-8582
Japan
genichi.hatakoshi@toshiba.co.jp

Petter Holmström   Ch. 12
Kishino Laboratory[a]
Dept. of Electrical and Electronics Engineering
Sophia University
7-1 Kioi-cho, Chiyoda-ku
Tokyo 102-8554
Japan
petterh@imit.kth.se

and

Department of Microelectronics and Applied Physics
Royal Institute of Technology (KTH)
Electrum 229
SE-164 40 Kista
Sweden

[a] Please use this address for postal communication.

Frank Jahnke   Ch. 9
Institute for Theoretical Physics
University of Bremen
P.O. Box 330 440
28334 Bremen
Germany
jahnke@itp.uni-bremen.de

Frank Jermann Ch. 15
Osram GmbH
Hellabrunner Str. 1
81536 München
Germany
f.jermann@osram.de

Sergey Yu. Karpov Ch. 14
Soft-Impact, Ltd.[a]
P.O. Box 83, 27 Engels av.
St. Petersburg, 194156
Russia
karpov@semitech.us

and

STR, Inc.
10404 Patterson Ave.
Suite 108
Richmond, VA 23238
USA

a) Please use this address for postal communication.

Stephan W. Koch Ch. 7
Department of Physics and Material Sciences Center
Philipps Universität Marburg
Renthof 6
35032 Marburg
Germany
stephan.w.koch@physik.uni-marburg.de

Yen-Kuang Kuo Ch. 13
Department of Physics
National Changhua University of Education
Changhua 500
Taiwan
ykuo@cc.ncue.edu.tw

Zhan-Ming (Simon) Li Ch. 19
Crosslight Software Inc.
206-3993 Henning Drive
Burnaby, BC,
Canada V5C 6P7
simon@crosslight.com

Norbert Linder Ch. 15
Osram Opto Semiconductors GmbH
Leibnizstr. 4
93055 Regensburg
Germany
norbert.linder@osram-os.com

Alexey Maslov Ch. 21
Center for Nanotechnology
NASA Ames Research Center
Mail Stop 229-1
Moffett Field, CA 94035
USA
amaslov@mail.arc.nasa.gov

Jerry R. Meyer Ch. 2
Code 5613
Naval Research Laboratory
Washington DC 20375
USA
jerry.meyer@nrl.navy.mil

Umesh K. Mishra Ch. 10
Department of Electrical and Computer Engineering
University of California
Santa Barbara, CA 93106-9560
USA
mishra@ece.ucsb.edu

Jerome V. Moloney Ch. 7
Nonlinear Control Strategies Inc.
5669 N. Oracle Rd.
Suite 2201
Tucson, AZ 85704
USA
j.moloney@nlcstr.com

and

Optical Sciences Center
University of Arizona
Tucson, AZ 85721
USA
jml@acms.arizona.edu

List of Contributors

Shuji Nakamura Ch. 19
Materials Department
University of California
Santa Barbara, CA 93106
USA

shuji@engineering.ucsb.edu

Włodzimierz Nakwaski Ch. 20
Laboratory of Computer Physics
Institute of Physics
Technical University of Łódź
ul. Wólczańska 219
93-005 Łódź
Poland

nakwaski@p.lodz.pl

Cun-Zheng Ning Ch. 21
Center for Nanotechnology
NASA Ames Research Center
Moffett Field, CA 94035

cning@arc.nasa.gov

and

Center for Nanophotonics
and Department of Electrical Engineering
Arizona State University
Tempe, AZ 85287
USA

ching@asu.edu

Marek Osiński Ch. 17
Center for High Technology Materials
University of New Mexico
1313 Goddard SE
Albuquerque
NM 87106-4343
USA

osinski@chtm.unm.edu

Roberto Paiella Ch. 6
Boston University
Electrical & Computer Engineering Dept.
8 Saint Mary's Street
Boston
MA 02215-2421
USA

rpaiella@bu.edu

Tomás Palacios Ch. 10
Department of Electrical Engineering and
Computer Science
Massachusetts Institute of Technology
77 Massachusetts Ave.
Cambridge, MA 02139
USA

tpalacios@mit.edu

Seoung-Hwan Park Ch. 8
Electronics Engineering
Hayang
Kyeongsan
Kyeongbuk 712-702
Korea

shpark@cu.ac.kr

Joachim Piprek Ch. 1, 19
NUSOD Institute
Newark, DE 19714-7204
USA

piprek@nusod.org

Robert P. Sarzała Ch. 20
Laboratory of Computer Physics
Institute of Physics
Technical University of Łódź
ul. Wólczańska 219
93-005 Łódź
Poland

rpsarzal@p.lodz.pl

Pascal Schley Ch. 5
Technische Universität Ilmenau
Institut für Physik
PF 100565
98684 Ilmenau
Germany

pascal.schley@tu-ilmenau.de

Ulrich T. Schwarz Ch. 18
Regensburg University
Department of Physics
93040 Regensburg
Germany

ulrich.schwarz@physik.uni-regensburg.de

Gennady A. Smolyakov   Ch. 17
Center for High Technology Materials
University of New Mexico
1313 Goddard SE
Albuquerque
NM 87106-4343
USA
gen@chtm.unm.edu

Nobuo Suzuki   Ch. 11
Advanced Electron Devices Laboratory
Corporate R&D Center
Toshiba Corp.
1 Komukai-Toshiba-cho
Saiwai-ku
Kawasaki 212-8582
JAPAN
nob.suzuki@toshiba.co.jp

Angela Thränhardt   Ch. 7
Department of Physics and Material
Sciences Center
Philipps Universität Marburg
Renthof 6
35032 Marburg
Germany
angela.thraenhardt@physik.uni-marburg.de

Igor Vurgaftman   Ch. 2
Code 5613
Naval Research Laboratory
Washington DC 20375
USA
vurgaftman@nrl.navy.mil

Hans Wenzel   Ch. 5
Ferdinand-Braun-Institut für Höchst-
frequenztechnik
Gustav-Kirchhoff-Straße 4
12489 Berlin
Germany
hans.wenzel@fbh-berlin.de

Andreas Theo Winzer   Ch. 5
Technische Universität Ilmenau
Institut für Physik
PF 100565
98684 Ilmenau
Germany
andreas.winzer@tu-ilmenau.de

Bernd Witzigmann   Ch. 18
ETH Zürich
Integrated Systems Laboratory
Gloriastrasse 35
8092 Zürich
Switzerland
bernd@iis.ee.ethz.ch

Sheng-Horng Yen   Ch. 13
Department of Physics
National Changhua University of Education
Changhua 500
Taiwan

Part 1 Material Properties

1
Introduction
Joachim Piprek

1.1
A Brief History

Considerable efforts to fabricate nitride devices began more than three decades ago. In 1971, Pankove *et al.* reported the first GaN-based light-emitting diode (LED) [1]. However, most of these early research programs were eventually abandoned due to fundamental materials problems. Since there was no suitable bulk-crystal technology for producing GaN substrates, epitaxy was done on highly lattice-mismatched substrates. The resulting heteroepitaxial films exhibited a high defect density and poor surface morphology. The high n-type background doping, coupled with the deep ionization levels of common acceptors, resulted in an inability to grow p-type materials.

It was not until the mid-1980s that these problems began to be overcome, due in large part to the work of Isamu Akasaki at Nagoya and Meijo Universities and Shuji Nakamura at Nichia Chemical Company in Japan. The use of AlN [2] or GaN [3] nucleation layers facilitated the growth of high-quality GaN films on sapphire substrates by metalorganic chemical vapor deposition (MOCVD). The first n-GaN/AlGaN transistor was demonstrated by Khan *et al.* in 1993 [4]. Another breakthrough was the first successful fabrication of p-type GaN by low-energy electron-beam irradiation (LEEBI) of Mg-doped GaN [5]. In 1992, Nakamura demonstrated that Mg-doped GaN could also be made conductive by thermal annealing in an N_2 ambient [6]. The development of high-quality InGaN films was the third main breakthrough towards the fabrication of InGaN/GaN high-brightness LEDs in 1994 [7]. Finally, Nakamura *et al.* succeeded, in 1995, in manufacturing the first nitride-based laser diode with continuous-wave room-temperature emission at 417 nm wavelength [8]. Soon thereafter, Nichia offered the first commercial GaN-based LEDs and laserdiodes. A detailed review of these developments can be found in [9].

Blue and green nitride LEDs are now widely used, for instance, in full-color displays and in traffic signals. Nitride laser diodes are key components in emerging high-definition DVD players. Other promising application areas

Nitride Semiconductor Devices: Principles and Simulation. Joachim Piprek (Ed.)
Copyright © 2007 WILEY-VCH Verlag GmbH & Co. KGaA, Weinheim
ISBN: 978-3-527-40667-8

are printing, sensors, communication, and medical equipment. However, despite intense research efforts worldwide, there still remains a strong need for a more detailed understanding of microscopic physical processes in nitride devices. Numerical simulation can help to investigate those processes and to establish quantitative links between material properties and measured device performance.

1.2
Unique Material Properties

The troubled history and the recent success of nitride semiconductor devices are both very much related to the unique material properties of GaN and its most relevant alloys InGaN and AlGaN. Depending on the alloy composition, the direct bandgap varies from about 0.7 eV to 6.2 eV, covering a wide wavelength range from red through yellow and green to blue and ultraviolet. While other compound semiconductors, such as GaAs and InP, are grown in the zinc blende crystal system, nitride devices are grown in the hexagonal (wurtzite) crystal system (Fig. 1.1). This leads to unique material properties, such as built-in electric fields due to spontaneous and piezoelectric polarization. Sapphire (Al_2O_3) or SiC are often used as substrates for GaN growth, which exhibit slightly different lattice constants a of 0.476 nm and 0.308 nm, respectively (GaN substrates became available only recently, $a = 0.319$ nm). The lattice-mismatched epitaxial growth causes a large number of dislocations in nitride devices, with dislocation densities that are more than five orders of magnitude higher than in other compound semiconductor devices. The surprisingly small impact of these defects on the performance of GaN-based light emitters is still not fully understood. Another unique property of nitrides is the high activation energy for acceptor (Mg) doping of about 170 meV. It requires high doping densities near 10^{20} cm^{-3} to achieve free hole concentrations of about 10^{18} cm^{-3}. The high doping density causes an extremely low hole mobility on the order of 10 cm^{2}V^{-1}s^{-1}. On the other hand, the high GaN electron mobility of up to 2000 cm^{2}V^{-1}s^{-1} and the large critical breakdown field of more than 3 MV cm^{-1} are advantageous in high-speed and high-power electronics. The thermal conductivity in GaN is more than three times higher than in GaAs.

Despite the recent commercial success of GaN-based devices, internal physical mechanisms are often not completely understood. Advanced models and numerical simulations are needed to support further performance enhancement as well as the emergence of new GaN devices. Sophisticated theories and physics-based software have been developed for previous generations of semiconductor devices (see, e.g., [10,11]). However, the unique material prop-

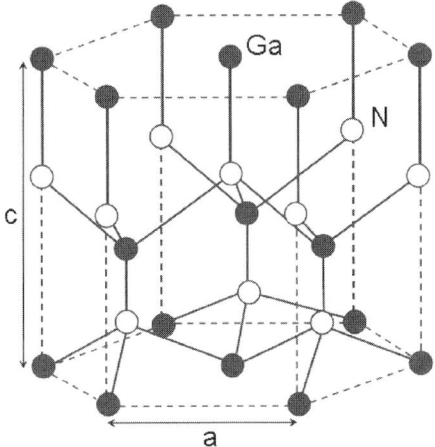

Fig. 1.1 Wurtzite crystal with lattice constants c and a. The structure is formed by two intertwined hexagonal sublattices of, for instance, Ga and N atoms.

erties of nitrides present a new challenge. Comprehensive device simulations require the knowledge of a large number of material parameters, including their variation with material composition, defect or carrier density, internal temperature, built-in electric field, or with other physical conditions. The reliability of simulation results strongly depends on the accuracy of the material parameters used. Measured device characteristics can be utilized to adjust specific parameters in the simulation, in particular those parameters that are affected by the fabrication process [12, 13].

The past decade of intense research effort has tremendously improved our knowledge of nitride material properties. However, many material parameters are still not exactly known. An example is the thermal conductivity of ternary and quaternary nitride alloys. The first thermal conductivity measurements for AlGaN have only recently been published.

Thermal properties of nitrides are briefly reviewed in the next section. Other material parameters are covered in the following chapters.

1.3
Thermal Parameters

Self-heating often limits the performance of semiconductor devices. Heat is generated when carriers transfer part of their energy to the crystal lattice. In consequence, the thermal (vibrational) energy of the lattice rises, which is measured as an increase in its temperature, T_L. We here assume a local thermal equilibrium between lattice and carriers, i.e., the lattice temperature and the carrier temperature are considered identical ($T = T_L = T_n = T_p$). Virtually all material properties like energy band gap and carrier mobility, change with rising temperature.

Within the crystal lattice, thermal energy is dissipated by traveling lattice vibrations. The smallest energy portions of lattice waves are phonons, which can be treated like particles. Microscopic theories of lattice heat generation and dissipation are based on the phonon picture outlined in many solid-state textbooks, e.g., [14, 16, 17].

In practical device simulation, the main thermal parameters are thermal conductivity κ_L and specific heat C_L of the crystal lattice. Electrons and holes also contribute to specific heat and thermal conductivity. However, those contributions are usually negligible. The lattice thermal conductivity κ_L controls the heat flux density (W cm^{-2})

$$\vec{J}_{\text{heat}} = -\kappa_L \nabla T \tag{1.1}$$

which is driven by the gradient of the temperature distribution $T(\vec{r})$. Conservation of energy requires that the temperature satisfies the heat flux equation

$$\rho_L C_L \frac{\partial T}{\partial t} = -\nabla \cdot \vec{J}_{\text{heat}} + H_{\text{heat}} \tag{1.2}$$

where ρ_L is the material's density and $H_{\text{heat}}(\vec{r}, t)$ is the heat power density (W cm^{-3}) generated by various sources. This equation relates the change in local temperature ($\partial T/\partial t$) to the local heat flux (in or out) and to the local heat generation. All parameters in Eq. (1.2) generally depend on the material composition and on the temperature itself. Near room temperature ($T = 300K$) [14]

$$C_L(T) = C_L(300\text{ K}) \frac{20 - (\Theta_D/T)^2}{20 - (\Theta_D/300\text{ K})^2} \tag{1.3}$$

with Θ_D giving the Debye temperature which itself may be temperature dependent. A recent review of Debye temperatures reported for GaN is given in [15]. Temperature effects on the thermal conductivity near room temperature can be described by a power law

$$\kappa_L(T) = \kappa_L(300\text{ K}) \left(\frac{T}{300\text{ K}}\right)^{\delta_\kappa} \tag{1.4}$$

Table 1.1 lists the above material parameters for binary nitride compounds and for some substrate materials.

The room-temperature thermal conductivity of bulk GaN was first measured in 1977 by Sichel and Pankove who reported a value of 130 Wm^{-1}K^{-1}

1.3 Thermal Parameters

Tab. 1.1 Thermal material parameters at room temperature for binary nitrides and typical substrates (density ρ_L, specific heat C_L, Debye temperature Θ_D, thermal conductivity κ_L and its temperature coefficient δ_κ) [18–21].

Parameter Unit	ρ_L g cm^{-3}	C_L J g^{-1}K^{-1}	Θ_D K	κ_L (W m^{-1}K^{-1})	δ_κ
GaN	6.15	0.49	600	160	−0.6 (250 K< T <370 K)
AlN	3.23	0.6	1150	210	−1.2 (150 K< T <300 K)
InN	6.81	0.32	660	45	—
SiC	3.21	0.69	1200	380	−1.4 (200 K< T <600 K)
Al$_2$O$_3$	3.97	0.88	1032	38	−1.5 (300 K< T <600 K)

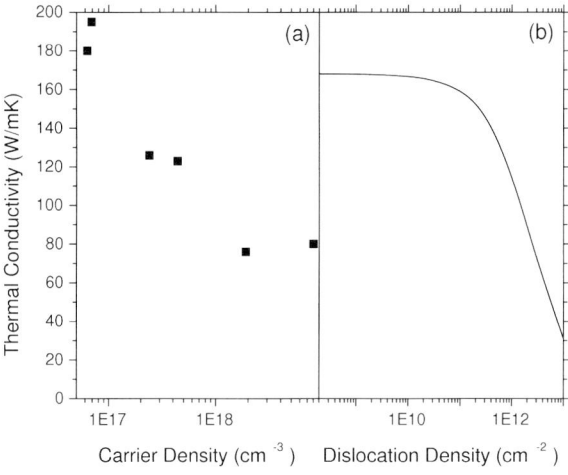

Fig. 1.2 GaN thermal conductivity as function of: (a) electron density of Si-doped samples (measured in [22]); (b) dislocation line density (calculated in [23]).

[24]. Although the crystal quality of their sample was not well known, this result is relatively close to the earlier theoretical prediction of 170 Wm^{-1}K^{-1} by Slack [25]. The recent success of GaN devices has motivated a number of new investigations of the GaN thermal conductivity. Typical results at room temperature are between 130 and 200 Wm^{-1}K^{-1}. The largest values are obtained for high crystal quality, with a maximum of 225 Wm^{-1}K^{-1} measured on free-standing GaN films [26]. The GaN thermal conductivity was found to drop significantly with increasing doping density (by about a factor of two per decade above 10^{17} cm^{-3} [22]) and for dislocation densities higher than 10^{11} cm^{-2} [23]. Both dependencies are plotted in Fig. 1.2. The temperature dependence is shown in Fig. 1.3 [19]. Near room temperature, it can be described by (1.4) with $\delta_\kappa = -0.6$ (dashed line in Fig. 1.3). This empirical parameter also depends on the defect density, it is $\delta_\kappa = -1.2$ for low defect density [20] and $\delta_\kappa = -0.2$ for polycrystalline GaN samples [27]. Over a wider temperature

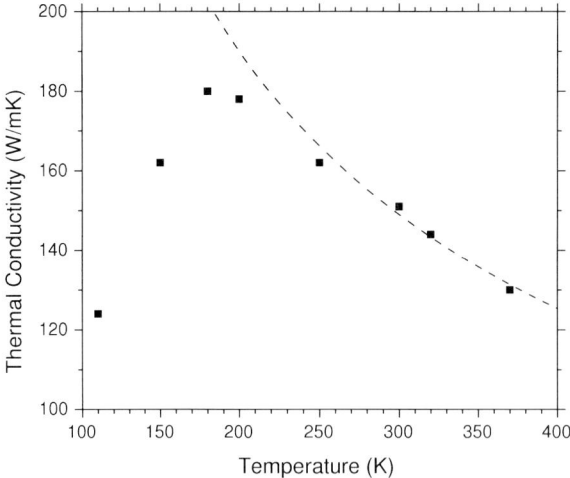

Fig. 1.3 GaN thermal conductivity vs. temperature. The dots show experimental results and the line plots Eq. (1.4) with $\delta_\kappa = -0.6$ [19].

range, the relation $\kappa_L(T)$ is more complex due to different dominating scattering mechanisms [23]. For doping densities of 10^{16}–10^{18} cm^{-3}, the electronic contribution to the GaN thermal conductivity is about 1000 times smaller than that of the lattice [22]. The thermal conductivity anisotropy in GaN at 300K is 1% or less [20].

Thermal parameters of AlN and InN are summarized in Table 1.1. However, these binary materials are commonly not employed in nitride devices. Alloys of GaN with InN and/or AlN are of higher importance. The random distribution of alloy atoms in ternary or quaternary semiconductor compounds causes strong alloy scattering of phonons, which leads to a significant reduction of the thermal conductivity. For ternary alloys, like $Al_xGa_{1-x}N$, the thermal conductivity is typically estimated from binary values using [28]

$$\frac{1}{\kappa_L(x)} = \frac{x}{\kappa_{AlN}} + \frac{1-x}{\kappa_{GaN}} + x(1-x)C_{AlGaN} \tag{1.5}$$

with the empirical bowing parameter C_{AlGaN}. The same bowing parameters can be employed for the quaternary alloy $Al_xIn_yGa_{1-x-y}N$

$$\frac{1}{\kappa_L(x,y)} = \frac{x}{\kappa_{AlN}} + \frac{y}{\kappa_{InN}} + \frac{1-x-y}{\kappa_{GaN}} + xyC_{AlInN} \\ + x(1-x-y)C_{AlGaN} + y(1-x-y)C_{InGaN} \tag{1.6}$$

Thus far, this bowing effect has only been measured for AlGaN [27, 29] (Fig. 1.4). The significant difference between both results may be due to the different measurement methods or to different quality of the AlGaN films,

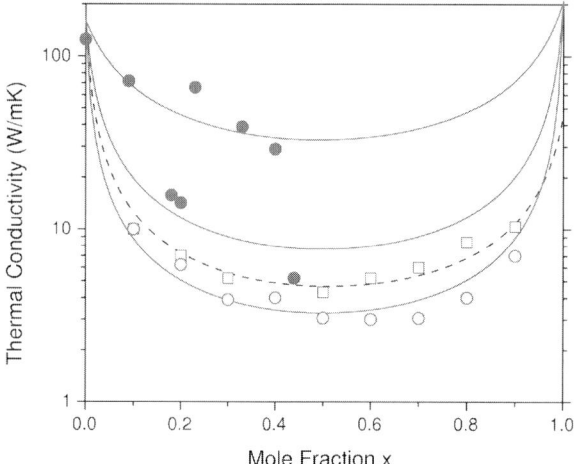

Fig. 1.4 Thermal conductivity of $Al_xGa_{1-x}N$ (circles, solid lines) and $In_xGa_{1-x}N$ (squares, dashed line). Open/solid symbols indicate theoretical/experimental results. Lines give the fit using Eq. (1.5).

that were both grown on sapphire (0001) substrates. It is interesting to note that the AlGaN thermal conductivity increases with temperature in one investigation [29] while it decreases in the other [27]. Using the bowing formula (1.5) and the binary data from Table 1.1, we obtain the room-temperature bowing parameters C_{AlGaN} of 0.1 Km W^{-1} and 0.6 Km W^{-1}, respectively (see lines in Fig. 1.4).

More sophisticated models have been applied to calculate the alloy thermal conductivity from fundamental material parameters, including a virtual-crystal model [29] and a molecular dynamics model [30]. The latter was also used to estimate the InGaN thermal conductivity [31] (open squares in Fig. 1.4). However, the accuracy of such fundamental models very much depends on the material parameters used and fits to measurements are often required to obtain reliable results. The bowing parameters estimated from the theoretical results in Fig. 1.4 are $C_{InGaN} = 0.8$ Km W^{-1} and $C_{AlGaN} = 1.2$ Km W^{-1}.

In practical device analysis, the internal temperature often needs to be known only for specific locations, for instance in the active region of a laser diode. If the heat power P_{heat} (W) is generated in the same location, then the heat flux from that location to the heat sink can be characterized by a thermal resistance R_{th} (K W^{-1}), giving the temperature difference

$$\Delta T = R_{th} P_{heat} \tag{1.7}$$

between the heat source and the heat sink. Similar to the electrical resistance, the thermal resistance depends not only on material properties (thermal conductivity) but also on the device geometry. The advantage of this approach is the thermal characterization of the device by one parameter R_{th} that usually can be measured. In cases with heat generation at different locations or multiple layers within the device, the single resistance R_{th} can be replaced by a thermal resistance network. In analogy to electrical circuits, simplified thermal models can be established this way. For homogeneous heat flux, the thermal resistance of a given layer of thickness d and cross-section A is linked to its thermal conductivity by

$$R_{th} = dA^{-1}\kappa_L^{-1} \tag{1.8}$$

Thermal resistances are also employed to characterize the heat flux through an interface. For instance, the thermal boundary resistance between GaN and the substrate depends on the transmission probability of phonons through that interface. Phonon scattering and reflection lead to a relatively large interface resistance value on the order of 10^{-3} Kcm^{2}W^{-1} which may have a considerable impact on the device performance [32].

Within thin semiconductor layers, interface scattering may reduce the phonon mean free path. As a consequence, the bulk thermal conductivity of this layer is reduced [33]. With an estimated GaN phonon mean free path of 88 nm at room temperature [27], nanometer-scale GaN layers are expected to exhibit a strongly reduced thermal conductivity.

References

1 J. I. Pankove, E. A. Miller and J. E. Berkeyheiser. *RCA Rev.*, 32:383, 1971.

2 H. Amano and N. Sawaki and I. Akasaki and T. Toyoda. *Appl.Phys.Lett.*, 48:353, 1986.

3 S. Nakamura *Japan. J. Appl.Phys., Part 2*, 30:L1705, 1991.

4 M. A. Khan, A. Bhattarai, J. N. Kuznia and D. T. Olson. *Appl. Phys. Lett.*, 63:1214, 1993.

5 H. Amano, M. Kito, K. Hiramatsu and I. Akasaki. *Japan. J. Appl. Phys., Part 1*, 28:L2112, 1989.

6 S. Nakamura, T. Mukai, M. Senoh and N. Iwasa. *Japan. J. Appl. Phys., Part 1*, 31:L139, 1992.

7 S. Nakamura, T. Mukai and M. Senoh. *Appl. Phys. Lett.*, 64:1687, 1994.

8 S. Nakamura, M. Senoh, S. Nagahama, N. Iwasa, T. Yamada, T. Matsushita, H. Kiyoku and Y. Sugimoto. *Japan. J. Appl. Phys., Part 1*, 35:L74, 1996.

9 S. Nakamura, S. Pearton and G. Fasol. *The Blue Laser Diode*. Springer-Verlag, Berlin, 2000.

10 V. Palankovski and R. Quay. *Analysis and Simulation of Heterostructure Devices*. Springer, Wien, 2004.

11 J. Piprek, editor. *Optoelectronic Devices: Advanced Simulation and Analysis*. Springer, New York, 2005.

12 J. Piprek and S. Nakamura. *IEE Proc., Part J: Optoelectron.*, 149:145, 2002.

13 J. Piprek, T. Katona, S. P. DenBaars and S. Li. 3d simulation and analysis of UV AlGaN/GaN LEDs. In *Light-Emitting Diodes: Research, Manufacturing and Applications VII*, volume 5366, Bellingham, 2004. SPIE-The International Society for Optical Engineering.

14 K. W. Boer. *Survey of Semiconductor Physics*, volume II. Van Nostrand Reinhold, New York, 1992.

15 C. Roder, S. Einfeldt, S. Figge and D. Hommel. *Phys. Rev. B*, 72:085218, 2005.

16 P. Y. Yu and M. Cardona. *Fundamentals of Semiconductors*. Springer-Verlag, Berlin, 1996.

17 Ch. Kittel. *Introduction to Solid State Physics*. Wiley, New York, 1996.

18 M. E. Levinshtein, S. L. Rumyantsev and M. S. Shur, editors. *Properties of Advanced Semiconductor Materials*. Wiley, New York, 2001.

19 M. Kamano, M. Haraguchi, T. Niwaki, M. Fukui, M. Kuwahara, T. Okamoto and T. Mukai. *Japan. J. Appl. Phys., Part 1*, 41:5034, 2002.

20 G. A. Slack, L. J. Schowalter, D. Morelli and Jaime A. Freitas. *J. Cryst. Growth*, 246:287, 2002.

21 Ch. Eichler. *Thermal Management of GaN-based Laserdiodes*. PhD thesis, University of Ulm, 2005.

22 D. I. Florescu, V. M. Asnin, F. H. Pollak, R. J. Molnar and C. E. C. Wood. *J. Appl. Phys.*, 88:3295, 2000.

23 J. Zou, D. Kotchetkov, A. A. Balandin, D. I. Florescu and F. H. Pollak. *J. Appl. Phys.*, 92:2534, 2002.

24 E. K. Sichel and J. I. Pankove. *J. Phys. Chem. Sol.*, 38:330, 1977.

25 G. A. Slack. *J. Phys. Chem. Sol.*, 34:321, 1973.

26 W. Liu, A. A. Balandin, C. Lee and H.-Y. Lee. *phys. stat. sol. (a)*, 202:R135, 2005.

27 B. C. Daly, H. J. Maris, A. V. Nurmikko, M. Kuball and J. Han. *J. Appl. Phys.*, 92:3820, 2002.

28 J. Piprek. *Semiconductor Optoelectronic Devices: Introduction to Physics and Simulation*. Academic Press, San Diego, 2003.

29 W. Liu and A. A. Balandin. *J. Appl. Phys.*, 97:073710, 2005.

30 T. Kawamura, Y. Kangawa and K. Kakimoto. *J. Cryst. Growth*, 284:197, 2005.

31 T. Kawamura, Y. Kangawa and K. Kakimoto. *phys. stat. sol. (c)*,3:1695, 2006.

32 V. O. Turin and A. A. Balandin. *Electron. Lett.*, 40:81, 2004.

33 J. Piprek, T. Troger, B. Schroter, J. Kolodzey and C. S. Ih. *IEEE Photon. Technol. Lett.*, 10:81, 1998.

2
Electron Bandstructure Parameters

Igor Vurgaftman and Jerry R. Meyer

2.1
Introduction

In response to the current intensive scientific and commercial interest in nitride semiconductors, several recent works have reviewed their material and physical properties [1–6]. However, the rapid pace of ongoing developments in the field have rendered obsolescent any review that does not incorporate results from the past five years. As an illustration, in 2000 we comprehensively reviewed the literature and recommended specific band parameters for all of the common III-V semiconductors and their alloys, including the nitrides [7]. While most of the tabulations presented in that work remain current, those for the nitride (and dilute-nitride) material systems needed to be updated as early as 2003 [8]. And now, only three years later, a large body of additional data has become available for inclusion in the tabulations.

While our earlier review [8] of the band parameters for nitrogen-containing III-V semiconductors will serve as the starting point for the present work, a number of the recommended band parameter values must now be revised. We also provide additions that, in our judgment, significantly enhance this work's utility as a reference. One example is the case of InN, where a number of in-depth arguments in favor and against the reduced value for its energy gap have been made recently. We also want the present work to be nearly self-contained, such that with a few minor exceptions the reader can understand our band parameter choices without referring to the earlier review [8].

Due to length constraints, the present chapter covers only the wurtzite form of the "conventional" nitrides (GaN, AlN, InN, and their alloys) and excludes their zinc-blende manifestations and dilute-nitride ternaries and quaternaries. We provide values of the energy gaps, spin–orbit and crystal-field splittings, electron and hole effective-mass parameters, conduction and valence band deformation potentials, elastic constants, and band offsets on an absolute scale. Once again, we maintain full internal consistency, e.g., between expressions for the alloy composition and temperature dependences and their end points.

For each parameter, the text briefly discusses the most salient results from the literature, followed by the specification of a recommended value. While in many cases it was impractical to cite every published work to treat a given parameter, enough references are included to provide a reasonable picture of the available knowledge base. We begin by summarizing the most useful band structure model in these materials, based on the $k \cdot p$ method, applicable to crystals with the wurtzite symmetry; and we also briefly describe how the various band parameters may be employed in a realistic band structure calculation for quantum-confined nitride heterostructures.

2.2
Band Structure Models

Energy gaps (E_g) for the wurtzite nitride semiconductors and their alloys span the wide range from 0.7 to 6.2 eV or, in optical terms, from the near-infrared to the ultraviolet (see Fig. 2.1). It is fortunately a very good approximation, even for the narrowest-gap material (InN), to reduce the 8-band description around the energy gap to separate models for the conduction and valence bands. In the following, we rely on the zone-center $k \cdot p$ method to accurately describe the relevant band structure. Since other available approaches, such as the tight-binding [9, 10] and pseudopotential [11, 12] methods, require either formidable computational resources or difficult-to-deduce parameter sets, they are less useful for routine band structure calculations. The $k \cdot p$ model of the relevant band structure incorporates the correct hexagonal symmetry and may be regarded as a convenient fit to some judiciously chosen combination of first-principles and experimental band energies. The band parameters specified below are the fitting variables in this model.

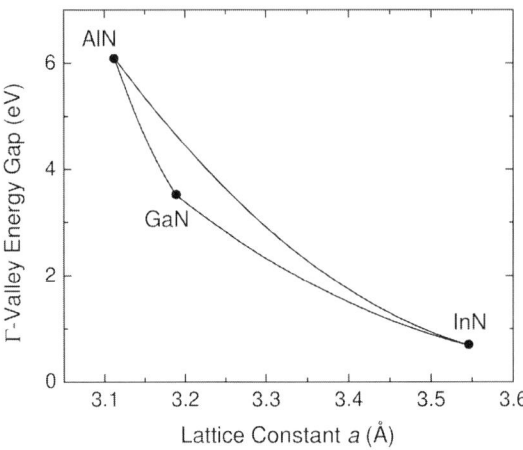

Fig. 2.1 Recommended energy gaps of wurtzite nitride semiconductor alloys (lines) and binaries (points) vs. lattice constant.

2.2 Band Structure Models

The spin-degenerate conduction band is described by the anisotropic, parabolic form with wavefunctions comprised of the s-orbital (angular momentum $l = 0$) states:

$$H_c = E_c + \frac{\hbar^2 k_\parallel^2}{2m_\parallel^*} + \frac{\hbar^2 k_z^2}{2m_\perp^*} + a_{c1}\varepsilon_{zz} + a_{c2}\left(\varepsilon_{xx} + \varepsilon_{yy}\right) \quad (2.1)$$

where $k_\parallel^*$, k_z^* are the wavevectors and $m_\parallel^*$, $m_\perp^*$ are the effective masses along the in-plane and c-axis ([0001]) directions, respectively, a_{c1} and a_{c2} are the conduction-band deformation potentials, and ε_{ij} are the strain tensor components. In the following, we cite the interband deformation potentials a_1 and a_2, from which a_{c1} and a_{c2} may be obtained by subtracting the hydrostatic shift in the valence band, as determined by the D parameters in Eq. (2.5). The anisotropy is a distinguishing feature of the hexagonal crystal structure. Since the crystal-field and spin–orbit splittings in wurtzite nitride materials are rather small, the momentum matrix element E_P and the F parameter are related to the effective mass and energy gap via:

$$\frac{m_0}{m^*} \approx 1 + 2F + \frac{E_P}{E_g} \quad (2.2)$$

where the anisotropy arises from that of E_P (and possibly F).

The valence band is described by a six-band Hamiltonian, originally derived by Rashba, Sheka, and Pikus in a basis-invariant form [13, 14]. For explicitness, we employ the following convenient basis:

$$\left\{|Y_{11}\uparrow\rangle = \frac{-1}{\sqrt{2}}|(X+iY)\uparrow\rangle, \quad |Y_{11}\downarrow\rangle = \frac{-1}{\sqrt{2}}|(X+iY)\downarrow\rangle \right.$$
$$|Y_{10}\uparrow\rangle = |Z\uparrow\rangle, \quad |Y_{10}\downarrow\rangle = |Z\downarrow\rangle \quad (2.3)$$
$$\left. |Y_{1-1}\uparrow\rangle = \frac{-1}{\sqrt{2}}|(X-iY)\uparrow\rangle, \quad |Y_{1-1}\downarrow\rangle = \frac{-1}{\sqrt{2}}|(X-iY)\downarrow\rangle \right\}$$

where the states corresponding to the spherical harmonics are also expressed in terms of the, perhaps more familiar, directed p (angular momentum $l = 1$) orbitals. Note that this basis is distinct from the total-angular-momentum ($J = L + S$) basis, in which the Kohn–Luttinger Hamiltonian applicable to zincblende semiconductors is typically cast. Including strain effects, the resulting wurtzite six-band Hamiltonian is then given by [15–17]:

$$H_v = \begin{bmatrix} F & 0 & -H^* & 0 & K^* & 0 \\ 0 & G & \Delta & -H^* & 0 & K^* \\ -H & \Delta & \lambda & 0 & I^* & 0 \\ 0 & -H & 0 & \lambda & \Delta & I^* \\ K & 0 & I & \Delta & G & 0 \\ 0 & K & 0 & I & 0 & F \end{bmatrix} \quad (2.4)$$

where

$$F = \Delta_1 + \Delta_2 + \lambda + \theta$$
$$G = \Delta_1 - \Delta_2 + \lambda + \theta$$
$$\lambda = \frac{\hbar^2}{2m_0}\left[A_1 k_z^2 + A_2(k_x^2 + k_y^2)\right] + D_1 \varepsilon_{zz} + D_2(\varepsilon_{xx} + \varepsilon_{yy})$$
$$\theta = \frac{\hbar^2}{2m_0}\left[A_3 k_z^2 + A_4(k_x^2 + k_y^2)\right] + D_3 \varepsilon_{zz} + D_4(\varepsilon_{xx} + \varepsilon_{yy})$$
$$K = \frac{\hbar^2}{2m_0} A_5 \left(k_x + ik_y\right)^2 + D_5 \left(\varepsilon_{xx} + 2i\varepsilon_{xy} - \varepsilon_{yy}\right) \quad (2.5)$$
$$H = \frac{\hbar^2}{2m_0} i A_6 k_z \left(k_x + ik_y\right) + i D_6 \left(\varepsilon_{zx} + i\varepsilon_{yz}\right) - A_7 \left(k_x + ik_y\right)$$
$$I = \frac{\hbar^2}{2m_0} i A_6 k_z \left(k_x + ik_y\right) + i D_6 \left(\varepsilon_{zx} + i\varepsilon_{yz}\right) + A_7 \left(k_x + ik_y\right)$$
$$\Delta = \sqrt{2}\Delta_3$$

The A_i parameters in Eq. (2.5) are related to the hole effective masses. The crystal-field splitting is $\Delta_{cr} = \Delta_1$, while the commonly employed approximation for the spin–orbit splitting is $\Delta_{so} = 3\Delta_2 = 3\Delta_3$ (although a small Δ_2/Δ_3 anisotropy has sometimes been reported [18,19]). Terms involving the valence-band deformation potentials, D_i, correspond to the analogous terms containing $\frac{\hbar^2}{2m_0} A_i$ with the simple substitution: $k_x^2 \to \varepsilon_{xx}$, $k_x k_y \to \varepsilon_{xy}$, etc. The sometimes-neglected A_7 parameter is needed to fully account for the spin–orbit interaction in an inversion-asymmetric crystalline structure, and to break the LH and CH spin degeneracies in the plane.

A basis transformation detailed by Ren et al. [17] transforms the 6×6 Hamiltonian of Eq. (2.4) into two 3×3 Hamiltonians. Although the two descriptions are equivalent, the transformation allows an analytical solution for bulk materials that may save considerable execution time when quantum-confined structures are modeled. Owing to space constraints, we do not reproduce the transformed basis vectors and Hamiltonian.

To evaluate the strain tensor components, we must specify the five distinct, nonvanishing elastic constants for wurtzite crystals: C_{11}, C_{12}, C_{13}, C_{33}, and C_{44}. In the important case of biaxial strain and growth along the c-axis [0001], the nondiagonal strain-tensor components vanish while $\varepsilon_{xx} = \varepsilon_{yy} = \frac{a_{sub} - a_{lc}}{a_{lc}}$, $\varepsilon_{zz} = \frac{c_{sub} - c_{lc}}{c_{lc}} = -2\frac{C_{13}}{C_{33}}\varepsilon_{xx}$.

The reduced symmetry of the wurtzite lattice leads to significant polarization even in the case of [0001] growth. The total polarization is the sum of the piezoelectric and spontaneous components. Using the customary compact form for the six distinct strain tensor components:

$$\{\varepsilon_{xx}, \varepsilon_{yy}, \varepsilon_{zz}, (\varepsilon_{yz}, \varepsilon_{zy}), (\varepsilon_{zx}, \varepsilon_{xz}), (\varepsilon_{xy}, \varepsilon_{yx})\} \equiv \{\varepsilon_1, \varepsilon_2, \varepsilon_3, \varepsilon_4, \varepsilon_5, \varepsilon_6\} \quad (2.6)$$

the piezoelectric polarization (proportional to strain) can be written:

$$P_i^{pz} = \sum_j e_{ij}\varepsilon_j = \sum_k d_{ik}\sigma_k = \sum_j \sum_k d_{ik}C_{kj}\varepsilon_j \qquad (2.7)$$

where the σ_i are components of the stress tensor, e_{ij} are piezoelectric constants, and d_{ij} are piezoelectric moduli. For fixed values of the elastic constants, only the nonvanishing, distinct piezoelectric moduli (d_{31}, d_{33}, and d_{15}) or piezoelectric constants (e_{31}, e_{33}, and e_{15}) need to be specified. For instance, in the case of [0001] growth (biaxial strain), the only nonvanishing polarization component is oriented along the growth direction:

$$P_3 = 2d_{31}\frac{a_{sub} - a_{lc}}{a_{lc}}\left(C_{11} + C_{12} - 2\frac{C_{13}^2}{C_{33}}\right) + P_{sp} \qquad (2.8)$$

where P_{sp} is the spontaneous polarization (independent of strain). A sheet charge (with density $\sigma = P_3^A - P_3^B$) is formed at A/B heterojunction interfaces, as demanded by the discontinuity of the polarization at those planes.

In quantum confinement situations where the translational invariance is relaxed, the conduction- and valence-band Hamiltonians of Eqs (2.1) and (2.4) can still be used to solve for the band structure if we make the well-known substitution $k \to -i\nabla$. To achieve acceptable accuracy, the band bending induced by polarization and the electrostatic potential due to the charge sheets must generally be included. The resulting set of differential equations can be solved by a variety of different methods, such as finite-difference [20], transfer-matrix [21, 22], finite-element [23, 24], spectral (reciprocal-space with a wave vector cut-off) [25], etc.

2.3 Band Parameters

Based on a critical review of the literature for each nitride material, we will recommend values for the following band parameters defined in the previous section: lattice constants a_{lc} and c_{lc}, Γ-valley energy gap and its temperature dependences as given by the Varshni [26] parameters α and β in $E_g(T) = E_g(T=0) - \frac{\alpha T^2}{\beta + T}$ (whereas other temperature-dependence models such as Bose–Einstein [27] or those described by Pässler [28–30] and others [31] are physically motivated and may provide slightly better fits, the Varshni form is accurate to within a few meV for the materials and temperature ranges of interest here), crystal-field splitting Δ_{cr}, spin–orbit splitting Δ_{so}, electron effective masses $m_\parallel^*$ and $m_\perp^*$, interband deformation potentials a_1 and a_2, valence-band deformation potentials $D_1 - D_6$, valence-band parameters $A_1 - A_7$, elastic constants $C_{11}, C_{12}, C_{13}, C_{33}$, and C_{44}, piezoelectric moduli d_{31},

d_{33}, and d_{15}, and spontaneous polarization P_{sp}. For alloys, it will be assumed that the parameters are given as:

$$P(A_{1-x}B_x) = (1-x)P(A) + xP(B) - C_P x(1-x) \qquad (2.9)$$

and only parameters P with nonzero bowing values C_P will be tabulated. The band offsets at the relevant heterojunctions will be discussed separately in Section 2.3.8.

The band structure of wurtzite GaN near the top of the valence band is plotted in Fig. 2.2. The bottom pair of crystal-hole (CH) bands result from the crystal-hole splitting and are formed at the zone center from a mixture of opposite-spin $|Y_{1(-)1}\rangle$ and $|Y_{10}\rangle$ states with predominant $|Z\rangle$ character. The top four bands are split by the spin–orbit interaction into two pairs with heavy and light effective masses in the plane (HH and LH, respectively), with their [0001] masses being similar. At the zone center, the HH states have exclusively $|Y_{1(-)1}\rangle$ character (that is, $|X \pm iY\rangle$), whereas the LH states are predominantly $|X \pm iY\rangle$, with $|Z\rangle$ admixture. In this sense, the HH band is analogous to heavy holes in zinc-blende semiconductors. At the zone center, the three pairs of bands have energies (with zero referenced to the valence-band maximum $E_0 = \Delta_1 + \Delta_2$, where $\Delta_2 = \Delta_3$ is assumed): $E_{HH}(k=0) = E_0$, $E_{LH}(k=0) = E_0 - \Delta_{cr}/2 - \Delta_{so}/2 + S/2$, and $E_{CH}(k=0) = E_0 - \Delta_{cr}/2 - \Delta_{so}/2 - S/2$, where $S = [(\Delta_{cr} - \Delta_{so}/3)^2 + 8/9\Delta_{so}^2]^{1/2}$. Note that in this paper the energy gap E_g is defined not as $E_c - E_0$, but as the difference between E_c and the topmost valence band (crystal-hole for the AlN-rich compounds). Inclusion of the A_7 term lifts the HH, LH, and CH degeneracy away from the zone center [17], in agreement with more sophisticated band structure computational methods.

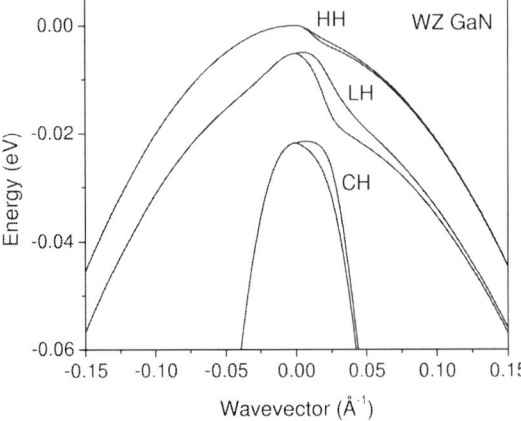

Fig. 2.2 Valence-band structure of wurtzite GaN using the band parameters recommended in this review. The energy zero is taken to be the top of the valence band.

2.3.1
GaN

The energy gap for GaN is most accurately determined from luminescence measurements of the exciton lines. The actual gap is recovered by adding the estimated binding energy to the observed exciton transition energy. For relatively pure GaN, perhaps the most reliable indicator is the peak associated with the free A exciton (E_A). Early measurements [32–34] led to a value $E_A = 3.475$ eV near $T = 0$, in conjunction with an estimate of 28 meV for the binding energy. Numerous other photoluminescence (PL) and absorption studies from the 1990s [35–48] broadened the range of reported A-exciton transition energies at 0 K to 3.474–3.496 eV, where some of the spread was apparently attributable to variations in the strain conditions [49]. Experimental A-exciton binding energies range from 18 meV to 28 meV [34, 35, 39, 40, 43, 46, 50–52]. Owing to uncertainty in the hole effective mass, the corresponding theoretical [53] range for the binding energy is 23–28 meV. Based on median values for both E_A and the binding energy, we recommend a value of 3.510 eV for the zero-temperature energy gap. Recent publications [54] have reported a significant dependence of the GaN energy gap on layer thickness, and attributed it to strain. Since the result $E_g(300K) \approx 3.38$ eV for the limit of very thick (several hundred µm) layers would require that the currently-accepted value be revised downwards by more than 50 meV, further confirmation is needed. Low-temperature optical transmission measurements [55] on single-crystal, highly-doped GaN yielded $E_g = 3.556$ eV, which considerably exceeds the adopted value at least in part because of the Moss–Burstein shift.

The Varshni coefficients for wurtzite GaN have also been studied extensively. From optical absorption measurements on bulk single crystals and epitaxial layers grown on sapphire, Teisseyre et al. [56] obtained $\alpha = 0.939$–1.08 meV K^{-1} and $\beta = 745$–772 K. For the temperature variation of the A exciton resonance, Shan et al. reported $\alpha = 0.832$ meV K^{-1} and $\beta = 836$ K [36], whereas Petalas et al. [57] fixed $\beta = 700$ K and found $\alpha = 0.858$ meV K^{-1} using spectroscopic ellipsometry. Salvador et al. [58] obtained $\alpha = 0.732$ meV K^{-1} and $\beta = 700$ K, based on PL results. Manasreh [41] reported $\alpha = 0.566$–1.156 meV K^{-1} and $\beta = 738$–1187 K from absorption measurements on samples grown by molecular beam epitaxy (MBE) and metalorganic chemical vapor deposition (MOCVD). The contactless electroreflectance study of Li et al. [44] led to $\alpha = 1.28$ meV K^{-1} and $\beta = 1190$ K for the A exciton transition energy, while Zubrilov et al. [59] suggested $\alpha = 0.74$ meV K^{-1} and $\beta = 600$ K based on exciton luminescence spectra. Photoluminescence data from a study of free and bound excitons by Reynolds et al. [60] were fitted to a modified Varshni-like form. Nam et al. [61] reported $\alpha = 0.84$ meV K^{-1} and $\beta = 789$ K from photoluminescence (PL) measurements, which were later updated in another work [62] by the same group to $\alpha = 0.94$ meV K^{-1} and $\beta = 791$ K. Su et al. [55] fixed $\beta = 600$ K and found $\alpha = 0.99$ meV K^{-1}. Our recommended Varshni parameters of $\alpha = 0.914$ meV K^{-1} and $\beta = 825$ K represent simple averages over the more credible reported values. Owing to the small relative change in the bandgap energy (only 72 meV between 0 and 300 K), GaN device characteristics tend to be relatively insensitive to the precise values of the Varshni parameters.

In the wurtzite phase, all conventional nitrides exhibit a direct energy gap, with the next conduction valleys being at least 2 eV higher than the Γ valley [63]. Therefore, we do not specify indirect gaps or other critical-point energies in this review, although various theoretical and experimental studies addressing this topic are cited in Ref. [8].

While early reflectivity and Faraday-rotation studies yielded $0.20m_0$ [64] and 0.24–$0.29m_0$, [65], respectively, for the electron effective mass in GaN, since then a considerable body of work has produced more precise evaluations. Meyer et al. [66] and Witowski et al. [67] obtained masses of $0.236m_0$ and $0.222m_0$, respectively, from shallow-donor transition energies. Pointing out that the strongly-polarized nature of GaN leads to a large polaron correction (8%), Drechsler et al. derived a bare mass of $0.20m_0$ from cyclotron resonance data [68]. Perlin et al. [69] obtained a similar result from infrared-reflectivity and Hall-effect measurements, and also found the anisotropy to be less than 1%. A slightly higher dressed mass of $0.23m_0$ was recently obtained by Wang et al. [70] and Knap et al. [71]. The former may require a small downward revision because the electrons were confined in a quantum well, whereas the latter authors apparently corrected for that effect. Infrared ellipsometry measurements on bulk n-doped GaN by Kasic et al. [72] yielded slightly-anisotropic electron masses of $0.237 \pm 0.006 m_0$ and $0.228 \pm 0.008 m_0$ along the two axes. The ellipsometry measurements of Shokhovets et al. [73] implied $0.201 \pm 0.005 m_0$, with the major uncertainty being the choice of valence-band parameters. Syed et al. [74] employed cyclotron resonance to estimate $0.208m_0$ for a 2D electron gas, and revised downward the polaronic enhancement for the 2D case. Finally, Elhamri et al. [75], Saxler et al. [76], Wong et al. [77], Wang et al. [78], and Hang et al. [79], determined masses ranging from $0.18m_0$ to $0.23m_0$ from Shubnikov–de Haas data for 2D electrons at a GaN/AlGaN heterojunction. It was suggested [75] that strain effects might have compromised somewhat the masses obtained by other studies. Our composite recommendation is $0.20m_0$ for the bare electron mass and $0.22m_0$ for the experimentally-relevant dressed mass. This bare mass agrees quite well with several theoretical estimates (see the list in Ref. [80]), whereas other calculations derived masses as low as 0.13–$0.16m_0$ [81,82]. The anisotropy magnitudes emerging from the published work allow us to suggest using $|m_{\parallel}^* - m_{\perp}^*| \approx 0.01 m_0$.

The momentum matrix element and F parameter were derived in a recent ellipsometric study [73]. We recommend using their $E_P = 19.8$ eV, in conjunction with $F = -0.82$ (slightly revised to fit our suggested effective mass and energy gap). While this large F-parameter is comparable to the values for other wider-gap III-V semiconductors; further confirmation of its unexpected significance is needed. The anisotropy of the momentum matrix element should match that of the effective mass. Despite the large mass, the electron dispersion relation becomes nonparabolic at energies somewhat above the conduction band minimum [73].

The spin–orbit and crystal-field splittings in GaN have been deduced from the A, B, and C free exciton energies. An early study of Dingle et al. found $\Delta_{cr} = 22$ meV and $\Delta_{so} = 11$ meV [32]. A recent and detailed analysis by Gil et al. yielded the values $\Delta_{cr} = 10$ meV and $\Delta_{so} = 18$ meV [38], although Chuang and Chang rederived $\Delta_{cr} = 16$ meV and $\Delta_{so} = 12$ meV from the same data using a more precise description of strain variations of the valence band-edge energies [15]. Reynolds et al. obtained $\Delta_{cr} = 25$ meV and $\Delta_{so} = 17$ meV from a fit to exciton data [51]. Using similar approaches, Δ_{cr}/Δ_{so} values of 22 meV/15 meV were obtained by Shikanai et al. [43], 37.5 meV/12 meV by Chen et al. [39], 9 meV/20 meV by Korona et al. [83], and 9–13 meV/17–18 meV by Campo et al. [84,85], while the values 10 meV/17 meV were determined by both Yamaguchi et al. [18] and Edwards et al. [86]. Finally, a recent detailed experimental investigation by Rodina et al. [47] found one of the smallest reported crystal-field splittings to date: $\Delta_{cr} = 9$ meV, along with $\Delta_{so} \approx 18$ meV. Ref. [8] reported first-principles calculations of the crystal-field and spin–orbit splittings, in conjunction with the A parameter sets as discussed below. Recent theoretical studies yielded crystal-field splittings from 12 to 43 meV [81,82,87] and spin–orbit splittings from 11 to 15 meV [87]. We recommend employing $\Delta_{cr} \approx 10$ meV and $\Delta_{so} \approx 17$ meV, since the recent experiments seem to converge on those values.

Since the valence-band structure in wurtzite GaN is strongly anisotropic and nonparabolic, directly determining the relevant A parameters from experiments is quite difficult. While early studies (overviewed in Ref. [8]) deduced a hole effective mass of $0.8m_0$, Orton [88] suggested a much smaller value of $0.4m_0$ based on the acceptor binding energies. Salvador et al. obtained a hole mass of $0.3m_0$ from a fit to PL spectra [58]. On the other hand, the absorption measurements of Im et al. yielded a rather heavy m_h^* of $2.2m_0$ [42]. Merz et al. [40] obtained an isotropically averaged heavy-hole bare mass of $0.54m_0$ from luminescence data, and also pointed out that the polaron correction for heavy holes in GaN is nearly 13%. Fits of the exciton binding energies yielded hole masses in the range 0.9–$1.2m_0$ [45, 89], while Kasic et al. [72] obtained $1.4m_0$ from an infrared ellipsometric study on p-doped GaN. A nonparabolic heavy-hole dispersion was reported by Shields et al. [90], with masses in the 0.75–$1.8m_0$ range. A bare heavy-hole mass of $0.52m_0$ was deduced from fits to experimental exciton-luminescence data by Chtchekine et al. [46]. Perhaps the most detailed such study to date was by Rodina et al. [47], whose fits produced bare and dressed hole masses in both directions, for excitons associated with all three valence bands. Santic examined the available theoretical and experimental evidence and showed that a density-of-states effective mass of $1.25m_0$ should be reasonable for approximating the carrier statistics, provided the hole density is not too high [91].

Since it does not appear possible to deduce a completely consistent set of A parameters from the experimental work, it seems advisable to employ instead the most reliable of the numerous theoretical sets that have appeared in the literature [15,17,39,81,82,92–95]. In particular, we recommend the parameter set proposed by Ren et al. [17], who derived the value of 93.7 meV·Å (note the unit error in the original paper) for the inversion parameter A_7. However, the reader must be cautioned that the parameters of Ren et al. [17] lead to spin–orbit and crystal-field splittings of 21.1 and 10.8 meV, respectively, which deviate slightly from the values recommended above based on the best experimental evidence (17 meV and 10 meV). Unfortunately, deriving a more consistent set of A parameters would require a new *ab initio* pseudopotential calculation. The valence-band structure of wurtzite GaN is shown in Fig. 2.3.

As in other III-V semiconductors, most of the energy shift induced by hydrostatic strain tends to occur in the conduction band. Here we focus on finding reliable values for the interband deformation potentials, a_1 and a_2, since these are better established than the conduction-band shifts. A note of caution is that the use of different elastic constants in the various studies may have significantly affected the derived deformation potentials. A recent calculation [96] yielded $a_1/a_2 = -4.09$ eV$/-8.87$ eV, while Shan et al. [97] reported $a_1/a_2 = -6.5$ eV$/-11.8$ eV. Fits to mobility data implied conduction-band deformation potentials approaching -9 eV [98,99], which is similar to the value of -9.43 eV obtained from pressure-dependent PL [100]. Other experimental studies obtained $a_1/a_2 = -3.1$ eV$/-11.2$ eV [101] and -5.22 eV$/-10.8$ eV or -6.85 eV$/-8.84$ eV, depending on the adopted elastic constants [102]. A very different set, $a_1/a_2 = -9.6$ eV$/-8.2$ eV, was recently obtained by studying optical absorption under uniaxial compression [103]. Our recommendation of $a_1/a_2 = -7.1$ eV$/-9.9$ eV represents an average of all the measured values.

Numerous values for the valence-band deformation potentials have been derived both from first-principles calculations [15, 104–108] and fits to experimental data [18,38,43,97,101–103,109,110]. There are considerable discrepancies between the reported data, and further work is needed to resolve which results are the most accurate. Our recommended values are obtained from an average of the available data and calculations for D_3, D_4, and D_5. Then we employ the quasi-cubic approximation [15] to estimate $D_6 = (D_3 + 4D_5)/\sqrt{2}$. Note that we do not require $D_3 = -2D_4$, although the determined values approximate that relation. After choosing D_1 to be half of the interband deformation potential a_1, the quasi-cubic approximation yields $D_2 = D_1 + D_3$. Our composite set of deformation potentials is then: $D_1 = -3.6$ eV, $D_2 = 1.7$ eV, $D_3 = 5.2$ eV, $D_4 = -2.7$ eV, $D_5 = -2.8$ eV, and $D_6 = -4.3$ eV.

Elastic constants for wurtzite GaN have been obtained from a number of experiments [111–116] and calculations [95, 96, 117–121]. The dependence of elasticity on applied pressure was recently explored by Lepkowski et al. [122]. It appears that the best agreement of theory and experiment is realized by adopting the data of Polian et al. [112]: $C_{11} = 390$ GPa, $C_{12} = 145$ GPa, $C_{13} = 106$ GPa, $C_{33} = 398$ GPa, and $C_{44} = 105$ GPa. The various experiments disagree considerably, however, so further investigation is needed to resolve the differences.

There have been a few theoretical and experimental attempts to determine the piezoelectric coefficients. Bykhovki et al. deduced e_{31} and e_{33} from the e_{14} coefficient in zinc-blende GaN, obtaining values of $e_{31} = -0.22$ C m^{-2} and $e_{33} = 0.43$ C m^{-2} [123]. Studies of polycrystalline GaN on zinc-blende Si and single-crystal GaN on wurtzite SiC by Guy et al. [124, 125] produced the results $d_{33} = 2.6$ pm V^{-1} and $d_{33} = 3.7$ pm V^{-1}, respectively. Measurements by Lueng et al. yielded a thin-film value (on an AlN buffer) of $d_{33} = 3.1$ pm V^{-1} [126, 127]. Since there was no way to determine d_{13} independently from those experiments, the relation $d_{13} = -d_{33}/2$ was used. A calculation of Shimada et al. yielded $e_{31} = -0.32$ C m^{-2} and $e_{33} = 0.63$ C m^{-2} [118], and Bernardini et al. derived $e_{31} = -0.49$ C m^{-2} and $e_{33} = 0.73$ C m^{-2} from first principles [128]. The theoretical work by Bernardini and Fiorentini [129] discussed the reliability of the experiments, and proposed values of $d_{33} = 2.7$ pm V^{-1} and $d_{13} = -1.4$ pm V^{-1}. Hangleiter et al. [130] estimated $d_{13} = -1.05 \pm 0.05$ pm V^{-1} from the PL energy in an InGaN/GaN quantum well. A similar approach was applied to other data by Christmas et al. [131] with the conclusion that the e coefficients of Shimada et al. [118] produced the best fit. Since the latest studies appear to favor smaller values, we recommend the piezoelectric coefficients (in light of our recommended elastic constants): $d_{33} = 1.9$ pm V^{-1} and $d_{13} = -1.0$ pm V^{-1}. Based on recent measurements [132] and calculations [129] of the shear piezoelectric coefficient, we also recommend $d_{15} = 3.1$ pm V^{-1}.

Several first-principles calculations have derived very different values for the spontaneous polarization P_{sp} in GaN, ranging from -0.034 C m^{-2} to -0.074 C m^{-2} [128, 133–135]. We recommend $P_{sp}(\text{GaN}) = -0.034$ C m^{-2}, although it should be emphasized that only the *differences* in spontaneous polarization are important in heterostructure band calculations. A full discussion is deferred until the AlN section.

A complete listing of the recommended band structure parameters for wurtzite GaN is compiled in Table 2.1.

Tab. 2.1 Recommended band structure parameters for wurtzite nitride binaries. The listed properties are lattice constants a_{lc} and c_{lc} at 300 K, energy gap E_g at 0 K and 300 K, Varshni parameters α and β, crystal-field and spin–orbit splittings Δ_{cr} and Δ_{so}, electron effective masses $m^*_{\parallel}$ and $m^*_{\perp}$ at 300 K, momentum matrix element E_P, F parameter, valence-band A parameters, interband deformation potentials a_1 and a_2, valence-band deformation potentials D, elastic constants C, piezoelectric coefficients d, and spontaneous polarization P_{sp}. Wherever a temperature dependence is not listed, no appreciable variation of that parameter should be assumed.

Parameters	GaN	AlN	InN
a_{lc} (Å) at $T = 300$ K	3.189	3.112	3.545
c_{lc} (Å) at $T = 300$ K	5.185	4.982	5.703
E_g (eV) at 0 K	3.510	6.10	0.69
E_g (eV) at 300 K	3.437	6.00	0.608
α (meV K^{-1})	0.914	2.63	0.414
β (K)	825	2082	154
Δ_{cr} (eV)	0.010	−0.227	0.024
Δ_{so} (eV)	0.017	0.036	0.005
$m^*_{\parallel}$ at 300 K	0.21	0.32	0.07
$m^*_{\perp}$ at 300 K	0.20	0.30	0.07
E_P (eV)	19.8	13.6	11.4
F	−0.82	0	−1.63
A_1	−7.21	−3.86	−8.21
A_2	−0.44	−0.25	−0.68
A_3	6.68	3.58	7.57
A_4	−3.46	−1.32	−5.23
A_5	−3.40	−1.47	−5.11
A_6	−4.90	−1.64	−5.96
A_7 (eV Å)	0.0937	0	0
a_1 (eV)	−7.1	−3.4	−4.2
a_2 (eV)	−9.9	−11.8	−4.2
D_1 (eV)	−3.6	−2.9	−3.6
D_2 (eV)	1.7	4.9	1.7
D_3 (eV)	5.2	9.4	5.2
D_4 (eV)	−2.7	−4.0	−2.7
D_5 (eV)	−2.8	−3.3	−2.8
D_6 (eV)	−4.3	−2.7	−4.3
C_{11} (GPa)	390	396	223
C_{12} (GPa)	145	137	115
C_{13} (GPa)	106	108	92
C_{33} (GPa)	398	373	224
C_{44} (GPa)	105	116	48
d_{13} (pm V^{-1})	−1.0	−2.1	−3.5
d_{33} (pm V^{-1})	1.9	5.4	7.6
d_{15} (pm V^{-1})	3.1	3.6	5.5
P_{sp} (C/m^2)	−0.034	−0.090	−0.042

2.3.2
AlN

The absorption measurements of Yim et al. [136] and Perry and Rutz [137] indicated that the energy gap in wurtzite AlN varied from 6.28 eV at 5 K to 6.2 eV at room temperature. Varshni parameters of $\alpha = 1.799$ meV K^{-1} and $\beta = 1462$ K were reported by Guo and Yoshida [138]. Their low-temperature gap of 6.13 eV is similar to that reported by Vispute et al. [139]. Tang et al. resolved what they believed to be the free or shallow-impurity-bound exciton in their cathodoluminescence data, at an energy of 6.11 eV at 300 K [140]. Wethkamp et al. used spectroscopic ellipsometry to determine that the energy gap decreases from 6.20 eV at 120 K to 6.13 eV at 300 K [141], while Brunner et al. reported a variation from 6.19 eV at 7 K to 6.13 eV at 300 K [142]. Kuokstis et al. resolved a low-temperature free-exciton transition at 6.07 eV [143]. Guo et al. reported the temperature dependence of the reflectance spectra, but fitted it to the Bose–Einstein expression [144]. Relatively large low-temperature energy gaps were also found by Jiang et al. [145] (6.280 eV, with $\alpha = 1.70$ meV K^{-1} and $\beta = 1480$ K), Chen et al. [146] (a transition attributed to the free A exciton at 6.18 eV), and Onuma et al. [147] ($E_g = 6.211$ eV). However, other workers reported lower values for the AlN energy gap. The differences between ultraviolet PL measurements and optical absorption/transmission measurements were discussed by Chen et al. [148]. Since the crystal-field splitting in AlN is negative (see below), the lowest-energy free A exciton is optically active only for polarization parallel to the c-axis. This may cause an overestimate of the energy gap in some passive optical experiments that are sensitive only to higher-energy transitions. Several groups have reported low-temperature free A exciton PL at 6.023–6.029 eV [61, 149–153], which we take to provide a better indication of the energy gap in AlN. Addition of the estimated binding energy of 71 meV [148] leads to our recommendation of $E_g = 6.10$ eV at $T = 0$. The most reliable Varshni parameters, extracted by examining the PL at temperatures up to 800 K, [61, 62] are $\alpha = 2.63$ meV K^{-1} and $\beta = 2082$ K.

Experimental investigations of the AlN electron effective mass [149, 152] have provided only a rather wide range of values based on estimates of the binding energy. Since a number of calculations are available [8, 81, 82, 93, 95, 154, 155], we obtained our recommend bare masses of $m_{\perp}^* = 0.30 m_0$ and $m_{\parallel}^* = 0.32 m_0$ by averaging the theoretical values. Note that this is more anisotropic than in wurtzite GaN [93]. Assuming $F = 0$, we obtain an average $E_P = 13.6$ eV. A valence-band density-of-states mass of $2.7 m_0$ was derived by Nam et al. [156] from an estimate of the binding energy for Mg acceptors. A number of theoretical valence-band parameter sets are available [15, 93–95]. An apparent disagreement in the signs for A_5 and A_6 is irrelevant, since only absolute values of those parameters enter the Hamiltonian [95]. In the absence of compelling experimental data, we continue to recommend the A parameters given by Kim et al. [95].

The crystal-field splitting in AlN is believed to be negative, which implies that the crystal hole is the topmost valence band. The available calculations listed in Ref. [8] give Δ_{cr} values ranging from -58 meV to -244 meV (see also the more recent results [81, 82]). The experimental investigations of the crystal-field splitting are tied in with measurements of the energy gap, insofar as the most reliable determinations are from separation of the free A exciton transition and the free B and C excitons (this separation approaches Δ_{cr}, provided $\Delta_{so} \ll \Delta_{cr}$). While Chen et al. [146] and Onuma et al. [147] reported values of -110 and -56 meV, respectively, our recommendation is to average the values of Chen et al. [148] (-230 meV) and Silveira et al. [149] (-225 meV). Spin–orbit splittings ranging from 11 meV to 20 meV have been cited in the literature [8]. We adopt the value of 36 meV derived by Silveira et al. [149] from PL measurements. The valence-band structure of wurtzite AlN is shown in Fig. 2.3.

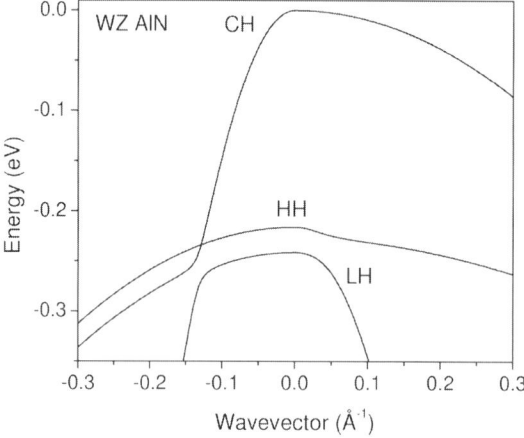

Fig. 2.3 Valence-band structure of wurtzite AlN using the band parameters recommended in this review. The energy zero is taken to be the top of the valence band.

The hydrostatic deformation potential for wurtzite AlN is believed to lie between -7.1 eV and -9.5 eV [63, 95], which is consistent with the observation that the bandgap pressure coefficients for AlGaN alloys have little dependence on composition [157]. Our recommended set ($a_1 = -3.4$ eV and $a_2 = -11.8$ eV) is taken from the recent work of Wagner and Bechstedt [96]. Theoretical values are also available for some of the valence-band deformation potentials [95, 118]. Our set is a composite of the recent calculations [158, 159]: $D_1 = -2.9$ eV, $D_2 = 4.9$ eV, $D_3 = 9.4$ eV, $D_4 = -4.0$ eV, $D_5 = -3.3$ eV, and $D_6 = -2.7$ eV, with the last value derived using the quasi-cubic approximation.

The elastic constants for wurtzite AlN were measured by Tsubouchi et al. [160], McNeil et al. [161], Deger et al. [116], and Bu et al. [162]. Theoretical results are also available [95, 120–122, 163–165]. We recommend the values $C_{11} = 396$ GPa, $C_{12} = 137$ GPa, $C_{13} = 108$ GPa, $C_{33} = 373$ GPa, $C_{44} = 116$ GPa suggested by Wright, who provides a detailed discussion of their expected accuracy [117].

The AlN piezoelectric coefficient $d_{33} = 5.6$ pm V^{-1} obtained in Ref. [125] agrees reasonably well with the early determinations [166, 167] reviewed in that work. Lueng et al. [127] measured $d_{33} = 5.1$ pm V^{-1}. Although these experiments found only d_{33}, both d_{33} and d_{13} can be calculated from first principles [118, 128, 129, 168, 169]. We recommend the recent theoretical values of Bernardini and Fiorentini: $d_{33} = 5.4$ pm V^{-1} and $d_{13} = -2.1$ pm V^{-1}, [129] although their elastic coefficients are somewhat larger than the values we recommend. Similar results were recently obtained for AlN thin films [170]. Based on recent measurements [132, 167] and a calculation [129] of the shear piezoelectric coefficient, we recommend $d_{15} = 3.6$ pm V^{-1}.

The difference between the GaN and AlN spontaneous polarizations strongly influences the band profiles and energy levels in GaN/AlN quantum heterostructures. Although rigorous calculations [128, 133–135] of the spontaneous polarization P_{sp} (AlN) have produced results spanning a fairly broad range, from -0.09 C m^{-2} to -0.12 C m^{-2}, values for the difference P_{sp}(AlN) $- P_{sp}$(GaN) tend to be more consistent, with most falling between 0.046 C m^{-2} and 0.056 C m^{-2}. For some time, most experimental studies of the GaN/AlGaN system have reported somewhat smaller P_{sp}(AlN) $- P_{sp}$(GaN). For example, Leroux et al. [171, 172] derived $-0.051 < P_{sp} < -0.036$ C m^{-2} for AlN. A study of the charging of GaN/AlGaN field-effect transistors led to a similar conclusion [173], and Hogg et al. were able to fit their luminescence data by assuming negligible spontaneous polarization [174]. Park and Chuang [175] required $P_{sp} = -0.040$ C m^{-2} to reproduce their GaN/AlGaN quantum-well data. On the other hand, Cingolani et al. [176] reported good agreement with experiment using the higher value from the original Bernardini et al. [128] calculation.

A significant step toward resolving this discrepancy has been the recent realization that the spontaneous polarization for AlGaN cannot be linearly interpolated between the binary endpoint values [177, 178]. In combination with an improved nonlinear-strain treatment of the piezoelectric effect, the discrepancy between theory and experiment for GaN/AlGaN quantum wells has largely been eliminated [134]. We adopt $P_{sp} = -0.090$ C m^{-2} as the recommended value for AlN, in conjunction with P_{sp}(GaN) $= -0.034$ C m^{-2}.

The recommended band structure parameters for wurtzite AlN are compiled in Table 2.1.

2.3.3
InN

Although the GaN and AlN energy gaps are now known to a high degree of accuracy, considerable controversy still surrounds the gap in InN. Early absorption studies on sputtered thin films found E_g in the 1.7–2.2 eV range [179–183]. Those polycrystalline or nanocrystalline films often had high electron densities and low mobilities, although low-density high-mobility samples were also reported. Subsequent investigations of epitaxially-grown wurtzite InN yielded a low-temperature gap of ≈ 1.99 eV [138, 184], which gained common acceptance even though the lack of luminescence at the gap energy remained puzzling. In 2001, however, optical characterizations of MBE-grown single-crystal InN indicated $E_g \approx 0.7$ eV [185–190], which represented a dramatic re-evaluation. Those results were confirmed by absorption and luminescence measurements on many samples grown by different groups [159,191–207]. Luminescence data for InGaN and InAlN alloys also supported the narrow-gap nature of the end-point InN [208, 209]. Theories similarly revised the gap downward, to 0.7–0.9 eV [210–212], and examined why it is anomalously narrow [213] in the context of trends for other semiconductor materials (violation of the common-cation rule).

The previous observations of a wide gap were variously attributed to a large Moss–Burstein shift [214], InN-In$_2$O$_3$ alloying due to significant oxygen incorporation into the sputtered samples [215], quantum confinement effects in nanoscale grains, and deviations from stoichiometry (excess nitrogen incorporation) [216]. While the first possibility may explain the results for highly-doped samples, Butcher and Tansley pointed out that it fails to account for reports of low-doped samples with large gaps [217]. InN-In$_2$O$_3$ alloying effects are not expected to be great enough [217], and in any case there is no hard evidence for such alloys forming (although oxygen incorporation may occur). Confinement effects cannot be ruled out in some cases, but there is no indication for gap variations with grain size. It is also unclear how excess N would impact the gap. Compounding the confusion are recent reports of strong PL at 1.87 eV [218], and of valence-electron energy-loss spectroscopy data indicating $E_g = 1.7$ eV [219]. Thus there there is no general explanation for the observation of $E_g \approx 2.0$ eV in some materials, unless one postulates that the determinations of low doping and high mobilities in some wide-gap samples were unreliable.

Attempts have also been made to explain away the 0.7 eV emission and apparent absorption band edge as artifacts. One possibility is a mid-gap defect (possibly an antisite with transition energy $\approx E_g/3$) [217, 218, 220, 221]. Monemar *et al.* [222] pointed out that this would require a neutral defect of thus far unestablished identity, which must invariably be present in very similar amounts. The apparent observation of low-energy conduction-band/acceptor transitions also defy the wide-gap model [222]. In a second line of attack, Schubina *et al.* suggested that excess In in the MBE growth may produce metallic In clusters [216], leading to a Mie-scattering effect on the absorption that would mask a true energy gap around 1.1–1.4 eV [223,224] (the same authors later appeared to accept the possibility of E_g as small as 0.9 eV [225]). Curiously enough, this "intermediate" gap was observed [226] or deduced [227] in several experiments. The emission near 0.7 eV would then be attributed to transitions at the interface between the InN matrix and the In inclusions. However, the presence of In clusters was ruled out for some of the MBE-grown materials displaying narrow-gap emission and absorption [192].

Several reviews of the InN material properties and energy-gap controversy have appeared [217,222,228–230], although a full understanding remains elusive at this stage. Since most of the materials with high optical quality display long-wavelength emission, we continue to recommend a 0 K gap of 0.69 eV, in conjunction with the Varshni parameters $\alpha = 0.414$ meV K^{-1} and $\beta = 154$ K determined by Wu *et al.* [206]. Future developments in this area are eagerly awaited.

The electron effective mass is intimately tied to the energy gap value [231]. Earlier measurements of the InN electron mass yielded $0.11m_0$ [181], $0.12m_0$ [232], and $0.14m_0$ [233], as well as $0.24m_0$ perpendicular to the c axis [234]. Masses in the 0.04–$0.11m_0$ range have been reported in the more recent literature, based on both experiments and calculations [81, 194, 200, 203, 207, 210, 211, 235–239]. Since only values in the lower half of this range are consistent with our adoption of small E_g, based on the recent experiments [238] and supported by theory [237] we recommend a band-edge mass of $0.07m_0$ (relatively isotropic). The presumed narrow-gap nature of the material also requires that nonparabolicity be taken into account. Assuming $F = -1.63$ (for want of any other information, taken from GaSb [7] with a similar gap), we obtain $E_P = 11.4$ eV.

Valence-band parameters were calculated by Yeo *et al.* [92] using the empirical pseudopotential method, and also by Pugh *et al.* [240] and Dugdale *et al.* [94] using essentially the same technique. The results of the first two studies are quite similar, and we recommend the parameters derived by Pugh *et al.* [240]. However, it should be pointed out that the lower InN energy gap may require a downward revision of the light-hole mass.

Spin–orbit splittings calculated for InN vary from 1 meV to 13 meV (see Ref. [8]), from which we recommend $\Delta_{so} = 5$ meV estimated by Carrier and Wei [210]. Although excitons cannot be resolved at present, owing to the small binding energy and marginal sample quality, assuming Δ_{so} is known, the crystal-field splitting can be deduced from the polarization dependence of the optical absorption [241]. That yields the recommended value of $\Delta_{cr} = 24$ meV. Theoretical estimates of Δ_{cr} range from 17 meV to 301 meV [8,81,92,242]. In any event, Δ_{cr} in InN is certain to be positive.

Christensen and Gorczyca predicted a hydrostatic deformation potential of −4.1 eV for wurtzite InN [63], although a smaller value of −2.8 eV was derived by Kim *et al.* [95]. More recently, Carrier and Wei calculated $a = -4.2$ eV, taking into account the downward revision of the energy gap [210]. Experimentally, the conduction and valence band deformation potentials were estimated as $a_c = -4.3$ eV and $a_v = 1.4$ eV [243]. We recommend the deformation potential of Carrier and Wei [210]. Since a complete set of the valence-band deformation potentials has not been reported, to the best of our knowledge, we recommend using those of GaN. While elastic constants were measured by Sheleg and Savastenko [111], we recommend the improved set of Wright [117]: $C_{11} = 223$ GPa, $C_{12} = 115$ GPa, $C_{13} = 92$ GPa, $C_{33} = 224$ GPa, $C_{44} = 48$ GPa. Kim *et al.* [95], Davydov [120], and Wang and Ye [121] calculated alternative sets.

The InN piezoelectric coefficients were calculated by Bernardini and Fiorentini [129]: $d_{33} = 7.6$ pm V^{-1}, $d_{13} = -3.5$ pm V^{-1}, and $d_{15} = 5.5$ pm V^{-1}. Several experimental reports appeared quite recently, with $d_{13} = -3.7 \pm 0.5$ pm V^{-1} obtained from fits of data for quantum-well devices [130]. Two measurements of d_{33} in wurtzite InN thin films yielded 3.12 ± 0.1 pm V^{-1} [244] and 3.9 ± 0.8 pm V^{-1} [245]. The latter work argued that the measurements possibly underestimate the true magnitude of d_{33}. This is supported by the above fit for d_{13}, which tends to be approximately half of d_{33}. We therefore continue to recommend the set proposed by Bernardini and Fiorentini [129]. While the spontaneous polarization data for GaN/GaInN structures remain inconclusive, most likely owing to material imperfections, the recommended value $P_{sp}(\text{InN}) = -0.042$ C m^{-2} is consistent with the most thorough comparison of experiment and theory [134].

Recommended band structure parameters for wurtzite InN are compiled in Table 2.1.

2.3.4
AlGaN

Initial studies of the energy gap's compositional dependence reported downward [246] as well as upward [247, 248] bowing. Subsequent PL [249] and absorption [250] measurements found a bowing parameter of $+1.0$ eV, which was often used in band structure calculations. Since then, many materials fabricated under a variety of growth conditions have been studied. In a good review of the results up to 1999, Lee *et al.* [251] divided the previous works into three general classes. First, the early findings of an upward bowing [247, 248] were generally not duplicated (with the somewhat inconclusive exception of Ref. [252]). The second class of materials, which were grown at high temperatures, generally exhibited a strong downward bowing of at least $+1.3$ eV [142, 253–258]. Often those results could not be fitted to a continuous parabolic curve, since they tended to jump to stronger bowing as the Al fraction increased [251]. It was proposed that the apparent strong bowing was actually an artifact resulting from defect- or impurity-related transitions at energies below the bandgap [259]. Lee et al. [251] further suggested that only samples fabricated by first growing a GaN buffer on sapphire at low temperature, followed by high-temperature growth of the alloy layer, may be expected to yield reliable energy gaps [260–265]. Residual anomalies for materials of the third class were attributed to incomplete strain relaxation. Based on these considerations, Lee et al. recommended the bowing parameter $C = 0.6$ eV [251].

Since then, there have been many other reports of energy gaps in $Al_xGa_{1-x}N$. Ochalski et al. [266] observed no detectable bowing for $x < 0.3$ (this work falls broadly into the third class of Lee et al.). A wider range of Al compositions was considered by Shan et al. [267], who deduced a bowing of $+1.33$ eV for alloy layers grown on AlN buffers. Meyer et al. [268] reported $C = 0.7$ eV for material that again falls into the third class as defined above. A similar value of $C = 0.8$ eV was reported by Omnes et al. [269]. Bergman et al. obtained $C = 1.2$ eV, and reported no evidence for x-dependent local bandgap variations induced by chemical ordering [270]. Cathodoluminescence, absorption, and reflectance measurements of epitaxial AlGaN grown on Si(111) suggested $C = 1.5$ eV [271–273]. Ebling et al. [274] found a large bowing parameter in AlGaN with partial chemical ordering. Other materials representing the third class were recently investigated by Jiang et al. [275] ($C = 0.53$ eV), Wagner et al. [276] ($C \approx 1$ eV), Zhou et al. [277] ($C = 0.85$ eV), Katz et al. [278] ($C = 1.38$ eV), and Yun et al. [279] ($C = 1.0$ eV). Paduano et al. [280] ($C = 0.70$ eV) suggested that the data are reliable only for growth on a specific sequence of buffer layers (giving narrow X-ray diffraction features), and special efforts are taken to compensate for the energy gap's strain dependence. Besides their own, they assigned two other studies [251, 268] to that category. Very recent reports include the PL measurements of Nepal et al. [62] ($C = 1.0$ eV), the spectroscopic ellipsometry data of Buchheim et al. [281] ($C = 0.9$ eV), PL on bulk crystals with Al fraction $x \leq 0.53$ [282] ($C = 1.3$ eV), PL and reflectance data [283] (showing that $C = 0.9$ eV provides a good fit until the crystal-field splitting becomes negative), optical transmission measurements [284] for $0.4 < x < 1$ ($C = 0.75$ eV), and the reflectance and PL data of Onuma et al. [285] ($C = 0.82$ eV).

Most theories project that the Γ-valley in wurtzite AlGaN should have a negligible to modest downward bowing ($C = 0$–0.75 eV) [80, 286–288]. Based on the consistent finding of a relatively small bowing parameter by the theories, as well as by some of the most authoritative experiments, we recommend $C = 0.8$ eV for AlGaN.

Nepal et al. found the Varshni parameters in AlGaN to exhibit bowing [62]. Although we recommend slightly different AlGaN bowing and GaN Varshni parameters, we expect their bowing parameters for α ($C = 2.15$ meV K^{-1}) and β ($C = 1561$ K) to be useful. Theory predicts that the energy gap's pressure dependence should be nonlinear [289], although the available data [290] (only for small Al fractions) does not allow a confirmation of this. The Al-GaN electron effective mass followed an approximately linear dependence on composition in the Shubnikov–de-Haas study by Jena et al. [291]. A bowing of $C(P_{sp}) = -0.021$ C m^{-2} for the spontaneous polarization is consistent with a detailed recent comparison of theory and experiment [134].

The recommended nonzero bowing parameters for AlGaN are summarized in Table 2.2.

Tab. 2.2 Nonzero bowing parameters for GaInN, AlGaN, and AlInN.

Parameters	GaInN	AlGaN	AlInN
E_g^Γ (eV)	1.4	0.8	3.4
α (meV K^{-1})	0	2.15	0
β (K)	0	1561	0
P_{sp} (C m^{-2})	−0.037	−0.021	−0.070

2.3.5
InGaN

Early determinations of the energy-gap bowing parameter for InGaN employed the ≈ 2 eV value for E_g. Further complications resulted from the frequent occurrence of chemical ordering and (partial) phase-decomposition effects [292], with clustering of In-rich regions [293]. However, recent work has produced considerable progress toward a fundamental understanding of this alloy gap.

A fit to the early data of Osamura et al. [179] yielded a bowing parameter C of ≈ 1.0 eV for wurtzite InGaN. That result was consistent with the theories of Wright and Nelson [294], Kassali and Bouarissa [295], Goano et al. [80], Ferhat et al. [296,297], and other early calculations [298,299]. Nakamura found that this bowing produced a good fit to the PL data for low-In compositions [300]. Li et al. obtained a slightly larger bowing [301], on the basis of PL from InGaN/GaN superlattices. Bellaiche and Zunger [302] established that a large apparent reduction in the InGaN bandgap could result from short-range atomic ordering. Wright et al. also considered strain and ordering effects [303].

A number of later works cast doubt on the picture of a small bandgap bowing parameter in InGaN alloys. For example, the experimental results of McCluskey et al. for In$_x$Ga$_{1-x}$N epilayers with $x < 0.12$ implied a bowing as large as 3.5 eV [304]. First-principles calculations by the same authors indicated that the bowing parameter itself may depend strongly on composition, at least for small In fractions. Kent et al. also calculated a strong variation of C

with x, and additionally determined that epitaxial layers may have a slightly smaller valence band offset than bulk materials [305]. Numerous subsequent studies reported similarly large bowings in the 2.4–4.5 eV range [306–323]. All of those works focused on In$_x$Ga$_{1-x}$N with $x < 0.20$. However, after noticing a weak temperature dependence of the alloy PL peak, as well as a Stokes shift between the PL and photoreflectance lines, Shan et al. [308] suggested that the PL may in fact be emitted primarily from material that is locally In-rich. Such ordering would naturally lead to an overestimate of the bowing parameter, which was disputed by McCluskey et al. [324]. Bellaiche et al. [325] suggested the interesting possibility that "clustering-like" electronic effects may occur without any actual chemical clustering, due to localization of the hole wavefunctions on the In sites. Stepanov et al. [326] noted that much of the scatter in the bowing parameter would be removed (to obtain C= 2.1–2.4 eV) were all of the studies to use the same value of Poisson's ratio.

Since we have adopted the lower InN band gap value, the data for In-rich In$_x$Ga$_{1-x}$N must be re-evaluated. The recent growth of high-quality epitaxial samples with large x has considerably broadened the compositional range over which bowing could be reliably determined [189,209,327]. Wu et al. [327] attributes a strong downward shift of the PL peak energy relative to the absorption edge to emission primarily from localized In-rich regions. A similar dependence on composition was observed by O'Donnell et al. [328]. Davydov et al. [189,228,329,330] arrived at a larger value of 2.5 eV from a PL absorption analysis that reduces to the narrow-gap InN end-point, and Hori et al. [208] obtained nearly the same curve. Bechstedt et al. [242] calculated a composition-dependent bowing parameter, with a large value for small In fractions and $C = 1.3$–1.4 eV for larger x. A variable C was also predicted by Sokeland et al. [331]. Other recent calculations indicate nearly constant bowing parameters of 1.7–1.8 eV [332,333] and 2.7 eV [334]. Our provisional recommendation of $C = 1.4$ eV is based on the work of Wu et al. [209,327].

Very little information is available on the other band parameters for InGaN. Effective-mass parameters for Ga-rich alloys were compiled by Pugh et al. [335]. Composition-dependent calculations [336,337] imply negligible bowing of the piezoelectric coefficients for wurtzite InGaN. We recommend a bowing of $C(P_{sp}) = -0.037$ C m^{-2} for the spontaneous polarization in the alloy, to be consistent with a recent detailed comparison of theory and experiment [134].

The recommended nonzero bowing parameters for InGaN are summarized in Table 2.2.

2.3.6
InAlN

$In_{1-x}Al_xN$ becomes lattice-matched to GaN at $x = 0.83$. The first experimental study of sputtered InAlN yielded bowing so strong as to imply a smaller energy gap for the lattice-matched alloy than for GaN [338]. Guo *et al.* [339] measured a wide gap in the InN limit of InN-rich InAlN alloys. For the opposite limit of Al-rich alloys, Kim *et al.* found consistency with downward bowing of at least 2.5 eV [340]. While Peng *et al.* gave a cubic expression for the energy gap that fit results spanning the entire range of compositions [341], that result also assumed a large InN gap. Furthermore, the strong bowing implied by their measurements, and also those of Yamaguchi *et al.* [342, 343], may be artifacts of polycrystallinity and clustering, by analogy to the effects already discussed for InGaN alloys. If combined with our recommended binary end-points, the substantially higher bandgap of 2.74 eV reported by Shubina *et al.* [344] for $In_{.68}Al_{0.32}N$ implies a slight upward bowing. Assuming wide-gap InN, Lukitsch *et al.* [345] reported strong downward bowing over a wide range of compositions. The absorption spectra fits of Guo *et al.* [346] also implied large gaps for In-rich InAlN. Recent studies measured the energy gaps in alloys exhibiting the narrow-gap InN behavior at the end-point [241, 347, 348]. Those works deduced bowing parameters of $C = 4.0$, 3.0, and 3.3 eV, respectively. Our recommended bowing is an average of those three: $C = 3.4$ eV. We should add that theoretical calculations for wurtzite InAlN produced bowing parameters of 2.38 eV [80], 4.1 eV [332], 3.7 eV [349], 3.3 eV [350] and 4.2 eV [334]. Some of those works employed an InN energy gap inconsistent with our recommendation.

A study [243] of pressure-dependent bandgaps in InAlN did not permit conclusions about the bowing, due to uncertain end-points. To be consistent with a recent detailed comparison of theory and experiment [134], we recommend a spontaneous-polarization bowing of $C(P_{sp}) = -0.070\,\mathrm{C\,m}^{-2}$.

The recommended nonzero bowing parameters for InAlN are summarized in Table 2.2.

2.3.7
AlGaInN

Energy gaps have also been reported for AlGaInN with rather small In fractions [342, 351–356]. Some results indicate a nearly linear bandgap reduction when $In < 2\%$ [357]. The cut-off wavelengths of AlGaInN (lattice-matched to GaN) ultraviolet photodetectors are also generally consistent with a linear interpolation [358]. Other studies [359] deduce a bowing coefficient of 2.5 eV associated with the In fraction.

2.3.8
Band Offsets

Bernardini and Fiorentini [360] have suggested that once the large electrostatic fields are included, even defining the band offset for a wurtzite system's polar interface becomes nontrivial. Confining our discussion to the (0001) orientation, the electrostatic potential takes on a characteristic sawtooth profile owing to the macroscopic polarization and corresponding interface charges [360, 361]. An additional complication is that the GaN/AlN and AlN/GaN cases are inequivalent [362], meaning that the two offsets must be specified separately [360, 361]. First-principles calculations found offsets in the rather narrow 0.7–0.8 eV range, although polarization and strain effects were not accounted for consistently [362]. From a detailed treatment of the strain-induced asymmetry at the (0001) polar heterojunction, Bernardini and Fiorentini obtained 0.2 eV for AlN/GaN and 0.85 eV for GaN/AlN [360].

On the experimental side, Baur et al. extracted a valence-band offset (VBO) of 0.5 eV from the difference between iron acceptor levels in GaN and AlN [363]. A fit to the PL spectrum for GaN/AlGaN quantum wells was consistent with VBO ≈ 0.9 eV [58]. A more recent fit by Nam et al. [364] implied ≈ 0.8 eV, and much the same value was obtained from deep-acceptor emission data [365]. X-ray photoemission spectroscopy yielded VBO $= 0.8 \pm 0.3$ eV at the wurtzite GaN/AlN junction [366], which was revised to 0.70 ± 0.24 eV in a later paper by the same authors [367]. Using the same approach, Waldrop and Grant found a considerably different value of 1.36 ± 0.07 eV [368]. A related study of $Al_xGa_{1-x}N$ alloys reported a nearly linear variation, with a positive bowing parameter of 0.59 eV [369]. Using X-ray and ultraviolet photoelectron spectroscopy, King et al. found 0.5 eV to 0.8 eV for the GaN/AlN VBO, depending on the growth temperature [370]. They attributed the differences to strain, defects, and film stoichiometry effects. Rizzi et al. [371] reported VBO $= 0.15$–0.4 eV for relaxed heterojunctions, and pointed out that in GaN the Ga $3d$ core level that has been used as a reference is in fact hybridized with other valence bands. From the exciton luminescence in AlGaN, Westmeyer et al. [372] found the conduction band offset to be 57% of the total energy gap discontinuity, which corresponds to a VBO of 1.1 eV. In the absence of VBO bowing, the VBO of 0.2 eV obtained by Foxon et al. for the $Al_{0.2}Ga_{0.8}N$/GaN junction [373] implies VBO $= 1.0$ eV for GaN/AlN. Considering the numerous pitfalls that can plague experiments such as these, we find no compelling reason to doubt the soundness of the theoretical evaluation by Bernardini and Fiorentini [360].

The theoretical work on valence band offsets at the important InN/GaN interface was summarized in Ref. [8]. X-ray photoemission spectroscopy measurements by Martin et al. gave a large VBO of 1.05 eV [367], and optical measurements on GaInN/GaN quantum wells were also consistent with a large value [374]. Martin *et al.* [367] corrected for the piezoelectric fields, and found no significant deviation from the transitivity rule when the InN/AlN heterojunction VBO was also determined. Although the InN bandgap was recently re-evaluated, we expect most of the energy difference to appear in the conduction band rather than the valence band. We therefore recommend VBO = 0.5 eV for both the InN/GaN and GaN/InN interfaces, since at present there is no compelling evidence for a significant difference. The InN/AlN case is more complicated, with growth-sequence asymmetries probably arising from the very different spontaneous polarizations in the two materials. One photoelectron spectroscopy study [375] obtained a very large 3.10 ± 0.04 eV VBO, which may not be consistent with the offsets proposed above for the other wurtzite nitride interfaces. The resolution of this issue must await further studies.

All of our band-offset recommendations should be considered provisional, and in particular cases involving InN, may be significantly altered by future developments. The recommended asymmetric wurtzite offsets are listed in Table 2.3, where, for lack of other information, we ignore any bowing of the VBOs. The conduction band minimum then bows by the amount of the energy-gap bowing, as is also typical of non-nitride III-V semiconductors [7].

Tab. 2.3 Recommended valence-band offsets (including strain and polarization effects) for the various binary wurtzite interface combinations. A positive VBO corresponds to higher valence band maximum in the first material than in the second.

Heterojunction	VBO (eV)
AlN/GaN (0001)	−0.20
GaN/AlN (0001)	0.85
InN/GaN (0001)	0.50
GaN/InN (0001)	−0.50

2.4
Conclusions

We have reviewed the literature for band structure parameters in wurtzite nitride semiconductors and their alloys and, based on this survey, recommended values for each parameter. We believe that future changes to some of the parameters will be minor. Assuming a fully satisfactory resolution of the InN energy-gap controversy, these include energy gaps for wurtzite GaN, AlN, and InN, as well as their electron effective masses. Agreement may also be at hand for the spin–orbit and crystal-field splittings in these materials, as well as the bowing parameters for InGaN, AlGaN, and InAlN. Interestingly, Wu *et al.* [347] noted that the bowing parameters for all three alloys, normalized by the difference in end-point energy gaps, fall within a fairly narrow range (0.3–0.6 for the parameters recommended here). The understanding of piezoelectric coefficients and spontaneous polarizations in GaN, AlN, InN, and their alloys has reached a new plateau, although further research is needed to fully confirm the proposed values.

However, a definitive valence parameter set, which is fully consistent with the consensus values for spin–orbit and crystal-field splittings, remains to be generated even for GaN. Other properties that are still somewhat or highly uncertain include the band offsets and deformation potentials for all of the binaries and their alloys. For those parameters, our recommendations should be considered provisional until more detailed and definitive experimental and theoretical evidence becomes available.

References

1 J. H. Edgar, editor, *Properties of Group-III Nitrides*, (INSPEC, IEE, London, 1994), EMIS Datareviews Series No. 11.

2 *Landolt–Börnstein Numerical Data and Functional Relationships in Science and Technology*, edited by O. Madelung, M. Schultz, and H. Weiss, New series, Vol. 17 (1982); reprinted in O. Madelung, editor, *Semiconductors-Basic Data*, 2nd ed. (Springer, New York, 1996).

3 F. A. Ponce, *III–V Nitrides*, edited by T. D. Moustakas, I. Akasaki, and B. A. Monemar, Mater. Res. Soc. Symp. Proc. **449** (1997).

4 *GaN and Related Materials*, edited by S. J. Pearton (Gordon and Breach, New York, 1997).

5 *Group III Nitride Semiconductor Compounds*, edited by B. Gil (Clarendon, Oxford 1998).

6 *GaN*, edited by J. I. Pankove and T. D. Moustakas (Academic, New York, 1998), Vol. 1.

7 I. Vurgaftman, J. R. Meyer, and L.-R. Ram-Mohan, J. Appl. Phys. **89**, 5815 (2001).

8 I. Vurgaftman and J. R. Meyer, J. Appl. Phys. **94**, 3675 (2003).

9 J. C. Slater and G. F. Koster, Phys. Rev. **94**, 1498 (1954).

10 P. Vogl, H. P. Hjalmarson, and J. D. Dow, J. Pys. Chem. Solids **44**, 365 (1983).

11 M. H. Cohen and V. Heine, Phys. Rev. **122**, 1821 (1961).

12 M. L. Cohen and J. R. Chelikowsky, *Electronic Structure and Optical Properties of Semiconductors*, Springer Series in Solid-State Sciences, ed. by M. Cardona (Springer, Berlin, 1988), Vol. 75.

13 E. I. Rashba, Fiz. Tverd. Tela (Leningrad) **1**, 407 (1959) [Sov. Phys. Solid State **1**, 368 (1959)]; E. I. Rashba and V. I. Sheka, *Fiz. Tverd. Tela (Leningrad), Collection of papers II* (Acad. Sci. USSR, Moscow, 1959), p. 162.

14 G. L. Bir and G. E. Pikus, *Symmetry and Strain-Induced Effects in Semiconductors* (Wiley, New York, 1974).

15 S. L. Chuang and C. S. Chang, Phys. Rev. B **54**, 2491 (1996).

16 Yu. M. Sirenko, J.-B. Jeon, K. W. Kim, M. A. Littlejohn, and M. A. Stroscio, Phys. Rev. B **53**, 1997 (1996).

17 G. B. Ren, Y. M. Liu, and P. Blood, Appl. Phys. Lett. **74**, 1117 (1999).

18 A. A. Yamaguchi, Y. Mochizuki, H. Sunakawa, and A. Usui, J. Appl. Phys. **83**, 4542 (1998).

19 A. V. Rodina and B. K. Meyer, Phys. Rev. B **64**, 245209 (2001).

20 X. Cartoixà, D. Z.-Y. Ting, and T. C. McGill, J. Appl. Phys. **93**, 3974 (2003).

21 B. Chen, M. Lazzouni, and L. R. Ram-Mohan, Phys. Rev. B **45**, 1204 (1992).

22. F. Szmulowicz, Phys. Rev. B **57**, 9081 (1998).
23. L. R. Ram-Mohan and J. R. Meyer, J. Nonlinear Opt. Phys. Mater. **4**, 191 (1995).
24. T. Li and K. J. Kuhn, J. Comput. Phys. **115**, 288 (1994).
25. B. Vinter, Phys. Rev. B **66**, 045324 (2002).
26. Y. P. Varshni, Physica (Amsterdam) **34**, 149 (1967).
27. L. Viña, S. Logothetidis, and M. Cardona, Phys. Rev. B **30**, 1979 (1984).
28. R. Pässler, Phys. Stat. Sol. B **216**, 975 (1999).
29. R. Pässler, J. Appl. Phys. **89**, 6235 (2001).
30. R. Pässler, Phys. Rev. B **66**, 085201 (2002).
31. F. J. Manjón, M. A. Hernandez-Fenollosa, B. Marí, S. F. Li, C. D. Poweleit, A. Bell, J. Menéndez, and M. Cardona, Eur. Phys. J. B **40**, 453 (2004).
32. R. Dingle, D. D. Sell, S. E. Stokowski, and M. Ilegems, Phys. Rev. B **4**, 1211 (1971).
33. O. Lagerstedt and B. Monemar, J. Appl. Phys. **45**, 2266 (1974).
34. B. Monemar, Phys. Rev. B **10**, 676 (1974).
35. C. Guenaud, E. Deleporte, M. Voos, C. Delalande, B. Beaumont, M. Leroux, P. Gibart, and J. P. Faurie, MRS Internet J. Nitride Semicond. Res. **2**, 10 (1997).
36. W. Shan, T. J. Schmidt, X. H. Yang, S. J. Hwang, J. J. Song, and B. Goldenberg, Appl. Phys. Lett. **66**, 985 (1995).
37. W. Shan, X. C. Xie, J. J. Song, and B. Goldenberg, Appl. Phys. Lett. **67**, 2512 (1995).
38. B. Gil, O. Briot, and R.-L. Aulombard, Phys. Rev. B **52**, R17028 (1995).
39. G. D. Chen, M. Smith, J. Y. Lin, H. X. Jiang, S.-H. Wei, M. A. Khan, and C. J. Sun, Appl. Phys. Lett. **68**, 2784 (1996).
40. C. Merz, M. Kunzer, U. Kaufmann, I. Akasaki, and H. Amano, Semicond. Sci. Technol. **11**, 712 (1996).
41. M. O. Manasreh, Phys. Rev. B **53**, 16425 (1996).
42. J. S. Im, A. Moritz, F. Steuber, V. Haerle, F. Scholz, and A. Hangleiter, Appl. Phys. Lett. **70**, 631 (1997).
43. A. Shikanai, T. Azuhata, T. Sota, S. Chichibu, A. Kuramata, K. Horino, and S. Nakamura, J. Appl. Phys. **81**, 417 (1997).
44. C. F. Li, Y. S. Huang, L. Malikova, F. H. Pollak, Phys. Rev. B **55**, 9251 (1997).
45. K. Reimann, M. Steube, D. Froelich, and S. J. Clarke, J. Cryst. Growth **189/190**, 652 (1998).
46. D. G. Chtchekine, Z. C. Feng, S. J. Chua, and G. D. Gilliland, Phys. Rev. B **63**, 125211 (2001).
47. A. V. Rodina, M. Dietrich, A. Göldner, L. Eckey, A. Hoffmann, Al. L. Efros, M. Rosen, and B. K. Meyer, Phys. Rev. B **64**, 115204 (2002).
48. H. B. Yu, H. Chen, D. Li, Y. J. Han, X. H. Zheng, Q. Huang, and J. M. Zhou, J. Cryst. Growth **263**, 94 (2004).
49. B. Monemar, J. P. Bergman, I. A. Buyanova, W. Li, H. Amano, and I. Akasaki, MRS Internet J. Nitride Semicond. Res. **1**, 2 (1996).
50. M. Smith, G. D. Chen, J. Y. Lin, H. X. Jiang, M. Asif Khan, C. J. Sun, Q. Chen, and J. W. Yang, J. Appl. Phys. **79**, 7001 (1996).
51. D. C. Reynolds, D. C. Look, W. Kim, O. Aktas, A. Botchkarev, A. Salvador, H. Morkoc, and D. N. Talwar, J. Appl. Phys. **80**, 594 (1996).
52. J. F. Muth, J. H. Lee, I. K. Shmagin, R. M. Kolbas, H. C. Casey, Jr., B. P. Keller, U. K. Mishra, and S. P. DenBaars, Appl. Phys. Lett. **71**, 2572 (1997).
53. S. Chichibu, T. Azuhata, T. Sota, and S. Nakamura, J. Appl. Phys. **79**, 2784 (1996).
54. A. Castaldini, A. Cavallini, and L. Polenta, Appl. Phys. Lett. **84**, 4851 (2004).
55. C.-H. Su, W. Palosz, S. Zhu, S. L. Lehoczky, I. Grzegory, P. Perlin, and T. Suski, J. Cryst. Growth **235**, 111 (2002).
56. H. Teisseyre, P. Perlin, T. Suski, I. Grzegory, S. Porowski, J. Jun, A. Pietraszko, and T. D. Moustakas, J. Appl. Phys. **76**, 2429 (1994).
57. J. Petalas, S. Logothetidis, S. Boultadakis, M. Alouani, and J. M. Wills, Phys. Rev. B **52**, 8082 (1995).
58. A. Salvador, G. Liu, W. Kim, O. Aktas, A. Botchkarev, and H. Morkoc, Appl. Phys. Lett. **67**, 3322 (1995).
59. A. S. Zubrilov, Yu. V. Melnik, A. E. Nikolaev, M. A. Jacobson, D. K. Nelson, and V. A. Dmitriev, Semicond. **33**, 1067 (1999) [Fiz. Tekh. Poluprov. **33**, 1173 (1999)].

60. D. C. Reynolds, J. Hoelscher, C. W. Litton, and T. C. Collins, J. Appl. Phys. **92**, 5596 (2002).
61. K. B. Nam, J. Li, J. Y. Lin, and H. X. Jiang, Appl. Phys. Lett. **85**, 3489 (2004).
62. N. Nepal, J. Li, M. L. Nakarmi, J. Y. Lin, and H. X. Jiang, Appl. Phys. Lett. **87**, 242104 (2005).
63. N. E. Christensen and I. Gorczyca, Phys. Rev. B **50**, 4397 (1994).
64. A. S. Barker, Jr. and M. Illegems, Phys. Rev. B **7**, 743 (1973).
65. B. Rheinlander and H. Neumann, Phys. Stat. Sol. B **64**, K123 (1974).
66. B. K. Meyer, D. Volm, A. Graber, H. C. Alt, T. Detchprohm, A. Amano, and I. Akasaki, Solid State Commun. **95**, 597 (1995).
67. A. M. Witowski, K. Pakula, J. M. Baranowski, M. L. Sadowski, and P. Wyder, Appl. Phys. Lett. **75**, 4154 (1999).
68. M. Drechsler, D. M. Hoffman, B. K. Meyer, T. Detchprohm, H. Amano, and I. Akasaki, Jpn. J. Appl. Phys. **34**, L1178 (1995).
69. P. Perlin, E. Litwin-Staszewska, B. Suchanek, W. Knap, J. Camassel, T. Suski, R. Piotrzkowski, I. Grzegory, S. Porowski, E. Kaminska, and J. C. Chervin, Appl. Phys. Lett. **68**, 1114 (1996).
70. Y. J. Wang, R. Kaplan, H. K. Ng, K. Doverspike, D. K. Gaskill, T. Ikedo, I. Akasaki, and H. Amono, J. Appl. Phys. **79**, 8007 (1996).
71. W. Knap, S. Contreras, H. Alause, C. Skierbiszewski, J. Camassel, M. Dyakonov, J. L. Robert, J. Yang, Q. Chen, M. Asif Khan, M. L. Sadowski, S. Huant, F. H. Yang, M. Goiran, J. Leotin, and M. S. Shur, Appl. Phys. Lett. **70**, 2123 (1997).
72. A. Kasic, M. Schubert, S. Einfeldt, D. Hommel, and T. E. Tiwald, Phys. Rev. B **62**, 7365 (2000).
73. S. Shokhovets, G. Gobsch, and O. Ambacher, Appl. Phys. Lett. **86**, 161908 (2005).
74. S. Syed, J. B. Heroux, Y. J. Wang, M. J. Manfra, R. J. Molnar, and H. L. Stormer, Appl. Phys. Lett. **83**, 4553 (2003).
75. S. Elhamri, R. S. Newrock, D. B. Mast, M. Ahoujja, W. C. Mitchel, J. M. Redwing, M. A. Tischler, and J. S. Flynn, Phys. Rev. B **57**, 1374 (1998).
76. A. Saxler, P. Debray, R. Perrin, S. Elhamri, W. C. Mitchel, C. R. Elsass, I. P. Smorchkova, B. Heying, E. Haus, P. Fini, J. P. Ibbetson, S. Keller, P. M. Petroff, S. P. DenBaars, U. K. Mishra, and J. S. Speck, J. Appl. Phys. **87**, 369 (2000).
77. L. W. Wong, S. J. Cai, R. Li, K. Wang, H. W. Jiang, and M. Chen, Appl. Phys. Lett. **73**, 1391 (1998).
78. T. Wang, J. Bai, S. Sakai, Y. Ohno, and H. Ohno, Appl. Phys. Lett. **76**, 2737 (2000).
79. D. R. Hang, C.-T. Liang, C. F. Huang, Y. H. Chang, Y. F. Chen, H. X. Jiang, and J. Y. Lin, Appl. Phys. Lett. **79**, 66 (2001).
80. M. Goano, E. Bellotti, E. Ghillino, G. Ghione, and K. F. Brennan, J. Appl. Phys. **88**, 6467 (2000).
81. D. Fritsch, H. Schmidt, and M. Grundmann, Phys. Rev. B **67**, 235205 (2003).
82. B. Rezaei, A. Asgari, and M. Kalafi, Physica B **371**, 107 (2006).
83. K. P. Korona, A. Wysmolek, K. Paluka, R. Stepniewski, J. M. Baranowski, I. Grzegory, B. Lucznik, M. Wroblewski, and S. Porowski, Appl. Phys. Lett. **69**, 788 (1996).
84. J. Campo, M. Julier, D. Coquillat, J. P. Lascaray, D. Scalbert, and O. Briot, Phys. Rev. B **56**, R7108 (1997).
85. M. Julier, J. Campo, B. Gil, J. P. Lascaray, and S. Nakamura, Phys. Rev. B **57**, R6791 (1998).
86. N. V. Edwards, S. D. Yoo, M. D. Bremser, T. W. Weeks, Jr., O. H. Nam, H. Liu, R. A. Stall, M. N. Horton, N. R. Perkins, T. F. Kuech, and D. E. Aspnes, Appl. Phys. Lett. **70**, 2001 (1997).
87. B. Bouhafs, F. Litimein, Z. Dridi, and P. Ruterana, Phys. Stat. Sol. B **236**, 61 (2003).
88. J. W. Orton, Semicond. Sci. Technol. **10**, 101 (1995).
89. M. Steube, K. Reimann, D. Froelich, and S. J. Clarke, Appl. Phys. Lett. **71**, 948 (1997).
90. P. A. Shields, R. J. Nicholas, F. M. Peeters, B. Beaumont, and P. Gibart, Phys. Rev. B **64**, 081203 (2001).

91 B. Santic, Semicond. Sci. Technol. **18**, 219 (2003).

92 Y. C. Yeo, T. C. Chong, and M. F. Li, J. Appl. Phys. **83**, 1429 (1998).

93 M. Suzuki, T. Uenoyama, and A. Yanase, Phys. Rev. B **52**, 8132 (1995).

94 D. J. Dugdale, S. Brand, and R. A. Abram, Phys. Rev. B **61**, 12933 (2000).

95 K. Kim, W. R. L. Lambrecht, B. Segall, and M. van Schilfgaarde, Phys. Rev. B **56**, 7363 (1997).

96 J.-M. Wagner and F. Bechstedt, Phys. Rev. B **66**, 115202 (2002).

97 W. Shan, R. J. Hauenstein, A. J. Fischer, J. J. Song, W. G. Perry, M. D. Bremser, R. F. Davis, and B. Goldenberg, Phys. Rev. B **54**, 13460 (1996).

98 L. Hsu and W. Walukiewicz, Phys. Rev. B **56**, 1520 (1997).

99 W. Knap, E. Borovitskaya, M. S. Shur, L. Hsu, W. Walukiewicz, E. Frayssinet, P. Lorenzini, N. Grandjean, C. Skierbiszewski, P. Prystawko, M. Leszczynski, and I. Grzegory, Appl. Phys. Lett. **80**, 1228 (2002).

100 P. Harmer, M. P. Halsall, D. Wolverson, P. J. Parbrook, and S. J. Henley, Semicond. Sci. Technol. **19**, L22 (2004).

101 S. Ghosh, P. Waltereit, O. Brandt, H. T. Grahn, and K. H. Ploog, Phys. Rev. B **65**, 075202 (2002).

102 B. Gil and A. Alemu, Phys. Rev. B **56**, 12446 (1997).

103 H. Y. Peng, M. D. McCluskey, Y. M. Gupta, M. Kneissl, and N. M. Johnson, Phys. Rev. B **71**, 115207 (2005).

104 S. Kamiyama, K. Ohnaka, M. Suzuki, and T. Uenoyama, Jpn. J. Appl. Phys. **34**, L822 (1995).

105 T. Ohtoshi, A. Niwa, and T. Kuroda, J. Appl. Phys. **82**, 1518 (1997).

106 M. Suzuki and T. Uenoyama, Jpn. J. Appl. Phys. **35**, 1420 (1996).

107 M. Suzuki and T. Uenoyama, J. Appl. Phys. **80**, 6868 (1996).

108 M. Kumagai, S. L. Chuang, and H. Ando, Phys. Rev. B **57**, 15303 (1998).

109 A. Alemu, B. Gil, M. Julier, and S. Nakamura, Phys. Rev. B **57**, 3761 (1998).

110 A. A. Yamaguchi, Y. Mochizuki, C. Sasaoka, A. Kimura, M. Nido, and A. Usui, Appl. Phys. Lett. **71**, 374 (1997).

111 A. U. Sheleg and A. V. Savastenko, Izv. Akad. Nauk SSSR, Neorg. Mater. **15**, 1598 (1979).

112 A. Polian, M. Grimsditch, and I. Grzegory, J. Appl. Phys. **79**, 3343 (1996).

113 Y. Takagi, M. Ahart, T. Azuhata, T. Sota, K. Suzuki, and S. Nakamura, Physica B **219–220**, 547 (1996).

114 R. B. Schwarz, K. Khachaturyan, and E. R. Weber, Appl. Phys. Lett. **70**, 1122 (1997).

115 M. Yamaguchi, T. Yagi, T. Azuhata, T. Sota, K. Suzuki, S. Chichibu, and S. Nakamura, J. Phys. Condens. Matter **9**, 241 (1997).

116 C. Deger, E. Born, H. Angerer, O. Ambacher, M. Stutzmann, J. Hornsteiner, E. Riha, and G. Fischerauer, Appl. Phys. Lett. **72**, 2400 (1998).

117 A. F. Wright, J. Appl. Phys. **82**, 2833 (1997).

118 K. Shimada, T. Sota, and K. Suzuki, J. Appl. Phys. **84**, 4951 (1998).

119 T. Deguchi, D. Ichiryu, K. Toshikawa, K. Sekiguchi, T. Sota, R. Matsuo, T. Azuhata, M. Yamaguchi, T. Yagi, S. Chichibu, and S. Nakamura, J. Appl. Phys. **86**, 1860 (1999).

120 S. Yu. Davydov, Semicond. **36**, 41 (2002).

121 S. Q. Wang and H. Q. Ye, Phys. Stat. Sol. B **240**, 45 (2003).

122 S. P. Lepkowski, J. A. Majewski, and G. Jurczak, Phys. Rev. B **72**, 245201 (2005).

123 A. D. Bykhovski, V. V. Kaminski, M. S. Shur, Q. C. Chen, and M. A. Khan, Appl. Phys. Lett. **68**, 818 (1996).

124 S. Muensit and I. L. Guy, Appl. Phys. Lett. **72**, 1896 (1998).

125 I. L. Guy, S. Muensit, and E. M. Goldys, Appl. Phys. Lett. **75**, 4133 (1999).

126 C. M. Lueng, H. L. W. Chan, C. Surya, W. K. Fong, C. L. Choy, P. Chow, and M. Rosamond, J. Non-Cryst. Solids **254**, 123 (1999).

127 C. M. Lueng, H. W. Chan, C. Surya, and C. L. Choy, J. Appl. Phys. **88**, 5360 (2000).

128 F. Bernardini, V. Fiorentini, and D. Vanderbilt, Phys. Rev. B **56**, R10024 (1997).

129 F. Bernardini and V. Fiorentini, Appl. Phys. Lett. **80**, 4145 (2002).

130 A. Hangleiter, F. Hitzel, S. Lahmann, and U. Rossow, Appl. Phys. Lett. **83**, 1169 (2003).

131 U. M. E. Christmas, A. D. Andreev, and D. A. Faux, J. Appl. Phys. **98**, 073522 (2005).

132 S. Muensit, E. M. Goldys, and I. L. Guy, Appl. Phys. Lett. **75**, 3965 (1999).

133 F. Bechstedt, U. Grossner, and J. Furthmüller, Phys. Rev. B **62**, 8003 (2000).

134 O. Ambacher, J. Majewski, C. Miskys, A. Link, M. Hermann, M. Eickhoff, M. Stutzmann, F. Bernardini, V. Fiorentini, V. Tilak, B. Schaff, and L. F. Eastman, J. Phys. Condens. Matter **14**, 3399 (2002).

135 M. B. Nardelli, K. Rapcewicz, and J. Bernholc, Phys. Rev. B **55**, R7323 (1997).

136 W. M. Yim, E. J. Stofko, P. J. Zanzucchi, J. I. Pankove, M. Ettenberg, and S. L. Gilbert, J. Appl. Phys. **44**, 292 (1973).

137 P. B. Perry and R. F. Rutz, Appl. Phys. Lett. **33**, 319 (1978).

138 Q. Guo and A. Yoshida, Jpn. J. Appl. Phys. **33**, 2454 (1994).

139 R. D. Vispute, H. Wu, and J. Narayan, Appl. Phys. Lett. **67**, 1549 (1995).

140 X. Tang, F. Hossain, K. Wongchotigul, and M. G. Spencer, Appl. Phys Lett. **72**, 1501 (1998).

141 T. Wethkamp, K. Wilmers, C. Cobet, N. Esser, W. Richter, O. Ambacher, M. Stutzmann, and M. Cardona, Phys. Rev. B **59**, 1845 (1999).

142 D. Brunner, H. Angerer, E. Bustarret, F. Freudenberg, R. Hoepler, R. Dimitrov, O. Ambacher, and M. Stutzmann, J. Appl. Phys. **82**, 5090 (1997).

143 E. Kuokstis, J. Zhang, Q. Fareed, J. W. Yang, G. Simin, M. Asif Khan, R. Gaska, M. Shur, C. Rojo, and L. Schowalter, Appl. Phys. Lett. **81**, 2755 (2002).

144 Q. Guo, M. Nishio, H. Ogawa, and A. Yoshida, Phys. Rev. B **64**, 113105 (2001).

145 L. F. Jiang, W. Z. Shen, H. Ogawa, and Q. X. Guo, J. Appl. Phys. **94**, 5704 (2003).

146 J. Chen, W. Z. Shen, H. Ogawa, and Q. X. Guo, Appl. Phys. Lett. **84**, 4866 (2004).

147 T. Onuma, S. F. Chichibu, T. Sota, K. Asai, S. Sumiya, T. Shibata, and M. Tomaka, Appl. Phys. Lett. **81**, 652 (2002).

148 L. Chen, B. J. Skromme, R. D. Dalmau, R. Schlesser, Z. Sitar, C. Chen, W. Sun, J. Yang, M. A. Khan, M. L. Nakarmi, J. Y. Lin, and H. X. Jiang, Appl. Phys. Lett. **85**, 4334 (2004).

149 E. Silveira, J. A. Freitas, O. J. Glembocki, G. A. Slack, and L. J. Schowalter, Phys. Rev. B **71**, 041201 (2005).

150 J. Li, K. B. Nam, M. L. Nakarmi, J. Y. Lin, H. X. Jiang, P. Carrier, and S. H. Wei, Appl. Phys. Lett. **83**, 5163 (2003).

151 J. A. Freitas, G. C. B. Braga, E. Silveira, J. G. Tischler, and M. Fatemi, Appl. Phys. Lett. **83**, 2584 (2003).

152 E. Silveira, J. A. Freitas, M. Kneissl, D. W. Treat, N. M. Johnson, G. A. Slack, and L. J. Schowalter, Appl. Phys. Lett. **84**, 3501 (2004).

153 N. Nepal, K. B. Nam, M. L. Nakarmi, J. Y. Lin, H. X. Jiang, J. M. Zavada, and R. G. Wilson, Appl. Phys. Lett. **84**, 1090 (2004).

154 M. Oshikiri, F. Aryasetiawan, Y. Imanaka, and G. Kido, Phys. Rev. B **66**, 125204 (2002).

155 C. Persson, Bo E. Sernelius, A. Ferreira da Silva, R. Ahuja, and B. Johansson, J. Phys. Condens. Matter **13**, 8915 (2001).

156 K. B. Nam, M. L. Nakarmi, J. Li, J. Y. Lin, and H. X. Jiang, Appl. Phys. Lett. **83**, 878 (2003).

157 W. Shan, J. W. Ager III, W. Walukiewicz, E. E. Haller, B. D. Little, J. J. Song, M. Schurman, Z. C. Feng, R. A. Stall, and B. Goldenberg, Appl. Phys. Lett. **72**, 2274 (1998).

158 S.-H. Park and S. L. Chuang, J. Appl. Phys. **87**, 353 (2000).

159 J.-M. Wagner and F. Bechstedt, Phys. Stat. Sol. B **234**, 965 (2002).

160 K. Tsubouchi, K. Sugai, and N. Mikoshiba, in *1981 Ultrasonics Symposium*, ed. by B. R. McAvoy (IEEE, New York, 1981), vol. 1, p. 375.

161 L. E. McNeil, M. Grimditch, and R. H. French, J. Am. Ceram. Soc. **76**, 1132 (1993).

162 G. Bu, D. Ciplys, M. Shur, L. J. Schowalter, S. Schujman, and R. Gaska, IEEE Trans. Ultrason. Ferroelect. Freq. Control **53**, 251 (2006).

163 V. Yu. Davydov, Yu. E. Kitayev, I. N. Goncharuk, A. N. Smirnov, J. Graul, O. Semchinova, D. Uffmann, M. B. Smirnov, A. P. Mirgorodsky, and R. A. Evarestov, Phys. Rev. B **58**, 12899 (1998).

164 E. Ruiz, S. Alvarez, and P. Alemany, Phys. Rev. B **49**, 7115 (1994).

165 R. Kato and J. Hama, J. Phys. Condens. Matter **6**, 7617 (1994).

166 A. R. Hutson, U. S. Patent No. 3,090,876 (May 21, 1963).

167 K. Tsubouchi, K. Sugai, and N. Mikoshiba, *Proceedings of the IEEE Ultrasonic Symposium* (IEEE, New York, 1982), p. 340.

168 T. Kamiya, Jpn. J. Appl. Phys. **35**, 4421 (1996).

169 A. Zoroddu, F. Bernardini, P. Ruggerone, and V. Fiorentini, Phys. Rev. B **64**, 045208 (2001).

170 V. Mortet, M. Nesladek, K. Haenen, A. Morel, M. D'Oelislaeger, M. Vanecek, Diamond and Related Mater. **13**, 1120 (2004).

171 M. Leroux, N. Grandjean, J. Massies, B. Gil, P. Lefebvre, and P. Bigenwald, Phys. Rev. B **60**, 1496 (1999).

172 N. Grandjean, B. Damilano, S. Dalmasso, M. Leroux, M. Laugt, and J. Massies, J. Appl. Phys. **86**, 3714 (1999).

173 J. A. Garrido, J. L. Sanchez-Rojas, A. Jimenez, E. Munoz, F. Omnes, and P. Gibart, Appl. Phys. Lett. **75**, 2407 (1999).

174 R. A. Hogg, C. E. Norman, A. J. Shields, M. Pepper, and M. Iizuka, Appl. Phys. Lett. **76**, 1428 (2000).

175 S.-H. Park and S.-L. Chuang, Appl. Phys. Lett. **76**, 1981 (2000).

176 R. Cingolani, A. Botchkarev, H. Tang, H. Morkoc, G. Traetta, G. Coli, M. Lomascolo, A. DiCarlo, F. Della Sala, and P. Lugli, Phys. Rev. B **61**, 2711 (2000).

177 F. Bernardini and V. Fiorentini, Phys. Rev. B **64**, 085207 (2001).

178 V. Fiorentini, F. Bernardini, and O. Ambacher, Appl. Phys. Lett. **80**, 1204 (2002).

179 K. Osamura, S. Naka, and Y. Murakami, J. Appl. Phys. **46**, 3432 (1975).

180 N. Puychevrier and M. Menoret, Thin Solid Films, **36**, 141 (1976).

181 V. A. Tyagai, A. M. Evstigneev, A. N. Krasiko, A. F. Andreeva, and V. Ya. Malakhov, Sov. Phys. Semicond. **11**, 1257 (1977) [Fiz. Tekh. Poluprov. **11**, 2142 (1977)].

182 T. L. Tansley and C. P. Foley, J. Appl. Phys. **59**, 3241 (1986).

183 K. L. Westra and M. J. Brett, Thin Solid Films **192**, 227 (1990).

184 L. F. Jiang, W. Z. Shen, H. F. Yang, H. Ogawa, and Q. X. Guo, Appl. Phys. A **78**, 89 (2004).

185 T. Inushima, V. V. Mamutin, V. A. Vekshin, S. V. Ivanov, T. Sakon, M. Motokawa, and S. Ohoya, J. Cryst. Growth **227–228**, 481 (2001).

186 V. Yu. Davydov, A. A. Klochikhin, R. P. Seisyan, V. V. Emtsev, S. V. Ivanov, F. Bechstedt, J. Furthmüller, H. Harima, A. V. Mudryi, J. Aderhold, O. Semchinova, and J. Graul, Phys. Stat. Sol. B **229**, R1 (2002).

187 J. Wu, W. W. Walukiewicz, K. M. Yu, J. W. Ager III, E. E. Haller, H. Lu, W. J. Schaff, Y. Saito, and Y. Nanishi, Appl. Phys. Lett. **80**, 3967 (2002).

188 T. Matsuoka, H. Okamoto, M. Nakao, H. Harima, and E. Kurimoto, Appl. Phys. Lett. **81**, 1246 (2002).

189 V. Yu. Davydov, A. A. Klochikhin, V. V. Emtsev, S. V. Ivanov, V. V. Vekshin, F. Bechstedt, J. Furthmüller, H. Harima, A. V. Mudryi, A. Hashimoto, A. Yamamoto, J. Aderhold, J. Graul, and E. E. Haller, Phys. Stat. Sol. B **230**, R4 (2002).

190 V. Yu. Davydov, A. A. Klochikhin, V. V. Emtsev, F. Bechstedt, A. V. Mudryi, and E. E. Haller, Phys. Stat. Sol. B **233**, R10 (2002).

191 H. Ahn, C.-H. Shen, C.-L. Wu, and S. Gwo, Appl. Phys. Lett. **86**, 201905 (2005).

192 K. M. Yu, Z. Liliental-Weber, W. Walukiewicz, W. Shan, J. W. Ager, S. X. Li, R. E. Jones, E. E. Haller, H. Lu, and W. J. Schaff, Appl. Phys. Lett. **86**, 071910 (2005).

193 R. Goldhahn, A. T. Winzer, V. Cimalla, O. Ambacher, C. Cobet, W. Richter, N. Esser, J. Furthmuller, F. Bechstedt, H. Lu, and W.

193 J. Schaff, Superlatt. Microstruct. **36**, 591 (2004).

194 W. Walukiewicz, S. X. Li, J. Wu, K. M. Yu, J. W. Ager, E. E. Haller, H. Lu, and W. J. Schaff, J. Cryst. Growth **269**, 119 (2004).

195 T. Matsuoka, H. Okamoto, H. Takahata, T. Mitate, S. Mizuno, Y. Uchiyama, and T. Makimoto, J. Cryst. Growth **269**, 139 (2004).

196 Y. Nanishi, Y. Saito, T. Yamaguchi, M. Hori, F. Matsuda, T. Araki, A. Suzuki, and T. Miyajima, Phys. Stat. Sol. A **200**, 202 (2003).

197 H. Lu, W. J. Schaff, L. F. Eastman, J. Wu, W. Walukiewicz, V. Cimalla, and O. Ambacher, Appl. Phys. Lett. **83**, 1136 (2003).

198 T. Matsuoka, M. Nakao, H. Okamoto, H. Harima, and E. Kurimoto, Jap. J. Appl. Phys. Part 1 **42**, 2288 (2003).

199 P. Trautman, K. Pakula, A. M. Witowski, and J. M. Barranowski, Acta Phys. Polonica A **108**, 903 (2005).

200 M. Higashiwaki and T. Matsui, J. Cryst. Growth **269**, 162 (2004).

201 J. W. Yoon, S. S. Kim, H. Cheong, H. C. Seo, S. Y. Kwon, H. J. Kim, Y. Shin, E. Yoon, and Y. S. Park, Semicond. Sci. Technol. **20**, 1068 (2005).

202 E. Trybus, G. Namkoong, W. Henderson, W. A. Doolittle, R. Liu, J. Mei, F. Ponce, M. Cheung, F. Chen, M. Furis, and A. Cartwright, J. Cryst. Growth **279**, 311 (2005).

203 B. Arnaudov, T. Paskova, P. P. Paskov, B. Magnusson, E. Valcheva, B. Monemar, H. Lu, W. J. Schaff, H. Amano, and I. Akasaki, Superlatt. Microstruct. **36**, 563 (2004).

204 Y. Ishitani, K. Xu, S. B. Che, H. Masuyama, W. Terashima, M. Yoshitani, N. Hashimoto, K. Akasaka, T. Ohkubo, and A. Yoshikawa, Phys. Stat. Sol. B **241**, 2849 (2004).

205 R. J. Kinsey, P. A. Anderson, C. E. Kendrick, R. J. Reeves, and S. M. Durbin, J. Cryst. Growth **269**, 167 (2004).

206 J. Wu, W. Walukiewicz, W. Shan, K. M. Yu, J. W. Ager, S. X. Li, E. E. Haller, H. Lu, and W. J. Schaff, J. Appl. Phys. **94**, 4457 (2003).

207 T.-C. P. Chen, C. Thomidis, J. Abell, W. Li, and T. D. Moustakas, J. Cryst. Growth **288**, 254 (2006).

208 M. Hori, K. Kano, T. Yamaguchi, Y. Saito, T. Araki, Y. Nanishi, N. Teraguchi, and A. Suzuki, Phys. Stat. Sol. B **234**, 750 (2002).

209 J. Wu, W. Walukiewicz, K. M. Yu, J. W. Ager, E. E. Haller, H. Lu, and W. J. Schaff, Phys. Stat. Sol. B **240**, 412 (2003).

210 P. Carrier and S. H. Wei, J. Appl. Phys. **97**, 033707 (2005).

211 D. Bagayoko and L. Franklin, J. Appl. Phys. **97**, 123708 (2005).

212 S. H. Wei, X. L. Nie, I. G. Batyrev, and S. B. Zhang, Phys. Rev. B **67**, 165209 (2003).

213 B. R. Nag, Phys. Stat. Sol. B **233**, R8 (2002).

214 J. Wu, W. Walukiewicz, S. X. Li, R. Armitage, J. C. Ho, E. R. Weber, E. E. Haller, H. Lu, W. J. Schaff, A. Barcz, and R. Jakiela, Appl. Phys. Lett. **84**, 2805 (2004).

215 A. G. Bhuiyan, K. Sugita, K. Kasashima, A. Hashimoto, A. Yamamoto, and V. Y. Davydov, Appl. Phys. Lett. **83**, 4788 (2003).

216 T. V. Shubina, S. V. Ivanov, V. N. Jmerik, M. M. Glazov, A. P. Kalvarskii, M. G. Tkachman, A. Vasson, J. Leymarie, A. Kavokin, H. Amano, I. Akasaki, K. S. A. Butcher, Q. Guo, B. Monemar, and P. S. Kop'ev, Phys. Stat. Sol. A **202**, 377 (2005).

217 K. S. A. Butcher and T. L. Tansley, Superlatt. Microstruct. **38**, 1 (2005).

218 Q. X. Guo, T. Tanaka, M. Nishio, H. Ogawa, X. D. Pu, and W. Z. Shen, Appl. Phys. Lett. **86**, 231913 (2005).

219 P. Specht, J. C. Ho, X. Xu, R. Armitage, E. R. Weber, R. Erni, and C. Kisielowski, Solid State Commun. **135**, 340 (2005).

220 K. S. A. Butcher, M. Wintrebert-Fouquet, P. P. T. Chen, H. Timmers, and S. K. Shrestha, Mat. Sci. Semicond. Proc. **6**, 351 (2003).

221 P. P.-T. Chen, K. S. A. Butcher, M. Wintrebert-Fouquet, R. Wuhrer, M. R. Phillips, K. E. Prince, H. Timmers, S. K. Shrestha, and B. F. Usher, J. Cryst. Growth **288**, 241 (2006).

222 B. Monemar, P. P. Paskov, and A. Kasic, Superlatt. Microstruct. **38**, 38 (2005).

223 T. V. Shubina, S. V. Ivanov, V. N. Jmerik, D. D. Solnyshkov, V. A. Vekshin, P. S.

Kop'ev, A. Vasson, J. Leymarie, A. Kavokin, H. Amano, K. Shimono, A. Kasic, and B. Monemar, Phys. Rev. Lett. **92**, 117407 (2004).

224 T. V. Shubina, S. V. Ivanov, V. N. Jmerik, P. S. Kop'ev, A. Vasson, J. Leymarie, A. Kavokin, H. Amano, B. Gil, O. Briot, and B. Monemar, Phys. Rev. Lett. **93**, 269702 (2004).

225 T. V. Shubina, J. Leymarie, V. N. Jmerik, A. A. Toropov, A. Vasson, H. Amano, W. J. Schaff, B. Monemar, and S. V. Ivanov, Phys. Stat. Sol. A **202**, 2633 (2005).

226 W. Z. Shen, X. D. Pu, J. Chen, H. Ogawa, and Q. X. Guo, Solid State Commun. **137**, 49 (2006).

227 J. C. Ho, P. Specht, Q. Yang, X. Xu, D. Hao, and E. R. Weber, J. Appl. Phys. **98**, 093712 (2005).

228 V. Y. Davydov and A. A. Klochikhin, Semicond. **38**, 861 (2004).

229 J. Wu and W. Walukiewicz, Superlatt. Microstruct. **34**, 63 (2003).

230 A. G. Bhuiyan, A. Hashimoto, and A. Yamamoto, J. Appl. Phys. **94**, 2779 (2003).

231 B. R. Nag, Phys. Stat. Sol. B **237**, R1 (2003).

232 T. Inushima, T. Yaguchi, A. Nagase, A. Iso, T. Shiraishi, S. Ooya, Proc. 7th International Conf. On InP and Related Materials, (1995), p. 187.

233 A. Kasic, M. Schubert, Y. Saito, Y. Nanishi, and G. Wagner, Phys. Rev. B **65**, 115206 (2002).

234 T. Inushima, T. Shiraishi, and V. Y. Davydov, Solid State Commun. **110**, 491 (1999).

235 T. Inushima, M. Higashiwaki, and T. Matsui, Phys. Rev. B **68**, 235204 (2003).

236 S. P. Fu and Y. F. Chen, Appl. Phys. Lett. **85**, 1523 (2004).

237 D. Fritsch, H. Schmidt, and M. Grundmann, Phys. Rev. B **69**, 165204 (2004).

238 A. Klochikhin, V. Davydov, V. Emtsev, A. Sakharov, V. Kapitonov, B. Andreev, H. Lu, and W. J. Schaff, Phys. Stat. Sol. A **203**, 50 (2006).

239 L. F. J. Piper, T. D. Veal, I. Mahboob, C. F. McConville, H. Lu, and W. J. Schaff, Phys. Rev. B **70**, 115333 (2004).

240 S. K. Pugh, D. J. Dugdale, S. Brand, and R. A. Abram, Semicond. Sci. Technol. **14**, 23 (1999).

241 R. Goldhahn, P. Schley, A. T. Winzer, G. Gobsch, V. Cimalla, O. Ambacher, M. Rakel, C. Cobet, N. Esser, H. Lu, and W. J. Schaff, Phys. Stat. Sol. A **203**, 42 (2006).

242 F. Bechstedt, J. Furthmüller, M. Ferhat, L. K. Teles, L. M. R. Scolfaro, J. R. Leite, V. Yu. Davydov, O. Ambacher, and R. Goldhahn, Phys. Stat. Sol. A **195**, 628 (2003).

243 S. X. Li, J. Wu, E. E. Haller, W. Walukiewicz, W. Shan, H. Lu, and W. J. Schaff, Appl. Phys. Lett. **83**, 4963 (2003).

244 C. B. Cao, H. L. W. Chan, and C. L. Choy, Thin Solid Films **441**, 287 (2003).

245 I. L. Guy, Z. Zheng, M. Wintrebert-Fouquet, K. S. A. Butcher, P. Chen, and T. L. Tansley, J. Cryst. Growth **269**, 72 (2004).

246 J. Hagen, R. D. Metcalfe, D. Wickenden, and W. Clark, J. Phys. C **11**, L143 (1978).

247 S. Yoshida, S. Misawa, and S. Gonda, J. Appl. Phys. **53**, 6844 (1982).

248 M. A. Khan, R. A. Skogman, R. G. Schulze, and M. Gershenzon, Appl. Phys. Lett. **43**, 492 (1983).

249 M. R. H. Khan, Y. Koide, H. Itoh, N. Sawaki, and I. Akasaki, Solid State Commun. **60**, 509 (1986).

250 Y. Koide, H. Itoh, M. R. H. Khan, K. Hiramatu, and I. Akasaki, J. Appl. Phys. **61**, 4540 (1987).

251 S. R. Lee, A. F. Wright, M. H. Crawford, G. A. Petersen, J. Han, and R. M. Biefeld, Appl. Phys. Lett. **74**, 3344 (1999).

252 G. Yu, H. Ishikawa, M. Umeno, T. Egawa, J. Watanabe, T. Jimbo, and T. Soga, Appl. Phys. Lett. **72**, 2202 (1998).

253 O. Ambacher, R. Dimitrov, D. Lentz, T. Metzger, W. Rieger, and M. Stutzmann, J. Cryst. Growth **170**, 335 (1997).

254 B. V. Baranov, V. B. Gutan, and U. Zhumakulov, Sov. Phys. Semicond. **16**, 819 (1982).

255 M. D. Bremser, W. G. Perry, T. Zheleva, N. V. Edwards, O. H. Nam, N. Parikh, D. E. Aspnes, and R. F. Davis, MRS Internet J. Nitride Semicond. Res. **1**, 8 (1996).

256 T. Huang and J. S. Harris, Jr., Appl. Phys. Lett. **72**, 1158 (1998).

257. G. A. Korkotashvili, A. N. Pikhtin, I. G. Pichugin, and A. M. Tsaregorodtsev, Sov. Phys. Semicond. **18**, 913 (1984).

258. T. Wethkamp, K. Wilmers, N. Esser, W. Richter, O. Ambacher, H. Angerer, G. Jungk, R. L. Johnson, and M. Cardona, Thin Solid Films **313**, 745 (1998).

259. W. Shan, J. W. Ager III, W. Walukiewicz, E. E. Haller, B. D. Little, J. J. Song, M. Shurman, Z. C. Feng, R. A. Stall, and B. Goldenberg, Appl. Phys. Lett. **72**, 2274 (1998).

260. T. Takeuchi, H. Takeuchi, S. Sota, H. Sakai, H. Amano, and I. Akasaki, Jpn. J. Appl. Phys. **36**, L177 (1997).

261. D. Korakakis, H. M. Ng, M. Misra, W. Grieshaber, and T. D. Moustakas, MRS Internet J. Nitride Semicond. Res. **1**, 10 (1996).

262. A. Y. Polyakov, M. Shin, J. A. Freitas, M. Skowronski, D. W. Greve, and R. G. Wilson, J. Appl. Phys. **80**, 6349 (1996).

263. G. Steude, D. M. Hofmann, B. K. Meyer, H. Amano, and I. Akasaki, Phys. Stat. Sol. B **205**, R7 (1998).

264. D. K. Wickenden, C. B. Bargeron, W. A. Bryden, J. Miragliotta, and T. J. Kistenmacher, Appl. Phys. Lett. **65**, 2024 (1994).

265. A. S. Zubrilov, D. V. Tsvetkov, V. I. Nikolaev, and I. P. Nikitina, Semicond. **30**, 1069 (1996).

266. T. J. Ochalski, B. Gil, P. Lefebvre, N. Granjean, M. Leroux, J. Massies, S. Nakamura, and H. Morkoc, Appl. Phys. Lett. **74**, 3353 (1999).

267. W. Shan, J. W. Ager III, K. M. Yu, W. Walukiewicz, E. E. Haller, M. C. Martin, W. R. McKinney, and W. Yang, J. Appl. Phys. **85**, 9505 (1999).

268. B. K. Meyer, G. Steude, A. Göldner, A. Hoffmann, H. Amano, and I. Akasaki, Phys. Stat. Sol. B **216**, 187 (1999).

269. F. Omnes, N. Marenco, B. Beaumont, Ph. de Mierry, E. Monroy, F. Calle, and E. Munoz, J. Appl. Phys. **86**, 5286 (1999).

270. L. Bergman, X.-B. Chen, D. McIlroy, and R. F. Davis, Appl. Phys. Lett. **81**, 4186 (2002).

271. S. A. Nikishin, N. N. Faleev, A. S. Zubrilov, V. G. Antipov, and H. Temkin, Appl. Phys. Lett. **76**, 3028 (2000).

272. S. Nikishin, G. Kipshidze, V. Kuryatkov, K. Choi, Iu. Gheraoiu, L. Grave de Peralta, A. Zubrilov, V. Tretyakov, K. Copeland, T. Prokofyeva, M. Holtz, R. Asomoza, Yu. Kudryavtsev, and H. Temkin, J. Vac. Sci. Technol. B **19**, 1409 (2001).

273. M. Holtz, T. Prkofyeva, M. Seon, K. Copeland, J. Vanbuskirk, S. Williams, S. A. Nikishin, V. Tretyakov, and H. Temkin, J. Appl. Phys. **89**, 7977 (2001).

274. D. G. Ebling, L. Kirste, K. W. Benz, N. Teofilov, K. Thonke, and R. Sauer, J. Cryst. Growth **227–228**, 453 (2001).

275. H. Jiang, G. Y. Zhao, H. Ishikawa, T. Egawa, T. Jimbo, and M. Umeno, J. Appl. Phys. **89**, 1046 (2001).

276. J. Wagner, H. Obloh, M. Kunzer, M. Maier, K. Köhler, and B. Johs, J. Appl. Phys. **89**, 2779 (2001).

277. Q. Zhou, M. O. Manasreh, M. Pophristic, S. Guo, and I. T. Ferguson, Appl. Phys. Lett. **79**, 2901 (2001).

278. O. Katz, B. Meyler, U. Tisch, and J. Salzman, Phys. Stat. Sol. A **188**, 789 (2001).

279. F. Yun, M. A. Reshchikov, L. He, T. King, H. Morkoc, S. W. Novak, and L. Wei, J. Appl. Phys. **92**, 4837 (2002).

280. Q. S. Paduano, D. W. Weyburne, L. O. Bouthillette, S.-Q. Wang, and M. N. Alexander, Jpn. J. Appl. Phys. **41**, 1936 (2002).

281. C. Buchheim, R. Goldhahn, M. Rakel, C. Cobet, N. Esser, U. Rossow, D. Fuhrmann, and A. Hangleiter, Phys. Stat. Sol. B **242**, 2610 (2005).

282. A. T. Winzer, R. Goldhahn, G. Gobsch, A. Link, M. Eickhoff, U. Rossow, and A. Hangleiter, Appl. Phys. Lett. **86**, 181912 (2005).

283. M. Leroux, S. Dalmasso, F. Natali, S. Helin, C. Touzi, S. Laugt, M. Passerel, F. Omnes, F. Semond, J. Massies, and R. Gibart, Phys. Stat. Sol. B **234**, 887 (2002).

284. F. Fedler, R. J. Hauenstein, H. Klausing, D. Mistele, O. Semchinova, J. Aderhold, and J. Graul, J. Cryst. Growth **241**, 535 (2002).

285. T. Onuma, S. F. Chichibu, A. Uedono, T. Sota, P. Cantu, T. M. Katona, J. F. Keading, S. Keller, U. K. Mishra, S. Nakamura, and

S. P. DenBaars, J. Appl. Phys. **95**, 2495 (2004).

286 V. G. Deibuk, A. V. Voznyi, and M. M. Sletov, Semicond. **34**, 35 (2000).

287 Y.-K. Kuo and W.-W. Lin, Jpn. J. Appl. Phys. **41**, 73 (2002).

288 L.-C. Duda, C. B. Stagarescu, J. Downes, K. E. Smith, D. Korakakis, T. D. Moustakas, J. Guo, and J. Nordgren, Phys. Rev. B **58**, 1928 (1998).

289 Z. Dridi, B. Bouhafs, and P. Ruterano, New J. Phys. **4**, 94 (2002).

290 P. Harmer, M. P. Halsall, D. Wolverson, P. J. Parbrook, and S. J. Henley, Semicond. Sci. Technol. **19**, L22 (2004).

291 D. Jena, S. Heikman, J. S. Speck, A. Gossard, U. K. Mishra, A. Link, and O. Ambacher, Phys. Rev. B **67**, 153306 (2003).

292 K. P. O'Donnell, R. W. Martin, and P. G. Middleton, Phys. Rev. Lett. **82**, 237 (1999).

293 L. K. Teles, J. Furthmüller, L. M. R. Scolfaro, A. Tabata, J. R. Leite, F. Bechstedt, T. Frey, D. J. As, and K. Lischka, Physica E **13**, 1086 (2002).

294 A. F. Wright and J. S. Nelson, Appl. Phys. Lett. **66**, 3051 (1995).

295 K. Kassali and N. Bouarissa, Solid-State Electron. **44**. 501 (2000).

296 M. Ferhat, J. Furthmüller, and F. Bechstedt, Appl. Phys. Lett. **80**, 1394 (2002).

297 M. Ferhat and F. Bechstedt, Phys. Rev. B **65**, 075213 (2002).

298 T. Nagatomo, T. Kuboyama, H. Minamino, and O. Omoto, Jpn. J. Appl. Phys. **28**, L1334 (1989).

299 N. Yoshimoto, T. Matsuoka, and A. Katsui, Appl. Phys. Lett. **59**, 2251 (1991).

300 S. Nakamura, J. Vac. Sci. Technol. A **13**, 705 (1995).

301 W. Li, P. Bergman, I. Ivanov, W.-X. Ni, H. Amano, and I. Akasa, Appl. Phys. Lett. **69**, 3390 (1996).

302 L. Bellaiche and A. Zunger, Phys. Rev. B **57**, 4425 (1998).

303 A. F. Wright, K. Leung, and M. van Schilfgaarde, Appl. Phys. Lett. **78**, 189 (2001).

304 M. D. McCluskey, C. G. Van de Walle, C. P. Master, L. T. Romano, and N. M. Johnson, Appl. Phys. Lett. **72**, 2725 (1998).

305 P. R. C. Kent, G. L. W. Hart, and A. Zunger, Appl. Phys. Lett. **81**, 4377 (2002).

306 T. Takeuchi, H. Takeuchi, S. Sota, H. Sakai, H. Amano, and I. Akasaki, Jpn. J. Appl. Phys. **36**, L177 (1997).

307 C. Wetzel, T. Takeuchi, S. Yamaguchi, H. Katoh, H. Amano, and I. Akasaki, Appl. Phys. Lett. **73**, 1994 (1998).

308 W. Shan, W. Walukiewicz, E. E. Haller, B. D. Little, J. J. Song, M. D. McCluskey, N. M. Johnson, Z. C. Feng, M. Schurman, and R. A. Stall, J. Appl. Phys. **84**, 4452 (1998).

309 C. A. Parker, J. C. Roberts, S. M. Bedair, M. J. Reed, S. X. Liu, N. A. El-Masry, and L. H. Robins, Appl. Phys. Lett. **75**, 2566 (1999).

310 H. P. D. Schrenk, P. de Mierry, M. Laugt, F. Omnes, M. Leroux, B. Beaumont, and P. Gibart, Appl. Phys. Lett. **75**, 2587 (1999).

311 M. E. Aumer, S. F. LeBoeuf, F. G. McIntosh, and S. M. Bedair, Appl. Phys. Lett. **75**, 3315 (1999).

312 B. Schineller, P. H. Lim, O. Schön, H. Protzmann, M. Heuken, and K. Heime, Phys. Stat. Sol. B **216**, 311 (1999).

313 L. Siozade, J. Leymarie, P. Disseix, A. Vasson, M. Mihailovic, N. Grandjean, M. Leroux, and J. Massies, Solid State Commun. **115**, 575 (2000).

314 S. Pereira, M. R. Correia, T. Monteiro, E. Pereira, E. Alves, A. D. Sequeira, and N. Franco, Appl. Phys. Lett. **78**, 2137 (2001).

315 P. Perlin, I. Gorczyca, T. Suski, P. Wisniewski, S. Lepkowski, N. E. Christensen, A. Svane, M. Hansen, and S. P. DenBaars, Phys. Rev. B **64**, 115319 (2001).

316 F. Scholz, J. Off, A. Sohmer, V. Sygnow, A. Dörnen, O. Ambacher, J. Cryst. Growth **189–190**, 8 (1998).

317 J. Wagner, A. Ramakrishnan, D. Behr, M. Maier, N. Herres, M. Kunzer, H. Ogloh, K.-H. Bachem, MRS Internet J. Nitride Semicond. Res. **4S1**, G2.8 (1999).

318 F. B. Naranjo, M. A. Sanchez-Garcia, F. Calle, E. Calleja, B. Jenichen, and K. H. Ploog, Appl. Phys. Lett. **80**, 231 (2002).

319 F. B. Naranjo, S. Fernandez, M. A. Sanchez-Garcia, F. Calle, E. Calleja, A. Trampert, and K. H. Ploog, Mat. Sci. Eng. **B93**, 131 (2002).

320 P. Ryan, C. McGuinness, J. E. Downes, K. E. Smith, D. Doppalapudi, and T. D. Moustakas, Phys. Rev. B **65**, 205201 (2002).

321 J. D. Beach, H. Al-Thani, S. McCray, R. T. Collins, and J. A. Turner, J. Appl. Phys. **91**, 5190 (2002).

322 M.-H. Kim, J.-K. Gho, I.-H. Lee, and S.-J. Park, Phys. Stat. Sol. A **176**, 269 (1999).

323 K. P. O'Donnell, J. F. W. Mosselmans, R. W. Martin, S. Pereira, and M. E. White, J. Phys. Condens. Matter **13**, 6977 (2001).

324 M. D. McCluskey, C. G. Van de Walle, L. T. Romano, B. S. Kruser, and N. M. Johnson, J. Appl. Phys. **94**, 4340 (2003).

325 L. Bellaiche, T. Mattila, L.-W. Wang, S.-H. Wei, and A. Zunger, Appl. Phys. Lett. **74**, 1842 (1999).

326 S. Stepanov, W. N. Wang, B. S. Yavich, V. Bougrov, Y. T. Rebane, and Y. G. Shreter, MRS Internet J. Semicond. Res. **6**, 6 (2001).

327 J. Wu, W. Walukiewicz, K. M. Yu, J. W. Ager III, E. E. Haller, H. Lu, and W. J. Schaff, Appl. Phys. Lett. **80**, 4741 (2002).

328 K. P. O'Donnell, I. Fernandez-Torrente, P. R. Edwards, and R. W. Martin, J. Cryst. Growth **269**, 100 (2004).

329 V. Y. Davydov, A. A. Klochikhin, V. V. Emtsev, D. A. Kurdyukov, S. V. Ivanov, V. A. Vekshin, F. Bechstedt, J. Furthmuller, J. Aderhold, J. Graul, A. V. Mudryi, H. Harima, A. Hashimoto, A. Yamamoto, and E. E. Haller, Phys. Stat. Sol. B **234**, 787 (2002).

330 V. Yu. Davydov, A. A. Klochikhin, V. V. Emtsev, A. N. Smirnov, I. N. Goncharuk, A. V. Sakharov, D. A. Kurdyukov, M. V. Baidakova, V. A. Vekshin, S. V. Ivanov, J. Aderhold, J. Graul, A. Hashimoto, and A. Yamamoto, Phys. Stat. Sol. B **240**, 425 (2003).

331 F. Sokeland, M. Rohlfing, P. Kruger, and J. Pollmann, Phys. Rev. B **68**, 075203 (2003).

332 Z. Dridi, B. Bouhafs, and P. Ruterano, Semicond. Sci. Technol. **18**, 850 (2003).

333 B. T. Liou, S. H. Yen, and Y. K. Kuo, Opt. Commun. **249**, 217 (2005).

334 A. V. Voznyy and V. G. Deibuk, Semicond. **38**, 304 (2004).

335 S. K. Pugh, D. J. Dugdale, S. Brand, and R. A. Abram, J. Appl. Phys. **86**, 3768 (1999).

336 A. Al-Yacoub and L. Bellaiche, Appl. Phys. Lett. **79**, 2166 (2001).

337 A. Al-Yacoub, L. Bellaiche, and S.-H. Wei, Phys. Rev. Lett. **89**, 057601 (2002).

338 K. Kubota, Y. Kobayashi, and K. Fujimoto, J. Appl. Phys. **66**, 2984 (1989).

339 Q. Guo, H. Ogawa, and A. Yoshida, J. Cryst. Growth **146**, 462 (1995).

340 K. S. Kim, A. Saxler, P. Kung, M. Razeghi, and K. Y. Lim, Appl. Phys. Lett. **71**, 800 (1997).

341 T. Peng, J. Piprek, G. Qiu, J. O. Olowolafe, K. M. Unruh, C. P. Swann, and E. F. Schubert, Appl. Phys. Lett. **71**, 2439 (1997).

342 S. Yamaguchi, M. Kariya, S. Nitta, H. Kato, T. Takeuchi, C. Wetzel, H. Amano, and I. Akasaki, J. Cryst. Growth **195**, 309 (1998).

343 S. Yamaguchi, M. Kariya, S. Nitta, T. Takeuchi, C. Wetzel, H. Amano, and I. Akasaki, Appl. Phys. Lett. **76**, 876 (2000).

344 T. V. Shubina, V. V. Mamutin, V. A. Vekshin, V. V. Ratnikov, A. A. Toropov, A. A. Sitnikova, S. V. Ivanov, M. Karlsteen, U. Söderwall, M. Willander, G. Pozina, J. P. Bergman, and B. Monemar, Phys. Stat. Sol. B **216**, 205 (1999).

345 M. J. Lukitsch, Y. V. Danylyuk, V. M. Naik, C. Huang, G. W. Auner, L. Rimai, and R. Naik, Appl. Phys. Lett. **79**, 632 (2001).

346 Q. X. Guo, T. Tanaka, M. Nishio, and H. Ogawa, Jap. J. Appl. Phys. Part 2 **42**, L141 (2003).

347 J. Wu, W. Walukiewicz, K. M. Yu, J. W. Ager, S. X. Li, E. E. Haller, H. Lu, and W. J. Schaff, Solid State Commun. **127**, 411 (2003).

348 T. Onuma, S. Chichibu, Y. Uchinuma, T. Sota, S. Yamaguchi, S. Kamiyama, H. Amano, and I. Akasaki, J. Appl. Phys. **94**, 2449 (2003).

349 B. T. Liou, S. H. Yen, and Y. K. Kuo, Appl. Phys. A **81**, 651 (2005).

350 Y. K. Kuo and W. W. Lin, Jap. J. Appl. Phys. Part 1 **41**, 5557 (2002).

351 M. E. Aumer, S. F. LeBoeuf, F. G. McIntosh, and S. M. Bedair, Appl. Phys. Lett. **75**, 3315 (1999).

352 G. Tamulaitis, K. Kazlauskas, S. Juršenas, A. Žukauskas, M. A. Khan, J. W. Yang, J. Zhang, G. Simin, M. S. Shur, and R. Gaska, Appl. Phys. Lett. **77**, 2136 (2000).

353 J. Li, K. B. Nam, K. H. Kim, J. Y. Lin, and H. X. Jiang, Appl. Phys. Lett. **78**, 61 (2001).

354 C. B. Soh, S. J. Chua, W. Liu, M. Y. Lai, and S. Tripathy, Solid State Commun. **136**, 421 (2005).

355 S. Nakazawa, T. Ueda, K. Inoue, T. Tanaka, H. Ishikawa, and T. Egawa, IEEE Trans. Electron Dev. **52**, 2124 (2005).

356 Y. Liu, T. Egawa, H. Ishikawa, H. Jiang, B. Zhang, M. Hao, and T. Jimbo, J. Cryst. Growth **264**, 159 (2004).

357 M. Asif Khan, J. W. Yang, G. Simin, R. Gaska, M. S. Shur, H.-C. zur Loye, G. Tamulaitis, A. Zukauskas, D. J. Smith, D. Chandrasekhar, and R. Bicknell-Tassius, Appl. Phys. Lett. **76**, 1161 (2000).

358 T. N. Oder, J. Li, J. Y. Lin, and H. X. Jiang, Appl. Phys. Lett. **77**, 791 (2000).

359 E. Monroy, N. Gogneau, F. Enjalbert, F. Fossard, D. Jalabert, E. Bellet-Amalric, L. S. Dang, and B. Daudin, J. Appl. Phys. **94**, 3121 (2003).

360 F. Bernardini and V. Fiorentini, Phys. Rev. B **57**, R9427 (1998).

361 N. Binggeli, P. Ferrara, and A. Baldereschi, Phys. Rev. B **63**, 245306 (2001).

362 A. Satta, V. Fiorentini, A. Bosin, and F. Meloni, Mat. Res. Soc. Symp. Proc. **395**, 515 (1996).

363 J. Baur, K. Maier, M. Kunzer, U. Kaufmann, and J. Schneider, Appl. Phys. Lett. **65**, 2211 (1994).

364 K. B. Nam, J. Li, K. H. Kim, J. Y. Lin, and H. X. Jiang, Appl. Phys. Lett. **78**, 3690 (2001).

365 D. R. Hang, C. H. Chen, Y. F. Chen, H. X. Jiang, and J. Y. Lin, J. Appl. Phys. **90**, 1887 (2001).

366 G. A. Martin, S. C. Strite, A. Botchkarev, A. Agarwal, A. Rockett, W. R. L. Lambrecht, B. Segall, and H. Morkoc, Appl. Phys. Lett. **67**, 3322 (1995).

367 G. A. Martin, A. Botchkarev, A. Rockett, and H. Morkoc, Appl. Phys. Lett. **68**, 2541 (1996).

368 J. R. Waldrop and R. W. Grant, Appl. Phys. Lett. **68**, 2879 (1996).

369 R. A. Beach, E. C. Piquette, R. W. Grant, and T. C. McGill, Mat. Res. Soc. Symp. Proc. **482**, 775 (1998).

370 S. W. King, C. Ronning, R. F. Davis, M. C. Benjamin, and R. J. Nemanich, J. Appl. Phys. **84**, 2086 (1998).

371 A. Rizzi, R. Lantier, F. Monti, H. Luth, F. Della Sala, A. Di Carlo, and P. Lugli, J. Vac. Sci. Technol. B **17**, 1674 (1999).

372 A. N. Westmeyer, S. Mahajan, K. K. Bajaj, J. Y. Lin, H. X. Jiang, D. D. Koleske, and R. T. Senger, J. Appl. Phys. **99**, 013705 (2006).

373 C. T. Foxon, S. V. Novikov, L. X. Zhao, and I. Harrison, Appl. Phys. Lett. **83**, 1166 (2003).

374 C.-C. Chen, H.-W. Chuang, G.-C. Chi, C.-C. Chuo, and J.-I. Chyi, Appl. Phys. Lett. **77**, 3758 (2000).

375 C.-L. Wu, C.-H. Shen, and S. Gwo, Appl. Phys. Lett. **88**, 032105 (2006).

3
Spontaneous and Piezoelectric Polarization: Basic Theory vs. Practical Recipes
Fabio Bernardini

3.1
Why Spontaneous Polarization in III-V Nitrides?

III-V nitrides and their alloys represent a very special class of Materials. They are best chosen for UV-blue light-emitting devices because of their favorable mechanical properties and the wide and direct band gap. At the same time, among III-V semiconductors, III-V nitrides, (III-Ns hereafter), are the only ones that show a property called spontaneous polarization ($\mathbf{P}_{sp}$), a built-in electric polarization absent in other well-known semiconductors used in optoelectronic devices such as GaAs and ZnSe. This property is of importance in applications since it tends to spoil the quantum efficiency of optoelectronic devices based on multi-quantum wells (MQWs) nanostructures. Also, $\mathbf{P}_{sp}$ prediction and control is part of the technological development of electronic devices like AlGaN/GaN High Electron Mobility Transistors (HEMT's). In nature there exist two types of materials carrying a $\mathbf{P}_{sp}$: the ferroelectrics and the pyroelectrics. In ferroelectrics the $\mathbf{P}_{sp}$ can be inverted by applying a suitably strong electrostatic field. This effect, known as *bistability*, is very important since it allows an accurate measurement of the $\mathbf{P}_{sp}$. In pyroelectrics $\mathbf{P}_{sp}$ cannot be directly measured because its direction and orientation cannot be altered and is always parallel to a low symmetry axis of the crystal, this being called the *pyroelectric axis*. Luckily, just after GaN became a target of technological applications, a big step forward in the theory of solids, called the Modern Theory of Polarization (MTP) [1], sometimes referred to as *Berry's phase* method, provided an easy and accurate way to compute $\mathbf{P}_{sp}$ for the first time. Within MTP the calculation of $\mathbf{P}_{sp}$ is performed using first-principles computational tools and does not require a previous experimental knowledge about the material structure.

Among the tetrahedrally coordinated solids, the most common pyroelectrics have a wurtzite structure; thus we find AlN, GaN, InN (the subject of this work), the hexagonal polytypes of SiC and well-known II-VI semiconductors such as ZnO and BeO [2]. In wurtzite crystals the pyroelectric axis is parallel to the [0001] direction and $\mathbf{P}_{sp}$ will be equally oriented. In pyroelectrics permanent polarization is an intrinsic property related to the bonding nature of the material, whose origin is somewhat subtle and can be ultimately attributed to the fact that the geometric center of the negative charges (electrons) in the solid does not coincide with the center for the positive charges (nuclei). Another less rigorous but more intuitive way to express this concept is to think that in the pyroelectrics the bonds connecting the atoms with their first neighbors are *not equivalent*, i.e., one of these bonds has a more (or less) ionic nature when compared to the others. This leads to an understanding of why most of the semiconductors do not show a $\mathbf{P}_{sp}$. Tetrahedrally coordinated semiconductors with cubic structure have four equivalent bonds featured by a *perfect sp^3 hybridization*. The equivalence of the bonds can be inferred easily by the band structure of the material having a triple degenerate valence band top at the Γ point. In this case the center of the electronic charge belonging to an atom coincides with the nucleus position. In lower symmetry crystals, such as hexagonal structure semiconductors, the split-off of the valence band top at Γ into a single and double degenerate band shows the existence of an asymmetry in the bonding, with an uneven bond among the four. This is the bond oriented in the [0001] direction whose ionicity is different from the other ones. This difference is reflected by the geometric structure of the crystal, as in most of these crystals this bond is longer than the others. Here the center of the electron charges will be displaced along the [0001] direction, i.e., the pyroelectric axis direction in the hexagonal lattices. This intuitive picture of the origin of the $\mathbf{P}_{sp}$ shows why elemental semiconductors (Si, C, Ge) and zincblende-structure semiconductors, such as most of the III-V and II-VI semiconductors, do not show a $\mathbf{P}_{sp}$.

The equivalence among the four bonds with the neighboring atoms in cubic semiconductors can be changed, applying a strain to the crystal structure, in particular along the [111] direction. In this case the bond along the [111] direction is shortened or elongated and the perfect symmetry of the sp^3 hybridization is broken: the ensuing polarization is called *piezoelectric* ($\mathbf{P}_{pz}$) because it is induced by a mechanical perturbation. It is important to stress, for the clarity of the following discussion, that the difference between piezoelectric and spontaneous polarization is due only to the effect originating the polarization: a mechanical stress for $\mathbf{P}_{pz}$; an intrinsic asymmetry of the bonding in the *equilibrium* crystal structure for $\mathbf{P}_{sp}$. Therefore the same kind of computation used for the determination of the $\mathbf{P}_{sp}$ can be performed for a crystal under the effect of a mechanical perturbation (strain) and the derivative of the polarization with respect to the strain field provides the piezoelectric coefficients listed in the literature.

The $\mathbf{P}_{sp}$ can arise in cubic semiconductors by the effect of alloying. In the case of InGaP alloys [3], materials formed by the substitution of a certain amount of In atoms at the Ga sublattice sites of a zincblende GaP crystal, the bond distortion induced by the atomic size mismatch between Ga and In, can induce a break in the symmetry among the four bonds of the tetrahedral structure. The ensuing polarization must be called $\mathbf{P}_{sp}$ since it is shown in the *equilibrium* structure of the alloy: this effect is fairly relevant in III-N alloys.

The mathematical aspects of MTP are discussed in a review by R. Resta [4] and will not be dealt with in this paper. Here we want to recall that, in order to establish an accurate value for the polarization by MTP, either the $\mathbf{P}_{sp}$ or the $\mathbf{P}_{pz}$, the correct determination of the crystal structure is of fundamental importance. Massidda *et al.* [5] showed that, for a given crystal geometry, the magnitude of the $\mathbf{P}_{sp}$ does not depend to any great degree on the technicalities used for the calculation of the wavefunctions, so Hartree–Fock, Density Functional Theory (DFT) based calculations and GW corrected wavefunctions provide very similar results when the same atomic positions are being used. This finding has consequentially oriented the choice of the computational method on the DFT based calculation, especially in combination with the generalized gradient approximation (GGA) functional for the exchange-correlation energy (hereafter referred to as GGA-DFT). Indeed, only the latter approach is able to provide a very accurate description of the structure of most semiconductors.

3.2
Theoretical Prediction of Polarization Properties in AlN, GaN and InN

Since 1997 several works have been published on the polarization effects in III-N binaries. It should be stressed that data from articles published at different times are to be cautiously compared. The first set of values for the $\mathbf{P}_{sp}$ released in 1997 [6] was published in order to show that electric fields arising in AlN/GaN superlattices [7,8] not only were of piezoelectric nature, but their explanation required and proved the existence of a $\mathbf{P}_{sp}$ in the nitrides too. Such values were obtained with the 1997 state-of-the-art methodology, using the DFT method within the local density approximation (LDA) for the exchange-correlation energy (hereafter referred to as LDA-DFT). Rapid advances in computational techniques allowed improvement of those data. Once the importance of accurate values of $\mathbf{P}_{sp}$ and $\mathbf{P}_{pz}$ for the scientific community was recognized, a new and more complete set of values was computed. Indeed, in 2001 we discovered that the GGA-DFT approach as implemented by Perdew and Wang [9] better reproduces the relevant properties of III-N binary compounds. To keep consistent values of the polarization, properties obtained by the GGA-DFT cannot be used together with LDA-DFT ones, therefore we strongly encourage the reader to use only those data coming from Refs. [10] and [11]. The values for the structural and elastic properties of AlN, GaN and InN computed with GGA-DFT approach are collected in Table 3.1

The values reported in the table refer to the wurtzite structure for the III-N binaries, whose conventional cell is an hexagonal prism described by the edge length a of the prism base, the height of the prism c, and an internal parameter u defined as the anion-cation bond length along the [0001] direction, expressed in units of c. We see that the structural data computed by the GGA-DFT approach are in excellent agreement with experiment.

Tab. 3.1 Structure of Group III-V binary nitrides. Values of the equilibrium structural parameters a, c, u, elastic constants $C_{11}+C_{12}$, C_{31}, C_{33} computed using the GGA exchange-correlation functional are compared with experimental values. All theoretical values are from Ref. [10] except the elastic constants $C_{11}+C_{12}$ quoted from Ref. [11]. (Note that because of a misprint the values of $C_{11}+C_{12}$ listed in Table I of Ref. [11] are erroneously labeled as C_{11}.)

AlN	a(Å)	c/a	u	$C_{11}+C_{12}$(GPa)	C_{13}(GPa)	C_{33}(GPa)
GGA	3.1079	1.6033	0.3814	506	94	377
Exp[a]	3.1106	1.6008	0.3821	550[b]	100[b]	390[b]
GaN	a(Å)	c/a	u	$C_{11}+C_{12}$(GPa)	C_{13}(GPa)	C_{33}(GPa)
GGA	3.1968	1.6297	0.3769	413	68	354
Exp[a]	3.1892	1.6258	0.3770	520[b]	110[b]	390[b]
InN	a(Å)	c/a	u	$C_{11}+C_{12}$(GPa)	C_{13}(GPa)	C_{33}(GPa)
GGA	3.5805	1.6180	0.3787	266	70	205
Exp[a]	3.538	1.6119	—	294[c]	121[c]	182[c]

a) Reference [12]
b) Reference [13]
c) Reference [14]

In Table 3.2 we list the polarization data computed in 2001. We suggest to use them, instead of the set we published in 1997, to compute the polarization-related quantities in the III-Ns. For instance the piezoelectric polarization field can be easily computed as a function of the applied strain η_j as

$$\mathbf{P}_{pz} = e_{33}\eta_3 + e_{31}(\eta_1 + \eta_2) \tag{3.1}$$

One can see that, in all the III-Ns, the $\mathbf{P}_{sp}$ has a negative value. Since $\mathbf{P}_{sp}$ in the wurtzite structure is parallel to the c-axis a negative value means that $\mathbf{P}_{sp}$ is antiparallel to the conventional direction chosen for the [0001] axis. The conventional direction taken as *positive* goes parallel along the bond to the c-axis and connects the Ga atoms with their first neighboring N atoms following the direction from the Ga to the N. This means that the electric dipole associated to the primitive cell is oriented along the c-axis directed from the N to the Ga atom.

A rationale for the negative value of the $\mathbf{P}_{sp}$ can be found in the chemical nature of the bonding in wurtzite crystals. As discussed in Section 3.1 in the

Tab. 3.2 Polarization properties of III-V nitrides. The spontaneous polarization P_{sp}, Born effective charges Z^*, piezoelectric coefficients e_{ij} and d_{ij} after Ref. [15] and [11]. The values were computed using the GGA exchange-correlation functional.

	P_{sp} (C m^{-2})	e_{33} (C m^{-2})	e_{31} (C m^{-2})	e_{31}^p (C m^{-2})	e_{15} (C m^{-2})
AlN	−0.0898	1.505	−0.533	−0.623	−0.351
GaN	−0.0339	0.667	−0.338	−0.372	−0.167
InN	−0.0413	0.815	−0.412	−0.454	−0.112
	Z^* (e)	d_{33} (pm V^{-1})	d_{31} (pm V^{-1})	d_{31}^p (pm V^{-1})	d_{15} (pm V^{-1})
AlN	2.653	5.386	−1.907	−2.103	2.945
GaN	2.670	2.718	−1.344	−1.432	1.808
InN	3.105	9.279	−3.331	−3.542	5.510

wurtzites, the bond oriented in the [0001] direction is inequivalent to the other three. It is longer and of more ionic nature than the others. This means that in this bond the effective center of the electronic charge is farther from the Ga atom. Therefore, the center of the electronic charges belonging to a Ga atom will be slightly off the nucleus position in the direction of the N atom. It can be said that each Ga atom is a carrier of an electric dipole moment antiparallel to the direction connecting it with the N atom along the [0001] axis.

When inspecting Table 3.2 we see that the $\mathbf{P}_{sp}$ value in the nitrides is very large, e.g., AlN has the largest $\mathbf{P}_{sp}$ known so far in wurtzite structure compounds. Piezoelectric coefficients are very large too, ten times larger than in typical III-V and II-IV compounds. Therefore, in these materials the polarization field will be larger than usual and of paramount relevance in predicting the properties of the nanostructures. In Table 3.2 the values for the Born charges are also listed. These quantities represent the effective ionic charge attached to each atom. In Table 3.2 two values of the e_{31} are listed, the so-called *improper* value of piezoelectric coefficient e_{31} and its *proper* value marked with e_{31}^p. The existence of two values for the constant e_{31} has been recently recognized and its justification is not trivial (a thorough technical discussion can be found in Ref. [16]). We reiterate that the improper constant e_{31} should be used whenever the value of the polarization is used to compute an electric field in a nanostructure. The proper constant e_{31}^p must be used when the polarization is used to compute a flow of electric current due to a piezoelectric field in an experimental setup such as that used to measure $\mathbf{P}_{sp}$ in the ferroelectrics. Since we are interested in the computation of the polarization-related electrostatic field, the improper piezoelectric constant e_{31} will be employed from now on. While the piezoelectric constants e_{ij} relate to linear order, the polarization-induced by a semiconductor structure strain, the piezoelectric moduli d_{ij} (sometimes referred to as extensional piezoelectric coefficients) listed in Table 3.2 give the strain induced by an applied electric field into a semiconductor, the so-called *converse* piezoelectric effect. The two sets of piezoelectric constants e_{ij} and d_{ij} are related each other (in Voigt notation) by the transformation

$$e_{ij} = \sum_k d_{ik} C_{kj} \quad (3.2)$$

where C_{kj} are the elastic constants shown in Table 3.1. In Ref. [11] we used MTP to compute directly the d_{ij} coefficients as derivatives of the polarization referred to the stress applied to the bulk structure. This allowed us to build a fully consistent set of data that favorably compares with experiments which directly access the d_{ij} measuring the strain caused by an applied electric field on a macroscopic sample.

3.3
Piezoelectric and Pyroelectric Effects in III-V Nitrides Nanostructures

The connection between electric polarization and the ensuing electrostatic fields in semiconductors nanostructures is by no means a trivial subject. For instance, the existence of a $\mathbf{P}_{sp}$ in a material does not automatically imply the formation of an electrostatic field $\mathbf{E}$, indeed in a massive homogeneous sample no electrostatic field will arise, regardless of the material $\mathbf{P}_{sp}$ strength. In this context it is useful to consider the electrostatic field inside a solid, seen as the superimposition of a microscopic and of a macroscopic component. Hereafter we will only take the macroscopic component into account, the so-called *macroscopic* electric field (MEF), this being the sole component of the electrostatic field affected by polarization. For a rigorous definition of MEF we suggest Ref. [17]. In short, we can say that MEFs are those fields that remain after the *periodic* bulk-like oscillation of the crystal potential is filtered out. To avoid cumbersome notation in the formulas we will refer to the MEF simply as $\mathbf{E}$. A MEF in a solid originates from an accumulation of charge in a localized region, where a discontinuity of the crystal structure takes place. The typical discontinuity we think about is represented by an interface between two different pyroelectric materials in a nanostructure, e.g., AlGaN/GaN multi-quantum wells (MQWs). According to the laws of electrostatics the macroscopic electric field $\mathbf{E}$ is related to the density of charge σ by the equation:

$$\varepsilon \, \text{div} \, \mathbf{E} = \sigma \quad (3.3)$$

where ε is the *static* (zero frequency) dielectric constant of the solid. For the sake of clarity, the anisotropic nature of the static dielectric response will be overlooked and a scalar notation for ε will be used, identifying its value with the ε_{33} component of the dielectric tensor in a wurtzite crystal. In turn, the density of the polarization-induced charge σ_{pol} localized at the interface between two materials carrying a polarization field is proportional to the divergence of the polarization as given by

$$\text{div}\,\mathbf{P}^T = \text{div}(\mathbf{P}_{sp} + \mathbf{P}_{pz}) = -\sigma_{pol} \tag{3.4}$$

where $\mathbf{P}^T$ is the so-called *transverse* polarization, i.e., the sum of the spontaneous and piezoelectric polarization the quantity accurately computed by the MTP. One must observe that an homogeneous mechanical strain does not induce a break of the periodicity in a crystal, as it does only change its structural parameters; therefore, piezoelectric polarization alone does not imply the existence of MEFs. Equations (3.3) and (3.4) provide the connection between MEF and polarization in a perfectly insulating material. To get an explicit value for the MEF the divergence field must be integrated and the effect of the free carriers must be included by the insertion of an additional field $\mathbf{D}$. The explicit expression for the MEF in a pyroelectric nanostructure therefore reads

$$\mathbf{D} = \varepsilon\mathbf{E} + \mathbf{P}_{sp} + \mathbf{P}_{pz} \tag{3.5}$$

Here $\mathbf{D}$ is the sum of the external field applied to the sample $\mathbf{D}^{ext} = \varepsilon_{vacuum}\mathbf{E}^{ext}$ and of the contribution of the free carriers (electrons and holes) $\mathbf{D}^{free}$. It should be pointed out that in Eq. (3.4) the negative sign in front of σ_{pol} is simply a consequence of the convention taken to define a polarization field, that is the polarization in a solid goes from the negatively charged to the positively charged interface, while the MEF does the opposite. So it is not surprising that in most cases $\mathbf{E}$ will be in the opposite direction with respect to $\mathbf{P}^T$.

We recommend splitting the contribution of the free carriers $\mathbf{D}^{free}$ into two terms: $\mathbf{D}^{bulk}$ due to the carriers distributed inside the volume of the sample and $\mathbf{D}^{surf}$ the field arising from the screening effect at the outer surface of the sample. The existence of the latter is experimentally confirmed by the presence of two-dimensional electron gas (2DEG) at the sample surfaces [18, 19] and can be inferred from the following theoretical argument: a *macroscopic* sample cannot sustain a MEF whose average on the sample volume does not vanish. Indeed, at finite temperature the presence of a MEF implies a drift of the free carriers generating a current that in a open circuit system cannot be constantly sustained. In massive homogeneous samples $\mathbf{D}^{surf}$ is trivially related to the transverse polarization by the relation

$$\mathbf{D}^{surf} = \mathbf{P}^T \tag{3.6}$$

In a multilayered nanostructure made of n layers of thickness l_k and transverse polarization $\mathbf{P}_k^T$, the expression for the surface screening is somewhat more complex and is given by

$$\mathbf{D}^{surf} = \frac{\sum_k l_k \mathbf{P}_k^T / \varepsilon_k}{\sum_k l_k / \varepsilon_k} \tag{3.7}$$

Plugging this back into Eq. (3.5) we obtain, for a sample where the contribution of the free carriers screening in the bulk is negligible, the following expression for the MEF

$$E_j = \frac{\sum_k l_k P_k^T / \varepsilon_k - P_j^T \sum_k l_k / \varepsilon_k}{\varepsilon_j \sum_k l_k / \varepsilon_k} \tag{3.8}$$

with sums running on all layers (including the j-th). This is the general expression for any thickness of layers (wells and barriers) in a generic nanostructure (e.g., a generic MQW), where the interfaces between the layers are oriented in the [0001] direction. Equation (3.8) has two remarkably limiting cases. The first case is that of a single-quantum well (SQW) having a well layer of thickness l_W embedded between two semi-infinite samples B_L and B_R. In this case the MEF inside the well layer is given by

$$E_W = \left[\tfrac{1}{2}(P_{B_L}^T + P_{B_R}^T) - P_W^T\right] / \varepsilon_W \tag{3.9}$$

where $P_{B_{L,R}}^T$ are the transverse polarizations in the SQW barriers at the left (L) and right (R) side of the well. Please notice that Eq. (3.9) is symmetric with respect to the exchange of the barriers positions, since only their average polarization matters. Outside of the well, in the barriers, a MEF arises at the interface with the well layer, whose magnitude is given by

$$E_{B_{L,R}} = \tfrac{1}{2}(P_{B_{R,L}}^T - P_{B_{L,R}}^T)/\varepsilon_{B_{L,R}} \tag{3.10}$$

This equation refers to a system devoid of free carriers screening. Actually, the farther the fields in the barriers are from the QW, the more they decrease because of the screening effect due to the accumulation of free carriers at the sides of the well layer.

The second limiting case of Eq. (3.8) is represented by an infinite superlattice (SL) structure where different epitaxially interfaced layers alternate in the direction of the pyroelectric axis. This is relevant, since it allows the calculation of the dielectric constants of a given solid within the MTP as discussed in Refs. [20] and [21]. In the simplest case represented by an alternation of two layers A and B we write the electrostatic fields as

$$E_A = l_B(P_B^T - P_A^T)/(l_A \varepsilon_B + l_B \varepsilon_A) \tag{3.11}$$

and

$$E_B = l_A(P_A^T - P_B^T)/(l_A \varepsilon_B + l_B \varepsilon_A) \tag{3.12}$$

where $P_{A,B}^T$ are the layer's transverse polarizations. It is worth noticing that Eqs (3.11) and (3.12) correctly predict the limiting case of a massive sample

($l_B = 0$) seen as a SL made of a single layer. In this case the MEF given by $\mathbf{E}_A$ vanishes, regardless of the transverse polarization value $\mathbf{P}_A^T$.

In the case of nanostructures with a continuous change of composition, simple rules cannot be given and the (self-consistent) solution of a differential equation is necessary in order to compute the macroscopic electric fields. In the simplified case in which the composition is a function of the distance z from the sample surface the equation reads

$$\varepsilon(z)\frac{d}{dz}\mathbf{E}(z) + \mathbf{E}(z)\frac{d}{dz}\varepsilon(z) = \frac{d}{dz}\left(\mathbf{D}(z) - \mathbf{P}^T(z)\right) \qquad (3.13)$$

Equations (3.8) and (3.13) show that the polarization-induced MEF in a system cannot be computed without the knowledge of the system geometry, i.e., it is the system inhomogeneity in the form of surfaces and interfaces that allows the polarization to show up as a MEF. The absence of MEFs does not mean absence of any polarization. The idea that electrostatic fields are linearly dependent on the polarization is only partly true: in most cases the electrostatic field does not need to be proportional, nor to be *parallel* to the polarization field. As we will see in Section 3.4, there are nanostructure geometries that are MEF-free in the presence of large polarization fields.

Equation (3.5) shows clearly that the calculation of the MEFs in the nanostructures requires the knowledge of both polarization properties and static dielectric response of the constituents. While MTP accurately and directly provides access to transverse polarization, for the dielectric response, the standard theoretical approach is represented by the Density Functional Perturbation Theory (DFPT) [22]. DFPT is a general approach used to compute the response of a solid to a perturbation. It allows the computation of dielectric and piezoelectric constants and Born charges. However, it cannot be applied to compute $\mathbf{P}_{sp}$, since this is a property of the equilibrium structure and not an effect of a perturbation. DFPT turns out to be useful for validation of the results of the MTP-made calculations of the piezoelectric constants and of Born charges, its results always being fairly consistent with those of DFPT [1]. While DFPT allows the computation of the dielectric constants in a given solid regardless of its polarization properties, we have been able to compute the dielectric tensor constant ε_{33} for wurtzite crystals using MTP by means of a simple procedure exploiting the relationship between MEFs and polarization in appropriate superlattices [7,8,23]. This procedure is described in Ref. [20] and makes use of the relation in Eqs (3.11) and (3.12) between polarization and MEF in a SL.

We tested our simplified approach on GaAs, AlAs, InN, SiC, ZnO, GaN, AlN, BeO, LiF, $PbTiO_3$, and $CaTiO_3$ [21]. The predicted dielectric constants agree well with those given by DFPT. In Table 3.3 we compare the values of the dielectric constants ε_{33} of III-Ns which we computed within the LDA and GGA functional, with experimental values, and a calculation performed within the DFPT using the LDA functional [24]. The set labeled with GGA-MTP is an unpublished set of numbers, while the LDA-MTP set was published in 1997 [20], when MTP within a GGA formalism was not yet available to us: therefore, the LDA functional was our standard choice. Calculated dielectric constants are generally 10–15% overestimated with respect to actual values, even in those cases where the structural properties are correctly described (e.g., GGA). III-Ns are not an exception, indeed we find that GGA-MTP values are between 8 and 14% larger than the experimental values. LDA-based calculations (second and third row) seem less reliable and in this context we suggest readers use the experimental values listed in Table 3.3.

Tab. 3.3 ε_{33} component of the static dielectric tensor of binary III-V nitrides compounds. First and second line values are obtained using MTP after Ref. [20].

	AlN	GaN	InN
GGA-MTP	9.90	11.85	16.77
LDA-MTP[a]	10.31	10.28	14.61
LDA-DFPT[b]	9.56	10.35	—
Exp.	9.18[c]	10.4[d]	15.3[e]

a) Reference [20]
b) Reference [24]
c) Reference [25]
d) Reference [26]
e) Reference [27]

3.4
Polarization Properties in Ternary and Quaternary Alloys: Nonlinear Compositional Dependence and Order vs. Disorder Effects

Device designers are interested in MQW systems made of ternary and quaternary alloys. In the absence of a direct investigation on the alloys, their electronic and structural properties are customarily obtained by linear interpolation formulas based on Vegard's law. This approach works well for the macroscopic structural parameters of alloys (e.g., a and c lattice parameters in wurtzites) and nitrides are not an exception. Conversely, microscopic degrees of freedom like bond lengths, and electronic properties like band gaps, significantly depart from linearity. If this is the case, interpolation formulas based on the concept of bowing are used. Within this interpolation model it is supposed that the relevant quantity (e.g., the spontaneous polarization $\mathbf{P}_{\text{sp}}$ in

3.4 Polarization Properties in Ternary and Quaternary Alloys

an $Al_xGa_{1-x}N$ alloy) is correctly reproduced by a second-order polynomial of the alloy composition

$$\mathbf{P}_{sp}(Al_xGa_{1-x}N) = x\,\mathbf{P}_{sp}(AlN) + (1-x)\,\mathbf{P}_{sp}(GaN) + b_{AlGaN}\,x(1-x) \quad (3.14)$$

where b_{AlGaN} is the so-called *bowing parameter*, defined as

$$b_{AlGaN} = 4\,\mathbf{P}_{sp}(Al_{0.5}Ga_{0.5}N) - 2\,[\mathbf{P}_{sp}(AlN) + \mathbf{P}_{sp}(GaN)] \quad (3.15)$$

The advantage of the bowing model is that it requires knowledge of the relevant quantity $\mathbf{P}_{sp}(Al_xGa_{1-x}N)$ only at a composition $x = 0.5$, together with the values for the binaries.

Beside their compositional dependence, the electronic properties (e.g., band gap) of alloys are found to depend strongly on the extent of order of their microscopic structure. Indeed, the distribution of the diluted elements can be random or ordered, depending on the thermodynamics and kinetics of growth. In the III-N alloys, ordered structures have been experimentally detected [28] and theoretically predicted [29]. Therefore, polarization in alloys must be investigated in both random (disordered) and ordered systems. Since the above-mentioned ordering shows remarkable similarities with the one found for InGaP alloys CuPt phase, we will label this structure as CP-ordered. Ordered alloys are periodic structures, being suitable for direct calculation of macroscopic polarization within MTP. Chemically disordered alloys, where Group-III elements are randomly distributed on the cation sites, are aperiodic structures. In our first work on polarization in III-N alloys [30] we showed that the problem of aperiodicity in disordered alloys can be solved by exploiting the so-called Special Quasi-random Structure (SQS) method [31] to model chemical disorder. The use of SQS makes the calculation of the macroscopic polarization in an alloy straightforward, as in the case of a binary compound. The results for the $\mathbf{P}_{sp}$ and average lattice parameters (a and c) of the hexagonal structure in disordered and CP-ordered alloys are listed in Table 3.4.

Tab. 3.4 Bowing parameters for the structural (a,c) and polarization properties ($\mathbf{P}_{sp}$) of disordered (odd columns) and CP-ordered (even columns) III-V nitride alloys, computed using the GGA functional (Ref. [32]).

	b_{AlGaN}		b_{InGaN}		b_{AlInN}	
	Random	CP-order	Random	CP-order	Random	CP-order
a (Å)	−0.002	−0.007	−0.003	+0.002	−0.031	−0.083
c (Å)	+0.028	+0.039	+0.024	+0.185	+0.054	+0.420
P_{sp} (C m^{-2})	+0.0191	+0.0176	+0.0378	+0.1934	+0.0709	+0.3336

We note that, in spite of the absence of sizable nonlinearity in the macroscopic structural parameters (a and c), the $\mathbf{P}_{sp}$ shows a pronounced upward bowing. The bowing is larger for highly mismatched components alloys (those containing In) and acts to reduce the absolute value of the $\mathbf{P}_{sp}$. One can see that alloy ordering greatly increases the bowing in InGaN and AlInN alloys, while ordering in AlGaN does not affect the bowing parameter. The rationale behind the existence of a nonlinear behavior in the compositional dependence of $\mathbf{P}_{sp}$ is discussed in detail in Ref. [30]. Computational evidence brings us to assert that deviations from the bowing model due to nonparabolic behavior should be tiny. The existence of a bowing in III-N alloys is primarily an effect of the internal strain in the alloy due to the atomic size mismatch among Group-III elements. Superimposed on this effect, a more genuine nonlinearity comes from the microscopic level hydrostatic stress, tensile for Al and compressive for In, due to the lattice mismatch between the alloy binary components. Since, in each of the alloys, binary components polarization has a nonlinear dependence on hydrostatic stress, this results in a nonlinear behavior in alloys.

Spontaneous polarization bowing values are not sufficient to predict the MEFs in nanostructures made of III-N alloys. So far few theoretical works [30, 33–35] dealt with the problem of nonlinear piezoelectricity in ternary alloys, focusing on different aspects of the widest problem. In Ref. [30] we narrowed our scope to the very common case of alloys pseudomorphically grown on a GaN substrate. In that work we computed the polarization for binaries and alloys, fully re-optimizing all structures, with the constraint that the basal lattice parameter would fit that of the GaN equilibrium structure. Calculations showed that nonlinear behavior of the piezoelectric polarizations in the alloys does exist. For clarity sake, we will first discuss the case of the disordered alloys, this being more technologically relevant and simpler to understand. In disordered alloys, opposite to the $\mathbf{P}_{sp}$ case, piezoelectric nonlinearity is not a genuine bowing, but it is simply the superimposition of the nonlinear piezoelectricity of the individual binary constituents of the alloys. Our calculations in Ref. [30] show that, at the typical strain values considered in usual III-N based MQW systems, $\mathbf{P}_{pz}$ in the binaries has quite a large nonlinear component. This nonlinearity is quite well reproduced by a second-order polynomial as:

$$\begin{aligned} P_{pz}^{AlN} &= -1.808\ \eta_1 + 5.624\ \eta_1^2 \quad &\text{for } \eta_1 < 0 \\ P_{pz}^{AlN} &= -1.808\ \eta_1 - 7.888\ \eta_1^2 \quad &\text{for } \eta_1 > 0 \\ P_{pz}^{GaN} &= -0.918\ \eta_1 + 9.541\ \eta_1^2 \\ P_{pz}^{InN} &= -1.373\ \eta_1 + 7.559\ \eta_1^2 \end{aligned} \quad (3.16)$$

where η_1 is the basal strain of the binary compound considered. Then the values for the alloys corresponding to a molar fraction x are reproduced by applying Vegard's interpolation formula to the piezoelectric component given for each binary III-N. The accounting procedure for the nonlinearity is shown as follows. For a given alloy composition (e.g., $Al_xGa_{1-x}N$) having an equilibrium lattice constant $a(x)$, the basal strain field $\eta_1(x)$ for the alloy matched to a GaN substrate defined as

$$\eta_1(x) = (a^{GaN} - a(x))/a(x) \tag{3.17}$$

is determined. The evaluation of the lattice constant does not require an explicit calculation since our work shows that Vegard's law is obeyed by alloys. Therefore Eq. (3.17) can be further simplified as:

$$\eta_1(x) = x(a^{GaN} - a^{XN})/[xa^{XN} + (1-x)a^{GaN}] \tag{3.18}$$

where X is the diluted element of the alloy. Then $\mathbf{P}_{pz}$ is computed for the two (or more) binary components of the alloy (e.g., AlN and GaN) for the above-mentioned strain field magnitude. The piezoelectric polarization values (e.g., $\mathbf{P}_{pz}(AlN,\eta_1)$ and $\mathbf{P}_{pz}(GaN,\eta_1)$) are then weighed using the Vegard's interpolation formula:

$$\mathbf{P}_{pz}(Al_xGa_{1-x}N, \eta_1) = x\mathbf{P}_{pz}(AlN, \eta_1) + (1-x)\mathbf{P}_{pz}(GaN, \eta_1) \tag{3.19}$$

The piezoelectric polarization thus obtained favorably compares with the outcome of the direct calculations by the SQS alloy model. As for ordered alloys, procedures feature a higher level of complexity. Our calculations in Ref. [30] already showed that in highly mismatched alloys, such as InGaN and AlInN, piezoelectricity is influenced by the presence of microscopic chemical order. The origin of this effect has been discussed by Al-Yacoub and Bellaiche [34] in the case of the CP-ordering in $In_{0.5}Ga_{0.5}N$. They found that the piezoelectric coefficients e_{33} and e_{31} for this alloy in its equilibrium structure are 15% and 28% smaller than the binary constituents averaged values after Vegard's law.

We agree with these results and suggest an explanation that equally accounts for both spontaneous and piezoelectric components of the polarization. Our understanding is that deviations of $\mathbf{P}_{sp}$ and $e_{33,31}$ from Vegard's law predictions are a consequence of: (i) the effect of the nonlinear response of the binary constituents of the alloys with respect to hydrostatic stress and epitaxial strain for the spontaneous and piezoelectric polarization, respectively; (ii) the effect of the ordering on the cation sublattice positions (internal strain) for spontaneous polarization, and their elastic response to epitaxial strain for the piezoelectric properties. The first effect is the dominant term in disordered alloys while the second accounts for the apparent *anomalies* found for the ordered alloys.

We wish to note an important consequence in the understanding of III-N alloys achieved by the work of Ref. [30]: nonlinearity in III-N alloys can be fully predicted by the knowledge of both alloy atomic level structure and the nonlinear response of the binary compounds on hydrostatic and epitaxial stress. No other nonlinear components for polarization seem to originate in the alloys due to the effects of disorder or chemistry. Such results provide the intriguing chance of predicting the polarization of alloys with arbitrary compositions by just computing their structure, with the aid of classical potential methods, such as the valence force field. The average structural parameters a, c and u, are then used to interpolate the polarization data of the binary compounds, in order to obtain the actual polarization in the alloys included the nonlinear dependence on stress and composition. This is indeed the strategy used by Al-Yacoub *et al.* [35] in their work on piezoelectricity in $In_xGa_{1-x}N$ alloys. They computed the piezoelectric constant e_{33} of InGaN alloys at different In concentrations combining first-principles calculations and classical potential simulation for very large alloy supercells. The overall piezoelectric response in *disordered* $In_xGa_{1-x}N$ was found to obey Vegard's scheme outlined in Ref. [30]. As a consequence of our work on polarization in alloys on the realization of III-Ns based devices, it is important to grow the nanostructures under suitable conditions so as control the amount of ordering in our system. Methods not enabling this control can easily lead to nonreproducible results on the production of devices since, as we saw, the polarization $\mathbf{P}^T$ source of the MEFs in the nanostructure dramatically depends on the amount of order in the microscopic structure of the employed materials.

We wish to end this section with a few considerations about polarization in quaternary alloys. Since McIntosh and co-workers successfully demonstrated the possibility of growing AlInGaN on sapphire [36], attempts where made to build AlInGaN based heterostructures. Thanks to the endurance of the researchers, AlInGaN/GaN lattice matched interfaces with a high 2DEG density and interesting transport properties have been obtained [37]. With respect to ternary alloys the AlInGaN system allows the possibility of playing with two degrees of freedom represented by two diluted elements, Al and In, instead of one. Moreover the effect of the two species on the quaternary alloy are very different. Al has little influence on the alloy lattice parameter and a large one on the $\mathbf{P}_{sp}$. Conversely, In incorporation only marginally influences $\mathbf{P}_{sp}$ and strongly influences the structural properties. Therefore, with the help of AlInGaN alloys it is possible to achieve the goal of finding an alloy lattice-matched to GaN having a quite large difference in the polarization so that at the AlInGaN/GaN interface a large 2DEG would accumulate. This would be of great importance for one of the most important applications of the polarization properties of III-N alloys, the so-called *piezoelectric doping* of a QW, i.e., the formation of a 2DEG at the well layer interfaces induced by the screening of the MEFs. This is normally accomplished by interfacing an AlGaN barrier to a GaN well. However, the high Al content AlGaN/GaN nanostructures required for HFETs poses important limits on the geometry of the system because of the strain imposed on the AlGaN barrier.

In this case, as for the ternary alloys, useful predictions from theoretical work are of great value for device design. As yet, there is no investigation on polarization properties focusing on the quaternary AlInGaN alloy. Here we suggest that the knowledge already acquired on ternaries can also provide useful interpolating formulas for quaternaries. The natural extension of the bowing formula of Eq. (3.14) to $Al_xIn_yGa_{1-x-y}N$ would be

$$\begin{aligned}\mathbf{P}_{sp}\left(Al_xIn_yGa_{1-x-y}N\right) &= x\,\mathbf{P}_{sp}\left(AlN\right) + y\,\mathbf{P}_{sp}\left(InN\right) \\ &+ (1-x-y)\,\mathbf{P}_{sp}\left(GaN\right) \\ &+ b_{AlGaN}\,x(1-x-y) + b_{InGaN}\,y(1-x-y) \\ &+ b_{AlInN}\,xy + b_{AlInGaN}\,xy(1-x-y)\end{aligned} \quad (3.20)$$

where the three bowing coefficients b_{AlGaN}, b_{InGaN} and b_{AlInN} are those of the ternary alloys listed in Table 3.4. Indeed, this equation simply reduces to Eq. (3.14) whenever the concentration of one of the Group-III elements vanishes. Now the sole unknown coefficient $b_{AlInGaN}$ in Eq. (3.20) appears in the last term. This coefficient is defined within the bowing model as

$$\begin{aligned}b_{AlInGaN} &= 27\,\mathbf{P}_{sp}\left(Al_{1/3}In_{1/3}Ga_{1/3}N\right) \\ &- 9\left(b_{AlGaN} + b_{InGaN} + b_{AlInN}\right) \\ &- 3\left(\mathbf{P}_{sp}\left(AlN\right) + \mathbf{P}_{sp}\left(GaN\right) + \mathbf{P}_{sp}\left(InN\right)\right)\end{aligned} \quad (3.21)$$

Since the Al and In concentrations of interest for device building are very low (typically 10% for Al and 2% for In) we may infer that the last term in Eq. (3.20) could be neglected. This proves to be true apart from the unlikely circumstance of a very high $b_{AlInGaN}$ value.

As for the piezoelectric component of the polarization, the approach to quaternary alloys is very simple when a disordered system is taken into account, if we suppose that, as in the case of the ternaries, only the binary components nonlinearity appears. Given this approximation, Eq. (3.19) is simply generalized for $Al_xIn_yGa_{1-x-y}N$ as

$$\begin{aligned}\mathbf{P}_{pz}\left(Al_xIn_yGa_{1-x-y}N, \eta_1\right) &= x\mathbf{P}_{pz}\left(AlN, \eta_1\right) + y\mathbf{P}_{pz}\left(InN, \eta_1\right) \\ &+ (1-x-y)\mathbf{P}_{pz}\left(GaN, \eta_1\right)\end{aligned} \quad (3.22)$$

where the basal strain η_1 is defined after Vegard's interpolation of the lattice parameter as

$$\eta_1\left(Al_xIn_yGa_{1-x-y}N\right) = \left[x(a^{GaN} - a^{AlN}) + y(a^{GaN} - a^{InN})\right] / \left[x\,a^{AlN} + y\,a^{InN} + (1-x-y)\,a^{GaN}\right] \quad (3.23)$$

and the functions $\mathbf{P}_{pz}\left(XN, \eta_1\right)$ represent the nonlinear response of the binary constituents mentioned above (Eq. (3.16)).

3.5
Orientational Dependence of Polarization

The origin of the polarization in III-Ns, its dependence on composition and strain, the relationship connecting polarization and MEF have been discussed so far. Now we want to deal with an intriguing aspect of polarization fields in nanostructures: the possibility of building a nanostructure where any electrostatic field is absent, in spite of the presence of strong polarization fields. The Quantum-Confined Stark Effect produced by the MEFs in the MQWs is one of the limiting factors for the use of III-Ns when applied in the field of optoelectronics (e.g., white LEDs and UV lasers diodes). Growing these materials in the zinc-blende phase (this being devoid of $\mathbf{P}_{sp}$, as in the case of GaAs) would be a straightforward solution. In spite of huge efforts, however, device-quality material has not yet been obtained.

As yet, nanostructures with suitable geometries (where MEFs are absent regardless of the magnitude of the polarization) seem to be the most likely solutions. Such geometries are not trivial outcomes of the electrostatic laws and deserve a thorough discussion. We remind the reader that Eq. (3.8) was obtained following the assumption that the interfaces between the nanostructure layers were oriented in the [0001] direction. If we see our system as a mosaic of different homogeneous pieces of material with an arbitrary shape (e.g., Quantum Dots) or that our sample is made of a constantly variable composition, a more general expression for the MEF is to be used. To get this, we have to step back to Eqs (3.4) and (3.3), where the polarization-induced charge density was related to MEF and $\mathbf{P}^T$. By replacing the definition for σ of Eq. (3.3) into Eq. (3.4) we get a somewhat relevant expression for our case:

$$\operatorname{div} \mathbf{P}^T(\mathbf{r}) = -\operatorname{div}(\varepsilon(\mathbf{r})\mathbf{E}(\mathbf{r})) = -\mathbf{E}(\mathbf{r}) \cdot \nabla \varepsilon(\mathbf{r}) - \varepsilon(\mathbf{r}) \operatorname{div} \mathbf{E}(\mathbf{r}) \quad (3.24)$$

where ε is the static dielectric constant of the material, and $\mathbf{r}$ underlines the spatial dependence of the above quantities in a nonhomogeneous system. Here the transverse polarization is to be computed from the value of $\mathbf{P}_{sp}$ and strain field η through this expression:

$$P_i^T(\mathbf{r}) = P_{i,\mathrm{sp}}(\mathbf{r}) + \sum_{jk} e_{ijk}(\mathbf{r}) \cdot \eta_{jk}(\mathbf{r}) \quad (3.25)$$

where indexes i, j, k run over the Cartesian components x, y and z of the polarization vectors P and the piezoelectric (strain) tensors e_{ijk} (η_{jk}). In most cases, a numerical solution of the problem is required.

This could be avoided if the nanostructure is simply a MQW system made of compositionally homogeneous layers whose interfaces are oriented towards the same direction; this implies that the polarization field will be homogeneous inside each layer. Since interfaces are not oriented along the pyroelectric axis [0001], $\mathbf{P}^T$ will not necessarily point to this direction and most likely both the transverse polarization magnitude and direction will vary from layer to layer. The formation of MEFs in these MQWs will follow a different relation than that in Eq. (3.8). Indeed, the MEFs expression has to take into account the effect due to the discontinuity in the *direction* of the polarization across the interfaces. In such case, the most general expression of Eq. (3.4) reduces to

$$\sigma = \hat{\mathbf{n}} \cdot (\mathbf{P}_B^T - \mathbf{P}_A^T) \tag{3.26}$$

where $\hat{\mathbf{n}}$ is a normalized vector perpendicular to the interface considered. Furthermore, in order to keep consistency with the sign of σ given in Eq. (3.4), the vector $\hat{\mathbf{n}}$ has to point from layer A to B. As for the MEFs it keeps the same expression used so far and given in Eq. (3.3). It should be noted that in a nanostructure with parallel interfaces Eq. (3.3) predicts that MEFs will always be perpendicularly oriented to the interface *regardless of* the polarization direction within each layer. Accordingly, Eq. (3.8) can be rewritten for MQWs with arbitrary orientation as

$$E_j = \frac{\sum_k l_k \mathbf{P}_{k,\perp}^T / \varepsilon_k - \mathbf{P}_{j,\perp}^T \sum_k l_k / \varepsilon_k}{\varepsilon_j \sum_k l_k / \varepsilon_k} \tag{3.27}$$

$\mathbf{P}_{k,\perp}^T$ being the k-th layer polarization component perpendicular to the interface

$$\mathbf{P}_{k,\perp}^T = (\hat{\mathbf{n}}_k \cdot \mathbf{P}_k^T) \hat{\mathbf{n}}_k \tag{3.28}$$

with the direction of the vector $\mathbf{n}_k$ pointing from the k-th to the $k+1$-th layer. An important consequence of Eq. (3.27) is that only the perpendicular polarization component matters. While in MQWs which are oriented in the [0001] direction the values for the dielectric constants could be safely identified with the ε_{33} component of the dielectric tensor, this is not the case in our study: an appropriate value of the dielectric constant $\varepsilon_\perp$ should be used. Such value is obtained from the knowledge of the dielectric tensor ε_{ij} as:

$$\varepsilon_\perp = \sum_{ij} \hat{\mathbf{n}}_i \varepsilon_{ij} \hat{\mathbf{n}}_j \tag{3.29}$$

When reconsidering the problem of finding a MEF-free structure for the MQWs, we may find a trivial solution by growing the layers perpendicular to the pyroelectric polarization axis so that $\mathbf{P}_\perp^T$ vanishes. Indeed in this case $\mathbf{P}_{sp}$ will be perpendicular to the MQW growth direction and the piezoelectric component of the polarization will be in the same direction as $\mathbf{P}_{sp}$ if no shear strain is present in the layers forming the nanostructure. Such a hypothesis has been accomplished by the group of H. Ploog [38, 39]. They grew an Al-GaN/GaN MQW on top of a γ-LiAlO$_2$ substrate. γ-LiAlO$_2$ is a tetragonal structure material whose (001) surface offers an ideal nucleation site for the GaN rectangular prism plane ('M'-plane). The MQW grown in the [1$\bar{1}$00] direction has the pyroelectric axis [0001] parallel to the growth plane. Therefore in the absence of shear strain, as in this case, no MEFs will originate from the existing polarization. Along with the difficulties of growing III-Ns on this unusual substrate, the authors showed the benefits of growing a MQW in a MEF-free geometry. They were able to show an emission of light from a GaN well 10 meV *above* the material bulk band gap, a result not achieved in [0001] oriented MQWs [38].

It is interesting to note that Seoung-Hwan Park and co-workers [40] discovered that vanishing MEFs can be obtained for AlInN/GaN MQWs at tilt angles different from 90°. This effect is not totally unknown for other materials: let us recall that in the case of the zincblende structure InGaAs/GaAs MQWs, interfaces grown in the [111] polar orientation (equivalent to the [0001] in the hexagonals) show large piezoelectric polarization discontinuities, while along the [001] and [110] nonpolar orientation piezoelectricity vanishes [41]. One can see that the piezoelectric axis [111] is not orthogonal to the nonpolar directions [001] and [110] as it forms angles of 54.7° and 35.2°, respectively. Clearly, since structures of III-N hexagonal and III-V cubic semiconductors are very similar, in MQWs made of hexagonal semiconductors a vanishing piezoelectric polarization could be found for tilt angles between 54.7° and 35.2°. Indeed this is the case for a large set of layer compositions in AlGaN/GaN and InGaN/GaN tilted systems where $\mathbf{P}_{pz}$ is oriented along the interface plane, so that $\mathbf{P}_{pz,\perp}$ vanishes. For instance, Takeuchi *et al.* [42] in their work on In$_{0.1}$Ga$_{0.9}$N/GaN MQWs grown on a tilted axis, proved it possible to obtain a vanishing piezoelectric effect for orientations of 39° off the [0001] pyroelectric axis. We should point out that, in this work, the effect of $\mathbf{P}_{sp}$ is overlooked, InN and GaN both being largely mismatched materials with similar $\mathbf{P}_{sp}$. Accurate calculation cannot rely on this approximation and we suggest using Eqs (3.25) and (3.27) instead. Details on the formulas used in the computation of $\mathbf{P}_{pz}$ in arbitrarily oriented MQWs can be found in Ref. [43]. We notice that for MQWs grown on the (10$\bar{1}$0) and (11$\bar{2}$0) planes the absence of a MEF is not related to the values assumed for $\mathbf{P}_{sp}$ and piezoelectric coefficients. In the case of tilted growth orientation, a delicate cancellation among $\mathbf{P}_{sp}$ and piezoelectricity due to basal plane and shear strain effects sets in. The actual values for the critical angle of the MEF vanishing condition can depend on the quantities involved. On the other hand, since high-quality multilayers cannot grow in an arbitrary orientation, the above-mentioned works suggest that almost MEF-free structures at least can be grown along the [11$\bar{2}$4] and [10$\bar{1}$2] directions of a hexagonal III-N crystal.

Acknowledgments

We acknowledge financial support from MIUR under project PON-CyberSar.

References

1. R. D. King-Smith and D. Vanderbilt, Phys. Rev. B **47**, 1651 (1993).
2. A. Dal Corso, M. Posternak, R. Resta, and A. Baldereschi, Phys. Rev. B **50**, 10715 (1994).
3. Sverre Froyen, Alex Zunger, and A. Mascarenhas, Appl. Phys. Lett. **68**, 2852 (1996).
4. R. Resta, Rev. Mod. Phys. **66**, 899 (1994).
5. S. Massidda, R. Resta, M. Posternak, and A. Baldereschi, Phys. Rev. B **52**, 16977 (1995).
6. F. Bernardini, V. Fiorentini, and D. Vanderbilt, Phys. Rev. B **56**, R10024 (1997).
7. F. Bernardini, V. Fiorentini, and D. Vanderbilt, in *III-V Nitrides*, edited by F. A. Ponce, T. D. Moustakas, I. Akasaki, and B. A. Monemar, MRS Symposia Proceedings, No. **449**, (Materials Research Society, Pittsburgh 1997), p. 923.
8. F. Bernardini, and V. Fiorentini, Phys. Rev. B **57**, R9427 (1998).
9. J. P. Perdew, in *Electronic Structure of Solids '91*, edited by P. Ziesche and H. Eschrig (Akademie-Verlag, Berlin, 1991), p. 11.
10. F. Bernardini, V. Fiorentini, and D. Vanderbilt, Phys. Rev. B **63**, 193201 (2001).
11. F. Bernardini, and V. Fiorentini, Appl. Phys. Lett. **80**, 4145 (2002).
12. K. Lawniczak-Jablonska, T. Suski, I. Gorczyca, N. E. Christensen, K. E. Attenkofer, R. C. C. Perera, E. M. Gullikson, J. H. Underwood, D. L. Ederer, and Z. Liliental Weber, Phys. Rev. B **61**, 16623 (2000).
13. C. Deger, E. Born, H. Angerer, O. Ambacher, M. Stutzmann, J. Hornsteiner, E. Riha, and G. Fischerauer, Appl. Phys. Lett. **72**, 2400 (1998).
14. A.U. Sheleg and V.A. Savastenko, Inorg. Mater. **15**, 1257 (1979).
15. A. Zoroddu, F. Bernardini, P. Ruggerone, and V. Fiorentini, Phys. Rev. B **64**, 45208 (2001).
16. D. Vanderbilt, J. Phys. Chem Solids **61**, 147 (2000).
17. A. Baldereschi, S. Baroni, and R. Resta, Phys. Rev. Lett. **61**, 734 (1988).
18. O. Ambacher, B. Foutz, J. Smart, J. R. Shealy, N. G. Weimann, K. Chu, M. Murphy, A. J. Sierakowski, W. J. Schaff, and L. F. Eastman, R. Dimitrov, A. Mitchell, and M. Stutzmann, J. Appl. Phys. **87**, 334 (2000).
19. M. Eickhoff, J. Schalwig, G. Steinhoff, O. Weidemann, L. Görgens, R. Neuberger, M. Hermann, B. Baur, G. Müller, O. Ambacher, M. Stutzmann, Phys. Stat. Sol. (c), Vol. **0**, No. 6, 1908 (2003).
20. F. Bernardini, V. Fiorentini, and D. Vanderbilt, Phys. Rev. Lett. **79**, 3958 (1997).
21. F. Bernardini and V. Fiorentini, Phys. Rev. B **58**, 15292 (1998).
22. S. Baroni, P. Giannozzi, and A. Testa, Phys. Rev. Lett. **58**, 1861 (1987); N.E. Zein, Sov. Phys. Sol. State, Vol. **26**, 1825 (1984).
23. D. Vanderbilt, and R. D. King-Smith, Phys. Rev. B **48**, 4442 (1993).
24. J.-M. Wagner and F. Bechstedt, Phys. Rev. B **66**, 115202 (2002).
25. F. Malengreau, M. Vermeersch, S. Hagege, R. Sporken, M.D. Lange, and R. Caudano, J. Mater. Res. **12**, 175 (1997).
26. A. S. Barker, Jr. and M. Ilegems, Phys. Rev. B **7**, 743 (1973).
27. V. W. L. Chin, T. L. Tansley, and T. Osotchan, J. Appl. Phys. **75**, 7365 (1994).
28. P. Ruterana, G. De Saint Jores, M. Laügt, F. Omnes, and E. Bellet-Amalric, Appl. Phys. Lett. **78**, 344 (2001); P. Ruterana, G. Nouet, W. Van der Stricht, I. Moerman, and L. Considine, Appl. Phys. Lett. **72**, 1742 (1998).
29. J. E. Northrup, L. T. Romano and J. Neugebauer, Appl. Phys. Lett. **74**, 2319 (1999).

30 F. Bernardini, and V. Fiorentini, Phys. Rev. B **64**, 085207 (2001).

31 S.-H. Wei, L.G. Ferreira, J.E. Bernard, and A. Zunger, Phys. Rev. B **42**, 9622 (1990).

32 F. Bernardini, and V. Fiorentini, Phys. Stat. Sol. (a) Vol. **190**, No. 1, 65 (2002).

33 V. Fiorentini, F. Bernardini, and O. Ambacher, Appl. Phys. Lett. **80**, 1204 (2002).

34 A. Al-Yacoub and L. Bellaiche, Appl. Phys. Lett. **79**, 2166 (2001).

35 A. Al-Yacoub, L. Bellaiche, and S.-H Wei, Phys. Rev. Lett. **89**, 57601 (2002).

36 F. G. McIntosh, K. S. Boutros, J. C. Roberts, S. M. Bedair, E. L. Piner, and N. A. El-Masray, Appl. Phys. Lett. **68**, 40 (1996). E.L. Piner, F.G. McIntosh, J.C. Roberts, M.E. Aumer, V.A. Joshkin, S.M. Bedair, and N.A. El-Masry, MRS Internet J. Nitride Semicond. Res. 1, **43** (1996).

37 M. Asif Khan, J. W. Yang, G. Simin, R. Gaska, M. S. Shur, Hans-Conrad zur Loye, G. Tamulaitis, A. Zukauskas, David J. Smith, D. Chandrasekhar and R. Bicknell-Tassius, Appl. Phys. Lett. **76**, 1161 (2000). M. Asif Khan, J. W. Yang, G. Simin, R. Gaska, M. S. Shur, and A. D. Bykhovski, Appl. Phys. Lett. **75**, 2806 (1999).

38 P. Waltereit, O. Brandt, A. Trampert, H. T. Grahn, J. Menniger, M. Ramsteiner, M. Reiche and K. H. Ploog, Nature **406**, 865 (2000).

39 P. Waltereit, O. Brandt, M. Ramsteiner, A. Trampert, H. T. Grahn, J. Menniger, M. Reiche, R. Uecker, P. Reiche and K. H. Ploog, Phys. Stat. Sol. (a) Vol. **180**, No. 1, 133 (2000).

40 Seoung-Hwan Park and Shun-Lien Chuang, Phys. Rev. B **59**, 4725 (1999).

41 D. Sun, and E. Towe, Jpn. J. of Appl. Phys. Vol. **33**, Part 1, No. 1B, pp. 702–708 (1994).

42 T. Takeuchi, H. Amano, and I. Akasaki, Jpn. J. Appl. Phys. Vol. **39**, Part 1, No. 2A, pp. 413–416 (2000).

43 Seoung-Hwan Park, and Shun-Lien Chuang, Semicond. Sci. Technol. **17**, 686 (2002).

4
Transport Parameters for Electrons and Holes
Enrico Bellotti and Francesco Bertazzi

4.1
Introduction

The technological importance of the III-Nitride semiconductor material system has rapidly increased in the last few years driven by a considerable number of commercial and defense applications. A substantial effort has been put into developing the material growth techniques and improving the film quality. At the same time electronic and optoelectronic devices have been fabricated with different levels of sophistication and performance. From the electronics standpoint, HEMTs and related microwave integrated circuits have become the focus of the main development thrust. Green, blue and UV LED device structures are currently being investigated for application in solid state lighting. Traffic lights, outdoor displays and DVD players are key commercial applications for GaN optoelectronics devices. Deep UV lasers and LED sources and detectors are of great interest for bio-agent detection and identification.

The design and optimization of all these III-Nitride semiconductor devices requires that the optical and electrical material properties be known and that reliable models to describe them be available. In spite of the increasing interest in the III-Nitride semiconductors, only a limited amount of information is available about their transport properties. This work is a contribution toward the goal of making a set of transport parameters available to the device designers. Specifically, we intend to provide a systematic description of the carrier velocity, mobility, energy and momentum relaxation times that can be directly used to build analytical or numerical device models. The analytical expressions for the transport parameters used in this work are those already implemented in physical numerical device simulators or previously employed to describe the properties of other compound semiconductors. Whenever experimental values are available we will directly employ them to set up the analytical model and augment the results obtained from more fundamental material modeling techniques. When no experimental information is avail-

Nitride Semiconductor Devices: Principles and Simulation. Joachim Piprek (Ed.)
Copyright © 2007 WILEY-VCH Verlag GmbH & Co. KGaA, Weinheim
ISBN: 978-3-527-40667-8

able, the calculated values of the transport parameters will be directly used to build the analytical models.

The rest of this chapter is organized as follows: in Section 4.2 we describe the numerical simulation model used to obtain the transport parameters. Section 4.3 presents the transport parameters' analytical models. Sections 4.4, 4.5 and 4.6 present the transport parameters for GaN, AlN and InN, respectively. Section 4.7 concludes the chapter.

4.2
Numerical Simulation Model

Within the framework of the semiclassical transport theory [1], the Boltzmann Transport Equation (BTE) describes the motion of carriers in phase space and can be considered as a continuity equation for the distribution function in phase space, defined by three momentum coordinates and three spatial coordinates. When the force field does not depend on the wavevector $\vec{k}$ and the velocity $\vec{v}$ is not functionally dependent on the space coordinate, that is in bulk material, the BTE has the form [2]

$$\frac{\partial f(\vec{r},\vec{k},t)}{\partial t} + \vec{v}\cdot\nabla f(\vec{r},\vec{k},t) + \frac{\vec{\mathcal{E}}}{\hbar}\cdot\nabla_k f(\vec{r},\vec{k},t) = \left(\frac{\partial f(\vec{r},\vec{k},t)}{\partial t}\right)_C \quad (4.1)$$

where $f(\vec{r},\vec{k},t)$ is the particle distribution function in phase space, $\vec{\mathcal{E}}$ is the applied electric field, $\vec{r}$ is the spatial coordinate, and t the time. The BTE treats electrons and holes as classical point particles with definite position and momentum. The carriers are also considered to be uncorrelated and the single-particle distribution functions are valid. The collision term is given by

$$\left(\frac{\partial f(\vec{r},\vec{k},t)}{\partial t}\right)_C = \int \left\{ f(\vec{r},\vec{k}',t)[1-f(\vec{r},\vec{k},t)]\, S(\vec{k}',\vec{k}) - f(\vec{r},\vec{k},t)[1-f(\vec{r},\vec{k}',t)]\, S(\vec{k},\vec{k}') \right\} d\vec{k}' \quad (4.2)$$

where $S(\vec{k},\vec{k}')$ is the scattering rate from the state $\vec{k}$ to $\vec{k}'$ due to several mechanisms, such as phonons, impurities, impact ionization and others. The collision integral in (4.2) contains the quantum description of the system, since the scattering rates are evaluated using quantum mechanical techniques. Clearly the description of the band structure enters in the BTE through the velocity computed as $\vec{v} = \nabla_k E(\vec{k})/\hbar$, and the momentum-energy relation for the particles as given by $E(\vec{k})$. In the present work, we have computed the full electronic structure of III-Nitride binaries using the semi-empirical pseudopotential method [3]. As an example, Fig. 4.1 presents the calculated GaN band

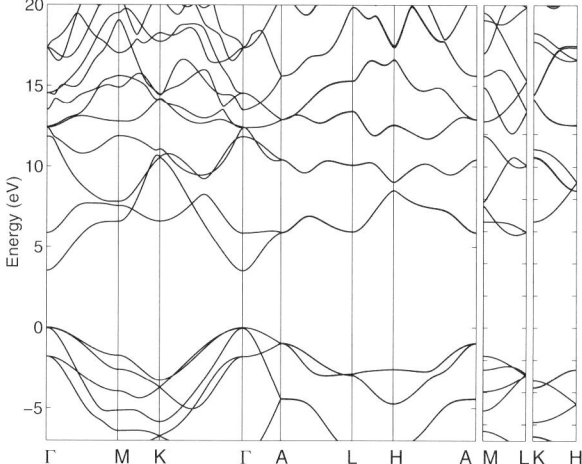

Fig. 4.1 Calculated GaN electronic structure [3]

structure. Eight conduction bands and six valence bands been have been considered in the present study.

The task of solving the BTE is made challenging by the presence of many obstacles. Among these are the complex description of scattering rates, the nonlinear response, and the boundary conditions in devices. Some analytical techniques are available. Assumptions can be made about the form of the distribution function, in this case a possible way is to use a drifted Maxwellian formulation [2]. Another possibility relies on the functional expansion of the distribution function in terms of spherical harmonics [4]. Although useful, these techniques provide only a limited answer to the study of transport in realistic semiconductors. Monte Carlo methods [5–8] certainly represent the most powerful tool in determining the transport properties in this situation. Rather than directly solving the BTE, the distribution function is built through direct simulation of the motion of an ensemble of particles in phase space. The ensemble is subjected to the action of external forces, such as an electric field, and scattering phenomena. In order to limit the computational requirements the motion is studied in the irreducible wedge of the Brillouin zone. The band structure and scattering rates have been computed for a set of 15936 points inside the irreducible wedge. Interpolation and symmetry operations enable calculation of the information needed outside the irreducible wedge. Before describing the results obtained for the material being studied a brief review of the scattering mechanisms is presented.

Tab. 4.1 Material parameters used to compute the scattering rates. $\hbar\omega_{LO}$ polar optical phonon energy, ϵ_0, ϵ_∞ static and high frequency permittivities, D_a deformation potential, D_{TK} optical coupling constant, ρ mass density, $\hbar\omega_{TO}$ nonpolar optical phonon energy, and e_{14}, e_{33}, e_{31}, e_{15} are the piezoelectric constants.

	GaN	InN	AlN
$\hbar\omega_{LO}$ [meV]	92.12	72.66	110.35
ϵ_0	9.7	13.52	8.5
ϵ_∞	5.28	5.8	4.46
D_a [eV]	8.3	7.1	9.5
D_{TK} [10^9 eV cm^{-1}]	1.0	1.0	1.0
ρ [g cm^{-3}]	6.087	6.810	3.230
$\hbar\omega_{TO}$ [meV]	69.55	59.02	83.16
$\hbar\omega_{max}$ [meV]	27.28	18.60	42.15
e_{14} [C m^{-2}]	0.375	0.375	0.92
e_{33} [C m^{-2}]	0.650	0.97	1.55
e_{31} [C m^{-2}]	−0.33	−0.57	−0.48
e_{15} [C m^{-2}]	−0.33	−0.48	−0.58

4.2.1
Scattering in the Semi-Classical Boltzmann Equation

The adiabatic approximation decouples the motion of the ions from the electronic part of the Hamiltonian. In this case the interaction with the phonons has to be introduced as a perturbation that forces a transition, from a one-electron Bloch state to another. The expression of the transition probability per unit time from a state $\vec{k}$ to a state $\vec{k}'$ induced by a perturbation Hamiltonian H_p, is given, to first order, by Fermi Golden Rule [2]

$$S(\vec{k},\vec{k}') = \frac{2\pi}{\hbar} |\langle \psi_k| H_p |\psi_{k'}\rangle|^2 \, \delta[E(\vec{k}) - E(\vec{k}')] \tag{4.3}$$

for elastic processes, and

$$S(\vec{k},\vec{k}') = \frac{2\pi}{\hbar} \left(N_q + \tfrac{1}{2} \pm \tfrac{1}{2}\right) |\langle \psi_k| H_p |\psi_{k'}\rangle|^2 \, \delta[E(\vec{k}) - E(\vec{k}') \mp \hbar\omega_q] \tag{4.4}$$

for phonon scattering processes. In (4.4) $\hbar\omega_q$ is the phonon energy, N_q is the phonon occupation number at the lattice temperature T, and the upper and lower signs correspond to emission and absorption, respectively. The total rate $1/\tau(\vec{k})$ out of the Bloch state ψ_k for each scattering mechanism can be evaluated numerically by integrating over the Brillouin zone using the full-band description

$$\frac{1}{\tau(\vec{k})} = \frac{V}{(2\pi)^3} \int S(\vec{k},\vec{k}') \, d\vec{k}' \tag{4.5}$$

The Fermi Golden Rule is derived under the assumption that the scattering event is instantaneous. This assumption breaks down in a few important cases. In the case of very fast transients, where the dynamics of the system has to be studied on a time scale comparable with the scattering rate, then the scattering rates have to be described in a different way [8]. In the case of very high scattering rates, the initial state may decay appreciably by the time the scattering is completed, and the determination of the rate has to be done taking this into account. Lastly, in the case of very high electric fields, the field can transfer a significant amount of energy to the particle during the scattering events (intracollisional field effect). Different attempts have been made to include such phenomena in transport simulators but no unified treatment is presently available. In cases where these effects can be neglected, the first-order transition rate provides a good description of the scattering events. Each scattering process can then be described by specializing (4.3), (4.4) using an appropriate expression for the perturbation Hamiltonian. The scattering events considered in the present transport study of wide band-gap semiconductor materials are polar scattering, deformation potential scattering, scattering with impurities, and dislocation scattering. Scattering with polar optical phonons can be described by using the Fröhlich formulation [1]

$$S_{po}(\vec{k},\vec{k}') = \frac{(2\pi e)^2 \omega_{LO}}{4\pi V} \left(\frac{1}{\epsilon_\infty} - \frac{1}{\epsilon_0} \right)$$
$$\times \left\{ N_q + \frac{1}{2} \pm \frac{1}{2} \right\} \frac{|I(\vec{k},\vec{k}')|^2}{q^2} \delta[E(\vec{k}) - E(\vec{k}') \mp \hbar\omega_{LO}] \quad (4.6)$$

where ϵ_∞, ϵ_0 are the optical and static dielectric functions, $I(\vec{k},\vec{k}')$ is the overlap integral between the initial $\vec{k}$ and final $\vec{k}'$ Bloch state, $\hbar\omega_{LO}$ is the longitudinal optical phonon energy and V the volume of the crystal. A common approximation is the use of a constant value for the polar optical energy associated with the longitudinal optical mode. The dependence of the matrix element on the inverse square of the phonon wave vector, and the angular dependence of the overlap integral makes this scattering mechanism anisotropic. In wurtzite crystals, the optical-phonon branches support mixed longitudinal and transverse modes due to the anisotropy [9]. Consequently, both the LO-like and the TO-like modes can contribute to the polar-optical phonon scattering. Moreover, the high-frequency dielectric constants are different within the basal plane and along the c-axis.

Nonpolar scattering rates are evaluated using the deformation potential formulation. In principle, by including the full description of the lattice dynamics in the computation of the matrix elements, it is possible to overcome the need of the empirically determined deformation potential [8]. Recently, the rigid pseudo-ion formalism has been applied to the scattering-rate calculation in wurtzite GaN [10]. Because of the problems associated with the selection of the phonon polarization vectors computed using empirical lattice dynamics models, it is not clear if the pseudo-ion formalism can be effectively used to determine the electron–phonon scattering rates for wide band-gap semiconductors. In the absence of better information about the deformation potentials, the nonpolar electron–phonon matrix element has been approximated by an isotropic coupling constant D_a. The resulting expression for the transition probability with acoustic phonons is given as,

$$S_{ac}(\vec{k},\vec{k}') = \frac{\pi D_a^2}{V\rho\omega_q}\left\{N_q + \frac{1}{2} \pm \frac{1}{2}\right\}\left|I(\vec{k},\vec{k}')\right|^2 q\,\delta[E(\vec{k}) - E(\vec{k}') \mp \hbar\omega_q] \quad (4.7)$$

where ρ is the density of the semiconductor, ω_q is the frequency of the phonon with wavevector $\vec{q}$, and $\vec{k}' = \vec{k} \mp \vec{q} + \vec{G}$ is the final electron wavevector which is mapped into the first BZ by adding a vector $\vec{G}$ of the reciprocal lattice. The acoustic phonon dispersion has been approximated by [11]

$$\hbar\omega_q = \begin{cases} \hbar\omega_{\max}\left[1 - \cos\left(\frac{qa}{4}\right)\right]^{1/2}, & q \leq 2\pi/(a\sqrt{3}) \\ \hbar\omega_{\max}, & q > 2\pi/(a\sqrt{3}) \end{cases} \quad (4.8)$$

Scattering with nonpolar optical phonons is modeled using a suitable deformation potential as,

$$S_{op}(\vec{k},\vec{k}') = \frac{\pi D_{TK}^2}{V\rho\omega_{TO}}\left\{N_q + \frac{1}{2} \pm \frac{1}{2}\right\}\left|I(\vec{k},\vec{k}')\right|^2 \delta[E(\vec{k}) - E(\vec{k}') \mp \hbar\omega_{TO}] \quad (4.9)$$

The parameters used in the scattering calculations are reported in Table 4.1 for convenience. For moderately doped semiconductors, ionized impurity scattering plays an important role and can dominate the total scattering rate for high doping concentrations. This type of collision is elastic in nature, and cannot alone control the transport process in the presence of an external field. Scattering with ionized impurities is modeled using the Brooks–Herring model [7]

$$S(\vec{k},\vec{k}') = \frac{32\pi^3 Z^2 N_I e^4}{\hbar V 4\pi\epsilon^2} \frac{|I(\vec{k},\vec{k}')|^2}{(\beta^2 + q^2)^2} \delta[E(\vec{k}) - E(\vec{k}')] \quad (4.10)$$

where β is the screening length, which depends on the free carrier density, and N_I is the density of ionized impurities of charge Ze. Numerical handling is more complicated than for phonon scattering, since the scattering probability also depends on the concentration of ionized impurities. Because of the difficulties in evaluating (4.10) numerically for a realistic band structure, an energy-dependent rate has been used for all the different materials studied. Notice, however, that impurity scattering is most important at low energies, where analytical models are most suitable. For a nonparabolic band the integrated scattering rate is [7]

$$P_{BH}(E) = \frac{N_I}{32\pi\sqrt{2m^*}} \left(\frac{Ze^2}{\epsilon_r\epsilon_0}\right)^2 \frac{(1+2\alpha E)}{\gamma^{3/2}} \frac{1}{(\beta/2k)^2[1+(\beta/2k)^2]} \quad (4.11)$$

where m^* is the density of states effective mass and $\gamma = \frac{\hbar^2 k^2}{2m^*}$.

In crystals whose lattice lacks inversion symmetry, such as semiconductors with wurtzite structure, the piezoelectric effect provides an additional coupling between electrons and acoustic vibrations. The piezoelectric scattering rate is given by

$$S_{piezo}(\vec{k},\vec{k}') = \frac{e^2 K_{av}^2 k_B T}{8\pi^2 \epsilon \hbar} \frac{q^2}{(q^2+\beta^2)^2} \delta[E(\vec{k}) - E(\vec{k}')] \quad (4.12)$$

The piezoelectric coupling constant has a complex directional dependence [12]. Ridley [12] introduced a spherical average of the piezoelectric constants separately for longitudinal and transverse modes in a wurtzite lattice. Longitudinal and transverse modes are lumped together by defining an average electromechanical coupling K_{av} such that

$$K_{av}^2 = \frac{\langle e_L^2 \rangle}{\epsilon c_L} + \frac{\langle e_T^2 \rangle}{\epsilon c_T} \quad (4.13)$$

where c_T, c_L are the transverse and longitudinal elastic constants and

$$\langle e_L^2 \rangle = \frac{1}{7}e_{33}^2 + \frac{4}{35}e_{33}(e_{31}+2e_{15}) + \frac{8}{105}(e_{31}+2e_{15})^2 \quad (4.14a)$$

$$\langle e_T^2 \rangle = \frac{2}{35}(e_{33}-e_{31}-e_{15})^2 + \frac{16}{105}e_{15}(e_{33}-e_{31}-e_{15}) + \frac{16}{35}e_{15}^2 \quad (4.14b)$$

Values of the piezoelectric constants for the nitrides are given in Table 4.1. Piezoelectric scattering weakens with increasing electron energy. The parabolic approximation is therefore appropriate for this scattering mechanism. After integrating we obtain, for a parabolic spherical band, the total scattering rate for absorption and emission [12]

$$P_{piezo}(E) = \frac{e^2 K_{av}^2 k_B T}{4\pi \epsilon \hbar^2 v} \left[\log\left(1 + \frac{8m^*E}{\hbar^2 \beta^2}\right) - \frac{1}{1+\hbar^2\beta^2/(8m^*E)}\right] \quad (4.15)$$

Defects can be conveniently categorized into three types: point (vacancies and interstitials), line (threading dislocations), and areal (stacking faults). In most cases, epitaxial layers of GaN contain a high density of defects with dislocations, which form because of the absence of a suitable lattice-matched substrate for epitaxial growth. These dislocations are predominantly oriented parallel to the c-axis of the material [13]. Epitaxial growth of (0001) GaN on sapphire (Al_2O_3) leads to high concentrations (typically 10^9–10^{11} cm^{-2}) of threading edge and screw dislocations which propagate vertically from the GaN/Al_2O_3 interface to the GaN surface. Recent works have shown that threading edge dislocations in GaN are electrically active [13]. Mobilities in molecular beam epitaxy (MBE) GaN are generally lower than those in metal-organic chemical vapor deposition (MOCVD) or hydride vapor phase epitaxy (HPVE) GaN. High MOCVD and HVPE mobilities (700–950 cm^{2}V^{-1}s^{-1}) have been correlated with low dislocation densities ($N_{dislo} < 5 \times 10^8$ cm^{-2}), while some of the best MBE mobilities (300–400 cm^{2}V^{-1}s^{-1} [13]) are from samples with $N_{dislo} > 5 \times 10^9$ cm^{-2}.

The scattering due to dislocation-line charges is two-dimensional, because only electrons moving perpendicular to the dislocation will be scattered. Thus, the relevant scattering wavevector is $\vec{q} = \vec{k}_\perp - \vec{k}'_\perp$, where $\vec{k}_\perp, \vec{k}'_\perp$ are the incoming and outgoing wavevectors in a direction perpendicular to the c-axis. It can be shown [14] that the scattering rate is given by the expression:

$$P_{dislo}(E) = \frac{N_{dislo}\, e^4 m^* \lambda^4}{\epsilon c^2 \hbar^3} \frac{1 + 2k_\perp^2 \lambda^2}{(1 + 4k_\perp^2 \lambda^2)^{3/2}} \tag{4.16}$$

where λ is the Debye length.

4.3
Analytical Models for the Transport Parameters

The dependence of the steady state electron drift velocity on the electric field of the III-nitride materials is similar to that of GaAs and other compound semiconductors. Typically, the velocity peaks for a given field value and then decreases. Because the electron velocity in the III-nitride binary compounds presents a dual-slope behavior [15, 16], the classical high-field mobility and velocity model employed for GaAs [17] cannot be used to fit the computed or measured velocity data. Recently, a different model has been proposed for III-nitride materials [15, 16]

$$v_e(\mathcal{E}) = \frac{\mu_{n0}\mathcal{E} + v_{e,sat}(\mathcal{E}/\mathcal{E}_{cr})^{\beta_1}}{1 + (\mathcal{E}/\mathcal{E}_{cr})^{\beta_1} + a(\mathcal{E}/\mathcal{E}_{cr})^{\beta_2}} \tag{4.17}$$

where $\mu_{n0} = \mu(T, N_i)$ is the electron temperature and doping dependent mobility, $v_{e,sat}$ is the saturation velocity and $\mathcal{E}_{cr}$ is the critical field. β_1 and β_2 are fitting parameters. The numerical values of these parameters, presented in Table 4.2, have been determined by fitting the calculated electron velocity-field characteristics of GaN, AlN, and InN for transport along the Γ-M (parallel to

Tab. 4.2 Fitting parameters to be used in Eq. (4.17) that approximate the electron steady state drift velocity in AlN, GaN and InN. The given parameters are valid for an applied electric field strength less than 800 kV cm^{-1}, an operating temperature of 300 K and doping concentration of $N_d = 10^{17}$ cm^{-3}.

	AlN (Γ-M)	AlN (Γ-A)	GaN (Γ-M)	GaN (Γ-A)	InN (Γ-M)	InN (Γ-A)
μ_{e0} [cm^2 V^{-1} s^{-1}]	400.0	450.0	1000.0	1000.0	3150.0	3150.0
$\mathcal{E}_{cr}$ [kV cm^{-1}]	420.0	400.0	215.0	200.0	90.0	90.0
$v_{e,sat}$ [10^7 cm s^{-1}]	1.5	1.5	1.3	1.3	1.0	1.0
a	6.75	6.75	7.0	5.5	5.75	5.5
β_1	6.0	6.0	4.5	4.25	3.2	3.1
β_2	0.825	0.825	0.7	0.7	0.85	0.9

the basal plane) and Γ-A (parallel to the c-axis) directions. In general, for layers grown on the wurtzite [0001] direction the transport coefficients along the Γ-A directions should be considered.

For the hole velocity-field characteristics, a simpler analytical model can be used. In this case we have employed the standard Caughey–Thomas formula [18] for the velocity

$$v_h(\mathcal{E}) = \frac{\mu_{h0}\mathcal{E}}{\left[1 + \left(\frac{\mu_{h0}\mathcal{E}}{v_{h,sat}}\right)^\beta\right]^{1/\beta}} \tag{4.18}$$

where $\mu_{h0}=\mu(T, N_i)$ (from Eq. (4.19)) is the hole temperature and doping dependent mobility, $v_{h,sat}$ is the hole saturation velocity, and β is a fitting parameter. As for the electrons, the numerical values of these parameters, presented in Table 4.3, have been determined by fitting the calculated hole drift velocity as a function of the electric field of GaN, AlN, and InN for transport along the Γ-M and Γ-A directions.

The temperature- and doping-dependent mobility obtained either from Monte Carlo simulation or experimental data is used to fit the widely used

Tab. 4.3 Fitting parameters to be used in Eq. (4.18) that approximate the hole steady state drift velocity in AlN, GaN and InN. The given parameters are valid for an applied electric field strength less than 800 kV cm^{-1} and an operating temperature of 300 K.

	AlN (Γ-M)	AlN (Γ-A)	GaN (Γ-M)	GaN (Γ-A)	InN (Γ-M)	InN (Γ-A)
β	1.0	1.9	0.85	0.725	1.0	0.75
μ_{h0} [cm^2 V^{-1} s^{-1}]	20.0	70.0	70.0	70.0	35.0	35.0
$v_{h,sat}$ [10^7 cm s^{-1}]	1.05	1.25	1.7	1.7	1.3	1.5

4 Transport Parameters for Electrons and Holes

Arora model [19]

$$\mu(T, N_i) = \mu_{min} + \frac{\mu_d}{1 + \left(\frac{N_i}{N_0}\right)^{A^*}} \quad (4.19)$$

where $N_i = N_a + N_d$ is the total impurity concentration and T the temperature. μ_{min}, μ_d, N_0 and A^* are temperature-dependent quantities given by:

$$\mu_{min} = A_{min} \left(\frac{T}{T_0}\right)^{\alpha_m} \qquad \mu_d = A_d \left(\frac{T}{T_0}\right)^{\alpha_d} \quad (4.20)$$

$$N_0 = A_N \left(\frac{T}{T_0}\right)^{\alpha_N} \qquad A^* = A_a \left(\frac{T}{T_0}\right)^{\alpha_a} \quad (4.21)$$

The quantities A_{min}, A_d, A_N, A_a, α_{min}, α_d, α_N, and α_a are fitting parameters and $T_0 = 300\,\text{K}$. For GaN two sets of fitting parameters that describe the electron mobility for two different values of dislocation density are given in Table 4.4 (columns three and four). The same table also presents three additional sets of parameters necessary to compute the GaN hole mobility and the AlN, InN electron mobility.

Tab. 4.4 Parameters for the Arora model (4.19) necessary to compute the temperature and doping dependent electron mobility of AlN, GaN, and InN. The fitting parameters are obtained assuming an areal dislocation density of $10^8\,\text{cm}^{-2}$ and $4\times10^8\,\text{cm}^{-2}$.

		GaN Electrons	GaN Electrons	GaN Holes	AlN Electrons	InN Electrons
N_{dislo}	cm^{-2}	4×10^8	10^8	0.0	10^8	10^8
A_{min}	[cm^{2}V^{-1}s^{-1}]	31.15	13.64	0.0	13.4	88.74
α_m		0.339	2.03	0.0	1.21	2.39
A_d	[cm^{2}V^{-1}s^{-1}]	1737.8	1662.2	41.7	725.1	6121.8
α_d		−2.96	−3.37	−2.34	−3.21	−3.70
A_N	[cm^{-3}]	1.69×10^{16}	1.38×10^{17}	1.97×10^{19}	2.25×10^{17}	7.65×10^{16}
α_N		7.57	6.27	0.869	7.45	6.95
A_a		0.299	0.50	0.309	0.41	0.366
α_a		−0.306	−0.144	−2.311	−0.18	−0.331

Hydrodynamic and energy balance simulation models require, as additional inputs, the energy and momentum relaxation times τ_E, τ_m as a function of the average carrier energy. This information has been obtained from bulk Monte Carlo simulation following the approach described in [20,21]. The energy relaxation time is given by:

$$\tau_E(\overline{E}) = \frac{\overline{E} - \overline{E}_0}{e\,\mathcal{E}(\overline{E})\,v_d(\overline{E})} \quad (4.22)$$

where $\bar{E}$ is the average electron/hole carrier energy, $\bar{E}_0$ is the thermal energy, $\mathcal{E}$ is the electric field and v_d the drift velocity. The momentum relaxation time is calculated using:

$$\tau_m(\bar{E}) = \frac{m(\bar{E})}{e} \frac{v_d(\bar{E})}{\mathcal{E}(\bar{E})} \qquad (4.23)$$

where e is the electron charge and $m(\bar{E})$ is the energy-dependent effective mass. The latter can be computed from the realistic density of states. A convenient analytical expression of the relaxation times is

$$\tau(\bar{E}) = \tau_0 \sum_n a_n \bar{E}^{p_n} / \sum_m b_m \bar{E}^{q_m} \qquad (4.24)$$

Where τ_0, a_n, p_n, b_m and q_m are fitting parameters. This parametrization is already implemented in commercial macroscopic device simulators [22, 23]. The specific coefficients for GaN, AlN, and InN will be given in the following sections of this chapter.

4.4 GaN Transport Parameters

Because of its potential applications GaN has been at the center of intense research activity. The results of a large number of theoretical [24–34] and experimental [13, 35–43] investigations of transport properties have been reported in the literature. Both electron and hole transport properties have been studied including the low-field mobility and the high-field coefficients. Because of the lack of a native substrate the low-field mobility is dominated by dislocation scattering [13] and defects. The experimental investigation of the high-field transport coefficients has not received much attention, in particular the impact ionization phenomena.

4.4.1 Electron Transport Coefficients

A number of experimental groups have reported measurements of the electron drift velocity in GaN [44–49]. Several discrepancies between the experimental data have emerged as a result of the different measurement techniques employed. In particular, the electric field strength value at which the electrons reach the maximum velocity is much higher, around 300 kV cm^{-1}, for measurements involving optical probing [44–47] than it is in other situations. On the other hand, in the case of measurement techniques using constriction-type devices [49], premature breakdown seems to limit the maximum field at which velocity can be measured and it is not clear if the maximum drift velocity is effectively reached. Experiments with a test device employing a metal-semiconductor-metal structure have shown a negative differential mobility region [48] for an electric field strength of the order of 200 kV cm^{-1}. Because of the complexity of the experiments performed and the test device geometry a careful analysis of the results is needed to extract the information on the electron drift velocity. In spite of these problems, all the experimental information seems to indicate that a peak velocity of the order of $2.0 \times 10^7 \sim 2.75 \times 10^7$ cm s^{-1} is reached for an electric field strength of the order of $200 \sim 300$ kV cm^{-1}. This is in general agreement with the prediction of Kolnik and co-workers [28], who were the first to employ a full-band Monte Carlo technique to compute the steady state electron drift velocity. In this work we use a similar full-band Monte Carlo simulation technique and employ an up-to-date electronic structure [3] to compute the steady state drift velocity. Figure 4.2 presents the calculated drift velocity for a applied electric

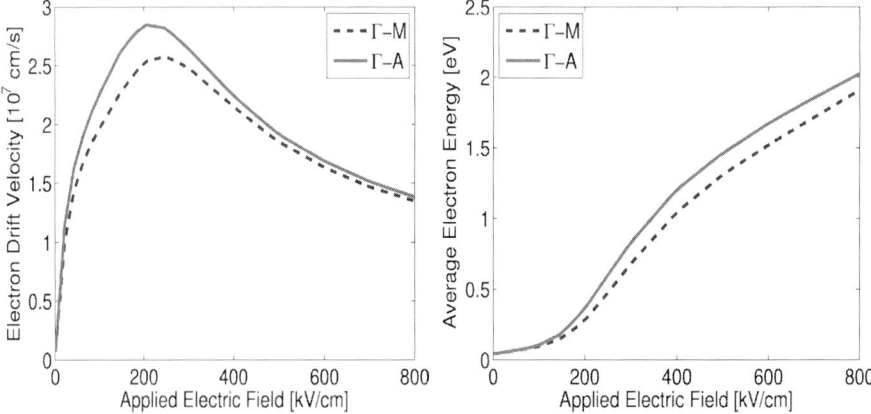

Fig. 4.2 Calculated steady state electron drift velocity for bulk GaN. A temperature of 300 K and a fully ionized doping concentration of $N_d = 10^{17}$ cm^{-3} are assumed. The solid line is for transport in the Γ-A direction and the dashed line along the Γ-M direction.

Fig. 4.3 Calculated steady average electron energy for bulk GaN. A temperature of 300 K and a fully ionized doping concentration of $N_d = 10^{17}$ cm^{-3} are assumed. The solid line is for transport in the Γ-A direction and the dashed line along the Γ-M direction.

field along the Γ-A and Γ-M crystallographic directions. The peak velocity for transport along the Γ-A direction is 2.85×10^7 cm s^{-1} reached at a field strength of 215 kV cm^{-1}. For transport in the Γ-M direction the peak velocity is slightly lower, 2.55×10^7 cm s^{-1}, and it is obtained for an applied field of 245 kV cm^{-1}. At higher electric field strength the velocity tends to a value close to 1.25×10^7 cm s^{-1}. The dependence of the electron drift velocity on the electric field applied along the two directions is accurately described by the analytical model (4.17) with the parameters reported in Table 4.2. A similar analytical expression of the drift velocity as a function both of the electric field and temperature is given in [50]. The calculated average electron energy is shown in Fig. 4.3. Also, in this case, we can observe a difference between the average carrier energy for the two crystallographic directions. Figure 4.4

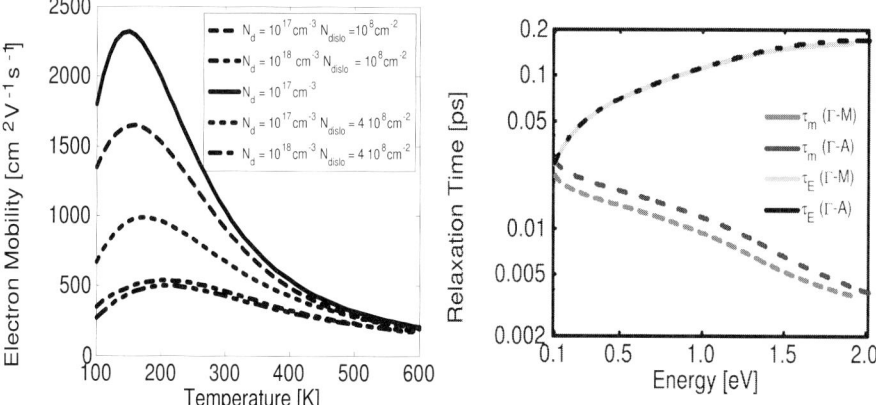

Fig. 4.4 Calculated low-field electron mobility in bulk GaN as a function of the lattice temperature, doping concentration and dislocation density. The effect of partial donor ionization is included.

Fig. 4.5 Calculated direction and energy dependent energy and momentum relaxation times for a lattice temperature of 300 K and a a fully ionized doping concentration of $N_d = 10^{17}$ cm^{-3}.

presents the calculated low-field electron mobility as a function of the doping concentration (including partial ionization), temperature and areal dislocation density. As can be noticed, the presence of the dislocations substantially reduces the mobility. This results from the Coulomb interaction of the carrier with the charged dislocation. The calculated mobility values obtained with a dislocation density of $N_{\mathrm{dislo}} = 4 \times 10^8$ cm^{-2} are also consistent with the experimental measurements in reference [13]. Hydrodynamic and energy balance models of GaN-based semiconductor devices require the energy and momentum relaxation times as input parameters [51]. The GaN electron relaxation times are computed according to Eqs (4.22) and (4.23) using the information on the average electron energy and drift velocity obtained from the Monte Carlo model. The relaxation times are then fitted for convenience to the analytical expression (4.24) employed in conventional macroscopic hydrodynamic simulation tools [22,23]. For GaN the fitting coefficients to be used in conjunction with Eq. (4.24) are reported in Table 4.5.

4.4.2
Hole Transport Coefficients

The development of a reliable doping technique necessary to obtain high-quality GaN p-type layers is still an open problem. As a result of the difficulty

Tab. 4.5 Fitting parameters needed to compute the GaN electron momentum (τ_m) and energy (τ_E) relaxation times as a function of the average electron energy and crystallographic direction by using Eq. (4.24).

	τ_m (Γ-M)	τ_m (Γ-A)	τ_E (Γ-M)	τ_E (Γ-A)
τ_0 [ps]	0.0045	0.0040	0.4432	0.0081
a_1	1.8982	0.8185	0.3139	−1.8162
p_1	−0.1593	−1.3858	−1.0605	1.1489
a_2	1.1903	0.1950	0.1630	7.6522
p_2	−1.1169	−0.4145	1.2191	0.1304
a_3	2.2649	1.9097	0.4238	−1.0803
p_3	−0.7078	−1.7917	−0.4381	−0.4110
b_1	0.9322	−0.2991	2.4656	1.2945
q_1	2.1524	0.5175	−1.4612	0.0231
b_2	1.9505	0.5492	0.1344	−1.0122
q_2	−0.5503	1.1801	2.6014	0.4498
b_3	−0.2746	0.7463	0.9425	0.0584
q_3	−0.4982	−1.4346	−0.6875	2.0144

in finding an efficient acceptor dopant, the present quality of the GaN p-type films, although improving, is still not adequate for high-performance bipolar transistors. The high activation energy of magnesium, in the range 150–250 meV above valence band edge, severely limits the doping efficiency to typically 1% at 300 K [52], although efficiency up to 10% has been reported [40]. As a consequence of this limited doping efficiency, an acceptor concentration of 10^{19}–10^{20} cm^{-3} is required to obtain a hole concentration of the order 10^{17}–10^{18} cm^{-3}. As several experimental investigations [40–43] have pointed out, this high doping concentration leads to a degradation of the transport properties. First the hole concentration exhibits a strong temperature dependence that has a significant effect on the device operation. Furthermore, the high doping concentration may lead to the formation of impurity bands and autocompensation that would limit the maximum attainable hole concentration. Consequently, modeling the hole transport in GaN is quite challenging and a satisfactory model that includes both low- and high-field transport, is not available yet. Using the measured data [40–43] it is possible to obtain an analytical approximation of the temperature and doping dependent hole mobility based on the Arora model. The dashed lines in Fig. 4.6 represent the hole mobility calculated using the analytical model for three experimental data sets (set 1,2 and 3) reported in [43]. Figure 4.7 shows the temperature-dependent hole concentration for the same five data sets [41].. The numerical values of the parameters to be used in Eqs (4.19)–(4.21) are reported in the fifth col-

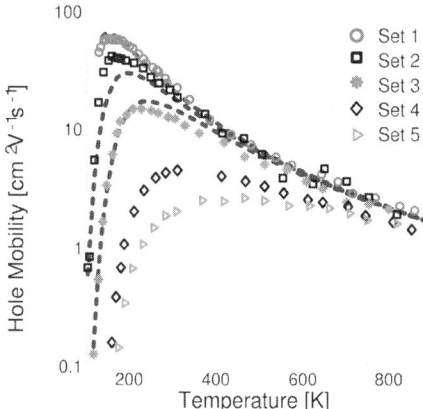

Fig. 4.6 Temperature-dependent mobility data for p-type GaN grown with MOCVD [43]. The dashed lines, corresponding to sets 1, 2 and 3, are obtained using the temperature and doping-dependent analytical model of Eq. (4.19) and the fitting parameters of Table 4.4. The corresponding hole concentration and doping levels are reported in Fig. 4.7.

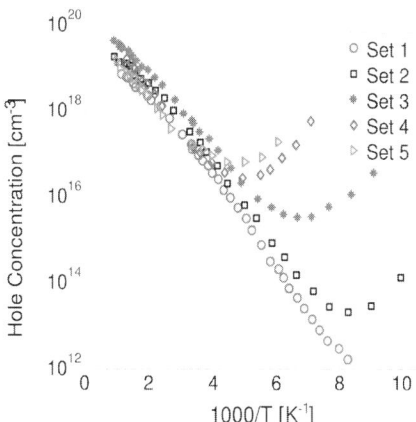

Fig. 4.7 Temperature-dependent carrier concentration data for p-type GaN grown with MOCVD [43]. The acceptor concentration for each data set is 1.8×10^{19}, 4.6×10^{19}, 1.4×10^{20}, 2.2×10^{20}, 8.6×10^{19} cm^{-3} for set 1, 2, 3, 4, 5, respectively. The corresponding temperature-dependent mobilities are reported in Fig. 4.6.

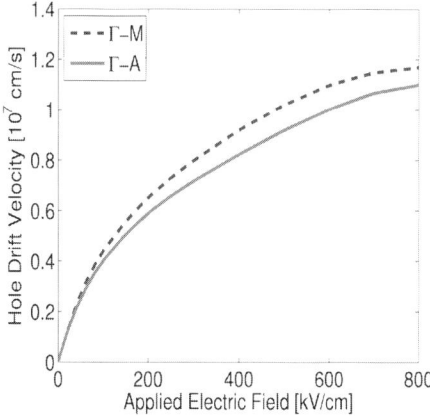

Fig. 4.8 Calculated steady state hole drift velocity for bulk GaN. A temperature of 300 K and a fully ionized doping concentration of $N_d = 10^{17}$ cm^{-3} are assumed. The solid line is for transport in the Γ-A direction and the dashed line along the Γ-M direction.

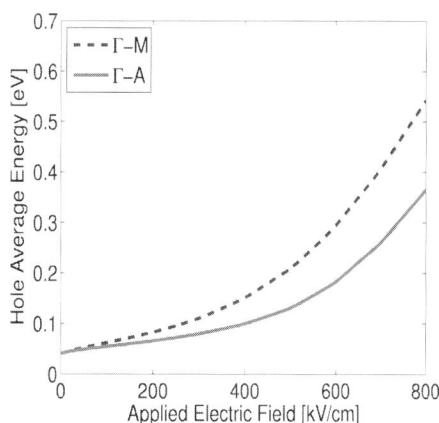

Fig. 4.9 Calculated steady state average hole energy for bulk GaN. A temperature of 300 K and a fully ionized doping concentration of $N_d = 10^{17}$ cm^{-3} are assumed. The solid line is for transport in the Γ-A direction and the dashed line along the Γ-M direction.

umn of Table 4.4. No experimental information is available on the GaN hole drift velocity. To date, only theoretical calculations, based on the full-band

Monte Carlo method [32, 33], have been performed to study the hole transport properties and hole-initiated impact ionization in GaN. In this work we have computed the direction-dependent steady state hole drift velocity using a full-band Monte Carlo method, employing the electronic structure presented in Fig. 4.1. Figure 4.8 shows the calculated steady state hole drift velocity in GaN. A small anisotropy can be noticed when comparing the velocity along the Γ-M and the Γ-A directions. For transport along the Γ-A direction the steady state hole velocity peaks at a value close to 1.2×10^7 cm s^{-1} for an applied electric field strength of 800 kV cm^{-1}. In the case of transport along the Γ-M direction the steady state hole velocity increases monotonically to a value of 1.1×10^7 cm s^{-1} for an applied electric field strength of 800 kV cm^{-1}. The fitting parameters to be used in Eq. (4.18) necessary to determine the hole velocity as a function of the applied electric field are reported in Table 4.3. The average hole energy is shown in Fig. 4.9.

4.5
AlN Transport Parameters

Among the binary compounds of the III-nitride material system, AlN has an important role in a variety of electronics and optoelectronics devices. AlN layers are normally used as buffer layers on sapphire substrates, as barrier layers in multi-quantum wells devices, for device surface passivation and for many other uses. AlN films grown with a variety of techniques are normally semi-insulating and this makes the measurement of the transport coefficients problematic.

4.5.1
Electron Transport Coefficients

The study of the carrier transport properties of AlN has received only limited attention. The theoretical prediction of the electron transport parameters has been mostly relegated to the use of simple transport models based on analytical bands [53–55] or even simpler models [25] based on the relaxation time approximation. To date, no information is available on the hole transport parameters in AlN, nor in any other of the AlGaN ternary alloys. For these materials, the complexity of the band structure requires the application of full-band Monte Carlo transport analysis. The material parameters employed are reported in Table 4.1. Figure 4.10 presents the calculated electron steady state drift velocity for AlN along two crystallographic directions, Γ-M and Γ-A. It is assumed a doping concentration $N_d = 10^{17}$ cm^{-3} and a temperature of 300 K. The numerical results indicate a small anisotropy, with the electron velocity being slightly lower for transport along the Γ-M direction. In this case a peak

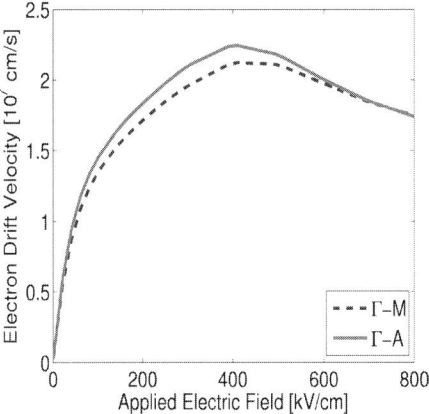 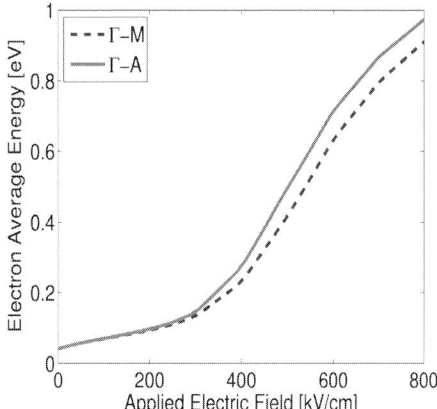

Fig. 4.10 Calculated steady state electron drift velocity for bulk AlN. A temperature of 300 K and a fully ionized doping concentration of $N_d = 10^{17}$ cm^{-3} are assumed. The solid line is for transport in the Γ-A direction and the dashed line along the Γ-M direction.

Fig. 4.11 Calculated steady state average electron energy for bulk AlN. A temperature of 300 K and a fully ionized doping concentration of $N_d = 10^{17}$ cm^{-3} are assumed. The solid line is for transport in the Γ-A direction and the dashed line along the Γ-M direction.

velocity of 2.1×10^7 cm s^{-1} is obtained for an applied electric field strength of 430 kV cm^{-1}. In the case of transport along the Γ-A a peak electron velocity of 2.25×10^7 cm s^{-1} is obtained for an applied electric field strength of 400 kV cm^{-1}. At higher field strength, the drift velocity will approach a value of 1.5×10^7 cm s^{-1} in both crystallographic directions. The two velocity-field characteristics have been fitted to the expression in Eq. (4.17), and the values of the relevant parameters are reported in Table 4.2. Figure 4.11 presents the calculated average electron energy. Figure 4.12 presents the calculated electron mobility as a function of the temperature, doping density and dislocation density. The partial ionization of the donors is included in the model. At each temperature the Fermi energy is computed and the resulting ionized dopant and electron/hole density are used to evaluate the impurity scattering. It can be seen that for a donor density of $N_d = 10^{17}$ cm^{-3} the calculated room temperature mobility is 470 cm^{2}V^{-1}s^{-1}. If the effect of the dislocation is included, assuming a density of $N_{dislo} = 10^8$ cm^{-2}, the electron mobility decreases to 430 cm^{2}V^{-1}s^{-1}. For higher doping density and the same dislocation density the electron mobility further decreases to the value of 260 cm^{2}V^{-1}s^{-1}. The fitting parameters in the analytical expression of Eq. (4.19) for the AlN temperature and doping dependent electron mobility are reported in Table 4.4. These values are valid for a dislocation density of $N_{dislo} = 10^8$ cm^{-2}. Following the same procedure described for GaN, the momentum and energy relaxation times for AlN have been evaluated (see Fig. 4.13). The corresponding fitting parameters in (4.24) are reported in Table 4.6.

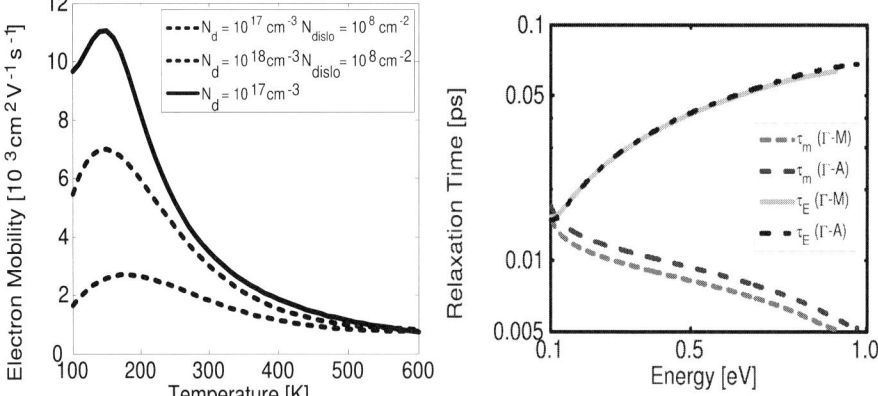

Fig. 4.12 Calculated low-field electron mobility in bulk AlN as a function of the lattice temperature, doping concentration and dislocation density. The effect of partial donor ionization is included.

Fig. 4.13 Calculated direction and energy-dependent energy and momentum relaxation times for a lattice temperature of 300 K and a fully ionized doping concentration of $N_d = 10^{17}$ cm^{-3}.

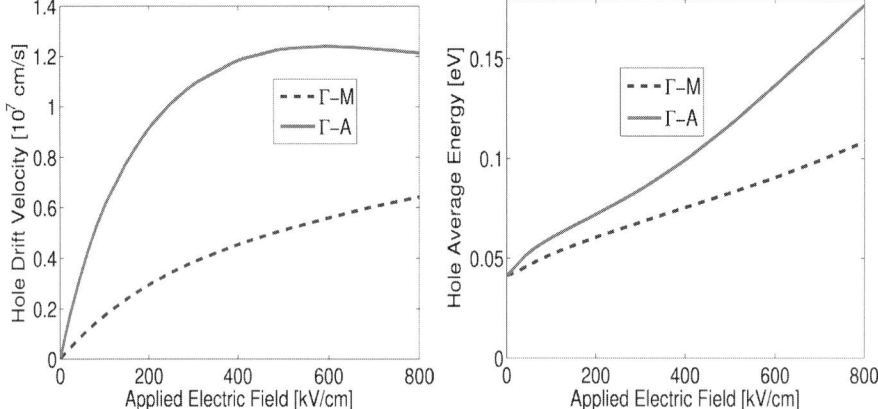

Fig. 4.14 Calculated steady state hole drift velocity for bulk AlN. A temperature of 300 K and a fully ionized doping concentration of $N_d = 10^{17}$ cm^{-3} are assumed. The solid line is for transport in the Γ-A direction and the dashed line along the Γ-M direction.

Fig. 4.15 Calculated steady state average hole energy for bulk AlN. A temperature of 300 K and a fully ionized doping concentration of $N_d = 10^{17}$ cm^{-3} are assumed. The solid line is for transport in the Γ-A direction and the dashed line along the Γ-M direction.

4.5.2
Hole Transport Coefficients

At present there is no information available on p-type AlN, and no experimental data on the transport coefficients is available. On the theoretical side,

Tab. 4.6 Fitting parameters needed to compute the AlN electron momentum (τ_m) and energy (τ_E) relaxation times as a function of the average electron energy and crystallographic direction by using Eq. (4.24).

	τ_m (Γ-M)	τ_m (Γ-A)	τ_E (Γ-M)	τ_E (Γ-A)
τ_0 [ps]	0.0005	0.0294	0.0007	0.0926
a_1	0.5122	0.0004	0.6843	0.3265
p_1	9.4227	−3.6235	3.2208	0.5902
a_2	3.0323	−0.7933	0.6986	1.2227
p_2	0.4459	1.6764	0.0882	0.4896
a_3	2.1434	2.3532	2.1900	0.4973
p_3	−1.5247	−0.4790	1.3451	−0.8956
b_1	−1.7153	1.0070	0.0495	−0.3133
q_1	−0.3644	−0.7284	1.2140	0.6085
b_2	0.7000	1.3461	−0.2747	1.6290
q_2	1.2064	3.7317	0.2959	0.5121
b_3	1.5851	7.0019	0.2606	1.3949
q_3	−0.6031	−0.1537	0.1935	−1.3109

there has been no attempt to compute the transport coefficients yet. As for the other III-nitride compounds, because of the complexity of the valence band structure, the full details of the electronic structure has to be included in a numerical model to determine the transport coefficients. Figure 4.14 shows the calculated steady state hole drift velocity in AlN. It can be seen that there is a substantial anisotropy as a result of the particular configuration of the valence band electronic structure that is different in the Γ-M and Γ-A directions. For transport along the Γ-A direction the steady state hole velocity peaks at a value of 1.2×10^7 cm s^{-1} for an applied electric field strength of 600 kV cm^{-1}. In the case of transport along the Γ-M direction the steady state hole velocity increases monotonically to a value of 6.4×10^6 cm s^{-1} for an applied electric field strength of 800 kV cm^{-1}. Table 4.3 presents the fitting parameters to be used in Eq. (4.18) necessary to determine the hole velocity as a function of the applied electric field. The dependence of the average hole energy on the electric field is shown in Fig. 4.15. As can be seen, the hole gas is significantly cooler if the electric field is applied along the Γ-M crystallographic direction.

4.6 InN Transport Parameters

InN has received much attention since new experimental evidence has shown that it has a smaller energy gap, around 0.8 eV [56], than what was thought

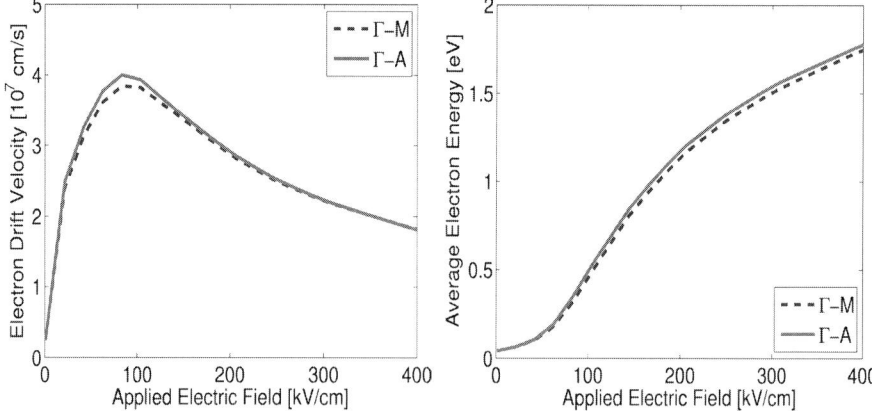

Fig. 4.16 Calculated steady state electron drift velocity for bulk InN. A temperature of 300 K and a fully ionized doping concentration of $N_d = 10^{17}$ cm^{-3} are assumed. The solid line is for transport in the Γ-A direction and the dashed line along the Γ-M direction.

Fig. 4.17 Calculated steady state average electron energy for bulk InN. A temperature of 300 K and a fully ionized doping concentration of $N_d = 10^{17}$ cm^{-3} are assumed. The solid line is for transport in the Γ-A direction and the dashed line along the Γ-M direction.

earlier ($\sim$ 2 eV). This new discovery opens the door for the III-nitride material system to be also used in the arena of infrared optoelectronics devices. As grown, InN is strongly n-type with a typical electron concentration in the range of $n = 10^{18} \sim 10^{20}$ cm^{-3} and to date there is no experimental evidence of p-type InN. The electron mobility in InN has been measured for material grown by different groups [57,58]. Several numerical studies were performed in the past when InN was thought to have a larger band gap. These included simple analytical band [55,59,60] and more complex full band [61] Monte-Carlo calculations. Recently, new theoretical investigations have been presented that included some of the new features of the electronic structure [62,63]. In this work we present the InN electron and hole transport parameters obtained from a full-band Monte Carlo calculation, based on the electronic structure that includes all information available to date.

4.6.1
Electron Transport Coefficients

Figure 4.16 presents the calculated steady state electron drift velocity for transport along the Γ-M and Γ-A crystallographic directions. Only a small anisotropy can be observed. The peak velocity in both cases is reached at an electric field strength of 90 kV cm^{-1} and it is 4.0×10^7 cm s^{-1} and 3.8×10^7 cm s^{-1} for the Γ-A and Γ-M directions, respectively. At higher field strengths the drift velocity approaches a value of 1.0×10^7 cm s^{-1} in both crystallographic directions. The two velocity-field characteristics have been fitted to the expression of Eq. (4.17), and the values of the relevant parameters are reported in Table 4.2. Figure 4.17 shows the calculated average electron energy. As can be seen, because of the small effective mass, the electron gas gains energy quickly and there is no appreciable anisotropy. The measured values of the electron

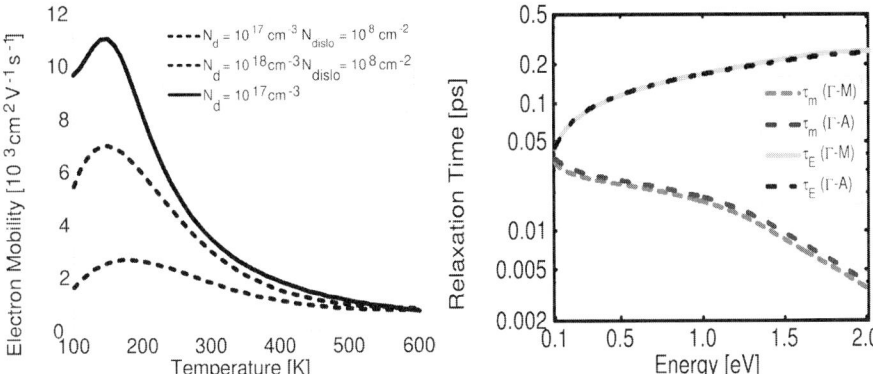

Fig. 4.18 Calculated low-field electron mobility in bulk InN as a function of the lattice temperature, doping concentration and dislocation density. The effect of partial donor ionization is included.

Fig. 4.19 Calculated direction and energy-dependent energy (τ_E) and momentum (τ_m) relaxation times for a lattice temperature of 300 K and a fully ionized doping concentration of $N_d = 10^{17}$ cm^{-3}.

mobility [57, 58] indicate a value in excess of $1000 \, \text{cm}^2 \text{V}^{-1}\text{s}^{-1}$ for carrier concentration below $10^{18} \, \text{cm}^{-3}$. Figure 4.18 presents the calculated InN electron mobility both without and with the effect of dislocations. The dislocation scattering clearly reduces the electron mobility, in particular at low temperatures. The value of the calculated electron mobility at 300 K, including the scattering due to an areal concentration of dislocation of $10^8 \, \text{cm}^{-2}$ are close to the measured valued in [58] (see Figure 1 of [58]) for an electron concentration of $10^{17} \, \text{cm}^{-3}$ and $10^{18} \, \text{cm}^{-3}$. Table 4.4 presents the fitting parameters for the Arora model (4.19).

The energy-dependent momentum and energy relaxation times are shown in Fig. 4.19. Table 4.7 presents the fitting parameters needed to compute the relaxation times as a function of the average electron energy according to Eq. (4.24).

4.6.2
Hole Transport Coefficients

At present no experimental data is available on p-type InN. The high electron concentration in the as-grown material makes p-type layers very challenging to fabricate. Nevertheless, to exploit the full potential of InN it will be neces-

Tab. 4.7 Fitting parameters needed to compute the momentum (τ_m) and energy (τ_E) relaxation times as a function of the average electron energy and crystallographic direction by using Eq. (4.24).

	τ_m (Γ-M)	τ_m (Γ-A)	τ_E (Γ-M)	τ_E (Γ-A)
τ_0 [ps]	0.0033	0.0054	0.0071	0.0066
a_1	−0.9579	7.7778	2.1680	2.4709
p_1	0.7050	−0.1306	−0.4225	−0.2218
a_2	1.7581	11.4953	2.1021	1.7485
p_2	0.1497	−1.9654	0.3685	0.4045
a_3	2.2140	−2.4579	0.0069	−0.0035
p_3	−0.2192	−2.3896	0.2600	1.1471
b_1	0.4200	1.3262	0.3604	0.2739
q_1	2.6350	3.0317	−0.4222	−0.3523
b_2	1.1886	8.4230	−0.1971	−0.1217
q_2	0.3538	−1.2180	0.0795	0.2631
b_3	−1.0271	−4.8988	0.0157	0.0133
q_3	1.0674	−1.1508	1.9963	1.9741

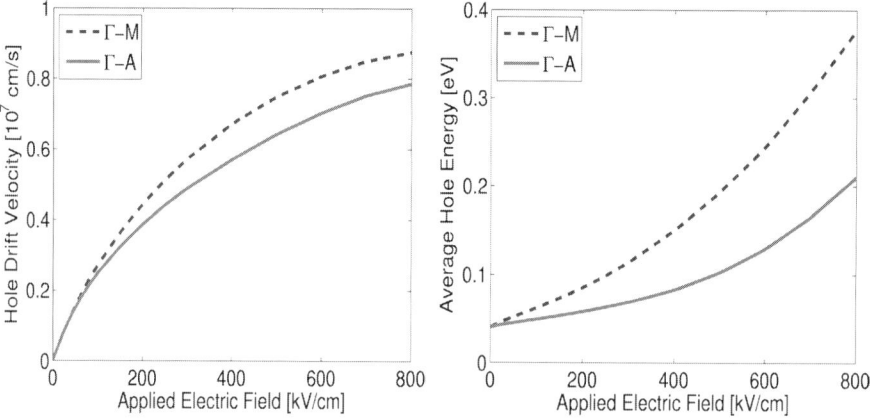

Fig. 4.20 Calculated steady state hole drift velocity for bulk InN. A temperature of 300 K and a fully ionized doping concentration of $N_d = 10^{17}$ cm^{-3} are assumed. The solid line is for transport in the Γ-A direction and the dashed line along the Γ-M direction.

Fig. 4.21 Calculated steady state average hole energy for bulk InN. A temperature of 300 K and a fully ionized doping concentration of $N_d = 10^{17}$ cm^{-3} are assumed. The solid line is for transport in the Γ-A direction and the dashed line along the Γ-M direction.

sary to develop techniques to obtain p-type InN. We have performed a preliminary calculation of the steady state hole velocity and average hole energy as a function of the electric field strength. Figure 4.20 presents the calculated steady state hole velocity. Only a small anisotropy is present when the transport along the *c*-axis and the basal plane is considered. The drift velocity saturates at a value between 8.0×10^6 cm s^{-1} and 9.0×10^6 cm s^{-1}. To complete the information on hole transport in InN, Fig. 4.21 presents the calculated average hole energy.

4.7 Conclusions

In this work we have provided a set of transport coefficients for the III-nitride binary compounds GaN, InN, and AlN. Analytical expressions for the steady state electron and hole drift velocities, mobilities, energy and momentum relaxation times have been chosen so that they can easily be incorporated in drift-diffusion and hydrodynamic simulation tools. All the parameters of the adopted analytical models have been determined from full-band Monte Carlo analysis and validated against available experimental information.

Acknowledgments

Funding for this work has been provided by the NSF 2005 Career Award ECS-0449232 made to E. Bellotti and by the Army Research Laboratory through funding provided by the Boston University Photonics Research Center. Funding for the computational resources was provided through the ONR Durip Award # N00014-05-1-03839. The authors would like to thank Prof. T.D. Moustakas, Prof. R. Paiella, Dr. A. Bhattacharyya, Dr. S. Rudin and Dr. M. Wraback for the information provided to us and for many useful discussions, and Dr. M. Goano for reading the manuscript.

References

1 K. F. Brennan, *The Physics of Semiconductors: with Applications to Optoelectronic Devices*. Cambridge, U.K.: Cambridge University Press, 1999.

2 K. Hess, *Advanced Theory of Semiconductor Devices*. Englewood Cliffs, NJ: Prentice-Hall, 1988.

3 M. Goano, E. Bellotti, E. Ghillino, G. Ghione, and K. F. Brennan, *J. Appl. Phys.*, vol. 88, no. 11, pp. 6467–6475, Dec. 2000.

4 S. Reggiani, M. C. Vecchi, and M. Rudan, *IEEE Trans. Electron Devices*, vol. ED-45, no. 9, pp. 2010–2017, Sept. 1998.

5 R. W. Hockney and J. W. Eastwood, *Computer Simulation Using Particles*. Bristol and New York: Adam Hilger, 1988.

6 C. Moglestue, *Monte Carlo Simulation of Semiconductor Devices*. London: Chapman & Hall, 1993.

7 C. Jacoboni and P. Lugli, *The Monte Carlo Method for Semiconductor Device Simulation*, ser. Computational Microelectronics. Wien: Springer-Verlag, 1989.

8. K. Hess, Ed., *Monte Carlo Device Simulation: Full Band and Beyond*. Boston: Kluver Academic Publishers, 1991.

9. B. C. Lee, K. W. Kim, M. Dutta, and M. A. Stroscio, *Phys. Rev. B*, vol. 56, no. 3, pp. 997–1000, July 1997.

10. S. Yamakawa, S. Aboud, M. Saraniti, and S. Goodnick, *Semiconductor Sci. Tech.*, vol. 19, pp. S475–S477, Apr. 2004.

11. M. V. Fischetti, *IEEE Trans. Electron Devices*, vol. ED-38, no. 3, pp. 634–649, Mar. 1991.

12. B. K. Ridley, *Quantum Processes in Semiconductors*, 4th ed. Oxford: Clarendon Press, 1999.

13. D. Look, *Phys. Rev. Lett.*, vol. 82, no. 6, pp. 1237–1240, 1999.

14. E. Bellotti, Unpublished, 2006.

15. E. Bellotti, M. Farahmand, M. Goano, E. Ghillino, C. Garetto, G. Ghione, H.-E. Nilsson, F. Breannan, and P. Ruden, in *Topics in High Field Transport in Semiconductors*, K. Brennan and P. Ruden, Eds. River Edge, NJ: World Scientific, 2001, pp. 163–222.

16. M. Farahmand, C. Garetto, E. Bellotti, K. F. Brennan, M. Goano, E. Ghillino, G. Ghione, J. D. Albrecht, and P. P. Ruden, *IEEE Trans. Electron Devices*, vol. ED-48, no. 3, pp. 535–542, Mar. 2001.

17. J. Barnes, R. Lomax, and G. Haddad, *IEEE Trans. Electron Devices*, vol. 23, pp. 1042–1048, Sept. 1976.

18. D. M. Caughey and R. E. Thomas, *Proc. IEEE*, pp. 2192–2195, Dec. 1967.

19. N. D. Arora, J. R. Hauser, and D. J. Roulston, *IEEE Trans. Electron Devices*, vol. 29, pp. 292–295, 1982.

20. M. Shur, *Electron. Lett.*, vol. 12, no. 23, pp. 615–616, Nov. 1976.

21. B. Carnez, A. Cappy, A. Kaszynski, E. Constant, and G. Salmar, *J. Appl. Phys.*, vol. 51, no. 1, pp. 784–790, Jan. 1980.

22. Integrated System Engineering, *Tcad 8.0 User's Manual*, Zurich, Switzerland, 2002.

23. Silvaco International, *ATLAS User's Manual*, Santa Clara, CA, 2000.

24. B. Gelmont, K. Kim, and M. Shur, *J. Appl. Phys.*, vol. 74, no. 3, pp. 1818–1821, Aug. 1993.

25. V. W. L. Chin, T. L. Tansley, and T. Osotchan, *J. Appl. Phys.*, vol. 75, no. 11, pp. 7365–7372, June 1994.

26. N. S. Mansour, K. W. Kim, and M. A. Littlejohn, *J. Appl. Phys.*, vol. 77, no. 6, pp. 2834–2836, Mar. 1995.

27. D. L. Rode and D. K. Gaskill, *Appl. Phys. Lett.*, vol. 66, no. 15, pp. 1972–1973, Apr. 1995.

28. J. Kolník, İ. H. Oğuzman, K. F. Brennan, R. P. Wang, P. P. Ruden, and Y. Wang, *J. Appl. Phys.*, vol. 78, no. 2, pp. 1033–1038, July 1995.

29. B. E. Foutz, L. F. Eastman, U. V. Bhapkar, and M. S. Shur, *Appl. Phys. Lett.*, vol. 70, no. 21, pp. 2849–2851, May 1997.

30. U. V. Bhapkar and M. S. Shur, *J. Appl. Phys.*, vol. 82, no. 4, pp. 1649–1655, Aug. 1997.

31. J. D. Albrecht, P. P. Ruden, E. Bellotti, and K. F. Brennan, *MRS Internet J. Nitride Semicond. Res.*, vol. 4S1, no. G6.6, 1999.

32. İ. H. Oğuzman, E. Bellotti, K. F. Brennan, J. K. R. Wang, and P. P. Ruden, *J. Appl. Phys.*, vol. 81, no. 12, pp. 7827–7834, June 1997.

33. İ. H. Oğuzman, J. Kolník, K. F. Brennan, R. Wang, T.-N. Fang, and P. P. Ruden, *J. Appl. Phys.*, vol. 80, no. 8, pp. 4429–4436, Oct. 1996.

34. J. Kolník, İ. H. Oğuzman, K. F. Brennan, R. Wang, and P. P. Ruden, *J. Appl. Phys.*, vol. 81, no. 2, pp. 726–733, Jan. 1997.

35. R. J. Molnar, T. Lei, and T. D. Moustakas, *Appl. Phys. Lett.*, vol. 62, no. 1, pp. 72–74, Jan. 1993.

36. D. C. Look, J. R. Sizelove, S. Keller, Y. F. Wu, U. K. Mishra, and S. P. DenBaars, *Solid State Commun.*, vol. 102, no. 4, pp. 297–300, 1997.

37. H. Eshghi, D. Lancefield, B. Beaumont, and P. Gibart, *Phys. Stat. Sol. (b)*, vol. 216, pp. 733–736, 1999.

38. M. Misra, A. V. Sampath, and T. D. Moustakas, *Appl. Phys. Lett.*, vol. 76, no. 8, pp. 1045–1047, Feb. 2000.

39. Z. Z. Bandić, P. M. Bridger, E. C. Piquette, and T. C. McGill, *Solid-State Electron.*, vol. 44, pp. 221–228, 2000.

40. A. Bhattacharyya, W. Li, J. Cabalu, T. Moustakas, D. Smith, and R. Hervig,

Appl. Phys. Lett., vol. 85, no. 21, pp. 4956–4958, 2004.

41 D. Lancefield and E. Eshghi, *J. Phys: Condens. Matter*, vol. 13, pp. 8939–8944, 2001.

42 K. Kim, M. Cheong, C. Hong, G. Yang, K. Lin, E. Suh, and H. Lee, *Appl. Phys. Lett.*, vol. 76, no. 9, pp. 1149–1151, 2000.

43 P. Kozodoy, H. Xing, S. DenBaars, U. Mishra, A. Saxler, R. Perrin, S. Elhamri, and W. Muitchel, *J. Appl. Phys.*, vol. 87, no. 4, pp. 1832–1835, 2000.

44 M. Wraback, H. Shen, J. C. Carrano, T. Li, J. C. Campbell, M. J. Schurman, and I. T. Ferguson, *Appl. Phys. Lett.*, vol. 76, no. 9, pp. 1155–1157, Feb. 2000.

45 M. Wraback, H. Shen, J. C. Carrano, C. J. Collins, J. C. Campbell, R. D. Dupuis, M. J. Schurman, and I. T. Ferguson, *Appl. Phys. Lett.*, vol. 79, no. 9, pp. 1303–1305, Aug. 2001.

46 M. Wraback, H. Shen, E. Bellotti, J. C. Carrano, C. J. Collins, J. C. Campbell, R. D. Dupuis, M. J. Schurman, and I. T. Ferguson, *Phys. Stat. Sol. (b)*, vol. 228, no. 2, pp. 585–588, 2001.

47 M. Wraback, H. Shen, S. Rudin, E. Bellotti, M. Goano, J. C. Carrano, C. J. Collins, J. C. Campbell, and R. D. Dupuis, *Appl. Phys. Lett.*, vol. 82, no. 21, pp. 3674–3676, May 2003.

48 Z. C. Huang, R. Goldberg, J. C. Chen, Y. Zheng, D. B. Mott, and P. Shu, *Appl. Phys. Lett.*, vol. 67, no. 19, pp. 2825–2826, Nov. 1995.

49 J. Barker, R. Akis, D. Ferry, S. Goodnick, T. Thornton, D. Koleske, A. Wickenden, and R. Henry, *Physica B*, vol. 314, no. 1, pp. 39–41, Mar. 2002.

50 V. Camarchia, M. Goano, G. Ghione, and E. Bellotti, *Semiconductor Sci. Tech.*, vol. 21, pp. 13–18, 2006.

51 V. Camarchia, E. Bellotti, M. Goano, and G. Ghione, *IEEE Electron Device Lett.*, vol. EDL-23, no. 6, pp. 303–305, June 2002.

52 D. Green, E. Haus, F. Wu, L. Chen, U. Mishra, and J. Speck, *J. Vac. Sci. Technol. B*, vol. 21, no. 4, pp. 1804–1811, 2003.

53 J. D. Albrecht, R. P. Wang, P. P. Ruden, M. Farahmand, and K. F. Brennan, *J. Appl. Phys.*, vol. 83, no. 3, pp. 1446–1449, Feb. 1998.

54 S. K. O'Leary, B. E. Foutz, M. S. Shur, U. V. Bhapkar, and L. F. Eastman, *Solid State Commun.*, vol. 105, no. 10, pp. 621–626, 1998.

55 B. E. Foutz, S. K. O'Leary, M. S. Shur, and L. F. Eastman, *J. Appl. Phys.*, vol. 85, no. 11, pp. 7727–7734, June 1999.

56 A. Bhuiyan, A. Hashimoto, and A. Yamamoto, *J. Appl. Phys.*, vol. 94, pp. 2779–2808, 2003.

57 T.-C. Chen, C. Thomidis, J. Abell, W. Li, and T. Moustakas, *J. Cryst. Growth*, vol. 288, pp. 254–260, 2006.

58 H. Lu, W. Schaff, L. Eastman, J. Wu, W. Walukievicz, D. Look, and R. Molnar, *Materials Research Society Symposium Proceedings*, vol. 743, p. L4.10.1, 2003.

59 S. K. O'Leary, B. E. Foutz, M. S. Shur, U. V. Bhapkar, and L. F. Eastman, *J. Appl. Phys.*, vol. 83, no. 2, pp. 826–829, Jan. 1998.

60 B. Nag, *J. Cryst. Growth*, vol. 269, pp. 35–40, 2004.

61 E. Bellotti, B. K. Doshi, K. F. Brennan, J. D. Albrecht, and P. P. Ruden, *J. Appl. Phys.*, vol. 85, no. 2, pp. 916–923, Jan. 1999.

62 C. Bultay and B. Ridley, *Superlatt. Microstruct.*, vol. 36, pp. 465–471, Oct. 2004.

63 V. Polyakov and F. Schwierz, *Appl. Phys. Lett.*, vol. 88, pp. 032 101–1, Jan. 2006.

5
Optical Constants of Bulk Nitrides

Rüdiger Goldhahn, Carsten Buchheim, Pascal Schley, Andreas Theo Winzer, and Hans Wenzel

5.1
Introduction

The development of nitride-based photonic devices requires detailed knowledge of the optical constants over an extended photon energy range. They are usually described in terms of the complex dielectric function (DF) or the complex index of refraction. Minor attention has been devoted so far to their reliable determination. A recent review showed that only the ordinary DF of wurtzite (hexagonal) GaN was known very precisely [1] until 2002, while for all other compounds pseudodielectric functions were reported. Those data do not provide a satisfactory basis for developing analytical DF models [2]. However, the situation has been improved considerably within the last few years. We succeed in determining the anisotropic DF of InN (with a band gap of 0.68 eV) [3,4], the ordinary DFs of In-rich InAlN [5] and InGaN [6] alloys as well as of AlGaN [7] compounds.

In this chapter, these latest experimental data are summarized. The fundamental properties influencing the dispersion of the DF are discussed in detail. Then a DF model will be presented which allows the description of the optical constants of the binary nitrides over a large photon energy range and the calculation of DFs for alloys. Finally, the influence of strong electric fields on the DF will be briefly addressed.

5.2
Dielectric Function and Band Structure

5.2.1
Fundamental Relations

Strain-free nitrides with wurtzite structure belong to the $P6_3mc(C_{6v}^4)$ space group and are optically uniaxial materials. It is typical for most epitaxially grown films that the optic axis (*c*-axis) is oriented normal to the surface (here

Nitride Semiconductor Devices: Principles and Simulation. Joachim Piprek (Ed.)
Copyright © 2007 WILEY-VCH Verlag GmbH & Co. KGaA, Weinheim
ISBN: 978-3-527-40667-8

the x-y plane). In this configuration the dielectric tensor takes the form

$$\overleftrightarrow{\varepsilon} = \begin{pmatrix} \varepsilon_x & 0 & 0 \\ 0 & \varepsilon_y & 0 \\ 0 & 0 & \varepsilon_z \end{pmatrix} = \begin{pmatrix} \varepsilon_o & 0 & 0 \\ 0 & \varepsilon_o & 0 \\ 0 & 0 & \varepsilon_e \end{pmatrix} \tag{5.1}$$

The principal tensor components ε_o (ordinary) and ε_e (extraordinary) describe the response to linearly polarized light, either perpendicular ($E \perp c$) or parallel ($E \parallel c$) to the optical axis, respectively. Their dependence on the photon energy ($E = \hbar\omega$), known as a complex dielectric function, is given in the form $\bar{\varepsilon}_j(\omega) = \varepsilon_{1,j}(\omega) + i \cdot \varepsilon_{2,j}(\omega)$ (j = o, e). Applying Fermi's Golden Rule, the imaginary part of the DF can be readily calculated from the single-particle band structure, representing the valence $E_v(\mathbf{k})$ and conduction $E_c(\mathbf{k})$ band energies as a function of the wave vector $\mathbf{k}$, by

$$\varepsilon_{2,j}(\omega) = \frac{\pi\hbar e^2}{\varepsilon_0 \omega m_0} \frac{1}{8\pi^3} \sum_v \sum_c \int_{BZ} f_{cv,j}(\mathbf{k}) \delta(E_c(\mathbf{k}) - E_v(\mathbf{k}) - \hbar\omega) d^3k \tag{5.2}$$

with the free electron mass (m_0), the elementary charge (e), and the permittivity of free space (ε_0). The optical anisotropy originates from the orientation-dependent dimensionless oscillator strength $f_{cv,j}(\mathbf{k})$. The integral runs over the whole Brillouin zone. In the vicinity of $\mathbf{k}$-points characterized by $\nabla_\mathbf{k}(E_c(\mathbf{k}) - E_v(\mathbf{k})) = 0$, the spectral shape of $\varepsilon_2(\omega)$ shows peculiarities. These points are called *critical points* (CP) [8] or Van Hove singularities. Since the real and imaginary part of the DF for each polarization direction obey the Kramers–Kronig (KK) relation

$$\varepsilon_{1,j}(\omega) = 1 + \frac{2}{\pi} \wp \int_0^{+\infty} \frac{\omega' \cdot \varepsilon_{2,j}(\omega')}{\omega'^2 - \omega^2} d\omega' \tag{5.3}$$

($\wp$ is the principal value of the integral), it is obvious that the singularities strongly influence the shape of $\varepsilon_{1,j}(\omega)$ far below the CP transition energy. Therefore, a discussion of nitride optical properties must not only focus on the band gap region but should also include the range of the high energetic CPs.

The complex refractive index $\tilde{N}_j(\omega)$ (j = o, e) is related to $\bar{\varepsilon}_j(\omega)$ via

$$\tilde{N}_j(\omega) = n_j(\omega) + i\kappa_j(\omega) = \sqrt{\bar{\varepsilon}_j(\omega)} \tag{5.4}$$

Here, $n_j = \{0.5[\varepsilon_{1,j} + (\varepsilon_{1,j}^2 + \varepsilon_{2,j}^2)^{1/2}]\}$ and $\kappa_j = \{0.5[\varepsilon_{1,j} - (\varepsilon_{1,j}^2 + \varepsilon_{2,j}^2)^{1/2}]\}$ denote the index of refraction and the extinction coefficient, respectively. Finally, for the estimation of the light penetration depth, for example, the knowledge of the absorption coefficient $\alpha_j(\omega)$ is important which is related to $\varepsilon_{2,j}$ by

$$\alpha_j(\omega) = \frac{\omega}{n_j(\omega)c_0} \varepsilon_{2,j}(\omega), \tag{5.5}$$

where c_0 is the velocity of light in a vacuum. It should be noticed that the spectral dependence of $\alpha_j(\omega)$ does not only depend on $\varepsilon_{2,j}(\omega)$ but also on $n_j(\omega)$ which is generally not constant and in particular not around the band gap! The effect is disregarded in most works.

5.2.2
Valence Band Ordering, Optical Selection Rules and Anisotropy

Wurtzite nitrides exhibit a pronounced absorption anisotropy in the vicinity of the fundamental band gap caused by the valence band ordering around the Γ point of the Brillouin zone and the symmetry of the corresponding wavefunctions. The valence bands are split into one Γ_9^v and two Γ_7^v bands due to crystal field (Δ_{cf}) and spin–orbit interaction (Δ_{so}). The energetic position of Γ_9^v relative to the other bands is given by the quasi-cubic model (Hopfield) with

$$\Gamma_9^v - \Gamma_{7\pm}^v = \frac{\Delta_{cf} + \Delta_{so}}{2} \pm \frac{1}{2}\sqrt{(\Delta_{cf} + \Delta_{so})^2 - \frac{8}{3}\Delta_{cf}\Delta_{so}} \tag{5.6}$$

The spin–orbit energies of all nitrides are positive; values of 5 meV [9], 18 meV [10], and 19 meV [9] are very likely for InN, GaN, and AlN, respectively. With these values, Δ_{cf} energies of 24 meV [5], 10 meV [10], and -230 meV [11] are estimated from experimental studies. The uppermost valence band will always be formed by Γ_9^v states if both, Δ_{so} and Δ_{cf}, are positive (InN, GaN, and their alloys). The $Al_xGa_{1-x}N$ alloys exhibit a different behavior. The Γ_{7-}^v band becomes the uppermost one with the sign change of Δ_{cf}. This is demonstrated in Fig. 5.1, where the relative energy position of the three valence bands is plotted as a function of Δ_{cf} for $\Delta_{so} = 18$ meV. Assuming a linear dependence of Δ_{cf} on the alloy composition (no data on the exact dependence have been reported so far), the band crossing occurs at an Al content of $x = 0.05$. It should be noted that similar behavior is expected for the InAlN system.

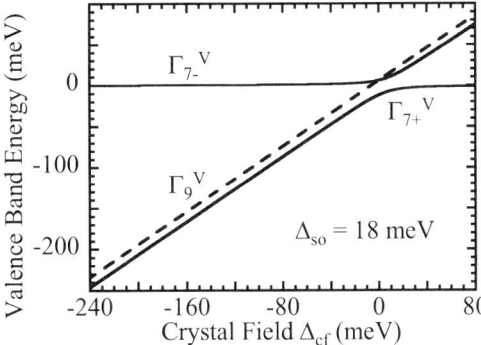

Fig. 5.1 Relative valence band energies at the Γ point of the Brillouin zone as a function of crystal field for a spin–orbit energy of 18 meV.

Optical transitions from Γ_9^v into the Γ_7^c conduction band (here labeled A) are only allowed in the configuration $E \perp c$, while $\Gamma_{7-}^v \to \Gamma_7^c$ (B) and $\Gamma_{7+}^v \to \Gamma_7^c$ (C) transitions contribute to both, ε_o and ε_e, but with a polarization dependent transition probability as depicted in Fig. 5.2. Together with the band energies of Fig. 5.1 the following fundamental properties should be noted: (i) A positive Δ_{cf} value results in a larger extraordinary absorption edge energy (B transition) with respect to the ordinary one (A transition). (ii) The relative contribution of B to ε_e becomes weaker with increasing Δ_{cf} which means that only a pronounced feature due to C is observed (for example in the case of InN). (iii) The lowest observable absorption edge for negative Δ_{cf} energies arises always from the B transition. It dominates for $E \parallel c$ but its contribution is only weak for $E \perp c$ compared to transition A (typical for AlN). In the following description E_0 will be used for denoting the fundamental band gap region if the different contributions (A, B, and C) are not discussed in detail.

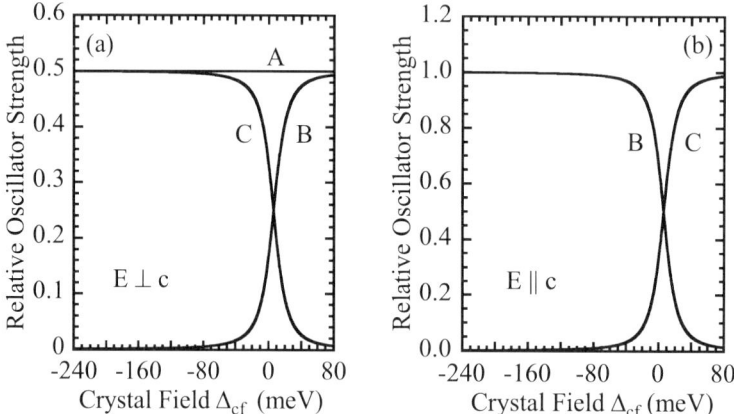

Fig. 5.2 Relative oscillator strength of transitions involving the three valence bands for light polarization $E \perp c$ (a) and $E \parallel c$ (b).

The previous considerations strictly apply only for strain-free nitrides. The pseudomorphic film growth on rather thick bulk-like layers or the mismatch in the thermal expansion coefficients introduces, however, in-plane biaxial strain. It shifts the transition energies and alters the transition probabilities in a characteristic manner (see Ref. [10], for example). Without going into details, tensile (compressive) strain has a similar effect as lowering (increasing) the crystal field energy. The inclusion of electron–hole interaction (excitonic effects), which is inevitable due to the large effective Rydberg energies of the nitrides, does not change the general dependence of the relative oscillator strength compared to the single-electron picture.

5.3 Experimental Results

Generally, the four quantities $\varepsilon_{1,o}$, $\varepsilon_{2,o}$, $\varepsilon_{1,e}$, and $\varepsilon_{2,e}$ have to be determined. It requires high-quality bulk samples or films with an optical axis lying in the surface plane (for example *a*-plane $(11\bar{2}0)$ or *m*-plane $(1\bar{1}00)$). Until now, those films have exhibited a rather poor surface quality compared to the smoother (0001) *c*-plane samples. Therefore, most studies were carried out on the latter materials. Those investigations allow only the determination of ordinary DFs.

Experimental methods, such as reflectance/transmission (R/T), prism coupling technique (PCT) and spectroscopic ellipsometry (SE) are applied for determining optical constants. R/T experiments below the band gap provide only reliable data if interface layers are taken into account [1, 12]. The propagating modes below the band gap are analyzed by PCT [13]. The method is very sensitive to the optical anisotropy in this range. However, only films with a refractive index larger than the substrate can be studied. The advantage of SE is that it provides at least two independent quantities, the ellipsometric parameters Ψ and Δ, at every wavelength. Reliable DFs are obtained from Ψ and Δ if appropriate multi-layer models will be used for the fitting procedure [1, 14]. They take both the formation of extended interface layers and surface roughness into account. With this approach it is not necessary to make any assumptions concerning the spectral shape of the DF. Only such data are presented here. They are based on measurements with a conventional lab ellipsometer (from 0.74 up to 5 eV) and the ellipsometer attached to the Berlin electron storage ring (BESSY II) in the spectral region from 4 up to 9.5 eV.

5.3.1 InN

With the recent progress in the growth of high-quality InN layers by plasma-induced molecular beam epitaxy (PI-MBE) and metal organic vapor phase epitaxy (MOVPE) single-crystalline films became available, which appreciably differ in their optical properties from the previously studied sputtered layers [14]. The results for two samples of this *low band gap* material, both grown by PI-MBE, will be discussed in detail. It exemplifies the behavior common to all hexagonal nitrides.

In order to get *a*- and *c*-plane InN orientation, $(10\bar{1}2)$ *r*-plane and *a*-plane sapphire substrates were used, respectively. The GaN buffer layer (on an AlN nucleation layer) was confirmed to be either $(10\bar{1}0)$ or (0001). The InN film was grown with an orientation following the GaN buffer. Room temperature Hall measurements yielded an electron concentration of $n = 6 \times 10^{18}$ cm^{-3} ($n = 1.5 \times 10^{18}$ cm^{-3}) and a mobility of 250 cm^{2}V^{-1}s^{-1} (1200 cm^{2}V^{-1}s^{-1}) for

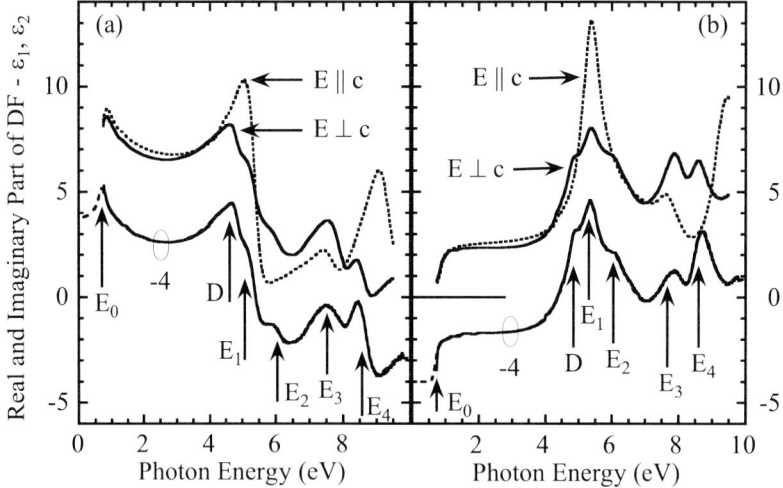

Fig. 5.3 Real (a) and imaginary (b) part of the dielectric function for wurtzite InN. Studies of an *a*-plane film yielded the upper curves representing the ordinary (solid lines) and extraordinary (dashed lines) tensor components. An effective ordinary DF (lower curves, shifted by -4 for the sake of clarity) is determined from *c*-plane measurements as explained in the text.

the *a*- (*c*-plane) films. The *a*-plane film was measured in two *c*-axis orientations with respect to the plane of incidence allowing the determination of both $\bar{\varepsilon}_o$ and $\bar{\varepsilon}_e$ [3]. The *c*-axis for the (0001) film is oriented normal to the surface, these data should nearly correspond to $\bar{\varepsilon}_o$ [4].

Figure 5.3 shows a comparison of the extracted real (a) and imaginary (b) parts of the DF for the fully relaxed InN layers. The data of the *c*-oriented film are vertically shifted (by -4) for the sake of clarity; similar results for this orientation were reported in Ref. [15] for energies up to 6.5 eV. The upper curves in Fig. 5.3 represent the ordinary (solid lines) and extraordinary (dashed lines) dielectric tensor components of hexagonal InN. The imaginary part for both polarizations increases sharply between 0.74 and 1 eV followed by a plateau-like behavior up to 4 eV and pronounced features in the high energy part of the spectra. The latter are related to the peculiarities of the joint density of states in the vicinity of CPs. A fit of the third derivatives of the DFs yields the CP transition energies; one gets 4.82 (D), 5.37 (E_1), 6.09 (E_2), 7.95 (E_3), and 8.56 eV (E_4) from $\bar{\varepsilon}_o$ with a refined analysis compared to previous studies [5]. Only three features dominate the lineshape of $\bar{\varepsilon}_e$ in the UV-VUV region, the CPs are found at 5.39 (E_1), 7.66, and 9.36 eV.

Typical for all hexagonal nitrides is the strong enhancement of ε_2 in the vicinity of E_1 for $\boldsymbol{E} \parallel \boldsymbol{c}$ compared to $\boldsymbol{E} \perp \boldsymbol{c}$; the ratio amounts to 13.1/8 for InN. Since ε_1 is related to the imaginary part by the KK relation, the enhance-

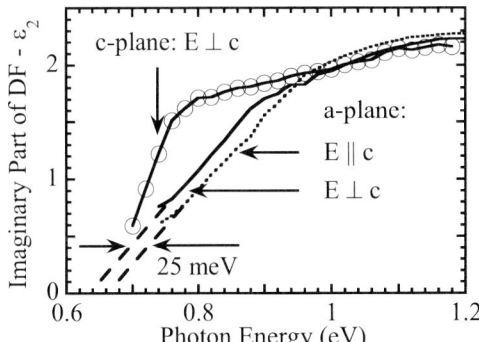

Fig. 5.4 Comparison of the imaginary parts of the DF for InN from Fig. 5.3 around the band gap.

ment effect is the major reason for the larger $\varepsilon_{1,e}$ values in the low-energy part of the spectra (except around the gap) compared to $\varepsilon_{1,o}$. Recent *ab initio* calculations [16] emphasize the experimental results on the anisotropy, and all calculated CP transition energies agree within a few 10 meV to the values given above if electron–hole interaction is consequently taken into account over the whole energy range. The excitonic effects influence the optical absorption by an overall redshift of the entire spectra and a redistribution of oscillator strength with respect to the single-particle band structure and DF.

If SE is employed to study *c*-oriented films, only an effective ordinary DF can be extracted because a small fraction of the *p*-polarized light is always aligned parallel to the *c*-axis. This influence is weak as long as the refractive index is large. The comparison of the ordinary (*a*-plane) with the effective ordinary DF of the *c*-plane film in Fig. 5.3 demonstrates the difference. The E_1/E_2 peak ratio differs slightly, and only the magnitude of E_4 is enhanced in the effective ordinary DF. Therefore, the effective DFs from *c*-plane studies are treated hereafter as equivalent to $\bar{\varepsilon}_o$ in the low-energy range.

The conclusion is corroborated by Fig. 5.4 where the ε_2 data of Fig. 5.3 are plotted on an enlarged scale. Due to the lower carrier density the onset of absorption is sharper for the (0001) film, but no difference between the $E \perp c$ data is found above 1 eV. It indicates, in addition, that the influence of conduction band filling is only restricted to the low-energy part. Furthermore, the slope of the $\varepsilon_{2,e}$ curve for the *a*-plane film with respect to $\varepsilon_{2,o}$ data is shifted by 25 meV to higher energies. The shift is the direct verification of the absorption anisotropy described theoretically in Section 5.2.2. The shape of ε_2 in the band gap region can be theoretically reproduced if conduction band nonparabolicity and band-filling are taken into account; examples are given in [5]. From the analysis the ordinary band gap at zero electron density $\Gamma_9^v \rightarrow \Gamma_7^c$ (A) is estimated with 0.68 eV. It should be noted that the determined optical anisotropy in this region is well confirmed by reflectance difference studies [3]. The influence of higher electron concentrations on $\bar{\varepsilon}_o$ has been discussed in detail in [15] and [17].

5.3.2
GaN and AlN

GaN is probably one of the most intensively studied semiconductors. Many investigations focus on the determination of the exciton transition energies. Reflectance and photoluminescence measurements [18] on homo-epitaxial nearly strain-free GaN films yielded values of 3.422 and 3.427 eV for the bound excitonic states FX^A and FX^B at room temperature, respectively. With an exciton binding energy of 25 meV it corresponds to a $\Gamma_9^v \rightarrow \Gamma_7^c$ (A) gap of 3.447 eV, while the B and C gaps are larger by 5 and 23 meV. A determination of the dielectric tensor components for GaN has not been reported so far.

Here, the properties of two unintentionally-doped (0001)-oriented films will be compared. They were grown by MOVPE (sample 1) and PI-MBE (sample 2) on the most common foreign substrates sapphire and 6H-SiC, respectively. The GaN lattice constants for sample 1 amount to $a = 3.186$ Å and $c = 5.187$ Å ($d = 1800$ nm, $n = 5 \times 10^{15}$ cm^{-3}). For sample 2, values of $n = 5 \times 10^{16}$ cm^{-3} ($d = 1000$ nm) and $c = 5.182$ Å were determined.

A comparison of the ordinary DFs in the vicinity of the gap is shown in Fig. 5.5(a), the solid (dashed) lines refer to sample 1 (2). Although studied at $T = 295$ K, a sharp exciton resonance below the exciton continuum is found for sample 1. The small splitting between FX^A and FX^B, however, did not allow us to resolve them as separate peaks. The absorption edge of sample 2

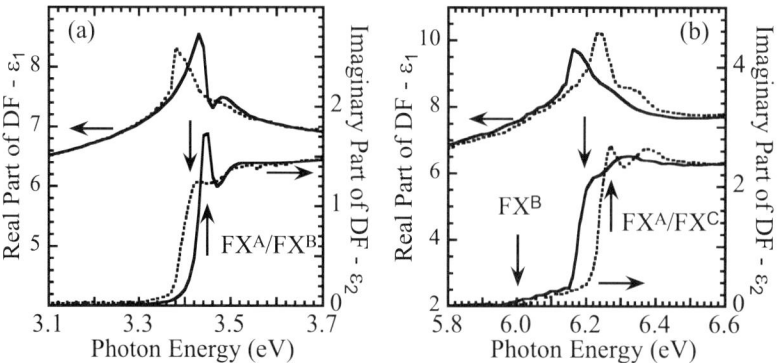

Fig. 5.5 Ordinary DFs of GaN (a) and AlN (b) around the band gap. The solid (dashed) lines in (a) refer to films grown on sapphire (6H-SiC) substrates and studied at $T = 295$ K. The solid and dashed lines in (b) represent data obtained at $T = 295$ K and $T = 155$ K, respectively.

is shifted to lower energies, and the exciton resonance is less pronounced. The latter is probably caused by the about three times larger surface electric field which influences the shape of the DF (Section 5.4.3). A fit yields the FX^A transition energies; they amount to 3.437 eV (3.405 eV) for sample 1 (2). The shift of 15 meV ($-$17 meV) with respect to strain-free GaN originates from the in-plane compressive (tensile) strain as indicated by the lattice constants.

Low-temperature ($T = 6$ K) polarized reflection measurements on AlN bulk crystals with *a*-plane [19] (or *m*-plane [11]) orientation revealed a sharp resonance at 6.029 eV (6.025 eV), which was unambiguously attributed to FX^B. The transition energies of FX^A and FX^C were determined with 6.243 eV and 6.268 eV (6.257 eV), respectively. With an estimated exciton binding energy of 48 meV (71 meV), a $\Gamma^v_{7-} \to \Gamma^c_7$ splitting of 6.077 eV (6.096 eV) was deduced at low temperature. The band gaps at room temperature are about 80 meV lower as evidenced by the reflectance studies of FX^B [19].

We studied a highly-resistive *c*-plane AlN film deposited by MBE on 6H-SiC substrate. The lattice constant of $c = 4.985$ Å corresponds to an in-plane compressive strain of 16×10^{-4}. The ordinary DFs, obtained at $T = 155$ K (dashed lines) and $T = 295$ K (solid lines), are compared in Fig. 5.5(b). The low-temperature $\bar{\varepsilon}_o$ data exhibit, well separated from the continuum, an excitonic resonance at 6.26 eV (roughly 20 meV higher than the strain-free value). For (0001) AlN, it arises from the superposition of FX^A and FX^C and undergoes a shift of 60 meV between both temperatures. The effective $\varepsilon_{2,o}$ spectra show an additional peculiarity. A weak plateau is observed below the dominating exciton resonance; its onset is found at about 5.99 eV for $T = 295$ K. This absorption is probably related to $\Gamma^v_{7-} \to \Gamma^c_7$ transitions. At present we cannot elucidate whether the observation of this feature originates from a spurious contribution of $\varepsilon_{2,e}$ (due to the SE set-up) or by the violation of selection rules at finite wave vectors as a result of valence band mixing.

Figure 5.6 shows the ordinary real (a) and imaginary (b) part of the DF at room temperature over the whole investigated energy range. The data for GaN (sample 1) and AlN are plotted by the solid and dashed lines, respectively. In contrast to InN, the D transition below the again dominating E_1 structure does not appear as a separate feature in ε_2, but E_2 is clearly resolved in both spectra. Again, a fit of the third derivatives of the ordinary DFs yields the CP transition energies, the corresponding values of all three binary nitrides are summarized in Table 5.1.

The data of Fig. 5.6 emphasize the strong impact of E_1 (and partially E_2) on the dispersion of $\varepsilon_{1,o}$. The spectral dependence at photon energies appreciably below the band gap region (E_0) is almost entirely determined by the transition probability at E_1. Obviously, similar arguments apply to $\varepsilon_{1,e}$. Benedict et al. [20] presented first-principles calculations of $\bar{\varepsilon}_o$ and $\bar{\varepsilon}_e$ for GaN and AlN that include electron–hole interaction. For both materials, the calculated

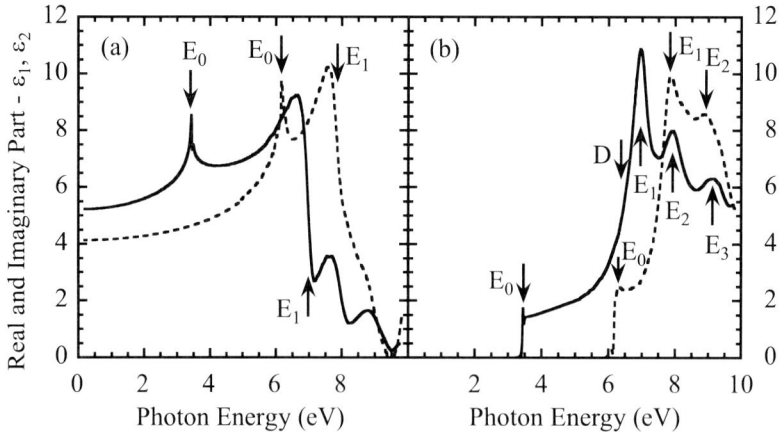

Fig. 5.6 Real (a) and imaginary (b) part of the ordinary DF for GaN (solid lines) and AlN (dashed lines) at room temperature.

Tab. 5.1 Critical point transition energies for the binary nitrides at room temperature as deduced from the ordinary dielectric function.

	$\Gamma_7^c - \Gamma_9^v$ (eV)	D (eV)	E_1 (eV)	E_2 (eV)	E_3 (eV)
InN	0.68	4.82	5.37	6.09	7.95
GaN	3.447	6.36	6.99	7.96	9.25
AlN	6.213		7.97	8.95	8.84

shape of the ordinary DF is in excellent agreement with the results presented in Fig. 5.6. The theoretically predicted enhancement of ε_2 for $E \parallel c$ compared to $E \perp c$ in the region of E_1 amounts to 13.5/9 (GaN) and 11.5/8.7 (AlN). The only experimental result concerning the anisotropy at high photon energies was reported by Cobet *et al.* [21]. They studied an *m*-plane $Al_{0.1}Ga_{0.9}N$ film. Already in the pseudo-DFs, a ratio of 10.6/8.6 has been found. Further experimental investigations are needed for this range in the future.

The aforementioned behavior is the major reason why $\varepsilon_{1,e}$ is larger than $\varepsilon_{1,o}$ below the fundamental gap. Figure 5.7 shows corresponding ellipsometry results for GaN and AlN [22]. The (0001)-oriented films were grown by PI-MBE on 6H-SiC substrates (for details of the SE data analysis for *c*-plane orientation see [22]). The accuracy of the method is corroborated by the excellent agreement with PCT data for GaN [13] and AlN [23]. The spectral dependence of the SE results is well described by the analytic formula [22]

$$\varepsilon_{1,j}(\omega) = 1 + \frac{A_{0,j}}{\pi}\left(\ln\frac{E_{p,j}^2 - \hbar^2\omega^2}{E_{0,j}^2 - \hbar^2\omega^2} + \frac{A_{p,j} \cdot E_{p,j}}{E_{p,j}^2 - \hbar^2\omega^2}\right) \tag{5.7}$$

which follows from a KK transformation of a constant step-like shape of

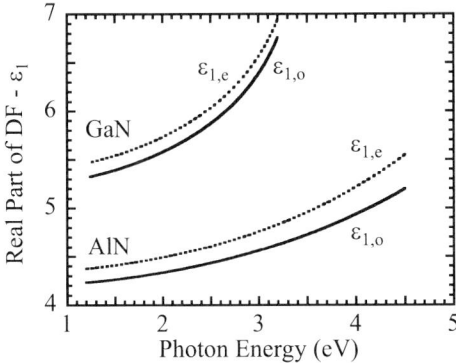

Fig. 5.7 Real part of the ordinary (solid lines) and extraordinary (dashed lines) dielectric function for GaN and AlN.

$\varepsilon_{2,j}(\omega)$ between $E_{0,j}$ and $E_{p,j}$ and a delta function at $E_{p,j}$. The former represents the exciton continuum and the latter summarizes contributions of all high-energy CPs. The fitted parameters describing the anisotropy of both GaN and AlN, are given in Table 5.2. Equation (5.7) can be applied in the photon energy range from 0.2 eV up to 3.2 eV (GaN) and 4.5 eV (AlN), respectively.

Tab. 5.2 Parameters of Eq. (5.7) for the real part $\varepsilon_{1,j}$ of the ordinary (j = o) and extraordinary (j = e) dielectric functions of wurtzite GaN and AlN in the transparent region [22].

Material	DF	$A_{0,j}$	$E_{0,j}$ (eV)	$A_{p,j}$ (eV)	$E_{p,j}$ (eV)
GaN	$\varepsilon_{1,o}$	1.837	3.450	40.65	8.175
	$\varepsilon_{1,e}$	1.929	3.504	41.89	8.164
AlN	$\varepsilon_{1,o}$	4.664	6.465	27.14	10.81
	$\varepsilon_{1,e}$	4.944	6.398	30.28	9.461

Finally, the curves in Fig. 5.7 emphasize that valence band ordering at the Γ point and selection rules become important when the anisotropy close to the gap is analyzed. While the splitting between $\varepsilon_{1,o}$ and $\varepsilon_{1,e}$ decreases towards the gap of GaN, it increases for AlN. The behavior mirrors the influence of positive and negative crystal field, respectively.

5.3.3
AlGaN Alloys

One of the prerequisites for the design and development of efficient blue-violet and UV emitters is the availability of reliable optical constants for $Al_xGa_{1-x}N$ alloys used as cladding and confinement layers. The data presented here were obtained from SE studies [7] on a series of undoped (0001) films ($d = 400 - 700$ nm) grown by MOVPE on sapphire substrate with and

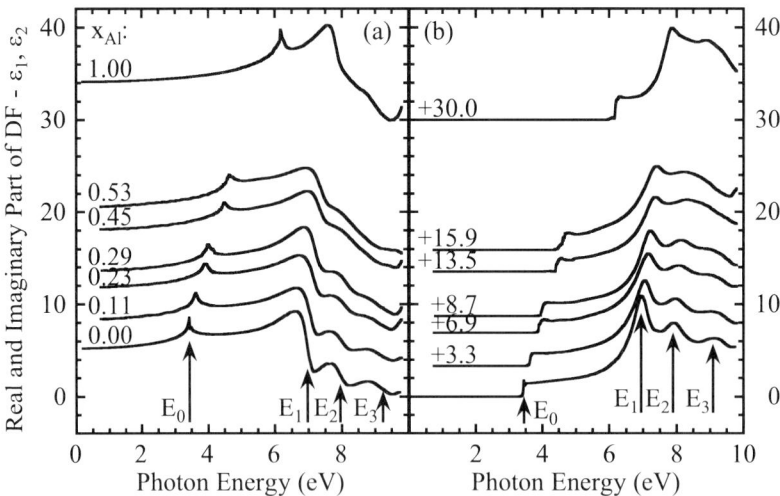

Fig. 5.8 Real (a) and imaginary (b) part of the ordinary dielectric function for $Al_xGa_{1-x}N$ alloys. All spectra are vertically shifted proportional to the Al-content as indicated by the numbers in (b).

without a GaN buffer layer ($d = 400$ nm). The real and imaginary parts of all samples are plotted in Fig. 5.8 (a) and (b), respectively. Although the Ψ and Δ spectra exhibit large interference oscillations below the gap, the point-by-point fitted DFs of the AlGaN layers are absolutely smooth. As demonstrated for a few Al compositions in [7], the spectral dependence in this range does not differ from the PCT results for $\varepsilon_{1,o}$ of Ref. [13]. The complicated layer structure did not allow us to determine $\varepsilon_{1,e}$ with a similar accuracy from the SE spectra. Therefore it is recommended to adopt the dependence reported in [13]. The ordinary and extraordinary tensor components can be approximated by two- and one-Sellmeier-term equations, respectively, of the form

$$\varepsilon_{1,o}(\lambda) = 1 + A^*_{1,o}\lambda^2/[\lambda^2 - (B^*_{1,o})^2] + A^*_{2,o}\lambda^2/[\lambda^2 - (B^*_{2,o})^2] \quad (5.8)$$

$$\varepsilon_{1,e}(\lambda) = 1 + A^*_e\lambda^2/[\lambda^2 - (B^*_e)^2] \quad (\lambda \geq 450 \text{ nm}) \quad (5.9)$$

with the coefficients summarized in Table 5.3. The validity range of Eq. (5.8) was not explicitly mentioned in [13]. The comparison with SE data [7] demonstrated, however, that no difference is found between both techniques up to 300 meV below the corresponding band gaps.

The absorption edges of the alloys are only slightly broader than for GaN confirming the compositional homogeneity. A fit of the DFs yields the $\Gamma_9^v \rightarrow \Gamma_7^c$ (A) band gaps which were confirmed by photoreflectance spectroscopy [7] for $x = 0.11$ and $x = 0.28$. The smooth dependence of the gap on the Al content can be approximated by the relation

$$A_{Al_xGa_{1-x}N} = xA_{AlN} + (1-x)A_{GaN} - b_A x(1-x) \quad (5.10)$$

with a bowing parameter of $b_A = 0.9$ eV and the binary gaps given in Table 5.1. The bowing energies for the high-energy critical points E_1, E_2 and E_3, indicated in Fig. 5.8(b), are small, they amount to $b_1 = 0.2$ eV, $b_2 = 0.1$ eV and $b_3 = 0.5$ eV only [7]. Note, E_3 undergoes a negative energy shift.

5.3.4
In-rich InGaN and InAlN Alloys

For the design of laser and light-emitting diodes working in the blue/green spectral region the properties of $In_xGa_{1-x}N$ alloys with high Ga content are particularly interesting. The $In_xAl_{1-x}N$ material system attracts much attention for the development of UV emitters because those layers can be deposited lattice-matched on GaN for a composition of about $x = 0.14$. However, most work has been carried out on In-rich alloys so far stimulated by the band gap revision for InN. The $In_xGa_{1-x}N$ results presented below are most suitable for demonstrating the principal behavior of these alloys.

The films were grown by PI-MBE on sapphire substrates. Their electron concentration amounts to $n = 2 - 6 \times 10^{18}$ cm^{-3} [6]. Figure 5.9 shows the real (a) and imaginary (b) part of the ordinary DFs for the $In_xGa_{1-x}N$ alloys with N-face ($x = 0.87$) and metal-face polarity ($x = 0.77$ and $x = 0.69$) in comparison to the results for InN. The step-like onset of ε_2 above the gap E_0 mirrors in a pronounced feature in ε_1 for all alloys. By fitting the shape of the imaginary part in this range with the procedure presented in [5] one can estimate the zero density band gaps (A) which are depicted in Fig. 5.10(a). The dependence on the In content is described by a bowing parameter of $b_A = 1.77$ eV. No influence of polarity on the optical properties was found. Despite larger alloy broadening, the D and E_1 to E_4 features in the high-energy part of ε_2 (Fig. 5.9(b)) are clearly resolved for all compositions. The CP energies shift smoothly with the Ga concentration to higher values as depicted in Fig. 5.10(a); the corresponding bowing parameters are summarized in Table 5.4.

Tab. 5.3 Coefficients taken from [13] to be used in Eq. (5.8) and Eq. (5.9) for representing $\varepsilon_{1,o}$ and $\varepsilon_{1,e}$ of $Al_xGa_{1-x}N$ alloys below the gap, respectively.

Al content	$\varepsilon_{1,o}$				$\varepsilon_{1,e}$	
	$A^*_{1,o}$	$B^*_{1,o}$ (nm)	$A^*_{2,o}$	$B^*_{2,o}$ (nm)	A^*_e	B^*_e (nm)
0	0.083	354.8	4.085	180.3	4.321	189.2
0.144	0.117	326.2	3.930	165.4	4.164	181.4
0.234	0.201	302.2	3.769	151.6	4.080	175.3
0.419	0.141	277.6	3.641	153.0	3.919	163.7
0.593	0.238	246.7	3.363	138.1	3.748	151.1
0.666	0.182	237.7	3.337	142.3	3.683	148.9

5 Optical Constants of Bulk Nitrides

Tab. 5.4 Bowing parameters for the CP energies of the ternary nitrides.

	b_A (eV)	b_D (eV)	b_1 (eV)	b_2 (eV)	b_3 (eV)
$In_xGa_{1-x}N$	1.77	0.97	1.03	1.10	0.79
$In_xAl_{1-x}N$	4.0		1.8	2.7	
$Al_xGa_{1-x}N$	0.9	0	0.2	0.1	0.5

The DFs presented in Ref. [5] for In-rich $In_xAl_{1-x}N$ alloys ($x \geq 0.71$) evidence a similar behavior as discussed above for InGaN. However, for an Al content of 29%, the spectra already become more AlN-like which means that transition D disappears. The fitted transition energies are plotted in Fig. 5.10(b). Note the weak dependence of E_3 on the In content which is expected by the observed merging of E_2 and E_3 for AlN. The bowing energies for the CPs are given in Table 5.4.

5.4
Modeling of the Dielectric Function

For the modeling of optoelectronic devices, an analytical representation of the real and imaginary parts of the DF below and above the absorption edge is needed. A few of the semi-empirical models were evaluated in [2]. The poor experimental data basis available at that time might be one of the reasons why

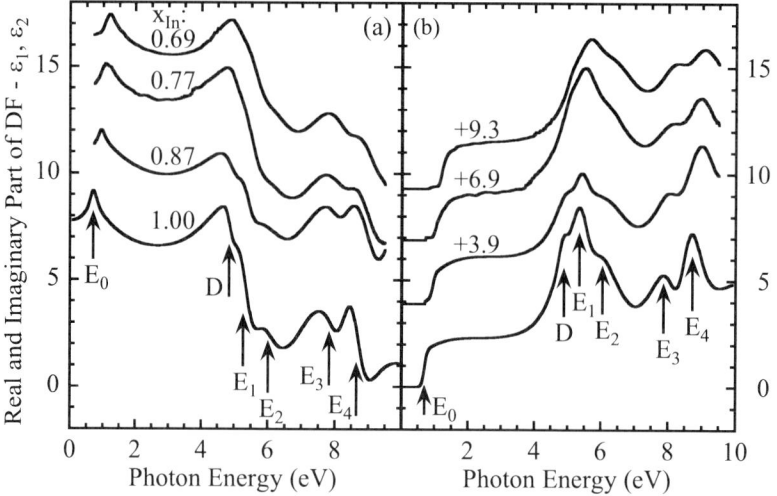

Fig. 5.9 Real (a) and imaginary (b) part of the ordinary dielectric function for $In_xGa_{1-x}N$ alloys. The alloy spectra are vertically shifted proportional to the Ga-content as indicated by the numbers in (b).

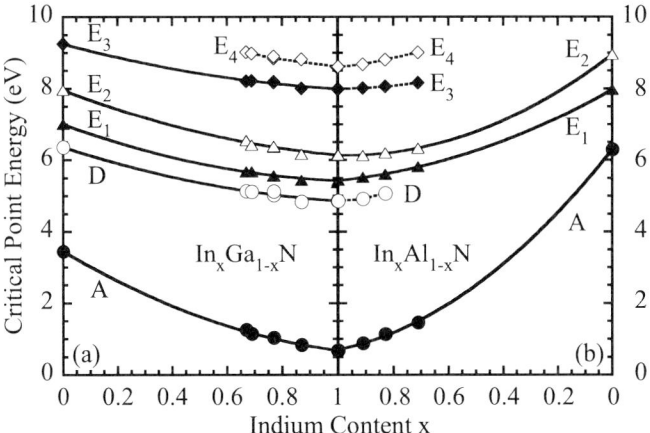

Fig. 5.10 Critical point transition energies of $In_xGa_{1-x}N$ (a) and $In_xAl_{1-x}N$ (b) alloys as a function of Indium content x.

no systematic variation of the alloy parameters was found. The second is that the well established single-particle CP models cannot be applied due to the strong excitonic enhancement found for the nitrides. Thirdly, all models derived for semiconductors assume parabolic bands. As demonstrated recently for InN [5] and GaN [24], at least the conduction band nonparabolicity has to be taken into account for fitting the shape of ε_2 above the band gap. A refined calculation should also include the nonmonotonous dispersion of the oscillator strength due to valence-band mixing around the Γ-point of the Brillouin zone. However, this effect is disregarded in the following.

5.4.1
Analytical Representation of the Dielectric Function

Although a simple expression for the DF cannot account for all peculiarities of the optical properties, one gets reasonable agreement over a limited frequency range if the real and imaginary parts are separately treated with the consequence that they cannot obey the KK relation. Due to lack of enough experimental data for the extraordinary tensor component, in the following only the ordinary one will be considered.

The imaginary part $\varepsilon_2(\omega)$ is split into two terms, $\varepsilon_2(\omega) = \varepsilon_{2,\text{low}}(\omega) + \varepsilon_{2,\text{high}}(\omega)$. The first term $\varepsilon_{2,\text{low}}(\omega)$ originates from contributions arising from the band gap and is given by

$$\varepsilon_{2,\text{low}}(\omega) = \sum_{j=A,(B\text{ or }C)} \frac{A_{BS}\,\hbar\omega\Gamma_{BS}}{((E_j - R)^2 - \hbar^2\omega^2)^2 + (\hbar\omega\Gamma_{BS})^2}$$

$$+ \sum_{j=A,(B\text{ or }C)} \frac{A_{CS}}{\hbar\omega}\,\frac{1 + \text{erf}[(\hbar\omega - E_j)/\Gamma_{CS}]}{1 - \exp(-2\pi\sqrt{R/|\hbar\omega - E_j|})}. \quad (5.11)$$

Transitions due to bound exciton states (BS) below the band gap are represented by real parts of damped harmonic oscillators (DHO), as indicated by the first term. Two DHOs for the transitions with the highest oscillator strength are sufficient, namely A and B (GaN) or C (AlN). The corresponding magnitudes and broadening energies for both contributing transitions are assumed to be equal. The exciton continuum is described by the second term (note, the usual term $A_{CS}/\hbar^2\omega^2$ is replaced here by $A_{CS}/\hbar\omega$ which accounts for band nonparabolicity).

The strong excitonic enhancement around the high-energy CPs allows us to describe them as well by DHOs in the form

$$\varepsilon_{2,\text{high}}(\omega) = \sum_{j=D,1,2,3} \frac{A_j\,\hbar\omega\Gamma_j}{(E_j^2 - \hbar^2\omega^2)^2 + (\hbar\omega\Gamma_j)^2} \times$$

$$\times \left[\Theta(E_j - \hbar\omega)\frac{\hbar\omega - E_A}{E_j - E_A} + \Theta(\hbar\omega - E_j)\right] \quad (5.12)$$

with the Heaviside step function Θ. The linear decay of the CP response (first term) towards the main band gap E_A is necessary in order to avoid artificial absorption below the gap. The procedure yields good agreement with the experimental data for GaN and AlN. For InN, we did not not succeed to represent the imaginary part of the DF in analytical form.

The real part $\varepsilon_1(\omega)$ follows from the real parts of the DHOs representing the bound excitonic states and the high-energy CPs. As already discussed in Section 5.3.2, the contribution of the exciton continuum can be modeled by a logarithmic function which results from a KK transformation of a constant-like shape of $\varepsilon_2(\omega)$ between E_A and a fictitious CP at E_P. Compared to Eq. (5.7), in order to get rid of the singularities, an imaginary part is added to the photon energy, so that $\hbar\omega$ is replaced by $\hbar\omega + i \cdot \Gamma_0$. The contributions due to CPs having higher energies than those taken into account as DHOs are summarized by a delta function at E_P. Finally, a constant term b is added. Then, one gets the following form:

$$\varepsilon_1(\omega) = b + \sum_{j=A,(B \text{ or } C)} \left(\frac{A_{BS}((E_j - R)^2 - \hbar^2\omega^2)}{((E_j - R)^2 - \hbar^2\omega^2)^2 + (\hbar\omega\Gamma_{BS})^2} \right)$$

$$+ \frac{1}{\pi} \Re \left(A_0 \cdot \ln \frac{E_P^2 - (\hbar\omega + i \cdot \Gamma_0)^2}{E_A^2 - (\hbar\omega + i \cdot \Gamma_0)^2} + \frac{A_P \cdot E_P}{E_P^2 - (\hbar\omega + i \cdot \Gamma_0)^2} \right)$$

$$+ \sum_{j=D,1,2,3} \frac{A_j (E_j^2 - \hbar^2\omega^2)}{(E_j^2 - \hbar^2\omega^2)^2 + (\hbar\omega\Gamma_j)^2}. \tag{5.13}$$

Here, for GaN and InN again transition B and for AlN transition C have to be included in the summation. With the parameters summarized in Table 5.5 the experimental data of GaN and AlN (Fig. 5.6) are well reproduced. It should be noted that the CP energies obtained by the fitting procedure differ slightly from those given in Table 5.1.

Tab. 5.5 Parameters to be used for calculating the ordinary DF of strain-free InN, GaN, and AlN.

		InN		GaN		AlN	
R (meV)		8		25		50	
E_A (eV)	E_B or E_C (eV)	0.68	0.685	3.447	3.452	6.213	6.227
A_{BS} (eV^2)	Γ_{BS} (eV)	–	–	0.082	0.037	0.462	0.080
A_{CS} (eV)	Γ_{CS} (eV)	0.290	0.030	1.163	0.033	3.411	0.055
A_0	Γ_0 (eV)	2.166	0.039	1.612	0.022	1.897	0.030
A_P (eV)	E_P (eV)	164.2	14.47	69.54	13.43	118.3	14.46
E_D (eV)	Γ_D (eV)	4.82	0.647	6.360	1.686	8.207	0.684
A_D (eV^2)		8.442		12.03		17.06	
E_1 (eV)	Γ_1 (eV)	5.37	0.725	6.940	0.626	7.800	0.510
A_1 (eV^2)		17.97		31.81		21.16	
E_2 (eV)	Γ_2 (eV)	6.09	0.983	7.907	0.924	8.950	1.335
A_2 (eV^2)		14.43		30.84		62.12	
E_3 (eV)	Γ_3 (eV)	7.95	1.167	9.214	1.838	8.841	1.838
A_3 (eV^2)		15.47		63.27		0	
b		−1.526		−0.041		−0.929	

5.4.2
Calculation of the Dielectric Function for Alloys

With the above developed DF model it becomes possible to calculate the optical response of alloys for all compositions. The example of $Al_xGa_{1-x}N$ demonstrates the procedure. The necessary CP energies to be inserted into

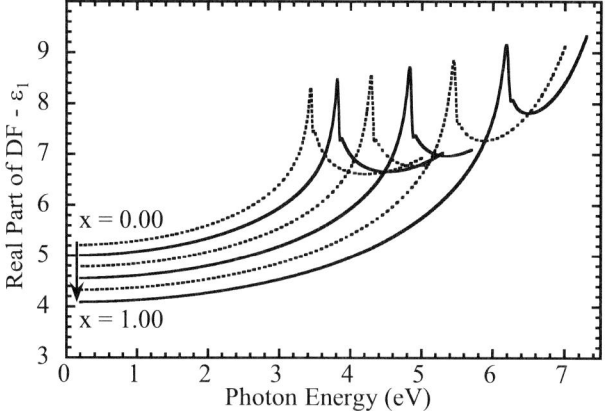

Fig. 5.11 Calculated real part of the ordinary DF for $Al_xGa_{1-x}N$ alloys in steps of $x = 0.20$.

Eqs (5.11)–(5.13) follow via Eq. (5.10) from the binary transition energies (Table 5.5) and the CP bowing parameters b_E given in Table 5.4. Only the band gap energies should be corrected for the influence of strain. For the magnitudes A and the broadening energies Γ we propose the following calculation scheme:

$$A_{Al_xGa_{1-x}N} = x \cdot A_{AlN} + (1-x) \cdot A_{GaN} - b_A x(1-x) \tag{5.14}$$

$$b_A = \frac{A_{AlN} - A_{GaN}}{E_{AlN} - E_{GaN}} \cdot b_E \tag{5.15}$$

$$\Gamma_{Al_xGa_{1-x}N} = x \cdot \Gamma_{AlN} + (1-x) \cdot \Gamma_{GaN} - b_\Gamma x(1-x) \tag{5.16}$$

$$b_\Gamma = \frac{\Gamma_{AlN} - \Gamma_{GaN}}{E_{AlN} - E_{GaN}} \cdot b_E . \tag{5.17}$$

Here, bowing parameters for the magnitude (b_A) and the broadening (b_Γ) are introduced, which are assumed to be proportional to the bowing parameters of the corresponding CPs (b_E). With this approach both, the real and imaginary parts of the ordinary DF data as shown in Fig. 5.8, are well reproduced. Results for $\varepsilon_1(\omega)$ covering the whole composition range of $Al_xGa_{1-x}N$ alloys are plotted in Fig. 5.11. The spectra demonstrate that a rigid shift of the curves is not sufficient to get the correct dispersion of the optical constants.

5.4.3
Influence of Electric Fields on the Dielectric Function

The gradients of the piezoelectric and spontaneous polarization at nitride hetero-boundaries lead to huge internal electric fields F, values between a few hundreds kV cm^{-1} up to several MV cm^{-1} are found typically. The fields alter the transition probability and therefore the shape of the DF around the band gap; as a result changes in the reflectivity are observed (photo- and electroreflectance spectroscopy makes use of the effect). If the field strength is reduced the absorption edge becomes much sharper and free excitonic lines are resolved [25, 26], even for material doped with a few 10^{17} cm^{-3}.

Numerical calculations [27] demonstrated that electron–hole interaction enhances the field-induced absorption coefficient below the gap by several orders of magnitude compared to the classical single-particle Franz–Keldysh effect. The most important parameter is the ionization field strength F_{ion}. With $R = 25$ meV and an exciton Bohr radius of $a = 2.8$ nm, one gets for GaN $F_{\text{ion}} = 90$ kV cm^{-1}, which is rather small compared to the internal fields. However, no closed analytical representation of the effect can be provided.

The calculations start from the field-free DF. Considering only one conduction and one valence band split by E_j at the center of the Brillouin zone, the light-induced creation of an electron–hole pair leads to an $\varepsilon_2(\omega)$ which is expressed in Elliots's theory [28] by

$$\varepsilon_{2,j}(\omega) = \frac{\pi \hbar e^2}{\varepsilon_0 m_0 \omega} f_{\text{cv},j} \sum_n |\phi_n(0)|^2 \delta(E_j + E_n - \hbar\omega). \qquad (5.18)$$

Here, $f_{\text{cv},j}$ denotes the oscillator strength. The sum is evaluated over all states n, characterized by the wave functions in relative motion coordinates $\phi_n(\mathbf{r})$ and the eigenenergies E_n. When a static homogeneous electric field is applied along the z direction one has to solve the effective mass Schrödinger equation:

$$\left(-\frac{\hbar^2}{2\mu m_0} \nabla^2 - \frac{e^2}{4\pi \varepsilon_0 \varepsilon_r r} - eFz \right) \phi_n(\mathbf{r}) = E_n \phi_n(\mathbf{r}). \qquad (5.19)$$

This problem is exactly solved by a parabolic coordinate transformation [29] allowing the direct insertion of the solutions $\phi_n(\mathbf{r})$ and E_n into Eq. (5.18) in order to calculate $\varepsilon_{2,j}(\omega, F)$. Then, the contributions from the three valence bands (weighted by the relative oscillator strengths) are added up and, for example, broadened by a Gaussian line shape function. To obtain the correct dispersion of $\varepsilon_1(\omega, F)$, the contribution of the high-energy CPs $\varepsilon_{2,\text{high}}$ must be added to $\varepsilon_2(\omega, F)$ prior to the KK transformation. Finally, $\alpha(\omega, F)$ is calculated by Eq. (5.5).

Figure 5.12 illustrates the typical behavior for GaN under tensile strain (contributions from A and C dominate). A band gap of A = 3.491 eV and a broadening energy of 2 meV correspond to the values found at $T = 5$ K. Pronounced excitonic resonances are only found at low fields ($F \ll F_{\text{ion}}$), which is consistent with the experimental observations [25,26], they completely disappear for $F > F_{\text{ion}}$. Unlike the classical picture, the ionization of the excitonic ground levels is not abrupt at F_{ion} because the increase of F causes a steady mixing

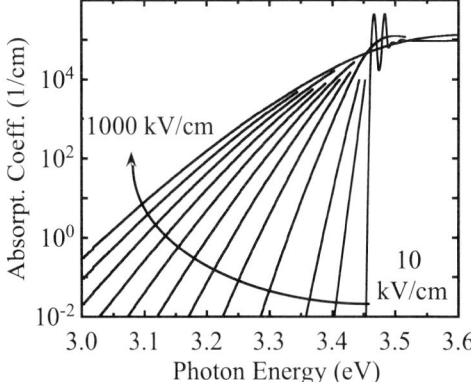

Fig. 5.12 Absorption coefficient of GaN near and below the band gap (A = 3.491 eV and a broadening energy of 2 meV) for electric field strengths of 10, 50, 100, 200, 300, 400, 500, 600, 700, 800, 900, and 1000 kV cm^{-1}.

of the bound state wave functions with the continuum states. Already in this picture it becomes clear that the alteration of the GaN optical properties due to sufficiently high electric fields can be superior to other broadening effects. Because F values were not reported in previous absorption studies, a detailed comparison with currently available absorption data is not possible.

References

1 R. Goldhahn, S. Shokhovets, in *III-Nitride Semiconductors: Optical Properties II, Series: Optoelectronic Properties of Semiconductors and Superlattices*, Vol. 14, (Eds.: M. O. Manasreh and H. X. Jiang), Taylor and Francis Books, New York, p. 73 (2002)

2 A. B. Djurišić, E. H. Li, in *III-Nitride Semiconductors: Optical Properties II, Series: Optoelectronic Properties of Semiconductors and Superlattices*, Vol. 14, (Eds.: M. O. Manasreh and H. X. Jiang), Taylor and Francis Books, New York, p. 115 (2002)

3 R. Goldhahn, A. T. Winzer, V. Cimalla, O. Ambacher, C. Cobet, W. Richter, N. Esser, J. Furthmüller, F. Bechstedt, H. Lu, W. J. Schaff, *Superlattices and Microstructures* **36**, 591 (2004)

4 R. Goldhahn, P. Schley, A. T. Winzer, M. Rakel, C. Cobet, N. Esser, H. Lu, W. J. Schaff, *J. Crystal Growth* **288**, 273 (2006)

5 R. Goldhahn, P. Schley, A. T. Winzer, G. Gobsch, V. Cimalla, O. Ambacher, M. Rakel, C. Cobet, N. Esser, H. Lu, W. J. Schaff, *phys. stat. sol. (a)* **203**, 42 (2006)

6 P. Schley, R. Goldhahn, A. T. Winzer, G. Gobsch, V. Cimalla, O. Ambacher, M. Rakel, C. Cobet, N. Esser, H. Lu, W. J. Schaff, *phys. stat. sol. (b)* **243**, 1572 (2006)

7 C. Buchheim, R. Goldhahn, M. Rakel, C. Cobet, N. Esser, U. Rossow, D. Fuhrmann, A. Hangleiter, *phys. stat. sol.(b)* **242**, 2610 (2005)

8 P. Y. Yu, M. Cardona, *Fundamentals of Semiconductors*, 3rd edn., Springer, Berlin, p. 261 (2001)

9 P. Carrier, S. H. Wei, *J. Appl. Phys.* **97**, 033707 (2005)

10 B. Gil, in *Gallium Nitride II, Series: Semiconductors and Semimetals* Vol. 57, (Eds.: J.I. Pankove and T. D. Moustakas), Academic Press, San Diego, p. 209 (1999)

11 L. Chen et al., *Appl. Phys. Lett.* **85**, 4334 (2004)

12 S. Shokhovets, R. Goldhahn, G. Gobsch, T. S. Cheng, C. T. Foxon, G. D. Kipshidze, W. Richter, *J. Appl. Phys.* **86**, 2602 (1999)

13 N. A. Sanford, L. H. Robins, A. V. Davydov, A. Shapiro, D. V. Tsvetkov, A. V. Dmitriev, S. Keller, U. K. Mishra, S. P. DenBaars, *J. Appl. Phys.* **94**, 2980 (2003)

14 R. Goldhahn, *Acta Physica Polonica* A **104**, 123 (2003)

15 A. Kasic, E. Valcheva, B. Monemar, H. Lu, W. J. Schaff, *Phys. Rev.* B **70**, 115217 (2004)

16 J. Furthmüller, P. H. Hahn, F. Fuchs, F. Bechstedt, *Phys. Rev.* B **72**, 205106 (2005)

17 H. Ahn, C. H. Shen, C. L. Wu, S. Gwo, *Appl. Phys. Lett.* **86**, 201905 (2005)

18 K. Thonke et al., in *Proc. Int. Workshop on Nitride Semiconductors, Nagoya 2000,* Conf. Ser. 1, The Institute of Pure and Applied Physics, p. 587 (2001).

19 E. Silveira, J. A. Freitas, Jr., O. J. Glembocki, G. A. Slack, J. A. Schowalter, *Phys. Rev.* B **71**, 041201 (2005)

20 L. X. Benedict, T. Wethkamp, K. Wilmers, C. Cobet, N. Esser, E. L. Shirley, W. Richter, M. Cardona *Solid State Commun.* **112**, 129 (1999)

21 C. Cobet et al., *Phys. Rev.* B **64**, 165203 (2001)

22 S. Shokhovets, R. Goldhahn, G. Gobsch, S. Piekh, R. Lantier, A. Rizzi, V. Lebedev, W. Richter, *J. Appl. Phys.* **94**, 307 (2003)

23 D. Blanc, A. M. Bouchoux, C. Plumerau, A. Cachard, J. F. Roux, *Appl. Phys. Lett.* **66**, 659 (1995)

24 S. Shokhovets, G. Gobsch, O. Ambacher, *Appl. Phys. Lett.* **86**, 161908 (2005)

25 S. Shokhovets, D. Fuhrmann, R. Goldhahn, G. Gobsch, O. Ambacher, M. Hermann, U. Karrer, *phys. stat. sol. (a)* **194**, 480 (2002)

26 P. Trautmann, K. Pakuła, R. Bożek, J. M. Baranowski, *Appl. Phys. Lett.* **83**, 3510 (2003)

27 J. D. Dow, D. Redfield, *Phys. Rev.* B **1**, 3358 (1970)

28 R. J. Elliott, *Phys. Rev.* **108**, 1384 (1957)

29 D. F. Blossey, *Phys. Rev.* B **2**, 3976 (1970)

6
Intersubband Absorption in AlGaN/GaN Quantum Wells

Sulakshana Gunna, Francesco Bertazzi, Roberto Paiella, and Enrico Bellotti[1]

6.1
Introduction

Intersubband transitions in semiconductor quantum wells, i.e., optical transitions between quantized states within the conduction band (or within the valence bands), have been the subject of extensive research over the past two decades, and are now used in several practical devices such as quantum-well infrared photodetectors (QWIPs) and quantum cascade lasers [1]. An important recent development in this field of research is the observation of near-infrared intersubband transitions in GaN-AlGaN quantum wells [2–10]. These heterostructures are characterized by extremely large conduction band discontinuities which can accommodate intersubband transition wavelengths as short as 1.1 µm [11]. Therefore III-nitride quantum wells can be used to extend the operation wavelength of intersubband devices into the technologically important fiber-optic spectral range.

Specific device applications that have already been demonstrated include QWIP-type photodetectors [4, 10] and nonlinear waveguides for all-optical switching [7]. In particular, the latter devices take advantage of the ultrafast (subpicosecond) relaxation times of intersubband transitions, and therefore are promising for all-optical signal processing at data rates of several hundred $Gb\,s^{-1}$. The use of III-nitride quantum wells has also been proposed for the development of near-infrared quantum cascade lasers [11], terahertz emitters [12, 13], and single-photon detectors at communication wavelengths based on a QWIP active material combined with an avalanche multiplication region [14].

In this work we present a comprehensive numerical study of intersubband absorption in GaN-AlGaN multiple-quantum-well (MQW) samples. In particular, we systematically investigate the dependence of the peak absorption energy on the main design parameters, such as well and barrier widths, barrier composition, and doping. To that purpose, the Schrödinger and Poisson

1) Corresponding author. bellotti@bu.edu

Nitride Semiconductor Devices: Principles and Simulation. Joachim Piprek (Ed.)
Copyright © 2007 WILEY-VCH Verlag GmbH & Co. KGaA, Weinheim
ISBN: 978-3-527-40667-8

equations are solved self-consistently to determine the conduction subband energy levels and envelope functions, from which the intersubband absorption spectrum is then computed. The numerical implementation of our theoretical model is based on the finite-element method, which ensures a high degree of efficiency and flexibility.

A distinctive feature of III-nitride heterostructures (e.g., compared to arsenide materials) is the presence of strong spontaneous and piezoelectric polarizations, leading to large interface charge densities [15, 16]. The resulting built-in electric fields strongly affect the intersubband absorption properties, leading to novel behaviors that are not observed in arsenic-based MQWs (e.g., a pronounced dependence of the intersubband absorption wavelength on barrier width). These effects are studied in detail in our simulations. We also investigate the role played in GaN-AlGaN quantum wells by the conduction-band nonparabolicity (which is important due to the large conduction-band offsets) and by many-body effects (which are important due to the large doping densities of order 10^{19} cm^{-3} typically employed in these near-infrared intersubband systems [2, 10]). While theoretical models of intersubband transitions in III-nitrides have already been presented [17, 18], no systematic investigation of the above-mentioned effects has yet been carried out. Finally, we present for the first time, simulations of complete MQW structures including buffer and cap layers, and investigate how the absorption spectrum is affected by the number of quantum wells.

6.2
Theoretical Model

Two important modeling aspects have to be considered. First, we need to specify a mathematical model to describe the physics of the semiconductor materials and the device, second it is necessary to use a suitable numerical technique to solve the coupled equations that describe the system. The theoretical model employed in this work is based on the self-consistent solution of Schrödinger and Poisson equations, within the finite-element approach framework. In the effective-mass approximation [19, 20], the electron energies and wavefunctions are obtained by solving the one-dimensional Schrödinger equation:

$$\left[-\frac{\hbar^2}{2} \frac{\partial}{\partial z} \frac{1}{m^*(z,E)} \frac{\partial}{\partial z} + V(z) \right] \psi_i(z) = E_i \psi_i(z) \tag{6.1}$$

where $m^*(z, E)$ is the energy and position-dependent electron effective mass and z is the distance along the growth axis. The total potential energy $V(z) = \Delta E(z) - q\phi(z) + V_{xc}$ includes the exchange correlation energy V_{xc}, the conduction band discontinuity $\Delta E(z)$, and the electrostatic potential $\phi(z)$, resulting from all the charges in the system. Once solved, the subbands energy eigenvalues E_i and wavefunctions $\psi_i(z)$, are available to compute the total charge density. The electrostatic potential $\phi(z)$ is obtained by solving the Poisson equation:

$$\frac{d}{dz}\epsilon(z)\frac{d}{dz}\phi(z) = q\left[n(z) - p(z) - N_D^+(z) + N_A^-(z) - \rho_{pol}\right] \quad (6.2)$$

where q is the electronic charge and ϵ is the dielectric constant. $N_D^+(z)$ and $N_A^-(z)$ are the local concentration of ionized donors and acceptors respectively, and $\rho_{pol}(z)$ is the bound polarization charge density at each well/barrier interface. The numerical model includes the partial ionization of acceptors and donors. This is particularly important for well-doped structures where a limited fraction of donors and acceptors can be ionized. The electron concentration is given by Fermi–Dirac statistics (a similar equation holds for holes)

$$n(z) = N \sum_i |\psi_i|^2 \ln\left[1 + \exp((E_f - E_i)/k_B T)\right] \quad (6.3)$$

where E_f is the Fermi energy for the system, E_i is the subband energy, $N = m^* k_B T / \pi \hbar^2$ and m^* is the electron effective mass in the well. At equilibrium, the Fermi level E_f is chosen to be constant and zero throughout the structure.

The intersubband absorption coefficient in the quantum well is calculated from the self-consistent solution [19]:

$$\alpha(\hbar\omega) = \left(\frac{e^2\omega}{n_r c\epsilon_0}\right)\frac{|\mu_{ji}|^2 (\Gamma/2)}{(E_j - E_i - \hbar\omega)^2 + (\Gamma/2)^2}(N_i - N_j) \quad (6.4)$$

$$\mu_{ji} = \langle \psi_j | z | \psi_i \rangle = \int \psi_j^* z \psi_i dz \quad (6.5)$$

where $\hbar\omega$ is the photon energy, n_r is the refractive index, c the speed of light, μ_{ji} is the intersubband dipole moment, ψ_j and ψ_i are the subband wavefunctions, Γ is the optical field linewidth (full width at half maximum, FWHM), and $N_{i(j)}$ is the number of electrons per unit volume in the $i(j)^{th}$ subband.

The linewidth Γ is an important parameter in many applications, in particular, a narrower Γ gives rise to a larger peak absorption in intersubband detectors or to a larger gain in intersubband lasers. The linewidth depends upon the elastic (impurity scattering, scattering by interface roughness) and inelastic (acoustic and optical phonon scattering) collisions and spontaneous photon emission and is given [21] by

$$\Gamma = 2\hbar\left(\frac{1}{2\tau_1} + \frac{1}{\tau_2}\right)$$

where τ_1 and τ_2 are the lifetimes corresponding to inelastic collisions and spontaneous photon emission and elastic collisions, respectively. The linewidth Γ for the MQW structures considered in this work will be discussed in Section 6.4.1.

In heterostructures made of materials with large conduction band offsets, as in the case of $Al_xGa_{1-x}N/GaN$ and $In_xGa_{1-x}N/GaN$, deep quantum wells can be fabricated. Electrons confined in these wells, as a result of an external excitation, can move in regions of the subband electronic structure that are significantly nonparabolic. The presence of a nonparabolic energy dispersion results in a shift of the energy eigenvalues of the confined states [22]. The higher subbands are shifted downwards and the ground state shifts upwards. In III-nitride materials this effect is compounded by the presence of polarization fields that further increase the well depth. The nonparabolicity effect is included in the Schrödinger equation through the energy dependent effective mass $m^*(z, E)$:

$$m^*(z, E) = m^*(z)\left[1 + \alpha(E - V(z))\right] \quad (6.6)$$

where E is the carrier energy, $qV(z)$ is the potential energy, and the nonparabolicity parameter α is given by [23]:

$$\alpha = \frac{1}{E_g(z)}\left[1 - \frac{m^*(0)}{m_0}\right]^2$$

where $E_g(z)$ is the energy gap. As a consequence of the inclusion of nonparabolicity in Eq. (6.1), the eigenfunctions have to be normalized accordingly [24]:

$$\left\langle \psi_i \left| \left[1 + p_z \frac{1/(2\alpha m^*)}{(E_i - V + \frac{1}{\alpha})^2} p_z \right] \right| \psi_i \right\rangle = 1$$

where $p_z = -i\hbar(\partial/\partial z)$ is the z-component of the momentum operator, and E_i is the energy eigenvalue corresponding to the ith subband with eigenfunction ψ_i. As a result of the inclusion of nonparabolic energy dispersion, the absorption coefficient needs to be computed taking this effect into account. It is given by:

$$\alpha(\hbar\omega) = \left(\frac{e^2\omega}{n_r c \epsilon_0}\right) |M_{ji}|^2 \frac{2}{(2\pi)^2} \times$$

$$\int_0^\infty d^2k \left[\frac{1}{1 + e^{(\frac{E_i - E_f}{k_B T})}} - \frac{1}{1 + e^{(\frac{E_j - E_f}{k_B T})}}\right] \frac{(\Gamma/2)}{(E_j - E_i - \hbar\omega)^2 + (\Gamma/2)^2}$$

and M_{ji} is the new dipole matrix element given by [17]:

$$M_{ji} = \frac{\hbar}{2(E_j - E_i)} \left\langle \psi_i \left| p_z \frac{1}{m^*(E_i, z)} + \frac{1}{m^*(E_j, z)} p_z \right| \psi_j \right\rangle$$

Due to the presence of doping and polarization fields at the AlGaN-GaN interfaces, which induce high sheet charge densities and hence high 2D concentrations in the wells, one cannot neglect the many-body effects on the electronic structure. From the local density approximation of density functional theory (DFT-LDA) [25], the exchange-correlation potential can be written as a functional of the electron density $V_{xc}[n(z)]$ and added as a contribution to the total potential term in the single-particle Schrödinger equation. The LDA correction is included using the Hedin–Lundqvist formulation, normally employed to study electrons in MQWs [20]:

$$V_{xc} = -\left(\frac{9\pi}{4}\right)^{1/3} \frac{2}{\pi r_s}\left[1 + \frac{0.6213}{21}r_s \ln\left(1 + \frac{21}{r_s}\right)\right]\frac{e^2}{8\pi\varepsilon\varepsilon_0 a^*} \tag{6.7}$$

where r_s is the dimensionless parameter characterizing the electron gas, given by $r_s = [(4\pi/3)a^{*3}n(z)]^{-1/3}$ corresponding to the mean electron distance normalized to the effective Bohr radius $a^* = (\varepsilon/m^*)a_B$. Furthermore, other many-body effects, such as depolarization and excitonic shift, also play a role in the absorption profile. Due to these phenomena the absorption peak does not occur at $E_{ji} = E_j - E_i$, but at a different value $\tilde{E}_{ji}$, given by

$$\tilde{E}_{21} = E_{21}\sqrt{(1+\gamma-\delta)}$$

where γ is the depolarization shift and δ is the excitonic shift. The depolarization effect [26–29] is due to resonant screening between the electron plasma in the quantum wells and the applied electromagnetic field. This leads to a shift in the frequency $\tilde{\omega}_{21}$, at which absorption occurs, different from the two-subband frequency ω_{21}. The modified absorption frequency is given by $\tilde{\omega}_{21}^2 = \omega_{21}^2 + f_{12}\omega_p^2$ with $\omega_p^2 = \frac{n_s e^2}{\epsilon_0 \epsilon m^* L_{eff}}$, where ω_p is the plasma frequency, f_{12} is the transition oscillator strength, $n_s = \int n(z)$ is the total areal electron concentration in the well and $L_{eff} = (\hbar^2 f_{12}/2m^* S E_{21})$. S is given as

$$S = \int_0^\infty dz \left[\int_0^z dz' \psi_2(z')\psi_1(z')\right]^2 \tag{6.8}$$

S can be readily computed once the self-consistent Poisson–Schrödinger problem has been solved. The parameter γ is given as

$$\gamma = \frac{2e^2 n_s S}{\varepsilon\varepsilon_0 E_{21}} \tag{6.9}$$

The excitonic effect is due to the Coulombic interaction between the excited electron and the quasi-hole left behind in the ground state which reduces the absorption frequency. The excitonic shift parameter δ is computed as [28, 30–32]:

$$\delta = -\frac{2n_s}{E_{21}}\int_{-\infty}^{\infty} dz\psi_2(z)^2\psi_1(z)^2\frac{\partial V_{xc}[n(z)]}{\partial n(z)} \tag{6.10}$$

6.2.1
Spontaneous and Piezoelectric Polarization

The stable crystalline phase of group III nitride semiconductors is wurtzite. Due to the comparative lack of symmetry of the hexagonal crystal structure, III-nitride materials exhibit a nonzero macroscopic polarization even at equilibrium. In other words, there is a spontaneous polarization [15] in the material layer. Because of strain due to the lattice mismatch between layers with different composition, there also exists a piezoelectric component. The total polarization $\vec{P}$ present in a GaN or AlGaN layer is the sum of the spontaneous polarization $\vec{P}_{SP}$ for the lattice at equilibrium, and the strain-induced or piezoelectric polarization $\vec{P}_{PE}$. Layers grown by different methods exhibit either N-face (or [000$\bar{1}$]) or Ga-face (or [0001]) polarity and the spontaneous polarization vector $\vec{P}_{SP}$ is directed from the nitrogen atomic layer to the gallium atomic layer. The spontaneous polarization along the c-axis for wurtzite crystal is given by $\vec{P}_{SP} = P_{SP}\vec{z}$. The piezoelectric component of the polarization can be calculated from:

$$P_{PE} = 2\frac{a - a_0}{a_0}\left(e_{31} - e_{33}\frac{c_{13}}{c_{33}}\right) \tag{6.11}$$

where a_0 is the lattice constant of the material layer at equilibrium and a is the lattice constant of the buffer layer or substrate. Furthermore, e_{31}, e_{33} are the elements of the piezoelectric strain matrix and c_{13} and c_{33} are the elastic constants of the material under consideration.

At the interface between two semiconductor layers with different compositions, a sheet charge is present as a result of the change in the total polarization in the layers. According to the electrostatic boundary conditions in MQWs and superlattices grown along the c-axis, the interface sheet charge is computed as:

$$\sigma = P^i - P^j = [P^i_{SP} + P^i_{PE}] - [P^j_{SP} + P^j_{PE}] \tag{6.12}$$

where i denotes the layer above layer j. This 2D sheet charge density is included in the Poisson equation as a source term.

An important practical consideration in the design of III-nitride-based MQWs or superlattices is that the buffer material composition should be chosen to minimize the overall stress in the structure. Therefore, the buffer layer composition has to be selected as a function of the characteristics of the barriers and wells in the system. In other words, for any MQW design there is a particular buffer layer composition that minimizes the global strain. The lattice constant of the buffer material, and consequently its composition, are determined once the well and barrier widths and their compositions are known [23]:

$$a = \frac{\sum_j \frac{C_j w_j}{a_j}}{\sum_j \frac{C_j w_j}{a_j^2}}$$

where w_j and a_j are the widths and lattice constants of each layer and $C_j = c_{11} + c_{12} - 2c_{13}^2/c_{33}$ is the elastic parameter for each layer. Once the lattice constant of the buffer is known, the molar fraction x can be determined using, for example, the Vegard law.

In this work we will primarily consider $Al_xGa_{1-x}N/GaN$ MQW structures grown on relaxed $Al_xGa_{1-x}N$ buffers, including AlN and GaN. The same formalism can be used to study $In_xGa_{1-x}N/GaN$-based MQWs. The material parameters needed for the numerical model have been obtained from references [33] and [34]. In particular, the conduction band edge discontinuity has been assumed to be $0.74\,\Delta E_g$ [17]. The dependence of the energy gap, lattice constant and effective mass on the molar fraction has been implemented using the Vegard law except when the relevant bowing parameter was available. In case of the $Al_xGa_{1-x}N$ energy gap we have used

$$E_g(x) = E_g^{GaN}(1-x) + E_g^{AlN} - bx(1-x)$$

with the bowing parameter b equal to 1.3 eV [17]. Table 6.1 presents all the relevant material parameters used in the numerical model.

6.3 Numerical Implementation

The efficient solution of the Schrödinger equation is becoming of growing interest for a large variety of semiconductor device applications, from MOSFETs [35] to advanced heterostructure devices [36]. Usually, a self-consistent solution is needed where the Poisson equation is also solved including the charge of quantized states. Among suitable analysis techniques for the self-consistent solution of the Schrödinger–Poisson problem (e.g., the transfer matrix technique [37] and the finite-difference method [38]), the finite-element method (FEM) is probably the most effective. Besides allowing for the use of nonuniform grids, the method can handle complex geometries, and can be readily extended to higher-order interpolation schemes. The physical model used for the description of bound states in a MQW consists of a nonlinear Poisson equation for the electrostatic potential ϕ coupled to an eigenvalue problem

Tab. 6.1 Material parameters used in the numerical model.

Parameter		GaN	AlN
Lattice Constant a	[Å]	3.189	3.122
Energy Gap Temperature Coefficient α	[meV K^{-1}]	0.909	1.799
Energy Gap Temperature Coefficient β	[K]	830.00	1462.00
Energy Gap (at 0K) E_g	[eV]	3.51	6.23
Energy Gap Bowing Parameter b	[eV]	1.3	1.3
Affinity χ	[eV]	3.5	2.76
Spin–Orbit Splitting E_{so}	[eV]	0.014	0.019
Relative Dielectric Constant ε_r		9.7	8.5
Spontaneous Polarization P_{SP}	[C m^{-2}]	−0.029	−0.081
Elastic Constant c_{11}	[GPA]	390.00	396.00
Elastic Constant c_{12}	[GPA]	145.00	137.00
Elastic Constant c_{13}	[GPA]	106.00	108.00
Elastic Constant c_{33}	[GPA]	398.00	373.00
Piezoelectric Constant e_{13}	[C m^{-2}]	−0.35	−0.5
Piezoelectric Constant e_{33}	[C m^{-2}]	1.27	1.79

for the Schrödinger equation in the effective-mass approximation

$$\frac{\partial}{\partial z} \epsilon(z) \frac{\partial}{\partial z} \phi(z) = -\rho(\phi) \tag{6.13a}$$

$$\left[-\frac{\hbar^2}{2} \frac{\partial}{\partial z} \frac{1}{m^*(z,E)} \frac{\partial}{\partial z} + V(z) \right] \psi(z) = E\psi(z) \tag{6.13b}$$

where ρ is the total charge density, including the quantum charge, and $m^*(z,E)$ is an energy-dependent effective mass accounting for nonparabolicity. Notice that Eq. (6.13b) is a nonlinear eigenvalue problem because the energy E appears also in the Hamiltonian through the effective mass m^*. We first consider the linear Schrödinger equation ($\alpha = 0$) with an impressed potential profile $V(z)$. The action integral can be written as [39]:

$$\mathcal{A} = \int dz\, \psi^*(z) \left[-\frac{\hbar^2}{2} \frac{\partial}{\partial z} \frac{1}{m^*} \frac{\partial}{\partial z} + V - E \right] \psi(z) \tag{6.14}$$

We start by discretizing the computational domain into line elements. In each element, we express the unknown wavefunction ψ in terms of Lagrange polynomials [39]

$$\psi(z) = \sum_i \psi_i N_i(z) \tag{6.15}$$

The simplest of the Lagrange interpolation schemes, beyond the piecewise constant interpolation, is linear interpolation within each element. In this case, we have two basis functions for each element, and the previous expansion reduces to

$$\psi(z) = \psi_1 \xi + \psi_2 (1 - \xi) \tag{6.16}$$

where ξ is a local coordinate in the element having a range $[0, 1]$, and ψ_1, ψ_2 are the values of ψ at the nodal points of the element. Substituting the expansion (6.15) in (6.14), the action within each element may be written, after some manipulation, as

$$\mathcal{A}_e = \sum_{i,j} \int_e dz\, \psi_i^* \left[-\frac{\hbar^2}{2} \frac{\partial N_i(z)}{\partial z} \frac{1}{m^*} \frac{\partial N_j(z)}{\partial z} + N_i(z)(V - E)N_j(z) \right] \psi_j \tag{6.17}$$

Summing the contributions from each element and applying the principle of stationary action, one obtains the generalized eigenmatrix equation

$$\sum_e \left(\frac{\hbar^2}{2m^*} [T]_e + V[U]_e \right) \{\psi\}_e = E \sum_e [U]_e \{\psi\}_e \tag{6.18}$$

with

$$[U]_e = \int_e \{N\}^T \{N\} dz = \frac{l_e}{6} \begin{bmatrix} 2 & 1 \\ 1 & 2 \end{bmatrix} \tag{6.19a}$$

$$[T]_e = \int_e \frac{\partial \{N\}^T}{\partial z} \frac{\partial \{N\}}{\partial z} dz = \frac{1}{l_e} \begin{bmatrix} 1 & -1 \\ -1 & 1 \end{bmatrix} \tag{6.19b}$$

where $\{\psi\}_e = \{\psi_1, \psi_2\}^T$, and l_e is the length of the element. In (6.18) we have assumed that the effective mass m^* and the potential energy V are piecewise constant over each element. The generalized eigenvalue problem (6.18) may be efficiently solved with the Implicitly Restarted Arnoldi Method [40] as implemented in ARPACK[2] [41]. The method allows for the efficient computation of a set of selected eigenstates taking full advantage of the sparsity of the problem.

For semiconductor heterostructures with relatively low barrier heights and low carrier densities, the electrons cluster around the subband minima. In this operation regime, the parabolic band model is indeed a good approximation. However, in conditions in which electrons are forced to higher energies, nonparabolic effects cannot be neglected, and a nonparabolic description of the

---

[2] ARPACK is available at http://www.caam.rice.edu/software/ARPACK/. ARPACK is a collection of FORTRAN subroutines designed to compute a few eigenvalues and, upon request, eigenvectors of large-scale sparse nonsymmetric matrices and pencils.

band structure should be included in the numerical model. It is often convenient to express nonparabolicity in terms of an energy-dependent effective mass

$$m^*(E) = m^*(0)[1 + \alpha(E - V)] \tag{6.20}$$

which leads to a nonlinear Schrödinger equation. Band nonparabolicity is usually handled with a modified shooting method in which the energy-dependent effective mass is adjusted for each energy level E [42]. The nonlinear Schrödinger equation can also be treated by setting up an iterative procedure to compute the energy levels; upon finding the eigenvalues, the corresponding eigenvectors are computed by solving a fictitious linear eigenproblem [18]. Nonparabolicity is included in our FEM model by solving directly the nonlinear eigenvalue problem. The action integral is

$$\mathcal{A}_e = \sum_{i,j} \int_e dz \times \tag{6.21}$$

$$\psi_i^* \left[-\frac{\hbar^2}{2} \frac{\partial N_i(z)}{\partial z} \frac{1}{m^*(0)} \frac{\partial N_j(z)}{\partial z} + (1 + \alpha(E - V))N_i(z)(V - E)N_j(z) \right] \psi_j$$

The FEM procedure leads to a quadratic eigenvalue problem

$$\sum_e \left(\frac{\hbar^2}{2m^*}[T]_e + V(1 - \alpha V)[U]_e \right) \{\psi\}_e$$
$$= E \sum_e (1 - 2\alpha V)[U]_e\{\psi\}_e + E^2 \sum_e [U]_e\{\psi\}_e \tag{6.22}$$

which is written for convenience as

$$[A]\{x\} = \lambda[B]\{x\} + \lambda^2[C]\{x\} \tag{6.23}$$

The $n \times n$ quadratic eigenvalue problem (6.23) can be linearized to a $2n \times 2n$ generalized eigenvalue problem [40]

$$\begin{bmatrix} 0 & I \\ A & -B \end{bmatrix} \begin{Bmatrix} x \\ \lambda x \end{Bmatrix} = \lambda \begin{bmatrix} I & 0 \\ 0 & C \end{bmatrix} \begin{Bmatrix} x \\ \lambda x \end{Bmatrix} \tag{6.24}$$

where I is a $n \times n$ identity matrix. The generalized eigenproblem (6.24) can be then solved by the Arnoldi method [3].

3) Notice that, since the linearization technique introduces additional eigenvalues, one has to check which of the computed eigenpairs satisfies the original polynomial equation by computing the residual error. Recently, numerical methods that directly handle the polynomial eigenvalue problem, e.g., by the Newton method, have been introduced, see [40] and references therein.

The Schrödinger equation has to be completed with a suitable set of boundary conditions. For bound states, the wavefunction is zero at the boundaries of the computational domain, far into the barrier regions. For free and quasibound states we have to account for the form of incoming and outgoing travelling waves [39]. The quasibound states are normalized to the length of the structure, the free propagating plane waves are launched at one end of the MQW and assigned a unitary amplitude.

The discretization of the Poisson equation (6.13a) is performed in a similar manner. Expressing the potential in terms of Lagrange polynomials $\phi(z) = \sum_i \phi_i N_i(z)$, the finite-element procedure leads to the discretized equation

$$\sum_e \epsilon [T]_e \{\phi\}_e = \sum_e \{\rho\}_e \tag{6.25}$$

where $\{\rho\}_e$ is the FEM representation of the total charge density within each element, including the quantum charge, polarization charges, and ionized donor and acceptor concentrations which depend on the electrostatic potential ϕ if incomplete ionization is assumed. The nonlinear equation (6.25), with appropriate boundary conditions, see [43] for a review, can be solved through the Newton method or its variants (e.g., the Newton–Richardson approach [43]).

6.3.1
Achieving Self-consistency: The Under-Relaxation Method

The Schrödinger and Poisson equations are coupled through the quantum charge n_q, which depends on the eigenfunctions and the energies of the bound states in the system. The self-consistent problem is usually approached by an iterative technique, where the Schrödinger and Poisson equations are solved in succession. Given the nonlinear nature of the calculation we separate the iterative process into two nested loops. In the outer iteration loop, the quantum electronic charge density is calculated by solving the Schrödinger equation with the latest value of the potential energy. In the inner loop, a nonlinear Poisson equation is solved for the electrostatic potential, with the ionized donor charge and the mobile charge density as source terms. The process is repeated until suitable convergence criteria are satisfied. It is well-known that this simple iterative procedure is intrinsically oscillating, and becomes unstable when the quantized charge substantially modifies the potential profile [44]. In order to achieve convergence it is necessary to under-relax the charge density using an adaptively determined relaxation parameter $\omega^{(k)}$. The standard approach is summarized as follows:

1. Solve the Poisson equation using the electron density $n_q^{(k)}$ from the previous iteration:

$$\frac{d}{dz}\epsilon\frac{d}{dz}\phi^{(k+1)} = q\left[n_q^{(k)} - N_D^+(\phi^{(k+1)}) - \rho_{pol}\right] \tag{6.26}$$

2. Solve the Schrödinger equation with the updated potential energy

$$-\frac{\hbar^2}{2}\frac{\partial}{\partial z}\left(\frac{1}{m^*}\frac{\partial}{\partial z}\psi_i^{(k+1)}\right) + [V^{(k+1)} + V_{xc}(n_q^{(k)})]\psi_i^{(k+1)} = E_i^{(k+1)}\psi_i^{(k+1)} \tag{6.27}$$

3. From the new eigenfunctions compute an intermediate new quantum electron density $\hat{n}_q^{(k+1)}$

$$\hat{n}_q^{(k+1)} = \frac{m^* k_B T}{\pi \hbar^2} \sum_i |\psi_i^{(k+1)}|^2 \ln\left[1 + \exp\left(\frac{E_f - E_i^{(k+1)}}{k_B T}\right)\right] \tag{6.28}$$

4. Under-relax in n_q to achieve convergence:

$$n_q^{(k+1)} = \omega^{(k+1)} \hat{n}_q^{(k+1)} + (1 - \omega^{(k+1)}) n_q^{(k)} \tag{6.29}$$

5. Repeat the iteration until n_q becomes stationary:

$$||n_q^{(k+1)} - n_q^{(k)}|| \leq \epsilon \tag{6.30}$$

The problem with this method is the inherent instability of the outer iteration which is controlled by the under-relaxation procedure only. The necessary relaxation parameter $\omega^{(k)}$ is not known in advance and needs to be readjusted during the iterative process. If $\omega^{(k)}$ is too large, the total quantized charge oscillates from one iteration step to the other without converging. If $\omega^{(k)}$ is too small, too many iteration steps are necessary for convergence. The physical reason for the numerical instability is the high sensitivity of the energy levels on the confining potential, and of the quantum charge n_q on the energy levels.

6.3.2
Predictor–Corrector Approach

Recently, Trellakis and co-workers have proposed a stable numerical technique based on a predictor–corrector approach for obtaining self-consistent solutions to the coupled Schrödinger and Poisson equations [44]. If we knew the exact dependence of the quantum electron density n_q on the electrostatic potential ϕ, we could solve a nonlinear Poisson equation without the need for a Schrödinger equation coupled by an outer iteration. The basic idea proposed in [44] is to find a suitable approximate expression $\tilde{n}_q$ for the quantum electron density n_q as a function of ϕ and use such an expression in a nonlinear Poisson equation of the type

$$\frac{d}{dz}\epsilon\frac{d}{dz}\phi^{(k+1)} = q\left[\tilde{n}_q(\phi^{(k+1)}) - N_D^+(\phi^{(k+1)}) - \rho_{pol}\right] \tag{6.31}$$

where the potential independent quantum electron density n_q is replaced by the potential dependent density (predictor) $\tilde{n}_q$. Using quantum mechanical perturbation theory, an approximate expression describing the dependence of the quantum electron density on the electrostatic potential can be derived:

$$\tilde{n}_q(\phi^{(k+1)}) = \frac{m^* k_B T}{\pi \hbar^2} \sum_i |\psi_i^{(k)}|^2 \times \qquad (6.32)$$

$$\ln\left[1 + \exp\left(\frac{E_f - E_i^{(k)} + q\left(\phi^{(k+1)} - \phi^{(k)}\right)}{k_B T}\right)\right]$$

where superscripts (k) denote quantities obtained in the previous outer iteration step. Comparing Eq. (6.32) to the original quantum electron density (6.28) we find that the only change made is that the energy levels $E_i^{(k)}$ are augmented by the position-dependent change in the electrostatic potential $q\delta\phi = q(\phi^{(k+1)} - \phi^{(k)})$, which disappears when convergence is reached. The Poisson equation can be easily solved by the Newton method, and the predicted results for n_q and ϕ are then corrected in the outer iteration loop by the exact solution of the Schrödinger equation

$$-\frac{\hbar^2}{2}\frac{\partial}{\partial z}\left(\frac{1}{m^*}\frac{\partial}{\partial z}\psi_i^{(k+1)}\right) + [V^{(k+1)} + V_{xc}(\tilde{n}_q^{(k+1)})]\psi_i^{(k+1)} = E_i^{(k+1)}\psi_i^{(k+1)} \qquad (6.33)$$

Unlike the standard algorithm, there is no need for under-relaxation and the outer iteration reduces to a simple alternation between the Poisson and Schrödinger equations until the quantum electron density becomes stationary.

In conclusion, in the predictor–corrector approach, numerical stability is obtained by moving the nonlinearity of the problem in the nonlinear Poisson equation. However, since the standard Poisson equation already includes nonlinear terms (e.g., the ionized impurity concentration), the addition of some extra terms increases the computational cost only minimally.

6.4
Absorption Energy in AlGaN-GaN MQWs

The numerical model presented in the previous section has been applied to the study of AlGaN-GaN MQW structures intended for a variety of applications where intersubband transitions are exploited in the device operation. Specifically we will focus our attention on two important aspects. First, we intend to study perfectly periodic AlGaN-GaN MQW structures; in other words, we will neglect the fact that they are in general grown on a given substrate and terminated with a cap layer. This approach has been followed by many researchers in designing MQWs and analyzing experimental data. Although this preliminary analysis is useful in understanding the general dependence of the intersubband transition energies on the design parameter space (i.e., layer thicknesses, composition and doping) and to evaluate second-order-effect corrections, it may introduce errors difficult to quantify since realistic structures are in general nonperiodic. The second aspect we set out to address is to understand the changes in intersubband transition energies when the nonperiodicity of the structure is considered. Furthermore, we want to investigate if and when the assumption of a periodic structure can be used. We will conclude by computing the optical response of MQW structures published in the literature and comparing the results with the experimental data.

It is important to point out that the applicability of the numerical models discussed here is not restricted to the case of AlGaN-GaN MQWs. In fact, GaN/InGaN MQWs can also be studied, as well as any other combination of binary, ternary and quaternary alloy MQWs. Since we are mainly interested in electron intersubband transitions, we will consider holes only by introducing a single parabolic valence band. It is clear that if interband transitions are of interest, the model presented in Section 6.2 has to include a more suitable description of the valence-band structure.

6.4.1
Numerical Analysis of Periodic AlGaN-GaN MQWs

In this section we study the intersubband transition energies of different MQW structures composed of GaN wells and $Al_xGa_{1-x}N$ barriers. The peak absorption energy has been calculated for various well and barrier width combinations and for several barrier molar fraction values. Furthermore, to evaluate the effect of the polarization charges on the intersubband energy, both cases, with and without polarization, have been considered. For each structure (characterized by well width, barrier width and molar fraction) the buffer layer composition has been chosen so that the overall stress in the device is minimized. Both well and barrier doping have also been investigated.

We note first that, for a rectangular well (in the absence of polarization), as the well width is decreased, the peak absorption energy between the ground and first excited state increases. As the well width is reduced, the subbands move toward higher energies, with higher subbands shifting more. The peak absorption energy increases as the well width is reduced until the second subband is pushed outside the confined region into the continuum. If the well width is further reduced, the ground state is shifted up and the peak absorption is observed at energies between the ground state and the continuum. As a result, the peak absorption energy will decrease. If we vary the barrier width instead, the peak absorbtion energy is almost unchanged for sufficiently thick barriers.

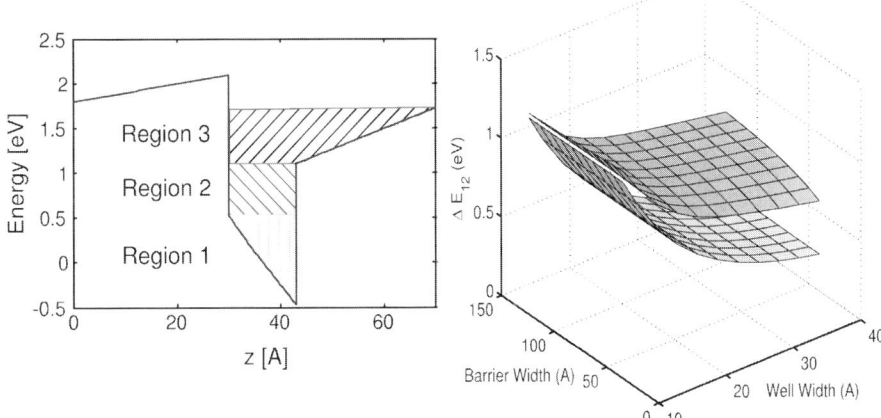

Fig. 6.1 Conduction band energy profile for a periodic AlGaN-GaN MQW structure.

Fig. 6.2 Calculated intersubband energy for an AlN-GaN MQW for several well barrier thicknesses. The lower surface is the calculated energy without polarization effects, the upper surface includes the effects of polarization.

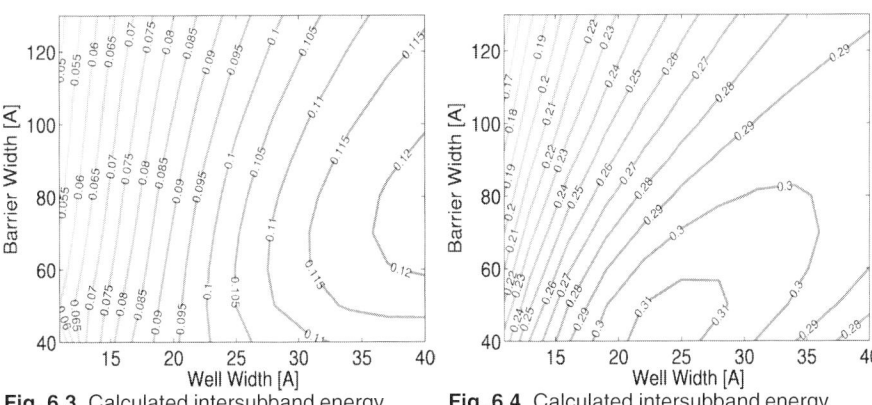

Fig. 6.3 Calculated intersubband energy (eV) for $Al_{0.1}Ga_{0.9}N$-GaN MQWs.

Fig. 6.4 Calculated intersubband energy (eV) for $Al_{0.3}Ga_{0.7}N$-GaN MQWs.

In the case of III-nitride-based MQWs, as a result of the spontaneous and induced polarization, a strong electric field is present in the wells and in the barriers. An example of the calculated conduction band profile in shown in Fig. 6.1. As can be observed, regions 1 and 3 resemble triangular wells and region 2 can still be considered a rectangular well. The overall different shape of the conduction band profile changes the dependence of the intersubband energy on the well width.

As in the case of rectangular wells, the peak absorption energy increases if the well width is reduced as long as the ground state and the first excited

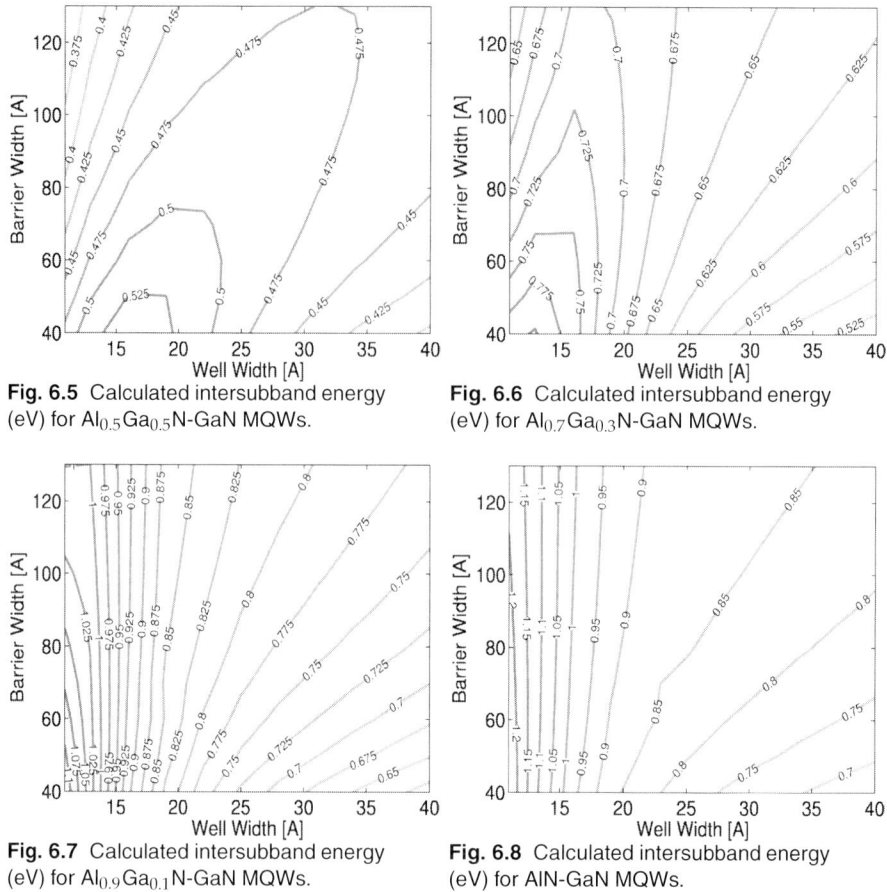

Fig. 6.5 Calculated intersubband energy (eV) for $Al_{0.5}Ga_{0.5}N$-GaN MQWs.

Fig. 6.6 Calculated intersubband energy (eV) for $Al_{0.7}Ga_{0.3}N$-GaN MQWs.

Fig. 6.7 Calculated intersubband energy (eV) for $Al_{0.9}Ga_{0.1}N$-GaN MQWs.

Fig. 6.8 Calculated intersubband energy (eV) for AlN-GaN MQWs.

state are in regions 1 and 2 of Fig. 6.1. If the first excited state crosses into region 3, then its energy eigenvalue increases progressively more slowly because of the widening well. The net result is that the peak absorption energy will eventually decrease.

It is also important to notice that, as opposed to the case of rectangular wells, in this situation the barrier width influences the peak absorption energy even for uncoupled wells. The piezoelectric fields present in the well and in the barrier depend on both the well and barrier thicknesses, and consequently so does the conduction band profile. In particular, the field strength in the well is directly proportional to the width of the barrier [23] (this is only true for periodic structures with periodic boundary conditions). As a result of this interdependence between wells and barriers, the peak absorption energy further depends in which region the subband energies are in Fig. 6.1.

If, for a given well barrier configuration, both subbands are in the rectangular region 2, then the barrier width does not significantly affect the absorption energy. If the ground state is in region 2 and the second higher state is in region 3, then the barrier width has little or no effect on the ground state. On the other hand, the higher subband in region 3 will shift down as the barrier width increases as a result of the reduction of the field in the barrier itself. If the ground state is in region 1, then a thicker barrier increases the field in the well and therefore its depth. Consequently, the ground state energy is shifted down and the peak absorption energy increases.

Figure 6.2 presents the calculated intersubband absorption energy for an AlN-GaN MQW with different well barrier width combinations and well doping of 2×10^{19} cm^{-3}. The lower surface is the calculated energy without polarization, the upper surface includes the effect of polarization. The relatively large difference between these two surfaces is a consequence of the greater effective well depth resulting from the presence of the polarization fields.

Figures 6.3–6.8 present the calculated intersubband energies for different well barrier width combinations and $Al_{0.1}Ga_{0.9}N$, $Al_{0.3}Ga_{0.7}N$, $Al_{0.5}Ga_{0.5}N$, $Al_{0.7}Ga_{0.3}N$, $Al_{0.9}Ga_{0.1}N$ and AlN barriers, respectively. As can be seen, for barriers with low AlGaN molar fraction, the intersubband energy first increases when the well width is reduced and subsequently decreases, as previously discussed. For barriers with higher AlGaN composition the intersubband energy increases monotonically in the range considered. The relation between the absorption energy and barrier width just described is also clearly observed in these figures.

The intersubband energies presented in Figs 6.4–6.7 have been calculated including both the nonparabolicity and the many-body effects described in Section 6.2. To investigate the relative importance of these second-order effects on the intersubband energy, we have computed the energy correction they introduce. The energy difference between the situation in which the nonparabolic effect is accounted for and the ideal case is presented in Figs 6.9–6.14 for various MQWs. It can be noticed that the nonparabolicity correction becomes progressively larger as the barrier height increases. In the case of $Al_{0.9}Ga_{0.1}N$-GaN MQWs the difference can be as high as 10% for narrow wells. In general, nonparabolicity mainly affects higher subbands, which are farther away from the band edge where the effective mass is higher, shifting them to a lower energy. This results in a lower absorption energy.

The relative importance of the many-body interactions, namely the exchange correlation energy and depolarization shift, is presented in Figs 6.15–6.20 for a well doping of 2×10^{19} cm^{-3}. As the well width is reduced, the subband energies move further away from the Fermi energy, consequently the electron population decreases and with it the many-body interaction strength. This is clearly noticeable for all the MQW configurations. Furthermore, an increase in many-body corrections is observed when the barrier width is increased for a given well width. This occurs because, as the barrier width in-

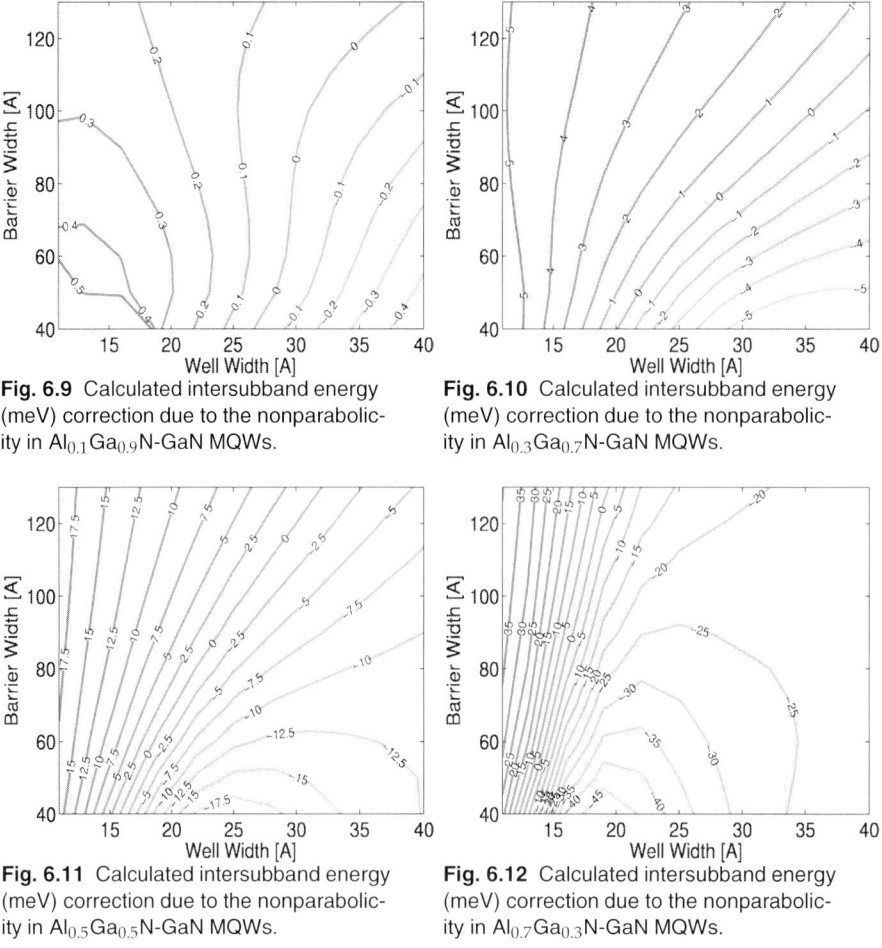

Fig. 6.9 Calculated intersubband energy (meV) correction due to the nonparabolicity in $Al_{0.1}Ga_{0.9}N$-GaN MQWs.

Fig. 6.10 Calculated intersubband energy (meV) correction due to the nonparabolicity in $Al_{0.3}Ga_{0.7}N$-GaN MQWs.

Fig. 6.11 Calculated intersubband energy (meV) correction due to the nonparabolicity in $Al_{0.5}Ga_{0.5}N$-GaN MQWs.

Fig. 6.12 Calculated intersubband energy (meV) correction due to the nonparabolicity in $Al_{0.7}Ga_{0.3}N$-GaN MQWs.

creases, the internal electric field in the well increases as well. As a result, the ground state moves closer to the Fermi energy and the electron population increases, resulting in a stronger many-body interaction. Notice that the magnitude of the many-body correction depends on the carrier density in the wells.

Figures 6.3–6.8 can be used to design several MQW structures of practical interest. For example, if one is interested in using these MQWs for QWIP application at optical communication wavelengths, MQWs with intersubband transition at $\lambda \sim 1.55$ μm and $\lambda \sim 1.3$ μm should be selected. Some of the possible choices (for a well doping of 2×10^{19} cm^{-3}) are reported in Tables 6.2–6.3.

As pointed out in Section 6.2 the linewidth Γ is an important parameter used to determine the absorption spectra. In general, both homogeneous and inhomogeneous broadening mechanisms are present. In MQWs, intrinsic pro-

6.4 Absorption Energy in AlGaN-GaN MQWs

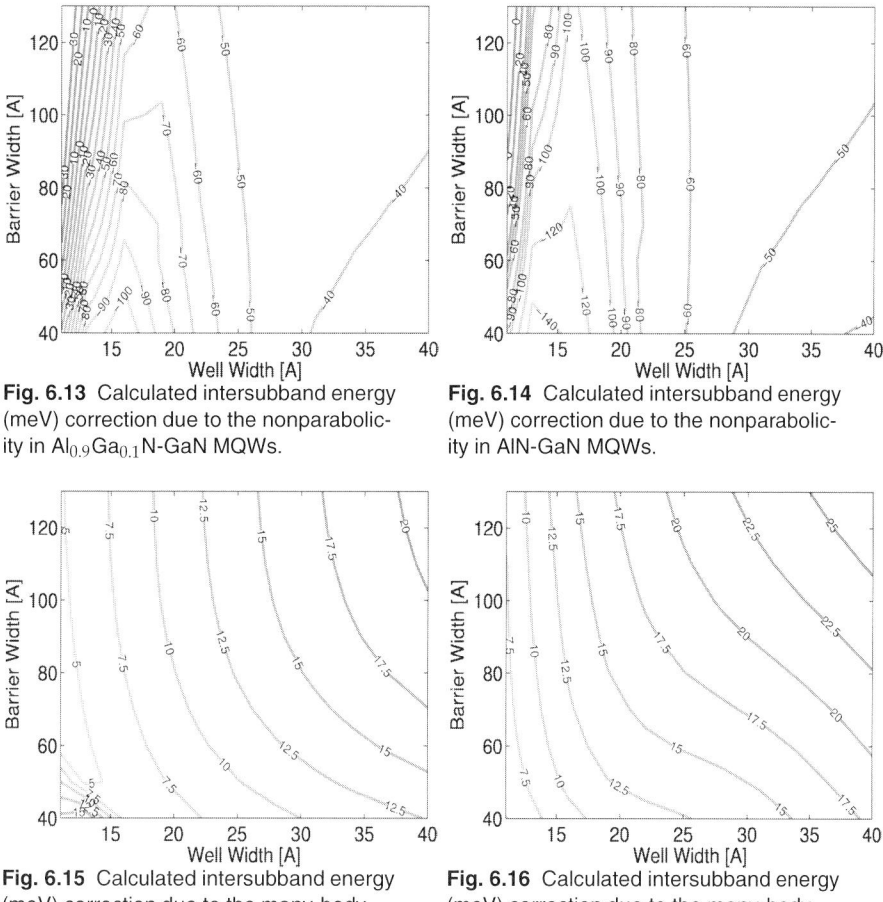

Fig. 6.13 Calculated intersubband energy (meV) correction due to the nonparabolicity in $Al_{0.9}Ga_{0.1}N$-GaN MQWs.

Fig. 6.14 Calculated intersubband energy (meV) correction due to the nonparabolicity in AlN-GaN MQWs.

Fig. 6.15 Calculated intersubband energy (meV) correction due to the many-body interaction in $Al_{0.1}Ga_{0.9}N$-GaN MQWs.

Fig. 6.16 Calculated intersubband energy (meV) correction due to the many-body interaction in $Al_{0.3}Ga_{0.7}N$-GaN MQWs.

cesses (namely phonon, impurity and surface roughness scattering) are responsible for homogeneous broadening [21]. Well-to-well thickness fluctua-

Tab. 6.2 MQW structures for $\lambda \sim 1.55$ μm.

$Al_xGa_{1-x}N$ Composition x	Barrier width [Å]	Well width [Å]
0.7	40	14
0.8	130	19
0.9	64	22
1.0	60	28

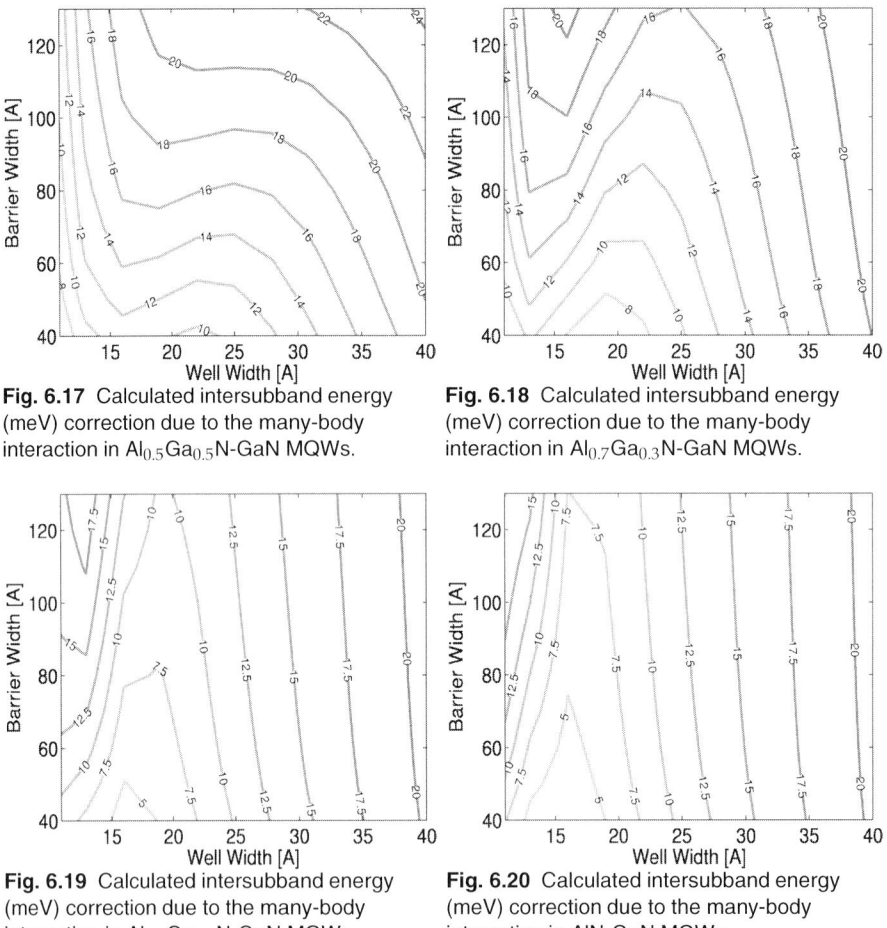

Fig. 6.17 Calculated intersubband energy (meV) correction due to the many-body interaction in $Al_{0.5}Ga_{0.5}$N-GaN MQWs.

Fig. 6.18 Calculated intersubband energy (meV) correction due to the many-body interaction in $Al_{0.7}Ga_{0.3}$N-GaN MQWs.

Fig. 6.19 Calculated intersubband energy (meV) correction due to the many-body interaction in $Al_{0.9}Ga_{0.1}$N-GaN MQWs.

Fig. 6.20 Calculated intersubband energy (meV) correction due to the many-body interaction in AlN-GaN MQWs.

tions are the most important source of inhomogeneous broadening and are responsible for most of the measured linewidth in typical AlGaN-GaN MQWs.

Tab. 6.3 MQW structures for $\lambda \sim 1.3$ μm.

Al_xGa_{1-x}N Composition x	Barrier width [Å]	Well width [Å]
0.8	50	11
0.9	40	15
1.0	30	16
1.0	130	19

To estimate the inhomogeneous contribution to Γ we have computed the maximum intersubband energy change for a barrier and well thickness fluctuation of one monolayer. The parameter Γ was computed as:

$$\Gamma = \max\left(\Delta E_{12}\left(Lb \pm \frac{ML}{2}, Lw \pm \frac{ML}{2}\right)\right) \\ - \min\left(\Delta E_{12}\left(Lb \pm \frac{ML}{2}, Lw \pm \frac{ML}{2}\right)\right) \quad (6.34)$$

where Lb is the barrier width, Lw is the well width and ML is one monolayer thickness. When applied to the MQW structures reported in Table 6.2, the linewidth value obtained was 64 meV. This value is similar to the smallest experimental values reported in the literature for similar structures [3,6].

For conventional GaAs/AlGaAs quantum wells, where no polarization charges are present, as the donor concentration increases, the absorption energy increases as well. This is due to the larger population of carriers in the ground state and the resulting lowering of the ground state energy caused by the exchange correlation effect. In the case of AlGaN-GaN MQWs, due to the presence of the polarization, the absorption energy also changes as a result of the polarization field screening caused by the free electrons. In fact, the carrier screening results in a lower effective electric field in the wells. This in turn increases the well effective width and decreases its depth, causing the absorption energy to decrease. For example, for the structures presented in Table 6.2, as the donor doping concentration in the wells increased from $10^{19}\,\text{cm}^{-3}$ to $2 \times 10^{20}\,\text{cm}^{-3}$ the absorption edge decreased from 0.8011 eV to 0.7813 eV. A significant change is also present when the barriers are doped instead of the wells. For example, for the structures of Table 6.2, if the barrier doping is $2 \times 10^{19}\,\text{cm}^{-3}$ the peak absorption energy is found to be 0.7648 eV as opposed to 0.80 eV with well doping. This shift is mainly due to the excitonic effect. Both the ionized impurity in the barriers and the enhanced electron population in the wells contribute to changing the peak absorption energy. As expected the barrier doping is more efficient in providing electrons in the wells since it leads to donor levels well above the MQW ground states. In fact, for the structures in Table 6.2, the sheet charge density in the wells was $2.659 \times 10^{13}\,\text{cm}^{-2}$ with barrier doping as opposed to $1.879 \times 10^{12}\,\text{cm}^{-2}$ with well doping, for the same donor concentration of $2 \times 10^{19}\,\text{cm}^{-3}$. It is clear that for efficient intersubband detectors the barrier doping option would be preferred.

When designing MQWs to be used as intersubband detectors, one needs to engineer the structure not only for the correct absorption energy, but also for the efficient extraction of the photo-excited electrons. The latter requirement can be achieved by using a bound-to-continuum transition where only one bound state is present in the wells, or a bound-to-quasibound transition,

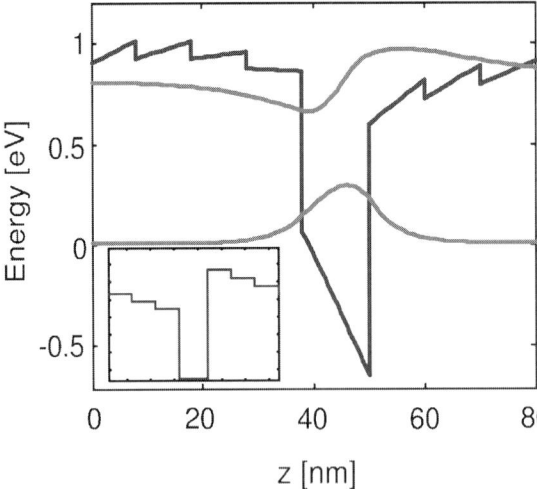

Fig. 6.21 Quantum well with graded barriers. The inset shows the flat band profile in the absence of polarization.

where the higher subband is very close the well edge. Vice versa, an excited state significantly below the well edge will result in a poor electron extraction to the continuum. Because of the presence of the polarization field in III-nitride-based MQWs, it is difficult to achieve this goal using barriers with uniform composition. In general, MQWs can be designed for intersubband absorption at $\lambda \sim 1.55$ μm and $\lambda \sim 1.3$ μm (see, for example, Tables 6.2 and 6.3) but the excited states are significantly below the well edge. As a result, although intersubband photocurrent generation in AlN-GaN MQWs has been experimentally observed [4, 5, 10] the electron extraction to the continuum is, in general, not optimal. As a possible solution, graded barriers can be used. The basic idea is to grade the AlGaN barrier composition in such a way as to compensate the piezoelectric field. An example of a structure designed to operate in the near-infrared region at $\lambda \sim 1.55$ μm (~ 800 meV) is shown in Fig. 6.21. In this case there are only two bound states in the well and the first excited state is close to the well edge. The inset in the bottom left corner shows the flat-band profile for the graded barrier. The structure consists of a 12 Å wide well doped with 2×10^{19} cm^{-3} donors and graded barriers with molar fraction of 0.72, 0.68, 0.64, 0.60, 0.56 and 0.52, each 10 Å wide.

6.4.2
Numerical Analysis of Non-periodic AlGaN-GaN MQWs and Comparison with Experimental Results

The numerical analysis and results presented in Section 6.4.1 were intended to study MQWs with a periodic structure. This not only has a mathematical implication on the numerical model, i.e., periodic boundary conditions, but also significantly changes the physics of the systems being investigated. In general,

MQW structures are grown on substrates and buffer layers that are selected according to given mechanical and electrical requirements. Furthermore, the whole structure can also be terminated with a cap layer with electrical characteristics suitable to form either Schottky or ohmic contacts. This type of arrangement leads to a structure that is not periodic. As opposed to MQWs based on conventional III-V semiconductors, the presence of the polarization fields in III-nitride MQWs further complicates the study of nonperiodic structures. To investigate the impact of the nonperiodicity, we consider a structure with a given substrate, cap layer and number of wells ranging from one to seventy. By using this prototype MQW structure we want to investigate the effects of different boundary conditions, and specifically understand if and when the structure can be studied under the assumption of a periodic system. We consider a MQW system at equilibrium, grown on sapphire and with of 1 µm AlN buffer layer and a 0.1 µm AlN cap layer. Both buffer and cap are n-type doped ($n \sim 10^{14}\,\text{cm}^{-3}$). The MQW portion of the sample consists of a 42 Å wide AlN barrier and 29 Å wide GaN wells. The GaN wells were n-type doped with $10^{20}\,\text{cm}^{-3}$ donors. The conduction band profile and the optical absorption spectrum have been computed for samples with 1, 5, 10, 20, 35, and 70 barrier-well pairs. This particular structure has been selected because a similar one with 35 pairs has been grown and characterized [6] and the experimental results are readily available for comparison.

In this study different boundary conditions have also been used. The simplest choice is to assume that the whole structure is itself a unit cell of a larger periodic system. Additionally, one can assume that on the top cap-layer and/or the substrate interface, either a Schottky or an ohmic contact is present.

As an example, the calculated band structure for a ten-well structure with buffer and cap (and global periodic boundary conditions) is shown in Fig. 6.22. It can be immediately noticed that, as a result of the presence of the polarization fields, the wells are not perfectly aligned to one another as they are in the periodic case. The rightmost well is closer to the Fermi energy (assumed to be constant and set to 0 eV) than the leftmost well. Consequently, the rightmost well is more populated than the wells closer to the buffer. The dependence of the conduction band profile with the number of wells results in a shift of the absorption energy. Furthermore, because the intersubband energy varies from well to well, a secondary absorption peak is in some cases present. Table 6.4 presents the calculated value of the main and secondary absorption peak energy. For the secondary peak, its magnitude is also indicated as a percentage of the primary. For a perfectly periodic structure (without buffer and cap), the peak absorption occurs at 0.656 eV. For a single well, the peak absorption occurs at 0.877 eV. In this case the conduction band profile is significantly different when compared with a perfectly periodic structure.

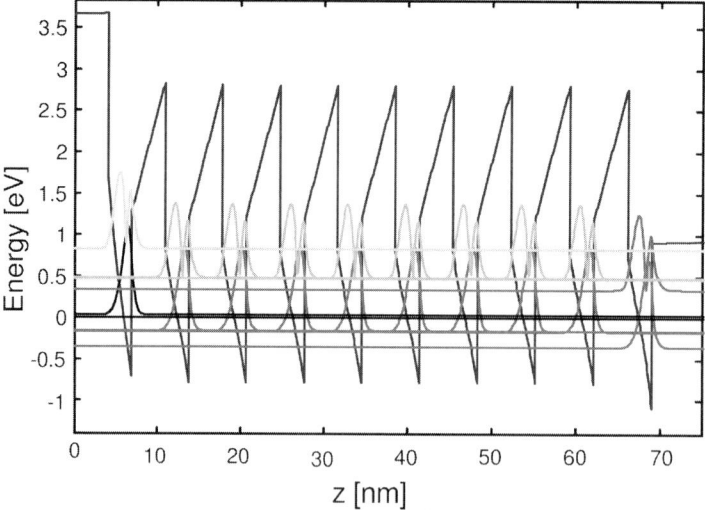

Fig. 6.22 Band structure for a ten-well system with 1 μm AlN buffer layer and 0.1 μm AlN cap layer.

This is caused by the polarization fields in the buffer and cap layers. As the number of wells increases, the center wells are progressively less and less affected by the polarization fields in the buffer and cap layers. Consequently the "inner" MQWs assume the perfectly periodic shape, as seen in Fig. 6.22. For a five-well device, the absorption profile has two peaks, one at 0.659 eV and the other at 0.771 eV, with 80% magnitude. This second peak is due to the rightmost well, which is heavily populated since it is closer to the Fermi energy. As the number of wells increases, the center wells dominate and this second peak diminishes, while the main peak comes closer and closer to that of the perfectly periodic structure. In particular, when the number of wells is in excess of 20, the periodic and nonperiodic results are similar. We also

Tab. 6.4 Variation of peak absorption energy with number of wells for a structure that includes buffer and cap layers.

Number of wells	Peak absorption (eV)	Secondary peak absorption (eV)	Experimental result (eV)
1	0.877		
5	0.659	0.771 (80%)	
10	0.658	0.771 (35%)	
20	0.656	0.771 (20%)	
35	0.656	0.771 (14%)	0.679 [6]
70	0.656	0.771 (10%)	
Periodic	0.656		

Tab. 6.5 Comparison between calculated and measured peak absorption energies for the AlN-GaN MQW structures of reference [6].

Periods	Barrier width (Å)	Well width (Å)	Calculated peak absorption edge (eV)	Experimental result (eV) [6]
35	42	20	0.829	0.815
35	42	29	0.706	0.701
35	42	40	0.63	0.65

notice that the calculated result for a 35-well structure is very close to the experimental value, within less than the absorption profile FWHM.

The absorption spectra for the MQW structures grown by Friel and co-workers [6] have been simulated using our numerical model and the results compared with the experimental values. The samples consisted of 35 periods of 42 Å AlN barriers and 20, 29 or 40 Å GaN wells grown on an unintentionally doped AlN buffer and with 2×10^{19} cm^{-3} donors in the GaN wells. The calculated peak absorption energies, reported in Table 6.5, for all the three structures are in agreement with the experiment results. Additionally, the superlattice structures fabricated by Kishino and co-workers [3] intended for intersubband absorption in the range from $\lambda \sim 1.08$ μm to 1.61 μm were simulated. The numerical results, reported in Table 6.6, are also in good agreement with the published experimental values.

Tab. 6.6 Comparison between calculated and measured peak absorption energies for the AlN-GaN MQW structures of reference [3].

Periods	Barrier width (Monolayers)	Well width (Monolayers)	Experimental result (μm) [3]	Calculated peak absorption energy (μm)
90	11	3.3	1.08	0.92
90	11	4.1	1.14	0.99
90	11	5.3	1.27	1.18
90	11	6.2	1.37	1.34
90	11	8.7	1.54	1.69
90	11	9.5	1.61	1.77

6.5 Conclusions

This work has presented a numerical study of the intersubband optical absorption in III-nitride AlGaN-GaN MQWs. The applicability of the numerical model is not restricted to the case of AlGaN structures: in fact, GaN/InGaN

MQWs can also be studied, as well as any other combination of binary, ternary and quaternary alloy MQWs. We have presented a systematic study of the dependence of the electron intersubband energy on geometrical parameters, barrier composition and doping. Furthermore, the role of conduction band nonparabolicity and many-body effects on the absorption spectra has been investigated. We have also studied the effect of the buffer and cap layers, normally neglected when the assumption of perfectly periodic MQWs is used. For the test cases that have been simulated, when the number of well barrier pairs is sufficiently large (typically greater than 20), the structure can be assumed periodic without loss of accuracy.

Acknowledgments

Funding for this work has been provided by the ONR through the 2003 YIP Award # N00014-03-1-0483 made to E. Bellotti. Funding for the computational resources was provided through the ONR Durip Award # N00014-05-1-03839. The authors would like to thank Prof. T.D. Moustakas, Dr. A. Bhattacharyya and Dr. M. Wraback for the information provided to us and many useful discussion, and also Dr. M. Goano for reading the manuscript.

References

1 R. Paiella, Ed., *Intersubband Transitions in Quantum Structures*. New York: McGraw-Hill, 2006.

2 C. Gmachl, H. Ng, S.-N. G. Chu, and A. Y. Cho, *Appl. Phys. Lett.*, vol. 77, no. 23, pp. 3722–3724, Dec. 2000.

3 K. Kishino, A. Kikuchi, H. Kanazawa, and T. Tachibana, *Appl. Phys. Lett.*, vol. 81, no. 23, pp. 1234–1236, 2000.

4 D. Hofstetter, S.-S. Schad, H. Wu, W. Shaff, and L. Eastman, *Appl. Phys. Lett.*, vol. 83, no. 3, pp. 572–574, July 2003.

5 E. Baumann, F. Giorgetta, D. Hofstetter, H. Wu, W. Schaff, L. Eastman, and L. Kirste, *Appl. Phys. Lett.*, vol. 86, p. 32110, 2005.

6 I. Friel, K. Driscoll, E. Kulenica, M. Dutta, R. Paiella, and T. Moustakas, *J. Cryst. Growth*, vol. 278, pp. 387–392, 2005.

7 N. Iizuka, K. Kaneko, and N. Suzuki, *Electron. Lett.*, vol. 40, pp. 962–963, 2004.

8 J. Hamazaki, S. M. ans H. Kunugita, K. Ema, H. Kanazawa, T. Tachibana, A. Kikuchi, and K. Kishino, *Appl. Phys. Lett.*, vol. 84, pp. 1102–1104, 2004.

9 M. Tchernycheva, L. Doyennette, L. Nevou, F. Julien, E. Warde, R. Colombelli, A. Lusson, F. Guillot, E. Monroy, S. Nicolay, E. Feltin, J. Carlin, N. Grandjean, T. Remmele, and M. Albrecht, in *Technical Digest for the 8th International Conference on Intersubband Transitions in Quantum Wells (ITQW 2005)(Cape Cod, 2005)*, 2005, p. 89.

10 H. Uchida, S. Matsui, T. Nakazato, P. Holmstrom, A. Kikuchi, and K. Kishino, in *Technical Digest for the 8th International Conference on Intersubband Transitions in Quantum Wells (ITQW 2005)(Cape Cod, 2005)*, 2005, p. 88.

11 A. Ishida, Y. Inoue, M. Kuwabara, H. Kan, and H. Fujiyasu, *Japan. J. Appl. Phys.*, vol. 41, pp. 1303–1305, 2002.

12 V. Jovanovic, D. Indjin, Z. Ikonic, and P. Harrison, *Appl. Phys. Lett.*, vol. 84, pp. 2995–2997, 2004.

13 G. Sun, R.A. Soref, and J. Khurgin, *Superlatt. Microstruct.*, vol. 37, pp. 107–113, 2005.

14 S. Gunna, E. Bellotti, and R. Paiella, in *Technical Digest for the 8th International Conference on Intersubband Transitions in Quantum Wells (ITQW 2005) (Cape Cod, 2005)*, 2005, p. 78.

15 F. Bernardini, V. Fiorentini, and D. Vanderbilt, *Phys. Rev. B*, vol. 56, p. R10024, 1997.

16 O. Ambacher, B. Foutz, J. Smart, J. R. Shealy, N. G. Weimann, K. Chu, M. Murphy, A. J. Sierakowski, W. J. Schaff, L. F. Eastman, R. Dimitrov, A. Mitchell, and M. Stutzmann, *J. Appl. Phys.*, vol. 87, no. 1, pp. 334–344, Jan. 2000.

17 V. Jovanovic, D. Indjin, Z. Ikonic, V. Milanovic, and J. Radovanovic, *Solid State Commun.*, vol. 121, p. 617, 2002.

18 V. Jovanovic, Z. Ikonic, D. Indjin, P. Harrison, V. Milanovic, and R. Soref, *J. Appl. Phys.*, vol. 93, pp. 3194–3167, 2003.

19 S. Chuang, *Physics of Optoelectronic Devices*. New York: Wiley Interscience, 1995.

20 M. O. Manasreh, *Semiconductor Quantum Wells and Superlattices for Long-Wavelength Infrared Detectors*. Boston: Artech House, 1993.

21 M. Helm, in *Intersubband Transitions in Quantum Wells: Physics and Device Applications I*, H. Liu and F. Capasso, Eds. San Diego: Academic Press, 2000, pp. 1–99.

22 H. Asai and Y. Kawamura, *Phys. Rev. B*, vol. 43, p. 4748, 1991.

23 B. K. Ridley, W. J. Schaff, and L. F. Eastman, *J. Appl. Phys.*, vol. 94, no. 6, p. 3972, 2003.

24 C. Sirtori, F. Capasso, and J. Faist, *Phys. Rev. B*, vol. 50, p. 8663, 1994.

25 W. Kohn and L. J. Sham, *Phys. Rev.*, vol. 140, no. 4A, pp. A1133–A1138, Nov. 1965.

26 B. Vinter, *Phys. Rev. B*, vol. 13, p. 4447, 1976.

27 B. Vinter, *Phys. Rev. B*, vol. 15, p. 3947, 1977.

28 T. Ando, A. B. Flower, and F. Stern, *Rev. Modern Phys.*, vol. 54, p. 437, 1982.

29 E. Batke, G. Weimann, and W. Schlapp, *Phys. Rev. B*, vol. 43, p. 6812, 1991.

30 W. Bloss, *J. Appl. Phys.*, vol. 66, p. 3639, 1989.

31 T. Ando, *Z. Phys. B*, vol. 26, p. 263, 1977.

32 T. Ando, *Solid State Commun.*, vol. 21, p. 133, 1977.

33 I. Vurgaftmana, J. R. Meyer, and L. R. Ram-Mohan, *J. Appl. Phys.*, vol. 89, p. 5815, 2001.

34 M. Goano, E. Bellotti, E. Ghillino, G. Ghione, and K. F. Brennan, *J. Appl. Phys.*, vol. 88, no. 11, pp. 6467–6475, Dec. 2000.

35 A. Abramo, A. Cardin, L. Selmi, and E. Sangiorgi, *IEEE Trans. Electron Devices*, vol. ED-47, no. 10, pp. 1858–1863, Oct. 2000.

36 C. Takano, Z. Yu, and R. W. Dutton, *IEEE Trans. Computer-Aided Design*, vol. CAD-9, no. 11, pp. 1217–1224, Nov. 1990.

37 B. Jonsson and S. T. Eng, *IEEE J. Quantum Electron.*, vol. QE-26, no. 11, pp. 2025–2035, Nov. 1990.

38 J. Y. Tang and K. Hess, *J. Appl. Phys.*, vol. 68, no. 8, pp. 4071–4076, Oct. 1990.

39 L. R. Ram-Mohan, *Finite Element and Boundary Element Applications in Quantum mechanics*. Oxford University Press, 2002.

40 Z. Bai, J. Demmel, J. Dongarra, A. Ruhe, and H. van der Vorst, Eds., *Templates for the Solution of Algebraic Eigenvalue Problems: A Practical Guide*. Philadelphia: SIAM, 2000.

41 R. B. Lehoucq, D. C. Sorensen, and C. Yang, *ARPACK Users' Guide: Solution of Large-Scale Eigenvalue Problems with Implicitly Restarted Arnoldi Methods*. Philadelphia: SIAM, 1998.

42 W. A. Harrison, *Electronic Structure and the Properties of Solids*. New York: Dover Publications, 1989.

43 S. Selberherr, *Analysis and Simulation of Semiconductor Devices*. Wien: Springer-Verlag, 1984.

44 A. Trellakis, A. T. Galick, A. Pacelli, and U. Ravaioli, *J. Appl. Phys.*, vol. 81, no. 12, pp. 7880–7884, June 1997.

7
Interband Transitions in InGaN Quantum Wells

Jörg Hader, Jerome V. Moloney, Angela Thränhardt, and Stephan W. Koch

7.1
Introduction

Group-III nitrides have been under intense investigation for their possible applications in light-emitting devices operating in the spectral range from the visible upto the ultraviolet range. Due to potential applications in the UV [1] and green [2] wavelength range, interest has soared again, triggered by the recent availability of GaN substrates.

For the theoretical description of GaN, Coulomb effects are exceedingly important due to the strong excitonic interaction. These have been factored into a free-carrier theory retroactively [3] or taken into account systematically at the level of the system Hamiltonian [4, 5]. Here, we present a microscopic theory able to predict a wide range of optical spectra in group-III nitrides, such as gain and refractive index, but also photoluminescence and radiative recombination as well as Auger rates. As input, the calculations only require the nominal structural layout (layer widths and material compositions) and well-known bulk material quantities like $\mathbf{k} \cdot \mathbf{p}$ bandstructure parameters, background refractive indices or phonon energies. They do not involve any fit-parameters, such as dephasing times or lineshape broadenings, which can be adjusted in order to get better agreement with experiment. As has been demonstrated in the past for zincblende-based structures (see e.g. Refs [6–8] and refs therein), the results of this approach show excellent quantitative agreement with experiment for gain/absorption and refractive index changes as well as luminescence (spontaneous emission) and for the carrier recombination rates due to spontaneous emission and Auger processes, yielding correct lineshapes, amplitudes and spectral positions as well as density and temperature dependencies. The deviations between theoretical results and experiment are usually found to be within the scattering range of the experimental data. Here, we show that this good theory–experiment agreement also extends to wurtzite-structured nitrides and we give an overview over their optical properties.

Nitride Semiconductor Devices: Principles and Simulation. Joachim Piprek (Ed.)
Copyright © 2007 WILEY-VCH Verlag GmbH & Co. KGaA, Weinheim
ISBN: 978-3-527-40667-8

On the other hand, attempts to simplify the calculations will generally fail [9]. Most prominently, the use of a dephasing time or lineshape broadening instead of explicitly calculating the scattering terms leads to wrong lineshapes, wrong density dependent spectral shifts and wrong amplitudes that can, e.g., lead to an error for the transparency density of fifty percent or more.

7.2
Theory

This section summarizes the microscopic model for the calculation of the optical and electronical properties of wide bandgap nitride materials. First, the $\mathbf{k} \cdot \mathbf{p}$ model for the single-particle energies and wavefunctions is described. This part also discusses the differences between the wurtzite-structured nitrides and more conventional zincblende-structured materials that are utilized for longer wavelength applications. The semiconductor Bloch equations used to calculate the absorption/gain and refractive index, the semiconductor luminescence equations for the spontaneous emission spectra and corresponding carrier lifetimes, and the model describing carrier losses due to Auger recombinations all take the same form, irrespective of the material or crystal structure under consideration.

7.2.1
Bandstructure and Wavefunctions

For the calculation of the single-particle wavefunctions and energies we use the $\mathbf{k} \cdot \mathbf{p}$ model for wurtzite crystals as outlined in [10]. However, since we need the coupling between conduction bands and valence bands for the evaluation of the Auger recombination processes, we use a fully coupled 8×8 model instead of the more commonly used 6×6 model in which only the coupling inbetween bulk valence bands is taken into account.

Assuming that the growth direction, z, is along the c (0001) axis, and using the cubic approximation, the $\mathbf{k} \cdot \mathbf{p}$ Hamiltonian takes the form:

$$\begin{pmatrix} E_c(\mathbf{k}) & 0 & -\frac{k_+ P_2}{\sqrt{2}} & \frac{k_- P_2}{\sqrt{2}} & k_z P_1 & 0 & 0 & 0 \\ 0 & E_c(\mathbf{k}) & 0 & 0 & 0 & \frac{k_- P_2}{\sqrt{2}} & -\frac{k_+ P_2}{\sqrt{2}} & k_z P_1 \\ -\frac{k_- P_2}{\sqrt{2}} & 0 & F & -K^\star & -H^\star & 0 & 0 & 0 \\ \frac{k_+ P_2}{\sqrt{2}} & 0 & -K & G & H & 0 & 0 & \sqrt{2}\Delta_2 \\ k_z P_1 & 0 & -H & H^\star & \lambda & 0 & \sqrt{2}\Delta_2 & 0 \\ 0 & \frac{k_+ P_2}{\sqrt{2}} & 0 & 0 & 0 & F & -K & H \\ 0 & -\frac{k_- P_2}{\sqrt{2}} & 0 & 0 & \sqrt{2}\Delta_2 & -K^\star & G & -H^\star \\ 0 & k_z P_1 & 0 & \sqrt{2}\Delta_2 & 0 & H^\star & -H & \lambda \end{pmatrix} \quad (7.1)$$

where we expanded in the following set of zone-center basis functions:

$$|1> = |iS \uparrow\rangle, \quad |2> = |iS \downarrow\rangle$$

$$|3> = \left|-\frac{1}{\sqrt{2}}(X+iY)\uparrow\right\rangle, \quad |4> = \left|\frac{1}{\sqrt{2}}(X-iY)\uparrow\right\rangle, \quad |5> = |Z\uparrow\rangle$$

$$|6> = \left|\frac{1}{\sqrt{2}}(X-iY)\downarrow\right\rangle, \quad |7> = \left|-\frac{1}{\sqrt{2}}(X+iY)\downarrow\right\rangle, \quad |8> = |Z\downarrow\rangle \quad (7.2)$$

In Eq. (7.1) we used the following definitions:

$$E_c(\mathbf{k}) = E_g + \Delta_1 + \Delta_2 + \Delta_\epsilon + \frac{\hbar^2(k_x^2 + k_y^2)}{2\tilde{m}_e^\perp} + \frac{\hbar^2 k_z^2}{2\tilde{m}_e^\parallel}$$

$$F = \Delta_1 + \Delta_2 + \lambda + \theta$$

$$G = \Delta_1 - \Delta_2 + \lambda + \theta$$

$$\lambda = \frac{\hbar^2}{2m_0}\left[\tilde{A}_1 k_z^2 + \tilde{A}_2\left(k_x^2 + k_y^2\right)\right] + \lambda_\epsilon$$

$$\theta = \frac{\hbar^2}{2m_0}\left[(\tilde{A}_2 - \tilde{A}_1)k_z^2 + \frac{1}{2}(\tilde{A}_1 - \tilde{A}_2)\left(k_x^2 + k_y^2\right)\right] + \theta_\epsilon$$

$$K = \frac{\hbar^2}{2m_0}\tilde{A}_5 k_+^2$$

$$H = \frac{\hbar^2}{2m_0}\frac{1}{\sqrt{2}}(\tilde{A}_2 - \tilde{A}_1 + 4\tilde{A}_5)k_+ k_z$$

$$\Delta_\epsilon = 2\epsilon_{xx}\left(a_2 - a_1 \frac{C_{13}}{C_{33}}\right)$$

$$\lambda_\epsilon = \left(-\frac{2D_1 C_{13}}{C_{33}} + 2D_2\right)\epsilon_{xx}$$

$$\theta_\epsilon = (D_1 - D_2)\left(\frac{2C_{13}}{C_{33}} + 1\right)\epsilon_{xx}$$

$$\epsilon_{xx} = 2\frac{a_{lc} - a_{lc}^r}{a_{lc} + a_{lc}^r}$$

$$k_\pm = k_x \pm i k_y \quad (7.3)$$

a_{lc}^r is the lattice constant of the neighboring material, the lattice mismatch of which induces strain in the material under consideration. m_0 is the bare electron mass. The definitions of all other symbols used here are as given by I. Vurgaftmann and J. R. Meyer in Chapter 2 of this book. Table 7.1 lists all parameter values that differ from those used there. For the fundamental bandgap we assume a bowing parameter of 1.4 eV. For all other parameters we use linear interpolation between the values for GaN and InN to obtain those for ternary mixtures.

Tab. 7.1 Band structure parameters that differ from those used by I. Vurgaftmann and J.R. Meyer in Chapter 2 of this book. α and β are the Varshni parameters describing the temperature dependence of the bandgap. $\Delta_{cr} = \Delta_1$ in the cubic approximation used here.

Parameter	GaN	InN
E_g (eV) at 0 K	3.510	0.780
E_g (eV) at 300 K	3.437	0.756
α (meV K^{-1})	0.909	0.245
β (K)	830	624
Δ_{cr} (eV)	0.010	0.040
$m_e^{\parallel}$ (m_0)	0.20	0.07
a_1 (eV)	−4.9	−3.5
a_2 (eV)	−11.3	−3.5
D_1 (eV)	−3.7	−3.7
D_2 (eV)	4.5	4.5
d_{13} (pm V^{-1})	−1.6	−3.5

Since the coupling between conduction bands and valence bands is included explicitly, the corresponding couplings have to be taken out of the parameters $A_{1,2,5}$ and $m_e^{\parallel,\perp}$ as they are used in the 6 × 6-Hamiltonian. The parameters $\tilde{A}_{1,2,5}$ and $\tilde{m}_e^{\parallel,\perp}$ are the corresponding versions after this Löwdin renormalization [10, 11].

The coupling parameters P_1 and P_2 are given by [10]:

$$P_1^2 = \frac{\hbar^2}{2m_0}\left(\frac{m_0}{m_e^{\parallel}} - 1\right) \frac{(E_g + \Delta_1 + \Delta_2)(E_g + 2\Delta_2) - 2\Delta_2^2}{E_g + 2\Delta_2}$$
$$P_2^2 = \frac{\hbar^2}{2m_0}\left(\frac{m_0}{m_e^{\perp}} - 1\right) \frac{E_g[(E_g + \Delta_1 + \Delta_2)(E_g + 2\Delta_2) - 2\Delta_2^2]}{(E_g + \Delta_1 + \Delta_2)(E_g + \Delta_2) - \Delta_2^2}$$

(7.4)

The Hamiltonian for quantum well calculations is obtained from Eq. (7.1) by replacing k_z by $-i\partial/\partial z$ in a hermitian way and considering all band structure parameters to be z dependent. The resulting eigenvalue problem is solved using the method described in [12].

In the excited state, nonzero local charge densities lead to a modification of the confinement potential due to screening. In this case, the charge-induced potentials have to be calculated by solving Poisson's equation. The bandstructure and the charge potential calculations have to be iterated until a self-consistent solution is found (see Ref. [13]).

In wurtzite heterostructures strong piezoelectric polarizations, Π_{pe}, and spontaneous polarizations, Π_{sp}, can be present:

$$\Pi_{sp} = P_{sp}$$
$$\Pi_{pe} = 2d_{31}\epsilon_{xx}\left(C_{11} + C_{12} - 2\frac{C_{13}^2}{C_{33}}\right)$$

(7.5)

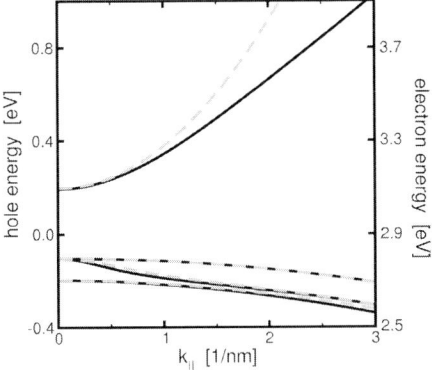

Fig. 7.1 Energetically lowest electron subband and highest three hole subbands for an $In_{0.1}Ga_{0.9}N/GaN$ structure with a 2 nm wide well (as in [5]) in the unexcited state. Solid black (dashed grey): with (without) coupling between conduction bands and valence bands.

Here, P_{sp} is the spontaneous polarization constant and d_{31} is the piezoelectric constant. These polarizations are accounted for by adding the corresponding potential V_p to the confinement potential.

$$V_p(z) = \begin{cases} ez\frac{\left(\Pi^b_{sp}+\Pi^b_{pe}-\Pi^w_{sp}-\Pi^w_{pe}\right)w^b}{\varepsilon^w_0 w^b + \varepsilon^b_0 w^w}, & z \text{ in well} \\ ez\frac{\left(\Pi^w_{sp}+\Pi^w_{pe}-\Pi^b_{sp}-\Pi^b_{pe}\right)w^w}{\varepsilon^w_0 w^b + \varepsilon^b_0 w^w}, & z \text{ in barrier} \end{cases} \quad (7.6)$$

ε_0 is the static dielectric constant, e the elementary charge, the superscript b/w stands for in the well/barrier and $w^{w/b}$ is the well/barrier width.

The coupling between conduction and valence bands is usually believed to be only significant for materials with narrow bandgaps since the corresponding coupling terms scale with the inverse of the energetic separation between the states. However, as can be seen in Fig. 7.1, it also leads to considerable changes in the bands for the wide bandgap materials considered here. Most importantly, it leads to a nonparabolicity of the conduction band.

7.2.2
Semiconductor Bloch Equations

The absorption/gain and carrier-induced refractive index changes are calculated by solving the equations of motion for the microscopic polarizations, $P^{ji}_\mathbf{k}$, i.e., the so-called semiconductor Bloch equations (SBE):

$$\frac{d}{dt}P^{ji}_\mathbf{k} = \frac{1}{i\hbar}\left\{\sum_{i',j'}\left[\mathcal{E}^{jj'}_\mathbf{k}\delta_{ii'}+\mathcal{E}^{ii'}_\mathbf{k}\delta_{jj'}\right]P^{j'i'}_\mathbf{k} + \left[1-f^i_\mathbf{k}-f^j_\mathbf{k}\right]U^{ij}_\mathbf{k}\right\} + \frac{d}{dt}P^{ji}_\mathbf{k}\bigg|_{corr} \quad (7.7)$$

where

$$\mathcal{E}_{\mathbf{k}}^{ii'} = \varepsilon_{\mathbf{k}}^{i}\delta_{ii'} - \sum_{i'',\mathbf{q}} V_{\mathbf{k}-\mathbf{q}}^{ii''i'i''} f_{\mathbf{q}}^{i''}, \qquad \mathcal{E}_{\mathbf{k}}^{jj'} = \varepsilon_{\mathbf{k}}^{j}\delta_{jj'} - \sum_{j'',\mathbf{q}} V_{\mathbf{k}-\mathbf{q}}^{j'j''jj''} f_{\mathbf{q}}^{j''}$$

$$\mathcal{U}_{\mathbf{k}}^{ij} = -\mu_{ij,\mathbf{k}}E(t) - \sum_{i',j',\mathbf{q}} V_{\mathbf{k}-\mathbf{q}}^{ij'ji'} P_{\mathbf{q}}^{j'i'} \qquad (7.8)$$

Here, $\varepsilon_{\mathbf{k}}^{i/j}$ are the electron/hole subbands as calculated with the single-particle bandstructure model from Section 7.2.1. The indices i, i', i'' (j, j', j'') label electron (hole) subbands; $\mathbf{k}$ and $\mathbf{q}$ are in-plane momentum wave vectors; f's are the carrier distribution functions; μ are dipole matrix elements and V are the Coulomb matrix elements.

$\mathcal{E}$ are the "renormalized energies". The Coulomb interaction leads to density dependent shifts of the transition energies, an effect usually called "bandgap renormalization". $\mathcal{U}$ is the "renormalized field" that includes the influence of the attractive Coulomb interaction between electrons and holes. As a consequence, the absorption is increased ("Coulomb enhancement") and quasi-bound states below the subband edges, excitonic resonances, appear.

In Eq. (7.7) the term $d/dt P_{\mathbf{k}}^{ji}|_{corr}$ comprises all higher order correlations due to the Coulomb and electron–phonon interactions. These correlations are responsible for the screening of the Coulomb interaction and the dephasing of the polarization due to electron–electron and electron–phonon scattering. Here, the effects of screening are included at the random phase approximation level (Lindhardt approximation) and the carrier scattering in the second Born–Markov limit. The explicit form of the resulting scattering equations, as well as more details about the treatment of the Coulomb interaction, can be found in [14]. More information about the derivation of Eq. (7.7) is given in [15].

Once the microscopic polarizations have been calculated, the macroscopic optical polarization, P, entering into Maxwell's equation is given by:

$$P(t) = \frac{1}{V} \sum_{i,j,\mathbf{k}} P_{\mathbf{k}}^{ji}(t)\mu_{\mathbf{k}}^{ji} \qquad (7.9)$$

where V is the optically active volume. From this, the linear optical susceptibility in frequency space $\chi(\omega)$ is obtained by a simple Fourier transform and division by the test field E:

$$\chi(\omega) = \frac{P(\omega)}{4\pi\varepsilon_0 \varepsilon E(\omega)} \qquad (7.10)$$

Here, ε is the permittivity. Finally, the linear (intensity) absorption/gain $\alpha(\omega)$ and the carrier-induced refractive index $\delta n(\omega)$ are given by:

$$\alpha(\omega) = -\frac{\omega}{n_b c}\mathcal{I}m\{\chi(\omega)\}, \quad \delta n(\omega) = \frac{1}{2n_b}\mathcal{R}e\{\chi(\omega)\} \qquad (7.11)$$

where $\mathcal{I}m$ and $\mathcal{R}e$ denote the imaginary and real parts, n_b is the background refractive index and c is the vacuum speed of light.

7.2.3
Semiconductor Luminescence Equations

For a calculation of the luminescence, i.e., spontaneous emission, the light field has to be quantized, which gives the semiconductor luminescence equations. Due to their high complexity, they will be presented in a schematic form here. Let $b_{\sigma\vec{q}}$ be an annihilator for a photon with wave number $\vec{q}$ and photon energy ω_q. Note that photons are characterized by a three-dimensional wave number $\vec{q}$ whereas electrons and holes have a subband index ν in addition to their in-plane (two-dimensional) wave number $\mathbf{k}$. Assuming a detector with infinite energy resolution and steady state conditions, the luminescence is given by the rate of spontaneously emitted photons [16],

$$S_\sigma(\omega_q) = \frac{\partial}{\partial t}\Delta\langle b^\dagger_{\sigma\vec{q}} b_{\sigma\vec{q}}\rangle$$

$\Delta\langle b^\dagger_{\sigma\vec{q}} b_{\sigma\vec{q}}\rangle$ denotes the incoherent spontaneous emission contribution,

$$\Delta\langle b^\dagger_{\sigma\vec{q}} b_{\sigma\vec{q}}\rangle = \langle b^\dagger_{\sigma\vec{q}} b_{\sigma\vec{q}}\rangle - \langle b^\dagger_{\sigma\vec{q}}\rangle\langle b_{\sigma\vec{q}}\rangle$$

and may be shown to be the real part of a weighted sum over the photon-assisted polarizations,

$$\Delta\langle b^\dagger_{\sigma\vec{q}} b_{\sigma\vec{q}}\rangle = \frac{2}{\hbar A}\mathrm{Re}\left[\sum_{\mathbf{k}\nu\lambda}\mathcal{E}_{\vec{q}}\mu^*_{\mathbf{k}\nu\lambda}\Delta\langle b_{\vec{q}\sigma}e^\dagger_{\nu\mathbf{k}-\mathbf{q}}h^\dagger_{\lambda\mathbf{k}}\rangle\right]$$

$\mathcal{E}_q$ is the vacuum field amplitude,

$$\mathcal{E}_q = \frac{\hbar\omega_q}{2\epsilon_0}$$

and A is the quantization area. The photoluminescence may thus be calculated from the photon-assisted interband polarization $\Pi_{\mathbf{k},\vec{q}} = \langle b^\dagger_{\vec{q}} v^\dagger_{\mathbf{k}-\vec{q}_\parallel} c_\mathbf{k}\rangle$ which adheres to a similar equation of motion as the polarization,

$$\left[i\frac{\partial}{\partial t} - (\mathcal{E}^e_k + \mathcal{E}^h_k) + \hbar\omega_q\right]\Pi_{\mathbf{k},\vec{q}} = \left[1 - f_{e,\mathbf{k}+\vec{q}_\parallel} - f_{h,\mathbf{k}}\right]\sum_{\mathbf{k}'}V_{\mathbf{k}-\mathbf{k}'}\Pi_{\mathbf{k}',\vec{q}}$$

$$+ \left.\frac{\partial}{\partial t}\Pi_{\mathbf{k},\vec{q}}\right|_{\mathrm{corr}} + \Omega^{\mathrm{stim}}_{\mathbf{k},\vec{q}_\parallel} + F_{\vec{q}}S_{\mathbf{k},\vec{q}_\parallel}$$

where the term marked "corr" is defined analogously to the corresponding one in the polarization equation of motion. S describes the incoherent source and Ω^{stim} denotes the feedback, e.g., of a cavity. Here, we consider photoluminescence in a quantum well structure where the feedback term is negligible. The incoherent source consists of Hartree–Fock and excitonic correlation terms,

$$S = \langle e^\dagger h^\dagger eh \rangle = f_e f_h + \Delta \langle e^\dagger h^\dagger eh \rangle$$

Even at high temperatures/densities, the omission of the excitonic correlation term causes an incorrect lineshape of the photoluminescence. Excitonic correlations

$$\left[i\hbar \frac{\partial}{\partial t} - E \right] \Delta \langle e^\dagger h^\dagger eh \rangle = S + M + O$$

are driven by products of singlet contributions $f_{e/h}$ (source term S) and are subject to phonon and carrier scattering (scattering term O). E contains the generalized renormalized eigenenergies and M gives the Coulomb-induced coupling between the excitonic correlations (by analogy with the field-renormalization terms in the polarization equation of motion).

The carrier recombination rate due to spontaneous emission R_{se} and the corresponding current density J_{se} are obtained by integrating over the spectrum:

$$J_{se} = eR_{se} = e \int d\omega_q S_\sigma(\omega_q) \tag{7.12}$$

and the carrier lifetime due to spontaneous emission is given by:

$$t_{se} = \frac{N^{2D}}{w^w R_{se}} \tag{7.13}$$

where N^{2D} is the sheet carrier density.

As pointed out earlier, the results of the semiconductor luminescence equations have been shown to yield very good quantitative agreement with experiment for zincblende-based structures [6,8]. Since the difference between these structures and the wurtzite-based wide bandgap nitrides is within the single-particle bandstructure, but not the model for the luminescence, one should expect similarly good results here.

7.2.4
Auger Recombination Processes

By analogy with the dephasing of the polarizations in the gain calculation, the Auger recombination processes are described by Boltzmann-type scattering equations in the second Born–Markov limit [8]. E.g., for the electron-loss, the relevant parts for the cases investigated here are:

$$J^e_{aug} = e \sum_{k,i,s} \frac{d f_k^{i,s}}{dt}$$

$$= -\frac{4\pi e}{\hbar} \sum_{k,k',q,i,s,s'} \text{Re} \left\{ \sum_{j_1,j_2,j_3} \left| \tilde{V}_q^{i\,j_3\,j_1\,j_2}_{s\,s'\,s'\,-s} \right|^2 \left(1 - f_{k'}^{j_3,s'}\right) f_{|q-k|}^{j_2,-s} f_{|k'-q|}^{j_1,s'} f_k^{i,s} \right.$$

$$\times \mathcal{D}\left(-\varepsilon_k^{i,s} - \varepsilon_{|k'-q|}^{j_1,s'} - \varepsilon_{|q-k|}^{j_2,-s} + \varepsilon_{k'}^{j_3,s'} \right)$$

$$+ \sum_{i_1,i_2,j_1} \left| \tilde{V}_q^{j_1\,i_1\,i_2\,i}_{-s\,s'\,s'\,s} \right|^2 \left(1 - f_{|k'-q|}^{i_1,s'}\right) f_{k'}^{i_2,s'} f_k^{i,s} f_{|k-q|}^{j_1,-s}$$

$$\left. \times \mathcal{D}\left(-\varepsilon_{|k'-q|}^{i_1,s'} + \varepsilon_{|k-q|}^{j_1,-s} + \varepsilon_{k'}^{i_2,s'} + \varepsilon_k^{i,s} \right) \right\} \qquad (7.14)$$

Here, s, s' are spin-indices. $\mathcal{D}$ is an energy-conserving function and $\tilde{V}$ are screened Coulomb matrix elements.

The second and third line in Eq. (7.14) describe Auger processes where an electron in subband i recombines with a hole in band j_2. The energy and momentum are transfered to a hole in band j_1, which is excited into band j_3. The last two lines describe recombinations where an electron is excited from band i_2 to i_1.

The opposite process to Auger recombination, impact ionization, is negligible here. The bandgap of the investigated materials is much larger than the thermal energy. In that case, the states high above the bandgap are empty and there are no carriers that have enough excess energy to produce an electron–hole pair by relaxing down towards the band-edge.

As has been shown in [8], Eq. (7.14) gives very good quantitative agreement with experiment if no uncontrolled approximations are made. Auger calculations in the past were based on the same fundamental equations. However, mostly due to numerical limitations, approximations like spin averaging or simplifications to the Coulomb matrix elements were made, that yield an uncertainty in the accuracy of the results by up to an order of magnitude or more. In contrast, here, all spin, subband and momentum dependencies are fully considered. E.g., the full k- and spin-dependence of the wavefunctions involved in the coupling matrix elements is taken into account. For zincblende-type structures operating in the 1.3 µm–1.5 µm range the calculations were found to agree with experiment to within an uncertainty of less than about 20%.

An 8×8-$k \cdot p$ band-structure model has been shown to be sufficient for the Auger calculations of those zincblende-type structures. For the wide-bandgap nitrides the involved states are even further detuned from the band-edge. Although this makes the validity of the 8×8-$k \cdot p$ model more questionable, one should not expect a fundamental breakdown in the accuracy.

7.3
Theory–Experiment Gain Comparison

Unfortunately, the amount of reliable experimental data in the literature for gain, absorption or luminescence in wide-bandgap nitrides is still quite limited. One publication that gives enough information to try to theoretically reproduce the results is Ref. [5]. The structure investigated there consists nominally of three 2 nm wide $In_{0.1}Ga_{0.9}N$-wells with 6 nm wide GaN barriers. The optical confinement factor is 0.017. For the comparisons shown below, only one well with cyclical boundary conditions was considered.

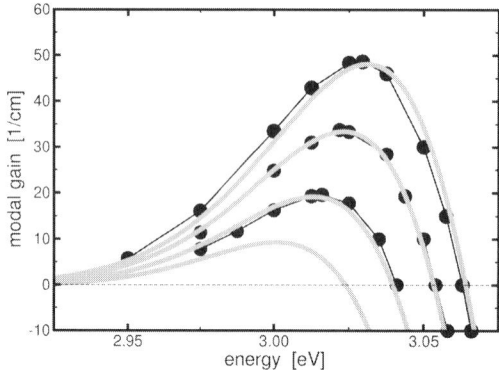

Fig. 7.2 Black dots: experimental gain spectra extracted from Ref. [5] for an $In_{0.1}Ga_{0.9}N$/GaN structure and pump currents of 60, 90 and 120 mA. Grey lines: theoretical gain spectra for sheet carrier densities of 7.00, 7.60, 8.25 and $8.85 \times 10^{12} cm^{-2}$ assuming an inhomogeneous broadening of 80 meV (FWHM).

Figure 7.2 shows a comparison between experimental data extracted from [5] and the results of the fully microscopic gain model described above. The theory agrees very well with experiment. The lineshapes, density-induced spectral shifts and amplitudes are all within the scattering range of the experiment. It is worth stressing that, because the confinement factor is known, the amplitudes of the theoretical spectra have to be correct.

As always in these comparisons two adjustments are necessary. First, an overall spectral shift of the theoretical spectra due either to a deviation between the nominal and actual well widths and/or compositions and/or discrepancies in the formula for the fundamental bandgap. Second, the theoretical spectra have to be inhomogeneously broadened in order to account for imperfect growth.

Here, all theoretical spectra had to be shifted by -110 meV. The most likely reason for this mismatch is that the theoretical value for the InGaN bandgap is incorrect. Published values for this bandgap show some uncertainty. However, the deviation found here lies outside the reasonable uncertainty range.

The commonly accepted bandgap energies as listed, e.g., in Table 7.1 are obtained from low-density luminescence measurements where the excitonic bandgap of the material is observed. The observed energies are then corrected for the exciton binding energy. However, the resulting values are not the bandgaps for interaction-free single particles. They still contain shifts due to the electron–phonon interaction. Since the electron–phonon interaction is taken into account explicitly here, this shift would have to be subtracted from the literature values first. As will be shown below, for this structure the phononic bandgap shift at 300 K and low densities is just about 110 meV, explaining the theory–experiment mismatch found here.

The shift could also be explained by an about 3% higher indium concentration in the wells than the nominal 10%. However, such a deviation is unreasonably high. Also, typical deviations from the nominal well width could only explain a fraction of maybe 20% of the mismatch. For all results shown below we corrected for that shift.

The theoretical spectra had to be broadened according to an inhomogeneous broadening of 80 meV (FWHM). This value is much higher than the typically 20 – 30 meV for InGaAs- or AlGaAs-based structures. However, it agrees with measurements of the Indium distribution in the wells by the authors of Ref. [5]. These measurements show strong fluctuations in the composition, of about ±2%. A variation by 2% leads to a change in the bandgap of about 80 meV.

As has been shown in [6], the inhomogeneous broadening can be determined rather easily through a comparison of theoretical and experimental low-excitation luminescence spectra. In Ref. [17] the authors did measure such spectra for a structure with nominally the same active region as the structure here. They found a total broadening of about 110 meV in good agreement with the 80 meV found here for just the inhomogeneous broadening.

As mentioned, since the confinement factor is known, the amplitude of the gain for a given density and broadening is fixed. Thus, for a known inhomogeneous broadening there is no ambiguity in the association of experimental pump currents and theoretical carrier densities. Changing the inhomogeneous broadening also changes the spectral shift of the gain maximum with increasing gain amplitude. For the case here, matching the experimentally found relation between spectral shift and gain amplitude within the scattering of the experiment determines the inhomogeneous broadening to within an uncertainty of less than 10%. This leaves less than about 5% uncertainty for the carrier density associated with a given pump current.

7.4
Absorption/Gain

All results shown below are for a temperature of 300 K and TE-polarized light.

7.4.1
General Trends

Figure 7.3 shows the absorption for the $In_{0.1}Ga_{0.9}N/GaN$-structure of Ref. [5] for various sheet carrier densities and two inhomogeneous broadenings.

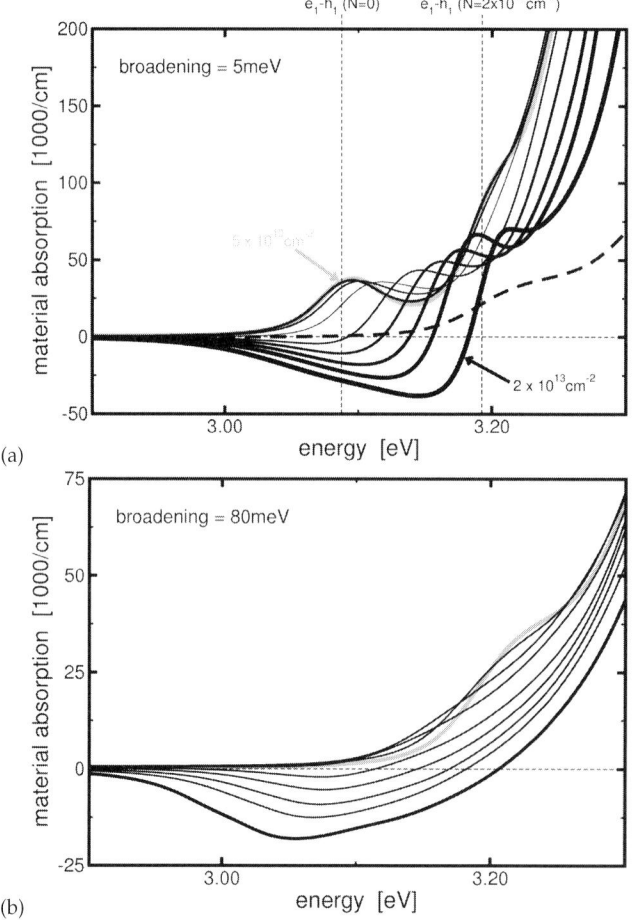

Fig. 7.3 Black, Grey: absorption for the $In_{0.1}Ga_{0.9}N/GaN$-structure of Ref. [5] for sheet carrier densities of 0.05, 0.50, 2.00, 4.0, 7.50, 10.00, 12.50, 15.00 and $20.00 \times 10^{12} cm^{-2}$ and inhomogeneous broadenings of 5 meV (a) and 80 meV (b). Thick dashed: absorption for 0.05 when Coulombic field renormalization is not included. Vertical dashed lines: single particle band-edge for the highest and lowest density.

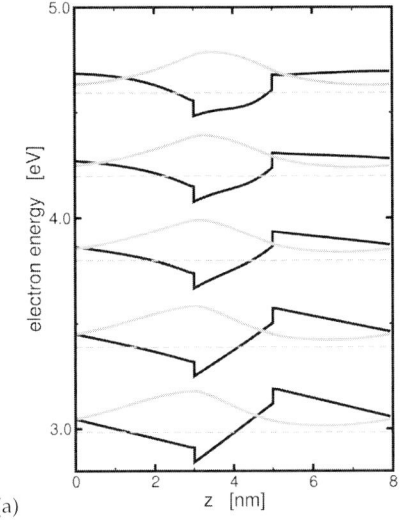 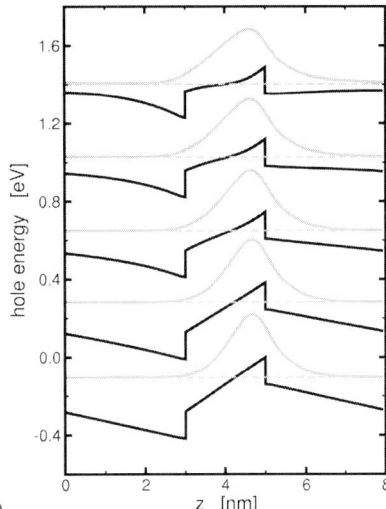

Fig. 7.4 Confinement potentials and wavefunctions (i.e., the dominant wavefunction spinor) of the energetically lowest electron (a) and hole (b) states for the $In_{0.1}Ga_{0.9}N/GaN$-structure of Ref. [5] and carrier densities of 0.05, 2.0, 7.5, 12.5 and $20.0 \times 10^{12} cm^{-2}$ (bottom to top). Dashed lines: energy of the state and the zero for the wavefunction amplitude. Results have been shifted by multiples of 0.4 eV for clarity. Here, piezoelectric and spontaneous polarizations lead to an internal electric field in the well of 1.39 MV cm^{-1}

With increasing carrier density several processes happen at the same time. The charge of the carriers leads to a screening of the internal electric fields. This reduces the quantum confined Stark Effect (QCSE) and results in a blue shift of the band-edge. It also reduces the local separation between electron and hole wavefunctions (see Fig. 7.4) increasing the wavefunction overlap. Meanwhile, the charges lead to a screening of the Coulomb interaction and reduce its effects. Finally, increasing the phase-space filling leads to a reduction of the absorption at and below the bandgap, and the formation of gain.

Several features are remarkable here. First, the bandgap renormalization due to electron–phonon scattering is very significant. The dashed spectrum in Fig. 7.3 shows the absorption when all renormalization effects are considered except for the Coulombic field renormalization (term involving V in the second line of Eq. (7.8)). This means that excitonic resonances are artificially turned off. Here, the absorption edge is about 110 meV higher than the lowest free-particle transition energy would suggest. Since the distributions are virtually zero for this density the bandgap renormalization effects in Eq. (7.8) are negligible and so is the electron–electron scattering. Thus, the shift is purely due to the electron–phonon interaction.

Second, the Coulomb effects are much more dramatic than in systems like InGaAs or AlGaAs. Here, the exciton binding energy is about 100 meV which is of the order of ten times higher than in those materials. The peak at the band-edge is the excitonic resonance associated with the $e1$–$hh1$ and $e1$–$lh1$ transitions. Due to the strong homogeneous broadening they cannot be resolved individually at this temperature.

Despite the strength of the excitonic features, they are virtually absent once the strong inhomogeneous broadening is considered. In materials with less inhomogenity the resonances should remain visible, but to the best of our knowledge no sharp excitonic resonances have ever been observed. Of course the absorption is still dominated by the excitonic features, even if the resonances are not apparent. They still determine the low-energy absorption and influence the density-dependent shifts and amplitudes.

With increasing density, the exciton binding energy and peak amplitude decrease due to screening which leads to a shift of the low-energy absorption edge. However, this shift coincides with the ones due to increased carrier scattering and due to the reduced QCSE. This demonstrates how the wide-bandgap nitride systems are dominated by many-particle interactions. Obviously, a proper description of the spectral positions and density-induced shifts requires a comprehensive microscopic model as outlined here. Any attempts based on simplified models are destined to fail.

Another noteworthy point here is that in this system the electron–hole overlap does not vary strongly with the density (see Fig. 7.4). Due to the strain-induced hydrostatic shift Δ_ε and the internal electric fields, the confinement for the electrons is very weak. It is not strong enough to lead to an electron state that is bound inside the well. Even the energetically lowest electron state is above the well edge. The electron is kept close to the well by the tilted barrier potential. Since the well and barriers are quite narrow in this case, the electron wavefunction still has good overlap with the hole wavefunction, leading to significant continuum absorption and very pronounced excitonic features. As will be demonstrated later, this can be significantly different in systems with other well and/or barrier configurations.

The carrier densities necessary to achieve gain are much higher than typically in systems for longer wavelengths like InGaAsP or AlGaAs. Whereas in the latter the transparency density is usually in the range of $2 - 4 \times 10^{12} \text{cm}^{-2}$, it is about twice that here. The most important reason for this is that here the effective electron mass is more than twice as high as the typical values in the latter systems. This means that the density of states for the electrons is higher and, thus, more carriers are required to achieve inversion.

7.4.2
Structural Dependence

Figure 7.5 shows the absorption for a structure similar to the one before, but where the well width was increased to 4 nm. Changing the well width while keeping the material composition the same also keeps the internal fields the same. Thus, at low densities, the confinement for the carriers is similar, but the local separation between electrons and holes is increased. This leads to a dramatic reduction in the absorption near the band-edge at low to intermediate densities. The excitonic resonances are almost completely absent and the absorption only becomes significant high above the band-edge, where the states are no longer influenced too much by the confinement potential. The gain is also reduced. The transparency density is about twice as high as in the 2 nm structure. At very high densities, where the internal field is strongly screened, the wavefunction overlap and gain are similar to the 2 nm case.

Figure 7.6 shows the absorption for an InGaN/GaN structure where the well width was kept at 2 nm, but the indium content was increased to 20 %. Due to the higher indium concentration the internal fields are twice as strong as in the case for only 10% indium, but the well is also much deeper. The higher field increases the wavefunction separation. The stronger confinement in the well decreases it. Overall, this leads to a somewhat smaller wavefunction overlap than in the 10% case for all densities. The excitonic resonances

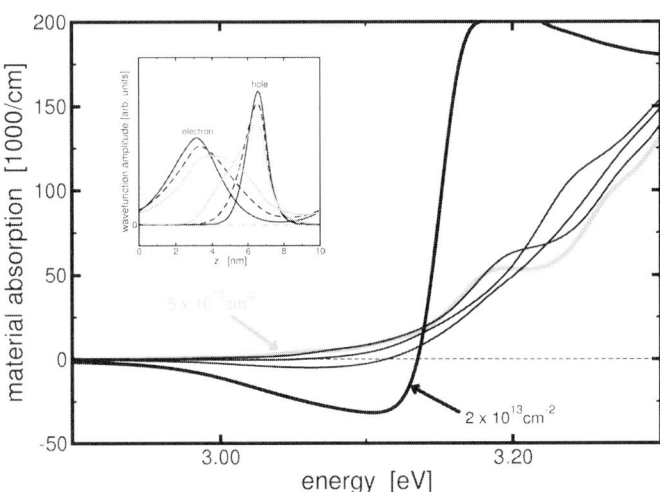

Fig. 7.5 Absorption for an $In_{0.1}Ga_{0.9}N$/GaN structure with a 4 nm wide well, carrier densities of 0.05, 2.00, 8.75, 15.00 and $20.00 \times 10^{12} cm^{-2}$ and an inhomogeneous broadening of 5 meV. Inset: wavefunctions of energetically lowest electron and hole states for densities of 0.05 (solid black), 8.75 (dashed black) and $20.00 \times 10^{12} cm^{-2}$ (grey).

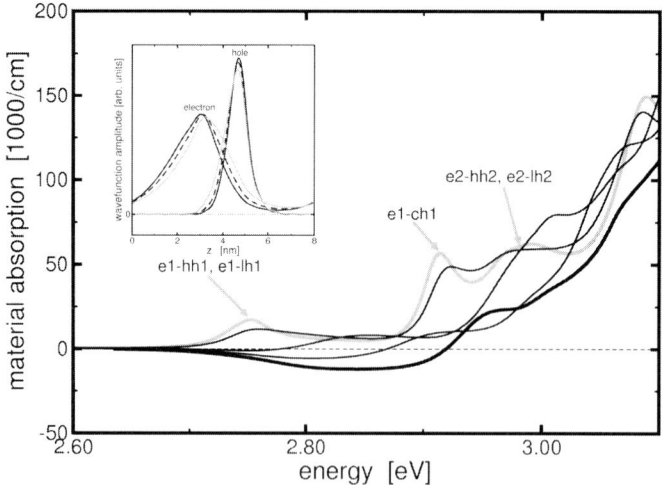

Fig. 7.6 Absorption for an $In_{0.2}Ga_{0.8}N/GaN$ structure with a 2 nm wide well, carrier densities of 0.05, 2.00, 8.75, 15.00 and $20.00 \times 10^{12} cm^{-2}$ and an inhomogeneous broadening of 5 meV. Inset: wavefunctions of energetically lowest electron and hole states for densities of 0.05 (solid black), 8.75 (dashed black) and $20.00 \times 10^{12} cm^{-2}$ (grey).

are still well pronounced, but less than for 10%, and the gain is reduced as well. The stronger confinement also leads to a stronger subband separation which allows to clearly identify not only the lowest heavy-hole and light-hole excitons, $e1$–$hh1$ and $e1$–$lh1$, but also the crystal-field split-hole, $e1$–$ch1$, and the second light-hole and heavy-hole excitons – at least for the small inhomogeneous broadening assumed here.

Since the barrier width influences the strength of the internal fields in the wells and barriers according to Eq. (7.6), it also influences the absorption and gain. E.g., if the width of the barriers is increased from 3 nm to 9 nm the internal field in the barriers is reduced to about one-third. The most dominant result of this is that the electrons are no longer localized as closely to the well as before. As mentioned earlier, in the structure of Ref. [5], the electrons are not confined in the well, but their confinement is provided by the tilted barriers. Thus, its localization is strongly influenced by the internal field in the barrier.

As can be seen from Fig. 7.7, the reduced overlap between electron and hole wavefunctions somewhat reduces the low-energy excitonic absorption and the gain. The smaller confinement for electrons also leads to a reduction of the band-edge at low densities by about 30 meV.

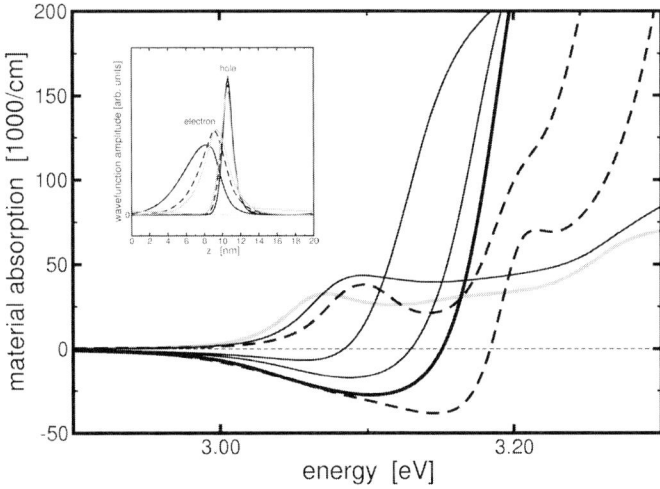

Fig. 7.7 Solid black (grey) lines: absorption for an $In_{0.2}Ga_{0.8}N/GaN$ structure with a 2 nm wide well and 9 nm wide barriers, carrier densities of $(0.05,) 2.00, 8.75, 15.00$ and $20.00 \times 10^{12} cm^{-2}$. Dashed lines: absorption of the same well, 3 nm wide barriers and densities of 0.05 and $20.00 \times 10^{12} cm^{-2}$. All results are for an inhomogeneous broadening of 5 meV. Inset: wavefunctions of energetically lowest electron and hole states for densities of 0.05 (solid black), 8.75 (dashed black) and $20.00 \times 10^{12} cm^{-2}$ (grey).

7.5 Spontaneous Emission

The spontaneous emission for the 2 nm $In_{0.1}Ga_{0.9}N/GaN$ structure of Ref. [5] is shown in Fig. 7.8. At low densities the main peak comes from the excitonic resonance as can be seen by comparing its energetic position to that of the absorption (Fig. 7.3). With increasing density the excitonic resonance becomes screened and the maximum shifts to higher energies.

As can be seen in Fig. 7.9, the spontaneous emission is strongly dependent on the structural details because its amplitude varies significantly with the wavefunction overlap. Thus, it is particularly small in the case of a (4 nm) wide well where electrons and holes are locally well separated.

In the low-density limit, spontaneous emission is assumed to scale with the density squared. As has been pointed out in Ref. [19], phase-space filling leads to a reduction in this density dependence at higher densities. In the limit of very high densities the dependence becomes only linear. Both the low and the high-density limits can be shown analytically if a series of assumptions is such as neglecting Coulomb interaction and all scattering processes as well as assuming single parabolic bands for electrons and holes. Another assumption

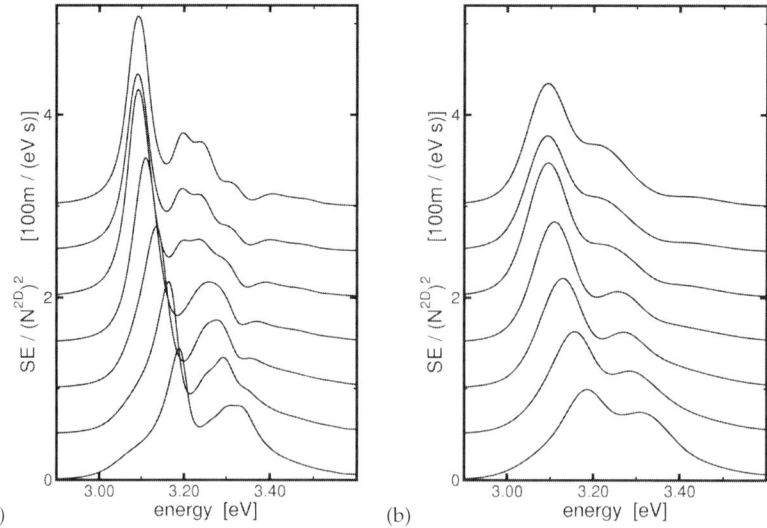

Fig. 7.8 Spontaneous emission divided by the density squared for the $In_{0.1}Ga_{0.9}N/GaN$ structure of Ref. [5] at densities of 0.05, 0.50, 2.00, 6.25, 10.00, 15.00 and $20.00 \times 10^{12} cm^{-2}$ (top to bottom) and inhomogeneous broadenings of 5 meV (a) and 80 meV (b). Spectra have been shifted by multiples of $50 \, m(eV \, s)^{-1}$ for clarity.

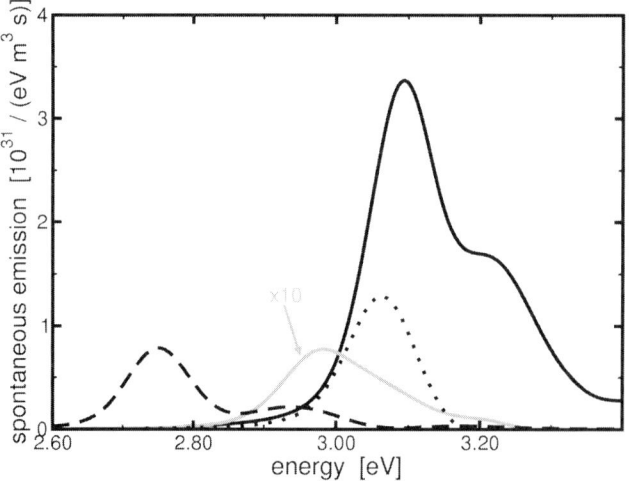

Fig. 7.9 Spontaneous emission for $In_xGa_{1-x}N/GaN$ structures at a density of $5 \times 10^{10} cm^{-2}$ and an inhomogeneous broadening of 80 meV. Black: $w^w = 2$ nm, $w^b = 3$ nm, $x = 0.1$; grey: $w^w = 4$ nm, $w^b = 3$ nm, $x = 0.1$, amplified by 10; dashed: $w^w = 2$ nm, $w^b = 3$ nm, $x = 0.2$; dotted: $w^w = 2$ nm, $w^b = 9$ nm, $x = 0.1$.

is particularly in question here: One has to assume that the bandstructure and dipole matrix elements are independent of the carrier density. Obviously,

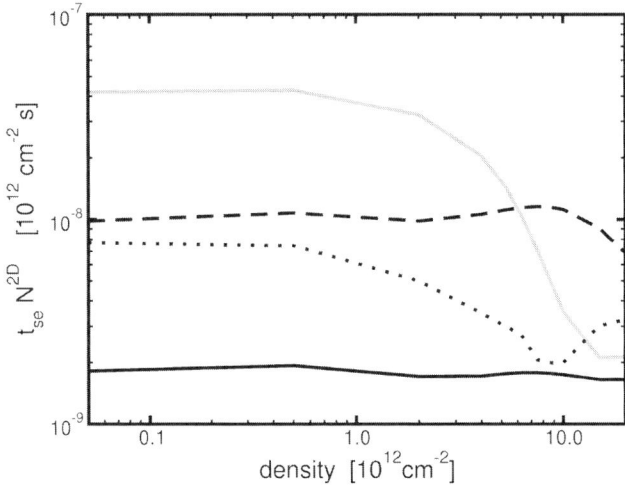

Fig. 7.10 Product of spontaneous emission lifetime and sheet carrier density for $In_xGa_{1-x}N/GaN$ structures as a function of the density. Black: $w^w = 2\,\text{nm}$, $w^b = 3\,\text{nm}$, $x = 0.1$; Grey: $w^w = 4\,\text{nm}$, $w^b = 3\,\text{nm}$, $x = 0.1$; Dashed: $w^w = 2\,\text{nm}$, $w^b = 3\,\text{nm}$, $x = 0.2$; Dotted: $w^w = 2\,\text{nm}$, $w^b = 9\,\text{nm}$, $x = 0.1$.

here, the latter assumption does not hold since the screening of the internal fields leads to strongly density-dependent wavefunction overlaps and thus dipole matrix elements. With increasing density, the overlap increases, which should lead to a stronger increase in the spontaneous emission than is the case in the absence of this effect.

The carrier lifetime due to spontaneous emission is shown in Fig. 7.10 for the structures discussed in Section 7.4. As can be seen from Eqs (7.12) and (7.13), the carrier lifetime due to spontaneous emission decreases linearly with the density if the spontaneous emission increases with the square of the density. This density dependence is indeed found for low densities in all structures investigated here.

For the structure with the 2 nm wide well and 10% indium, the overlap of the wavefunctions does not vary much with the carrier density. Thus, this effect does not lead to a reduction in the lifetime. For the structure with the 4 nm wide well the wavefunction overlap is strongly dependent on the density. Here the lifetime decreases substantially faster with density than linearly. A similar effect is also visible for the structure with 9 nm wide barriers.

Once the internal field is substantially screened, the wavefunction overlap will not improve any further and the resulting decreasing effect on the lifetimes ends.

The phase-space filling does not appear to have a significant impact on the carrier lifetime in these systems. It would lead to an increase of the ratios

t_{se}/N^{2D} in Fig. 7.10. Only for the structure with the 4 nm wide well and the one with the 9 nm wide barriers and the highest densities shown here, is such an effect observable. Obviously the formation of gain shows that phase-space filling occurs. However, at the same time, the absorption energetically right above the gain increases at the same time, off-setting the effect of the filling of the states below.

7.6
Auger Recombinations

Auger recombination is traditionally considered to be negligible in wide-bandgap materials. The matrix elements that allow for the electron–hole coupling necessary for Auger transitions, scale about inversely with the bandgap. This leads to Coulomb matrix elements V in Eq. (7.14) that scale roughly like $1/E_g^2$. Since the square of the Coulomb matrix element enters in the Auger rates, the rates should scale roughly with $1/E_g^4$. For wide-bandgap materials as investigated here, this should lead to a reduction in the Auger rates by almost two orders of magnitude when compared to those in devices operating near 1.3 µm. In the latter devices, carrier losses due to Auger recombinations are known to be of similar or greater importance to those due to spontaneous emission [20].

However, modern laser diodes based on the wide-bandgap nitrides operate under very high excitation conditions. The pump powers can be five or more times the threshold. Auger losses (at least at low densities) increase with about the third power of the carrier density. Losses due to spontaneous emission scale only with the square of the density at low densities and, as pointed out in Ref. [8], only linearly at high densities. Thus, the Auger rates might become important at these high excitation powers.

Also, the Coulomb interaction is much stronger in these materials due to the smaller dielectric constant.

However, as can be seen from Fig 7.11, for all situations investigated here and all densities, the Auger losses turn out to be negligibly small as compared with the losses due to spontaneous emission. Even at the highest densities the difference is still about two orders of magnitude.

7.7
Internal Field Effects

In the comparison between theoretical and experimental gain spectra (Fig. 7.2), we found good agreement using standard literature values for the internal piezoelectric and spontaneous polarization fields. However, there is still an

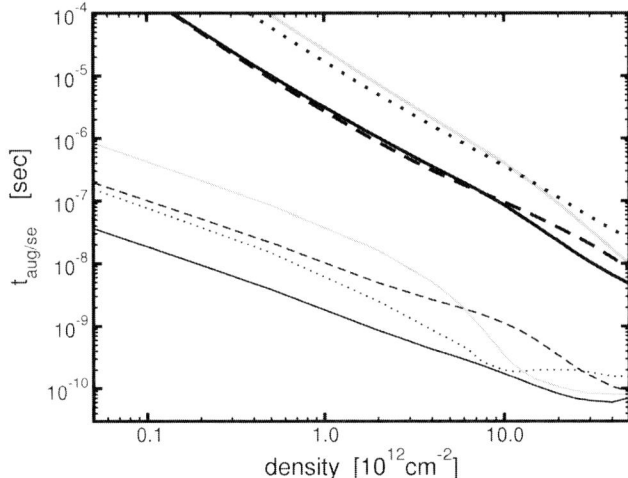

Fig. 7.11 Auger (thick lines) and spontaneous emission (thin lines) carrier lifetimes for $In_xGa_{1-x}N/GaN$ structures as a function of the density. Black: $w^w = 2\,nm$, $w^b = 3\,nm$, $x = 0.1$; grey: $w^w = 4\,nm$, $w^b = 3\,nm$, $x = 0.1$; dashed: $w^w = 2\,nm$, $w^b = 3\,nm$, $x = 0.2$; dotted: $w^w = 2\,nm$, $w^b = 9\,nm$, $x = 0.1$.

ongoing debate about the correct strength of these fields (see, e.g., Ref. [18] and Refs therein).

Typically the strength of the fields is determined measuring spectral positions of the luminescence as a function of varying external fields. However, as shown in Ref. [18], determining the correct strength of the field in the active region as a function of the external field is not trivial and can lead to significant errors in the analysis.

Figure 7.12 demonstrates the strong field dependence of the amplitude and spectral position of the luminescence and gain for a 4 nm wide $In_{0.1}Ga_{0.9}N/GaN$ well. Here, the strength of the internal field has been scaled artificially. Reducing the field by one-half leads to an increase in the luminescence peak by one order of magnitude. It also leads to a spectral shift of it by about 150 meV. For the gain, such a reduction in the field leads to an amplification by more than two (for this density, broadening and temperature) while the spectral shift is only about 20 meV.

As long as the excitation is kept in the low-density regime, where phase-space filling and screening of the internal field are negligible, changes in the carrier density only lead to an overall scaling of the luminescence but do not change its spectral position or lineshape. Thus, the field dependence of the luminescence, as shown in Fig. 7.12, is insensitive to the exact excitation conditions.

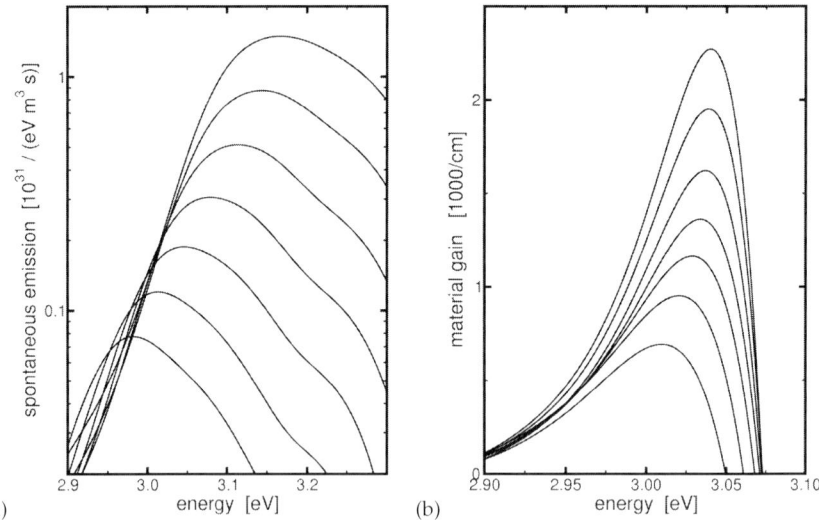

Fig. 7.12 (a): spontaneous emission and (b) gain for an $In_{0.1}Ga_{0.9}N$/GaN-structure with a 4 nm wide well, a density of 5×10^{10} cm^{-2} (1×10^{13} cm^{-2}) and an inhomogeneous broadening of 80 meV. From top to bottom the internal fields have been scaled by a factor $\eta = 0.4, 0.5, 0.6, 0.7, 0.8, 0.9$ and 1.0.

7.8
Summary

Fully microscopic models are used to investigate the absorption/gain and spontaneous emission as well as Auger processes in wide-bandgap nitride materials. The models only involve basic input like bulk bandstructure parameters and the nominal structural layout, but no fit parameters. Very good agreement with experiment is demonstrated.

The calculations demonstrate that the properties of wide-bandgap nitride materials are dominated by manybody interactions. The Coulomb interaction leads to very pronounced excitonic signatures and strong Coulomb enhancement of the absorption. The bandgap shifts with excitation density due to a screening of the internal fields as well as due to a reduction in the Coulomb interaction, the Coulomb induced bandgap renormalization and phononic energy renormalizations. All effects have to be taken into account on a level as presented here for a correct understanding of the system. Simplified models will not yield correct spectral positions, lineshapes, amplitudes and/or density dependencies.

Due to the strong internal fields, all properties are strongly dependent on the structural details like well widths and depths as well as barrier widths. In particular, the amplitude and spectral position of the spontaneous emission is extremely sensitive to any structural variations.

Acknowledgments

This work was supported by the U.S. Air Force Office of Scientific Research under contract F49620-02-1-0380, by the Deutsche Forschungsgemeinschaft and the Humboldt Foundation. NLCST acknowledges support through a Phase I STTR contract No. FA9550-05-C-0069.

References

1 M. Kneissl, D. W. Treat, M. Teepe, N. Miyashita, N. M. Johnson, Phys. Stat. Sol. A **200**, 118 (2003).

2 B. Neubert, P. Brückner, F. Habel, F. Scholz, T. Riemann, J. Christen, M. Beer, J. Zweck, Appl. Phys. Lett. **87**, 182111 (2005).

3 F. Jain, W. Huang, IEEE J. Quant. El. **32**, 859 (1996).

4 W. W. Chow, A. Girndt, S. W. Koch, Optics Express **2**, 119 (1998).

5 B. Witzigmann, V. Laino, M. Luisier, *et al.*: Appl. Phys. Lett. **88**, 021104 (2006).

6 J. Hader, A.R. Zakharian, J.V. Moloney, *et al.*: IEEE Photon. Technol. Lett. **14**, 762 (2002).

7 J. Hader, J.V. Moloney, S.W. Koch, and W.W. Chow: IEEE J. Sel. Topics Quantum Electron. **9**, 688 (2003).

8 J. Hader, J.V. Moloney, and S.W. Koch: IEEE J. Quantum Electron. **41**, 1217 (2005).

9 S.W. Koch, J. Hader, A. Thränhardt, and J.V. Moloney: *Gain and Absorption: Many-Body Effects*, in *Optoelectronic Devices; Advanced Simulation and Analysis*, ed. by J. Piprek (Springer, New York, 2005).

10 S.L. Chuang, and C.S. Chang: Phys. Rev. **B 54**, 2491 (1996).

11 P. Löwdin: J. Chem. Phys. **19**, 1396 (1951).

12 R. Winkler, and U. Rössler: Phys. Rev. **B 48**, 8918 (1993).

13 D. Ahn, and S.L. Chuang: J. Appl. Phys. **64**, 6143 (1988).

14 J. Hader, S.W. Koch, J. V. Moloney: Sol. Stat. Electron. **47**, 513 (2003).

15 W.W. Chow, S.W. Koch: *Semiconductor-Laser Fundamentals; Physics of the Gain Materials*, (Springer, Berlin Heidelberg New York 1999).

16 M. Kira, F. Jahnke, W. Hoyer, S. W. Koch, Progress in Quantum Electronics **23**, 189 (1999).

17 U.T. Schwarz, E. Sturm, W. Wegscheider, *et al.*: Appl. Phys. Lett. **85**, 1475 (2004).

18 I.H. Brown, I.A. Pope, P.M. Smowton, *et al.*: Appl. Phys. Lett. **86**, 131108 (2005).

19 J. Hader, J.V. Moloney, and S.W. Koch: Appl. Phys. Lett. **87**, 201112 (2005).

20 C. Schlichenmaier, A. Thränhardt, T. Meier, *et al.*: Appl. Phys. Lett. **87**, 261109 (2005).

8
Electronic and Optical Properties of GaN-based Quantum Wells with (10$\bar{1}$0) Crystal Orientation

Seoung-Hwan Park and Shun-Lien Chuang

8.1
Introduction

Wurtzite (WZ) GaN-based quantum wells (QWs) have received a great deal of attention in recent years for their potential application in blue-green optoelectronic devices [1]. The Hamiltonian based on the **k.p** method, for the conduction and valence-band structures of the GaN-based quantum wells have been derived [2] and applied to strained GaN quantum-well lasers [3]. The hole effective mass of WZ GaN-based QWs with the (0001) crystal orientation is significantly larger than that of conventional zincblende (ZB) crystals used in semiconductor lasers, such as GaAs and InP. In addition, it has been shown that the introduction of biaxial strain into the (0001)-plane of WZ GaN-based QWs does not effectively reduce the effective masses in the transverse direction [2–8]. In the conventional ZB crystals, on the other hand, a biaxial compressive strain in the QW structure reduces the in-plane effective mass of the top heavy-hole band and the threshold current density of a semiconductor laser [9]. WZ GaN-based QWs with the (0001) crystal orientation require higher carrier densities to generate optical gain compared to ZB GaAs-based QWs [10,11]. In addition, it was found that the GaN-based QW structures have a large internal electric field due to piezoelectric (PZ) and spontaneous (SP) polarizations [12–15]. This internal field in the QW structures causes the intrinsic quantum confined Stark effect, resulting in a red shift of the transition energy and a decrease in the radiative transition probability. The large internal field is expected to strongly affect the electronic and optical properties of GaN-based QW structures. Thus, the control of the internal field and the reduction of the effective mass is very important for the realization of high-performance, GaN-based devices.

As an additional parameter for band-structure engineering, the crystal orientation effect on characteristics of WZ GaN-based QW lasers has been studied for the realization of high-performance devices [16–20]. In particular, the

WZ GaN-based QW structures with a ($10\bar{1}0$) crystal orientation is interesting because the polarization field component is expected to be zero along the growth direction [$10\bar{1}0$] [18]. For a ($10\bar{1}0$) crystal orientation, Domen *et al.* [21] investigated the band structure and optical gain of bulk ($10\bar{1}0$) GaN. Niwa *et al.* [22] obtained the valence-subband structures by using the tight-binding method, and Yeo *et al.* [23] studied the valence-subband structure and optical gain by using the multiband effective-mass theory. They were compared with those for the (0001)-oriented QW lasers. However, most works have been studied under the free-carrier (FC) model without considering the many-body effects and under the flat-band (FB) model without considering the internal field due to the PZ and SP polarizations. The electronic and the optical properties of GaN-based QW lasers can be influenced significantly by the internal field. Thus, the inclusion of the internal field is important for an exact comparison with (0001) QW structures. Also, it is well-known that the many-body effects are significant due to the large exciton binding energy in GaN-based QWs [24]. More recently, effects of ($10\bar{1}0$) crystal orientation on the electronic and optical properties of wurtzite GaN/AlGaN QW structures, in particular, comparisons with those with (0001) crystal orientation have been extensively investigated with the inclusion of the internal electric field and the many-body effect [25–32]. On the experimental side, there has been a renewed interest in the growth of nonpolar III-nitride films for devices with polarization-free active regions which should increase their efficiency. Several groups have reported the deposition of planar m, and a-plane GaN films [33–38]. These QWs indeed gave experimental indications for the absence of the internal, built-in electric fields.

In this chapter, we review the electronic and the optical properties of ($10\bar{1}0$)-oriented WZ GaN-based QW lasers by using multiband effective-mass theory and the non-Markovian gain model with many-body effects. We compare the results with those of (0001)-oriented QW structures. The valence-band structures for the ($10\bar{1}0$)-oriented QW structure are calculated by using a 6×6 Hamiltonian based on the $\mathbf{k}\cdot\mathbf{p}$ method [17]. For the (0001)-oriented QW structure, self-consistent (SC) solutions are obtained by solving the Schrödinger equation for electrons, the block-diagonalized 3×3 Hamiltonian for holes, and Poisson's equation, iteratively [14, 39].

8.2
Theory

The Hamiltonian for the valence-band structure has been derived by using the $\mathbf{k} \cdot \mathbf{p}$ method [2,3]. The full Hamiltonian for the (0001)-oriented WZ crystal can be written as

$$H(\mathbf{k}, \epsilon) = \begin{pmatrix} F & -K^* & -H^* & 0 & 0 & 0 \\ -K & G & H & 0 & 0 & \Delta \\ -H & H^* & \lambda & 0 & \Delta & 0 \\ 0 & 0 & 0 & F & -K & H \\ 0 & 0 & \Delta & -K^* & G & -H^* \\ 0 & \Delta & 0 & H^* & -H & \lambda \end{pmatrix} \begin{matrix} |U_1> \\ |U_2> \\ |U_3> \\ |U_4> \\ |U_5> \\ |U_6> \end{matrix} \quad (8.1)$$

where

$$F = \Delta_1 + \Delta_2 + \lambda + \theta$$
$$G = \Delta_1 - \Delta_2 + \lambda + \theta$$
$$\lambda = \frac{\hbar^2}{2m_o}[A_1 k_z^2 + A_2(k_x^2 + k_y^2)] + \lambda_\epsilon$$
$$\theta = \frac{\hbar^2}{2m_o}[A_3 k_z^2 + A_4(k_x^2 + k_y^2)] + \theta_\epsilon$$
$$K = \frac{\hbar^2}{2m_o}A_5(k_x + ik_y)^2 + D_5\epsilon_+$$
$$H = \frac{\hbar^2}{2m_o}A_6(k_x + ik_y)(k_z) + D_6\epsilon_{z+}$$
$$\lambda_\epsilon = D_1(\epsilon_{zz}) + D_2(\epsilon_{xx} + \epsilon_{yy})$$
$$\theta_\epsilon = D_3(\epsilon_{zz}) + D_4(\epsilon_{xx} + \epsilon_{yy})$$
$$\epsilon_+ = \epsilon_{xx} - \epsilon_{yy} + 2i\epsilon_{xy}$$
$$\epsilon_{z+} = \epsilon_{xz} + i\epsilon_{yz}$$
$$\Delta = \sqrt{2}\Delta_3 \quad (8.2)$$

Here, the A_is are the valence-band effective-mass parameters which are similar to the Luttinger parameters in a ZB crystal, the D_is are the deformation potentials for WZ crystals, k_i is the wave vector, ϵ_{ij} is the strain tensor, Δ_1 is the crystal-field split energy, and Δ_2 and Δ_3 account for spin–orbit interactions. The bases for the Hamiltonian are defined as

$$|U_1> = -\frac{1}{\sqrt{2}}|(X+iY)\uparrow>$$
$$|U_2> = \frac{1}{\sqrt{2}}|(X-iY)\uparrow>$$
$$|U_3> = |Z\uparrow>$$
$$|U_4> = \frac{1}{\sqrt{2}}|(X-iY)\downarrow>$$
$$|U_5> = -\frac{1}{\sqrt{2}}|(X+iY)\downarrow>$$
$$|U_6> = |Z\downarrow> \quad (8.3)$$

The Hamiltonian for an arbitrary crystal orientation can be obtained using a rotation matrix

$$U = \begin{pmatrix} \cos\theta\cos\phi & \cos\theta\sin\phi & -\sin\theta \\ -\sin\phi & \cos\phi & 0 \\ \sin\theta\cos\phi & \sin\theta\sin\phi & \cos\theta \end{pmatrix} \quad (8.4)$$

Rotations of the Euler angles θ and ϕ transform the physical quantities from (x,y,z) coordinates to (x',y',z') coordinates. The z-axis corresponds to the c-axis [0001], and the growth axis (defined as the z'-axis) is normal to the QW plane $(hkil)$. The relation between the coordinate systems for vectors and tensors is expressed as [40,41]

$$\begin{aligned} k'_i &= U_{i\alpha} k_\alpha \\ \epsilon'_{ij} &= U_{i\alpha} U_{j\beta} \epsilon_{\alpha\beta} \\ C'_{ijkl} &= U_{i\alpha} U_{j\beta} U_{i\gamma} U_{j\delta} C_{\alpha\beta\gamma\delta} \end{aligned} \quad (8.5)$$

where summation over repeated indices is indicated.

The strain coefficients in the (x,y,z) coordinates for a general crystal orientation are determined from the condition that the layer is grown pseudomorphically and these strain coefficients should minimize the strain energy of the layer simultaneously [42]. We consider only the θ dependence of the physical quantities in the following, due to the hexagonal symmetry [17,43]. Here, θ is defined as the angle between the growth direction in the general crystal orientation and that (c-axis) in the (0001) crystal orientation; $\theta = 0$ corresponds to the (0001) growth direction and $\theta = \pi/2$ corresponds to the $(10\bar{1}0)$ growth direction. Then, the following relations for the strain tensors are obtained:

$$\begin{aligned} \epsilon_{xx} &= \epsilon_{xx}^{(0)} + \epsilon_{xz}\frac{\sin\theta}{\cos\theta} \\ \epsilon_{yy} &= \epsilon_{xx}^{(0)} \\ \epsilon_{zz} &= \epsilon_{xz}\frac{\cos\theta}{\sin\theta} + \epsilon_{zz}^{(0)} \\ \epsilon_{xy} &= \epsilon_{yz} = 0 \end{aligned} \quad (8.6)$$

where $\epsilon_{xx}^{(0)} = (a_s - a_e)/a_e$ and $\epsilon_{zz}^{(0)} = (c_s - c_e)/c_e$, which are due to mismatches between the lattice constants of the well (a_e and c_e) and the substrate (a_s and c_s). By minimizing the strain energy with respect to the variable ϵ_{xz}, we can obtain the following expression for ϵ_{xz}:

$$\epsilon_{xz} = -\frac{\left[\left(\dfrac{c_{11}+c_{12}+c_{13}\epsilon_{zz}^{(0)}}{\epsilon_{xx}^{(0)}}\right)\sin^2\theta + \left(\dfrac{2c_{13}+c_{33}\epsilon_{zz}^{(0)}}{\epsilon_{xx}^{(0)}}\right)\cos^2\theta\right]\epsilon_{xx}^{(0)}\cos\theta\sin\theta}{c_{11}\sin^4\theta + 2(c_{13}+2c_{44})\sin^2\theta\cos^2\theta + c_{33}\cos^4\theta} \tag{8.7}$$

where c_{ij} are the stiffness constants for the WZ structure. Hence, we obtain a 6×6 Hamiltonian for the $(10\bar{1}0)$ orientation by substituting the transformation relation with $\theta = \pi/2$ for the vector k in Eq. (8.5). The strain coefficients for the (0001) and the $(10\bar{1}0)$ orientations are given by substituting $\theta = 0$ and $\pi/2$ into Eq. (8.7), respectively. That is, the strain coefficients for the (0001) orientation are given by

$$\begin{aligned}\epsilon_{xx} &= \epsilon_{yy} = \epsilon_{xx}^{(0)}\\ \epsilon_{zz} &= -2\frac{c_{13}}{c_{33}}\epsilon_{zz}^{(0)}\\ \epsilon_{xy} &= \epsilon_{yz} = \epsilon_{zx} = 0\end{aligned} \tag{8.8}$$

Then, the strain-induced PZ field for the (0001)-WZ structure is written as [12]

$$F_z = \frac{2d_{31}}{\epsilon}\left(c_{11}+c_{12}-\frac{2c_{13}^2}{c_{33}}\right)\epsilon_{||} \tag{8.9}$$

where d_{31} is the piezoelectric constant, ϵ is the dielectric constant, and $\epsilon_{||} = \epsilon_{xx}^{(0)} = \epsilon_{yy}^{(0)}$.

Similarly, we can obtain the strain coefficients for the $(10\bar{1}0)$ orientation:

$$\begin{aligned}\epsilon_{xx} &= -\frac{c_{12}}{c_{11}}\epsilon_{xx}^{(0)} - \frac{c_{13}}{c_{11}}\epsilon_{zz}^{(0)}\\ \epsilon_{yy} &= \epsilon_{xx}^{(0)}\\ \epsilon_{zz} &= \epsilon_{zz}^{(0)}\\ \epsilon_{xy} &= \epsilon_{yz} = \epsilon_{zx} = 0\end{aligned} \tag{8.10}$$

where c_{11} and c_{12} are the stiffness constants for the WZ structure.

In general, the strain-induced PZ field is given as a function of the crystal orientation and can be expressed as [17, 43]

$$\begin{aligned}P_{PZ} &= P_x \sin\theta + P_z \cos\theta\\ P_x &= 2d_{15}c_{44}\epsilon_{xz}\\ P_z &= [d_{31}(c_{11}+c_{12}) + d_{33}c_{13}](\epsilon_{xx}+\epsilon_{yy}) + [2d_{31}c_{13}+d_{33}c_{33}]\epsilon_{zz}\end{aligned} \tag{8.11}$$

and d_{ij} are the piezoelectric constants and $d_{33}=-0.2d_{31}$. The strain-induced PZ field becomes zero for the $(10\bar{1}0)$ orientation ($\theta = \pi/2$). Thus, we use the flat-band (FB) model without the PZ field effect for the $(10\bar{1}0)$ oriented structure. The spontaneous polarization P_{SP} along the growth direction in the arbitrary crystal orientation is also estimated by the relation $P_{SP} = P_{SP}^{(0001)} \cos\theta$ since the spontaneous polarization in WZ (0001) GaN-based QW structures exists along the (0001) axis [18]. Here, $P_{SP}^{(0001)}$ is the spontaneous polarization in the (0001) crystal orientation, which is given in Table 8.1.

The optical momentum matrix elements in (x', y', z') coordinates for a general crystal orientation are given by

$$|\hat{e}' \cdot \mathbf{M}'^{\eta}|^2 = |<\Psi_l'^{c\eta}|\hat{e}' \cdot \mathbf{p}'|\Psi_m'^{v}>|^2 \tag{8.12}$$

where Ψ'^{c} and Ψ'^{v} are wavefunctions for the conduction and the valence bands, respectively, and $\eta = \uparrow$ and $\downarrow$ for both electron spins. The polarization-dependent interband momentum-matrix elements can be written as
TE-polarization ($\hat{e}' = \cos\phi'\hat{x}' + \sin\phi'\hat{y}'$):

$$\begin{aligned}|\hat{e}' \cdot \mathbf{M}'^{\uparrow}|^2 &= \Big|\cos\phi'\Big\{-\frac{1}{\sqrt{2}}\cos\theta P_x <g_m'^{(1)}|\phi_l> +\frac{1}{\sqrt{2}}\cos\theta P_x <g_m'^{(2)}|\phi_l>\\ &\quad -\sin\theta P_z <g_m'^{(3)}|\phi_l>\Big\}\\ &\quad +\sin\phi'\Big\{-i\frac{1}{\sqrt{2}}P_x <g_m'^{(1)}|\phi_l> -i\frac{1}{\sqrt{2}}P_x <g_m'^{(2)}|\phi_l>\Big\}\Big|^2 \end{aligned} \tag{8.13}$$

$$\begin{aligned}|\hat{e}' \cdot \mathbf{M}'^{\downarrow}|^2 &= \Big|\cos\phi'\Big\{\frac{1}{\sqrt{2}}\cos\theta P_x <g_m'^{(4)}|\phi_l> -\frac{1}{\sqrt{2}}\cos\theta P_x <g_m'^{(5)}|\phi_l>\\ &\quad -\sin\theta P_z <g_m'^{(6)}|\phi_l>\Big\}\\ &\quad +\sin\phi'\Big\{-i\frac{1}{\sqrt{2}}P_x <g_m'^{(4)}|\phi_l> -i\frac{1}{\sqrt{2}}P_x <g_m'^{(5)}|\phi_l>\Big\}\Big|^2 \end{aligned} \tag{8.14}$$

TM-polarization ($\hat{e}' = \hat{z}'$):

$$\begin{aligned}|\hat{e}' \cdot \mathbf{M}'^{\uparrow}|^2 &= \Big|-\frac{1}{\sqrt{2}}\sin\theta P_x <g_m'^{(1)}|\phi_l> +\frac{1}{\sqrt{2}}\sin\theta P_x <g_m'^{(2)}|\phi_l>\\ &\quad +\cos\theta P_z <g_m'^{(3)}|\phi_l>\Big|^2 \end{aligned} \tag{8.15}$$

$$\begin{aligned}|\hat{e}' \cdot \mathbf{M}'^{\downarrow}|^2 &= \Big|-\frac{1}{\sqrt{2}}\sin\theta P_x <g_m'^{(1)}|\phi_l> +\frac{1}{\sqrt{2}}\sin\theta P_x <g_m'^{(2)}|\phi_l>\\ &\quad +\cos\theta P_z <g_m'^{(3)}|\phi_l>\Big|^2 \end{aligned} \tag{8.16}$$

where $g'^{(\nu)}_m$ ($\nu = 1, 2, 3, 4, 5,$ and 6) is the wavefunction for the mth subband in (x', y', z') coordinates. Also,

$$P_x = P_y = <S|p_x|X> = <S|p_y|Y> = \frac{m_o}{\hbar} P_2$$

$$P_z = <S|p_z|Z> = \frac{m_o}{\hbar} P_1$$

$$P_1^2 = \frac{\hbar^2}{2m_o} \left(\frac{m_o}{m_e^z} - 1\right) \frac{(E_g + \Delta_1 + \Delta_2)(E_g + 2\Delta_2) - 2\Delta_3^2}{E_g + 2\Delta_2}$$

$$P_2^2 = \frac{\hbar^2}{2m_o} \left(\frac{m_o}{m_e^t} - 1\right) \frac{E_g[(E_g + \Delta_1 + \Delta_2)(E_g + 2\Delta_2) - 2\Delta_3^2]}{(E_g + \Delta_1 + \Delta_2)(E_g + \Delta_2) - \Delta_3^2} \quad (8.17)$$

Then, the optical momentum matrix elements for the (0001) and (10$\bar{1}$0) orientation are given by substituting $\theta = 0$ and $\pi/2$ into Eqs (8.13)–(8.16), respectively.

8.2.1
Non-Markovian gain model with many-body effects

The optical polarization-dependent gain spectra are calculated by the non-Markovian gain model with many-body effects [44, 45]. The optical gain with many-body effects including the effects of anisotropy on the valence band dispersion is given by

$$g(\omega) = \sqrt{\frac{\mu_o}{\epsilon}} \left(\frac{e^2}{m_o^2 \omega}\right) \int_0^{2\pi} d\phi \int_0^\infty dk_{||} \frac{2k_{||}}{(2\pi)^2 L_w} |M_{nm}(k_{||}, \phi)|^2$$
$$\times [f_n^c(k_{||}, \phi) - f_m^v(k_{||}, \phi)] L(\omega, k_{||}, \phi) \quad (8.18)$$

where ΔF is the quasi-Fermi level separation, ω is the angular frequency, μ_o is the vacuum permeability, ϵ is the dielectric constant, $k_{||}$ and ϕ are the magnitude and angle of the in-plane wave vector in the QW plane, respectively; L_w is the well thickness, $|M_{nm}|^2$ is the momentum matrix element in the strained QW, f_n^c and f_m^v are the Fermi functions for the conduction band states and the valence band states, and $\hbar$ is the Planck constant. The Gaussian lineshape function $L(\omega, k_{||}, \phi)$ renormalized with many-body effects is given by

$$L(\omega, k_{||}, \phi)$$
$$= \frac{(1 - \text{Re}Q(k_{||}, \hbar\omega))\text{Re}\Xi(E_{lm}(k_{||}, \hbar\omega)) - \text{Im}Q(k_{||}, \hbar\omega)\text{Im}\Xi(E_{lm}(k_{||}, \hbar\omega))}{(1 - \text{Re}Q(k_{||}, \hbar\omega))^2 + (\text{Im}Q(k_{||}, \hbar\omega))^2}$$
$$(8.19)$$

where [44, 45]

$$\mathrm{Re}\Xi(E_{lm}(k_{||},\hbar\omega)) = \sqrt{\frac{\pi\tau_{co}}{2\hbar\Gamma_{cv}(k_{||},\hbar\omega)}}$$

$$\times \exp\left(-\frac{\tau_{co}}{2\hbar\Gamma_{cv}(k_{||},\hbar\omega)}E_{lm}^2(k_{||},\hbar\omega)\right) \quad (8.20)$$

and

$$\mathrm{Im}\Xi(E_{lm}(k_{||},\hbar\omega)) = \frac{\tau_{co}}{\hbar}\int_0^\infty \exp\left(-\frac{\Gamma_{cv}(k_{||},\hbar\omega)\tau_{co}}{2\hbar}t^2\right)$$

$$\times \sin\left(\frac{\tau_{co}E_{lm}(k_{||},\hbar\omega)}{\hbar}t\right)dt \quad (8.21)$$

In the above, $E_{lm}(k_{||},\hbar\omega) = E_l^c(k_{||}) - E_m^v(k_{||}) + E_g + \Delta E_{SX} + \Delta E_{CH} - \hbar\omega$ is the renormalized transition energy between electrons and holes, where E_g is the bandgap of the material. ΔE_{SX} and ΔE_{CH} are the screened exchange and Coulomb-hole contributions to the bandgap renormalization. $Q(k_{||},\hbar\omega)$ is the term related to the excitonic or Coulomb enhancement of the interband transition probability [46]. In the above, $\Gamma_{cv}(k_{||},\hbar\omega)$ are broadening factors related to the decay rate of the electron and hole wavefunctions. The correlation time τ_{co} is related to the non-Markovian enhancement of optical gain [44] and is assumed to be constant. We used the bandgap bowing value of -3.2 eV for InGaN.

The bandgap renormalization is given as a summation of the Coulomb-hole self-energy and the screened-exchange shift [46]. The ϕ dependence of the bandgap renormalization is very small. Here, the ϕ dependence of the bandgap renormalization is neglected for simplicity. The Coulomb-hole contribution to the bandgap renormalization is written as

$$\Delta E_{CH} = -2E_R a_o \lambda_s \ln\left(1 + \sqrt{\frac{32\pi N L_w}{C\lambda_s^3 a_o}}\right) \quad (8.22)$$

where N is the carrier density, λ_s is the inverse screening length, and C is a constant usually taken between 1 and 4. The Rydberg constant E_R and the exciton Bohr radius a_o are given by

$$\Delta E_R(eV) = 13.6\frac{m_o/m_r}{(\epsilon/\epsilon_o)^2} \quad (8.23)$$

and

$$a_o(\text{Å}) = 0.53\frac{\epsilon/\epsilon_o}{m_o/m_r} \quad (8.24)$$

where m_r is the reduced electron–hole mass defined by $1/m_r = 1/m_e + 1/m_h$. The exchange contribution to the bandgap renormalization is given by

$$\Delta E_{SX} = -\frac{2E_R a_o}{\lambda_s} \int_0^\infty dk_{||} k_{||} \frac{1 + \frac{C\lambda_s a_o k_{||}^2}{32\pi NL_w}}{1 + \frac{k_{||}}{\lambda_s} + \frac{C a_o k_{||}^3}{32\pi NL_w}} \left[f_n^c(k_{||}) + 1 - f_m^v(k_{||}) \right] \quad (8.25)$$

The factor $1/(1 - Q(k_{||}, \hbar\omega))$ represents the Coulomb enhancement in the Padé approximation. Here, the factor $Q(k_{||}, \hbar\omega)$ is given by [45]

$$Q(k_{||}, \hbar\omega) = i \frac{a_o E_R}{\pi \lambda_s |M_{nm}(k_{||})|} \int_0^\infty dk'_{||} k'_{||} |M_{nm}(k'_{||})| (f_n^c(k'_{||}) - f_m^v(k'_{||})) \quad (8.26)$$
$$\times \Xi(E_{lm}(k'_{||}, \hbar\omega)) \Theta(k_{||}, k'_{||})$$

where

$$\Theta(k_{||}, k'_{||}) = \int_0^\infty d\theta \left(1 + \frac{C\lambda_s a_o q_{||}^2}{32\pi NL_w}\right) \left(1 + \frac{q_{||}}{\lambda_s} + \frac{C a_o q_{||}^3}{32\pi NL_w}\right)^{-1} \quad (8.27)$$

and

$$q_{||} = |\mathbf{k}_{||} - \mathbf{k}'_{||}|$$

The radiative current density can be related to the spontaneous emission rate as

$$J_{\text{rad}} = e N_w L_w \int_0^\infty r_{\text{spon}}(\lambda) d\lambda \quad (8.28)$$

where N_w is the number of wells. The spontaneous emission rate r_{spon} is given by

$$r_{\text{spon}}(\lambda) = \frac{8\pi c n^2}{\lambda^2} g_{\text{sp}}(\lambda) \quad (8.29)$$

where $\lambda = 2\pi c/\omega$, c is the speed of light, and n is the refractive index of the QW. The spontaneous emission coefficient $g_{\text{sp}}(\omega)$, taking into account the non-Markovian relaxation and the many-body effects, is given by

$$g_{\text{sp}}(\omega) = \sqrt{\frac{\mu_o}{\epsilon}} \left(\frac{e^2}{m_o^2 \omega}\right) \int_0^{2\pi} d\phi \int_0^\infty dk_{||} \frac{2k_{||}}{(2\pi)^2 L_w} \quad (8.30)$$
$$\cdot |M_{nm}(k_{||}, \phi)|^2 f_n^c(k_{||}, \phi) [1 - f_m^v(k_{||}, \phi)] L(\omega, k_{||}, \phi)$$

8.3 Results and Discussion

In this section, we present the numerical results for the crystal orientation effects on important electronic and optical properties for $(10\bar{1}0)$-oriented WZ InGaN/GaN and GaN/AlGaN QW lasers.

Tab. 8.1 Physical parameters of GaN, AlN, and InN.

Parameter	GaN		AlN		InN	
Lattice constant (Å)						
a	3.1892	[47]	3.112	[48]	3.53	[49]
c	5.1850	[47]	4.982	[48]	5.54	[49]
Energy parameters						
E_g (eV)	3.44	[50]	6.28	[50]	1.89	[51]
$\Delta_{cr} = \Delta_1$ (meV)	22.0	[52]	−58.5	[53]	41.0	[54]
$\Delta_{so} = 3\Delta_2$ (meV)	15.0	[52]	20.4	[53]	1.0	[54]
$\Delta_3 = \Delta_2$ (meV)						
Conduction band effective masses						
m_{ez}^w/m_o	0.20	[53]	0.33	[53]	0.11	[55]
m_{et}^w/m_o	0.18	[53]	0.25	[53]	0.11	[55]
Valence band effective-mass parameters						
A_1	−6.56	[53]	−3.95	[53]	−9.09	[55]
A_2	−0.91	[53]	−0.27	[53]	−0.63	[55]
$A_3 = A_2 - A_1$						
$A_4 = A_3/2$						
A_5	−3.13	[53]	−1.95	[53]	−4.36	[55]
$A_6 = (A_3 + 4A_5)/\sqrt{2}$						
Deformation potentials (eV)						
$a_c = -6.4 + a_v$ [49]	−4.60	[56]	−4.50	[49]	−1.40	[56]
D_1	−1.70	[56]	−2.89	[57]	−1.76	[57]
D_2	6.30	[56]	4.89	[57]	3.43	[57]
$D_3 = D_2 - D_1$						
$D_4 = D_3/2$						
D_5	−4.00	[58]	−3.34	[57]	−2.33	[57]
Dielectric constant						
ϵ	10.0	[59]	8.5	[59]	15.3	[12]
Elastic stiffness constant (10^{11} dyn cm^{-2})						
c_{11}	39.0	[60]	39.8	[59]	27.1	[12]
c_{12}	14.5	[60]	14.0	[59]	12.4	[12]
c_{13}	10.6	[60]	12.7	[59]	9.4	[12]
c_{33}	39.8	[60]	38.2	[59]	20.0	[12]
c_{44}	10.5	[60]	9.6	[59]	4.6	[49]
c_{66}	12.3	[60]	12.9	[59]	7.4	[49]
Piezoelectric constant						
d_{31} ($\times 10^{-12}$ m/V)	−1.7	[12]	−2.0	[12]	−1.1	[12]

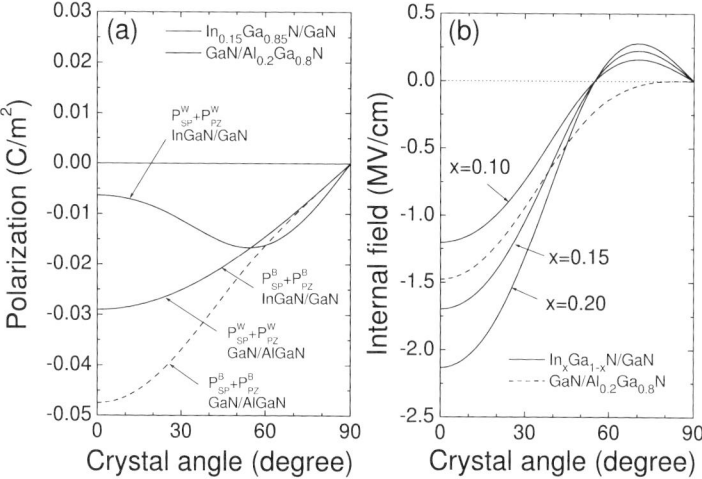

Fig. 8.1 (a) Sum of the strain-induced PZ and SP polarizations normal with respect to the growth plane for the well and the barrier and (b) the internal field as a function of the angle θ between the growth direction and the c-axis of $In_xGa_{1-x}N/GaN$ and $GaN/Al_{0.2}Ga_{0.8}N$ QW structures.

Figure 8.1 shows (a) the sum of the strain-induced PZ and SP polarizations normal with respect to the growth plane for the well and the barrier and (b) the internal field as a function of the angle θ between the growth direction and the c-axis of $In_xGa_{1-x}N/GaN$ and $GaN/Al_{0.2}Ga_{0.8}N$ QW structures. We consider a GaN/AlGaN MQW structure with the AlGaN barrier under tensile strain

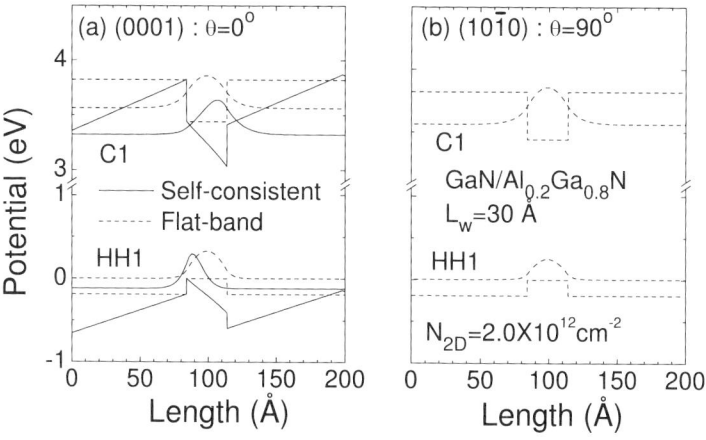

Fig. 8.2 Potential profiles and the wavefunctions (C1 and HH1) at the zone center for the self-consistent (SC) model with the PZ and SP polarization of: (a), (0001) and (b), (10$\bar{1}$0)-oriented 30 Å $GaN/Al_{0.4}Ga_{0.6}N$ QW structures. Here, the barrier width is set to be 84 Å.

and the GaN well lattice-matched to the GaN substrate. The InGaN/GaN QW structure has the compressively strained InGaN well and the unstrained GaN barrier. Note that the strain-induced normal PZ polarization P_{PZ} is zero for the barrier in the InGaN/GaN structure and the well in the GaN/AlGaN structure. The normal PZ polarization is important because the QW structures have quantized energy levels along the growth direction. The normal PZ polarization leads to the accumulation or depletion of carriers at the interfaces and creates a piezoelectric field [43]. In the case of the GaN/AlGaN QW structure, it is shown that the internal field gradually decreases with increasing crystal angle. On the other hand, the InGaN/GaN QW structure shows that the internal field becomes zero near the crystal angle of 55°, irrespective of the In composition in the well. This is because the sum of the PZ and SP polarization in the barrier is equal to that in the well. Also, it is observed that the internal field changes its sign when the crystal angle further increases. The internal field for both QW structures becomes zero for the $(10\bar{1}0)$ crystal orientation ($\theta = 90°$) because the SP and strain-induced PZ polarization is zero.

Figure 8.2 shows the potential profiles and the wavefunctions (C1 and HH1) at the zone center for the self-consistent (SC) model with the PZ and SP polarization of (a) (0001)-and (b) $(10\bar{1}0)$-oriented 30 Å GaN/Al$_{0.2}$Ga$_{0.8}$N QW structures. Here, the barrier width is set to be 84 Å and C1 and HH1 are acronyms for the first subband and the first heavy-hole subband. The SC solutions are obtained at a surface carrier density of $N_{2D} = 2 \times 10^{12}$ cm^{-2} and the dashed line corresponds to potential profiles for the flat-band (FB) model without the SP and PZ polarization. The AlGaN barrier is under 0.484% tensile strain and we assume that the layers have the Ga(A1)-face. For the Ga-face, the SP polarization is pointing towards the substrate because the SP polarization for GaN and AlN is found to be negative [61]. Then, the alignment of the PZ and SP polarization is parallel in the case of tensile strain. If the barrier is under compressive strain, the alignment of the PZ and SP polarization becomes antiparallel.

Potential profiles show that there is a large internal field in the well, due to the difference in the SP polarization between the well and barrier, although the PZ field in the QW is zero. The estimated field in the well is -1.55 MV cm^{-1} for a (0001)-oriented GaN/AlGaN QW with $L_w = 30$ Å. The FB model shows symmetric electron and hole wavefunctions with respect to the well center, while the SC model shows asymmetric electron and hole wavefunctions. In particular, a (0001)-oriented QW structure shows large spatial separation of the electron and hole wavefunctions. Also, the ground-state transition energy C1-HH1 is redshifted compared to the FB model due to the quantum-confined Stark effect. On the other hand, it is observed that a $(10\bar{1}0)$-oriented QW structure shows symmetric wavefunctions because of the disappearance of the PZ and SP polarization. Thus, we expect that important properties such as the

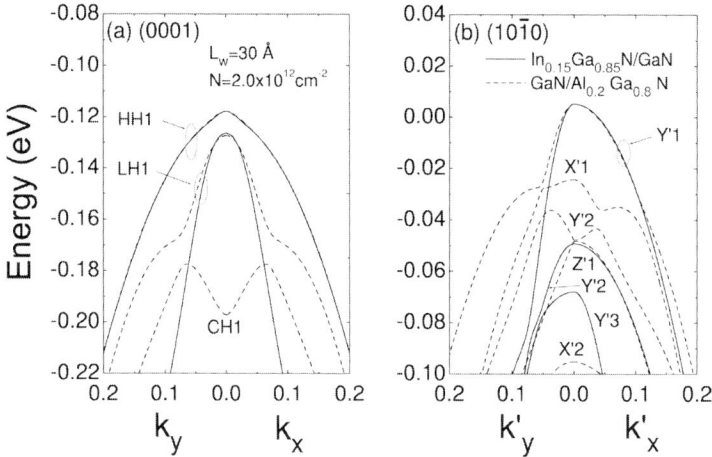

Fig. 8.3 Valence-band structures of: (a), (0001)- and (b), (10$\bar{1}$0)-oriented In$_{0.15}$Ga$_{0.85}$N/GaN QWs ($L_w = 30$ Å). For comparison, results for GaN/Al$_{0.2}$Ga$_{0.8}$N QWs are also plotted.

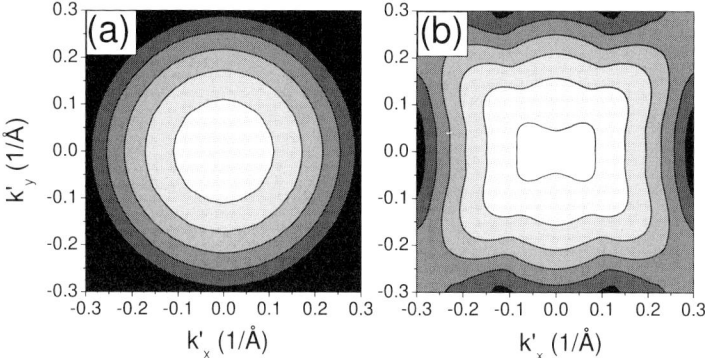

Fig. 8.4 Equienergy patterns in the in-plane wave vector space of: (a), (0001)- and (b), (10$\bar{1}$0)-oriented WZ GaN/Al$_{0.2}$Ga$_{0.8}$N QWs ($L_w = 30$ Å). Here, k'_x and k'_y are the in-plane wave vectors in the (x', y', z') coordinate system.

effective mass and the optical moment matrix elements are affected by the crystal orientation.

Figure 8.3 shows the valence-band structures of (a) (0001)- and (b) (10$\bar{1}$0)-oriented In$_{0.15}$Ga$_{0.85}$N/GaN QWs ($L_w = 30$ Å). For comparison, results for GaN/Al$_{0.2}$Ga$_{0.8}$N QWs are also plotted. The valence-band structures were calculated at a carrier density of 2×10^{12} cm^{-2}. The energies of the In-GaN/GaN QWs are rescaled to give the same Γ point energy as that of the GaN/InGaN QWs for the first subband. The added energies are 26 and -20 meV for the (0001) and the (10$\bar{1}$0) orientations, respectively. Here, the naming of the subbands for the (10$\bar{1}$0)-oriented structure follows the dominant composition of the wavefunction at the Γ point in terms of the $|X'\rangle$, $|Y'\rangle$, and $|Z'\rangle$ bases. The components $P_m^{i(=X',Y',Z')}$ of each wavefunction are given by

$$P_m^{X'} = \left\langle g_m'^{(3)} | g_m'^{(3)} \right\rangle + \left\langle g_m'^{(6)} | g_m'^{(6)} \right\rangle \quad (8.31)$$

$$P_m^{Y'} = \left(\langle g_m'^{(1)} + g_m'^{(2)} | g_m'^{(1)} + g_m'^{(2)} \rangle + \langle g_m'^{(4)} + g_m'^{(5)} | g_m'^{(4)} + g_m'^{(5)} \rangle \right)/2$$

$$P_m^{Z'} = \left(\langle g_m'^{(2)} - g_m'^{(1)} | g_m'^{(2)} - g_m'^{(1)} \rangle + \langle g_m'^{(4)} - g_m'^{(5)} | g_m'^{(4)} - g_m'^{(5)} \rangle \right)/2$$

The valence-band structure of the $(10\bar{1}0)$-oriented structure shows an anisotropy in the QW plane, unlike the (0001)-oriented structure. The subband energy spacing (∼10 meV) between the first two subbands for the (0001)-oriented QWs is shown to be very small (∼70 meV) compared to that of InP-based QWs due to the heavy-hole effective masses of GaN-based QWs. However, the subband energy spacing between the first two subbands increases with the inclusion of the $(10\bar{1}0)$ crystal orientation. In particular, the increase in the subband energy spacing is significant in the case of the $(10\bar{1}0)$-oriented InGaN/GaN QW. In addition, effective masses of the holes for the $(10\bar{1}0)$-oriented QWs are shown to be greatly reduced compared to the value for the (0001)-oriented QWs. We know that the effective mass of the topmost valence band along k'_y is much smaller than that along k'_x.

Figure 8.4 shows equi-energy patterns in the in-plane wave vector space of (a) (0001)- and (b) $(10\bar{1}0)$-oriented WZ GaN/Al$_{0.2}$Ga$_{0.8}$N QWs ($L_w = 30$ Å). Here, k'_x and k'_y are the in-plane wave vector in the (x', y', z') coordinate system. The k'_x axes for (0001) and $(10\bar{1}0)$ orientations are $[10\bar{1}0]$ and $[000\bar{1}]$, respectively. The k'_y axes for (0001) and $(10\bar{1}0)$ orientations are both $[\bar{1}2\bar{1}0]$. They are related to $k_\parallel$ and ϕ in Eq. (8.30) by $k'_x = k_\parallel \cos\phi$ and $k'_x = k_\parallel \sin\phi$. The energy pattern of the (0001) orientation is completely isotropic. On the other hand, the valence-band structure of the $(10\bar{1}0)$-oriented GaN/AlGaN QW shows anisotropy in the QW plane, unlike the (0001)-oriented structure. That is, the energy pattern of the $(10\bar{1}0)$ orientation shows two-fold symmetry. It is observed that the effective mass along k'_y around the topmost valence band is reduced for the $(10\bar{1}0)$-oriented structure. This is mainly attributed to the fact that the energy difference between the first and second subbands increases with the angle between the growth and the c-axis. The optical matrix elements also showed two-fold symmetry for the $(10\bar{1}0)$-oriented QW structures.

Figure 8.5 shows the hole density of states of the valence-band structures for: (a), (0001)- and (b), $(10\bar{1}0)$-oriented WZ In$_{0.15}$Ga$_{0.85}$N/GaN QW structures ($L_w = 30$ Å). For comparison, the results for GaN/Al$_{0.2}$Ga$_{0.8}$N QWs are also plotted. The energies of InGaN/GaN QWs are rescaled to give the same Γ point energy as that of the GaN/InGaN QWs for the first subband

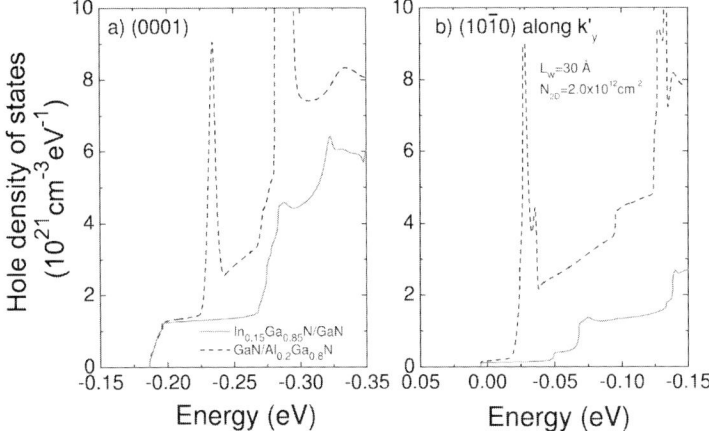

Fig. 8.5 Hole density of states of the valence-band structures for (a), (0001)- and (b), ($10\bar{1}0$)-oriented WZ $In_{0.15}Ga_{0.85}N$/GaN QW structures ($L_w = 30$ Å). For comparison, the results for GaN/$Al_{0.2}Ga_{0.8}$N QWs are also plotted.

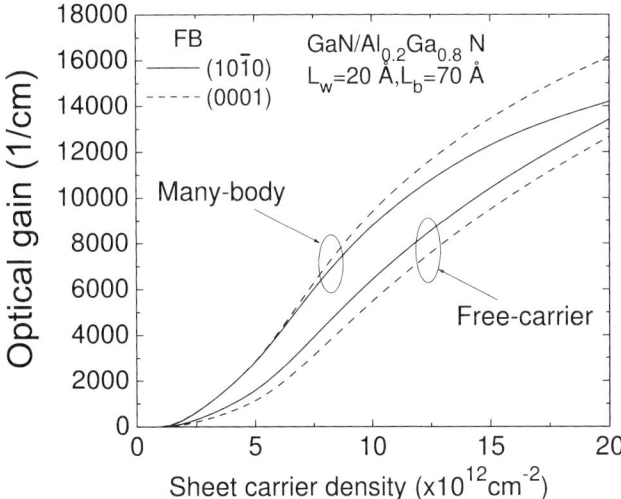

Fig. 8.6 y'-polarized TE optical gains as a function of carrier density for the free-carrier and many-body models of (0001)- and ($10\bar{1}0$)-oriented WZ GaN/$Al_{0.2}Ga_{0.8}$N QWs ($L_w = 20$ Å). Here, the FB model without the PZ and SP polarization is used for the (0001) orientation, to investigate many-body effects.

band, as shown in Fig. 8.3. It is evident that the density of states for the ($10\bar{1}0$)-oriented QW structures is greatly reduced compared with the (0001)-oriented QW structures. In particular, the reduction in the density of states is much more dominant in the case of the InGaN/GaN QW structure. This is mainly attributed to the fact that the energy spacing between the first two subbands of the InGaN/GaN QW structure is much larger than that of the GaN/AlGaN QW structure. The reduction in the density of states near the subband edge is important for improving the laser characteristics because the electrons and the holes are mostly near the subband edge. On the other hand, the (0001)-oriented structures show a higher density of states in the overall range of energy. Also, the densities of states on the lower energy side of GaN/AlGaN QWs are observed to be higher than those of the InGaN/GaN QWs. This is because the energy spacings between the subbands on the lower energy side of the GaN/AlGaN QWs are smaller than those of the InGaN/GaN QWs [30].

Figure 8.6 shows y'-polarized TE optical gains as a function of carrier density for the free-carrier and many-body models of (0001)- and (10$\bar{1}$0)-oriented WZ GaN/Al$_{0.2}$Ga$_{0.8}$N QWs (L_w = 20 Å). Here, the optical gain means the peak optical gain and the FB model without the PZ and SP polarization is used for the (0001) orientation to investigate many-body effects. In the case of the (10$\bar{1}$0) orientation, we always use the FB model since the PZ and SP polarization is expected to be zero at this orientation. The free-carrier model shows that the optical gain of the (10$\bar{1}$0)-oriented QW is enhanced over that of the (0001)-oriented QW. This is because, in the case of the (10$\bar{1}$0)-oriented QW structure, the states constituting the topmost valence subband near the band edge are predominantly $|Y'>$-like. This results in significantly enhanced y'-polarized optical matrix elements for the C1-Y'1 transition in the (10$\bar{1}$0)-oriented QW structure, in comparison with the y'-polarized optical matrix elements in the (0001)-oriented QW structure where $|X'>$ and $|Y'>$ are equally mixed. In addition, the average hole effective mass of the (10$\bar{1}$0) crystal orientation is smaller than that of the (0001) crystal orientation. However, considering many-body effects, we know that the optical gain of the (10$\bar{1}$0)-oriented QW is smaller than that of the (0001)-oriented QW. That is, the Coulomb enhancement contribution to the optical gain of the (0001)-oriented QW is larger than that for the (10$\bar{1}$0)-oriented QW. This is because the average hole effective mass of the (0001)-oriented QW is larger than that of the (10$\bar{1}$0)-oriented QW. These results suggest that the inclusion of the many-body effects is very important in order to give a guideline on device design issues.

Figure 8.7 shows y'-polarized TE optical gain spectra with many-body effects for: (a), (0001)- and (b), (10$\bar{1}$0)-oriented WZ In$_{0.15}$Ga$_{0.85}$N/GaN QW structures (L_w = 30 Å). For comparison, the results (dashed line) for GaN/Al$_{0.2}$Ga$_{0.8}$N QWs are also plotted. The optical gain spectra were calculated at a carrier density of 20×10^{12} cm^{-2}. The peak energy corresponds to the energy difference between the first conduction subband (C1) and the first valence subband. The peak energy of the GaN/AlGaN QW is shown to be larger than that of the InGaN/GaN QW because of its larger bandgap energy. We know that, in the case of the (0001)-oriented QW structure, the difference in the peak gain between InGaN/GaN and GaN/AlGaN QWs is

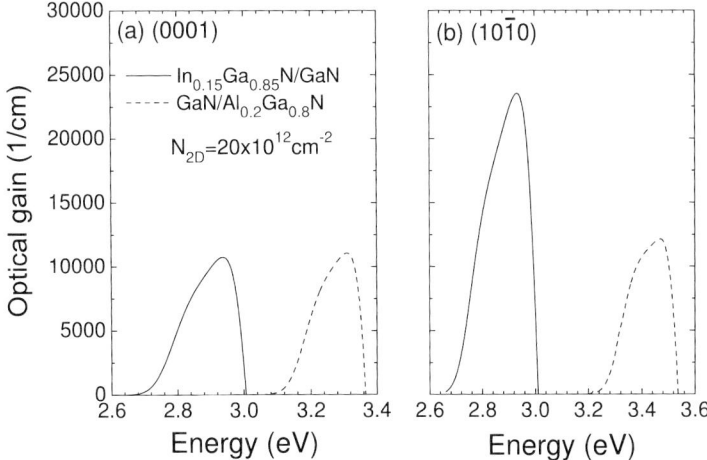

Fig. 8.7 y'-polarized TE optical gain spectra with many-body effects for: (a), (0001)- and (b), $(10\bar{1}0)$-oriented WZ $In_{0.15}Ga_{0.85}N$/GaN QW structures. For comparison, the results (dashed line) for the GaN/$Al_{0.2}Ga_{0.8}N$ QWs are also plotted.

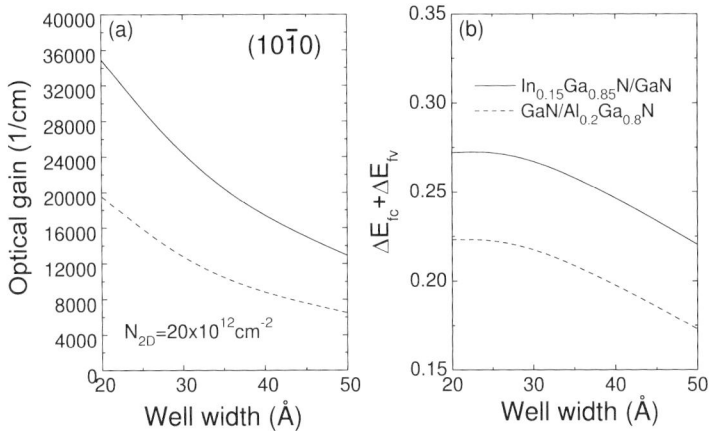

Fig. 8.8 (a) Many-body optical gain and (b) the quasi-Fermi-level separation $\Delta E_{fc} + \Delta E_{fv}$ as functions of the well width of $(10\bar{1}0)$-oriented WZ $In_{0.15}Ga_{0.85}N$/GaN QW structures. For comparison, the results (dashed line) for the GaN/$Al_{0.2}Ga_{0.8}N$ QWs are also plotted.

very small. This can be explained by the fact that hole effective masses of both QW structures are similar to each other. On the other hand, with the inclusion of the $(10\bar{1}0)$ crystal orientation, the optical gain of the InGaN/GaN QW structure becomes significantly larger than that of the GaN/AlGaN QW structure because the reduction in the density of states with the inclusion of the $(10\bar{1}0)$ crystal orientation is much more dominant in the case of the InGaN/GaN QW structure, as shown in Fig. 8.1.

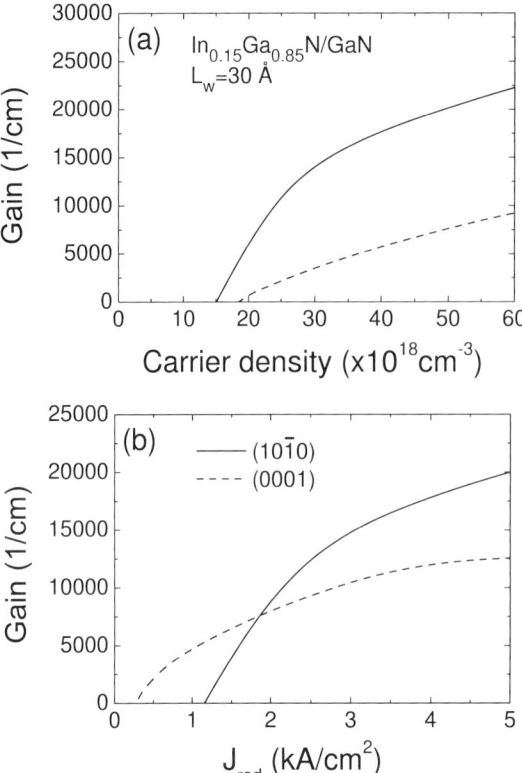

Fig. 8.9 y'-polarized TE optical gains as a function of carrier density for the free-carrier and many-body models of (0001)- and (10$\bar{1}$0)-oriented WZ GaN/Al$_{0.2}$Ga$_{0.8}$N QWs ($L_w = 30$ Å). Here, the FB model without the PZ and SP polarization is used for the (0001) orientation to investigate many-body effects.

Figure 8.8 shows: (a), the many-body optical gain and (b), the quasi-Fermi-level separation $\Delta E_{fc} + \Delta E_{fv}$ as functions of the well width of (10$\bar{1}$0)-oriented WZ In$_{0.15}$Ga$_{0.85}$N/GaN QW structures. For comparison, the results (dashed line) for the GaN/Al$_{0.2}$Ga$_{0.8}$N QWs are also plotted. The optical gain was calculated at a carrier density of 20×10^{12} cm^{-2}. The quasi-Fermi-level separation ΔE_{fc} (ΔE_{fv}) is defined as the energy difference between the quasi-Fermi level and the ground-state energy in the conduction band (the valence band). The optical gain is observed to decrease gradually with increasing well width for both QW structures because the subband energy spacing decreases with increasing well width. The decrease in the subband energy spacing results in a reduction in the quasi-Fermi level separation with increasing well width, as shown in Fig. 8.4(b). Also, the optical gain of the InGaN/GaN QW structure is found to be significantly larger than that of the GaN/AlGaN QW struc-

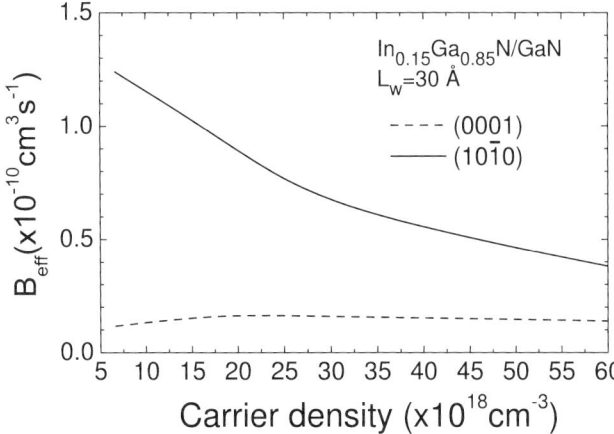

Fig. 8.10 Radiative recombination coefficient B_{eff} as a function of the carrier density for (0001)- and ($10\bar{1}0$)-oriented WZ $In_{0.15}Ga_{0.85}N$/GaN QW structures ($L_w = 30$ Å).

ture in the investigated range of well widths because the quasi-Fermi-level separation of the InGaN/GaN QW structure is much larger than that of the GaN/AlGaN QW structure. The increase in the quasi-Fermi-level separation observed in the InGaN/GaN QW structure is responsible for the reduction in the hole effective mass.

Figure 8.9 shows y'-polarized TE optical gains as a function of: (a) carrier density and (b) radiative current density for the many-body model of (0001)- and ($10\bar{1}0$)-oriented WZ $In_{0.15}Ga_{0.85}N$/GaN QW structures ($L_w = 30$ Å). In the case of the ($10\bar{1}0$) orientation, we use the flat-band (FB) model since the PZ and SP polarization is expected to be zero at this orientation [45]. The optical gain of the ($10\bar{1}0$)-oriented QW is largely enhanced over that of the (0001)-oriented QW in the investigated range of the carrier density. This is because, in the case of the ($10\bar{1}0$)-oriented QW structure, the states constituting the topmost valence subband near the band edge are predominantly $|Y'>$-like and the internal field due to the PZ and SP polarization is negligible. This results in significantly enhanced y'-polarized optical matrix elements in the ($10\bar{1}0$)-oriented QW structure, in comparison with the y'-polarized optical matrix elements in the (0001)-oriented QW structure where $|X'>$ and $|Y'>$ are equally mixed. Assuming a threshold current density J_{th} of about 4.3 kA cm^{-2} [54], the threshold optical gain is estimated to be about 13000 cm^{-1} for the (0001)-oriented QW structure. If we take this value as a threshold optical gain for the ($10\bar{1}0$)-oriented QW structure, we can obtain the threshold current density J_{th} of about 2.4 kA cm^{-2}, which is about half that of the (0001)-oriented QW structure. Thus, we expect that improved lasing characteristics are obtained by using the ($10\bar{1}0$) crystal orientation. The low gain for ($10\bar{1}0$)-oriented structures at low radiative current values is attributed to the fact that the radiative recombination coefficient of $(10\bar{1}0)$-oriented structures is much larger than that of the (0001)-oriented structures at low radiative current values, as discussed below.

Figure 8.10 shows the radiative recombination coefficient B_{eff} as a function of the carrier density for (0001)- and $(10\bar{1}0)$-oriented WZ $In_{0.15}Ga_{0.85}N/GaN$ QW structures ($L_w = 30$ Å). The radiative recombination coefficient is obtained from the calculated J_{rad} by $B_{\text{eff}} = J_{\text{rad}}/(eL_wN^2)$. The B_{eff} of the $(10\bar{1}0)$-oriented QW structure rapidly decreases with increasing carrier density. On the other hand, in the case of the (0001)-oriented QW structure, the B_{eff} is nearly independent of the carrier density. It is observed that the B_{eff} of the $(10\bar{1}0)$-oriented QW structure is much larger than the (0001)-oriented QW structure in the range of the lower carrier density. This results in significantly enhanced optical gain in the $(10\bar{1}0)$-oriented QW structure, in comparison with that in the (0001)-oriented QW structure, as shown in Fig. 8.3(a).

8.4
Summary

In summary, the electronic and the optical properties of $(10\bar{1}0)$-oriented WZ GaN-based QW lasers were reviewed by using multiband effective-mass theory and the non-Markovian gain model with many-body effects. We compare the results with those of (0001)-oriented QW structures with SP and PZ polarizations taken into account. The density of states of the $(10\bar{1}0)$ InGaN/GaN QW structures are shown to be largely reduced compared to those of the $(10\bar{1}0)$ GaN/AlGaN or (0001) QW structures. This is because the subband energy spacing between the first two subbands of the former is larger than it is for the $(10\bar{1}0)$ GaN/AlGaN or (0001) QW structures. The optical gain of the $(10\bar{1}0)$ InGaN/GaN QW structure is significantly larger than that of the $(10\bar{1}0)$ GaN/AlGaN QW structure in the investigated range of well-widths. This can be explained by the fact that the quasi-Fermi-level separation of the $(10\bar{1}0)$ InGaN/GaN QW structure is larger than that of the $(10\bar{1}0)$ GaN/AlGaN QW structure. Also, the increase in the quasi-Fermi-level separation is due to the fact that the $(10\bar{1}0)$ InGaN/GaN QW structures have much smaller hole effective masses than the $(10\bar{1}0)$ GaN/AlGaN QW structure. The optical gain of the $(10\bar{1}0)$-oriented QW is shown to be largely enhanced over that of the (0001)-oriented QW in the investigated range of the carrier density. This is because the $(10\bar{1}0)$-oriented QW has much larger optical matrix elements than the (0001)-oriented QW, in addition to the reduction of the hole effective mass. It is expected that the threshold current density J_{th} of the $(10\bar{1}0)$-oriented InGaN/GaN QW can be reduced to be about a half of that of the (0001)-oriented InGaN/GaN QW structure.

Acknowledgments

The authors wish to thank Dr. D. Ahn for his encouragement and support.

References

1. S. Nakamura and G. Fasol, *The Blue Green Diode*, Chapter 1, Springer, Berlin, 1997.
2. S. L. Chuang and C. S. Chang, *Phys. Rev. B* **54**, 2491(1996).
3. S. L. Chuang, *IEEE J. Quantum Electron.* **32**, 1791 (1996).
4. W. Fang and S. L.Chuang, *Appl. Phys. Lett.* **67**, 751 (1995).
5. M. Suzuki and T. Uenoyama, *Jpn. J. Appl. Phys.* **35**, L953 (1996).
6. S. H. Park and D. Ahn, *J. Korean Phys. Soc.* **30**, 345 (1997).
7. S. H. Park and D. Ahn, *J. Korean Phys. Soc.* **30**, 661 (1997).
8. S. H. Park, S. L. Chuang, and D. Ahn, *Appl. Phys. Lett.* **75**, 1354 (1999).
9. S. L. Chuang, *Physics of Optoelectronic Devices*, Chapter 4, Wiley, New York, 1995.
10. S. H. Park and S. L. Chuang, *J. Appl. Phys.* **87**, 353 (2000).
11. S. H. Park and S. L. Chuang, *Appl. Phys. Lett.* **72**, 287 (1998).
12. G. Martin, A. Botchkarev, A. Rockett, and H. Morkoç, *Appl. Phys. Lett.* **68**, 2541 (1996).
13. F. Bernardini, V. Fiorentini, and D. Vanderbilt, *Phys. Rev. B* **56**, 10024 (1997).
14. S. H. Park and S. L. Chuang, *Appl. Phys. Lett.* **72**, 3103 (1998).
15. S. H. Park and S. L. Chuang, *Appl. Phys. Lett.* **76**, 1981 (2000).
16. T. Ohtoshi, A. Niwa, and T. Kuroda, *Japan. J. Appl. Phys.* **35**, L1566 (1996).
17. S. H. Park and S. L. Chuang, *Phys. Rev. B* **59**, 4725 (1999).
18. S. H. Park, *Jpn. J. Appl. Phys.* **39**, L3478 (2000).
19. T. Takeuchi, H. Amano and I. Akasaki, *Jpn. J. Appl. Phys.* **39**, 413 (2000).
20. F. Mireles and S. E. Ulloa, *Phys. Rev. B* **62**, 2562 (2000).
21. K. Domen, K. Horino, A. Kuramata and T. Tanahashi, *Appl. Phys. Lett.* **70**, 987 (1997).
22. A. Niwa, T. Ohtoshi and T. Kuroda, *Appl. Phys. Lett.* **70**, 2159 (1997).
23. Y. C. Yeo, T. C. Chong and M. F. Li, *IEEE J. Quantum Electron.* **34**, 1270 (1998).
24. W. W. Chow, A. Knorr, and S. W. Koch, *Appl. Phys. Lett.* **67**, 754 (1995).
25. S. H. Park and H. M. Kim, *J. Korean Phys. Soc.* **36**, 78 (2000).
26. S. H. Park, *Appl. Phys. Lett.* **77**, 4095 (2000).
27. S. H. Park, *Appl. Phys. Lett.* **80**, 2830 (2002).
28. S. H. Park, *Jpn. J. Appl. Phys.* **41**, 2084 (2002).
29. S. H. Park, *J. Appl. Phys.* **93**, 9665 (2003).
30. S. H. Park, *J. Korean Phys. Soc.* **42**, 696 (2003).
31. S. H. Park, *J. Korean Phys. Soc.* **42**, 344 (2003).
32. R. Sharma, P. M. Pattison, H. Masui, R. M. Farrell, T. J. Baker, B. A. Haskell, F. Wu, S. P. DenBaars, J. S. Speck, and S. Nakamura, *Appl. Phys. Lett.* **87**, 231110 (2005).
33. P. Waltereit, O. Brandt, A. Trampert, H. T. Grahn, J. Menniger, M. Ramsteiner, M. Reiche, and K. H. Ploog, *Nature (London)* **406**, 865 (2000).
34. M. D. Craven, S. H. Lim, F. Wu, J. S. Speck, and S. P. DenBaars, *Appl. Phys. Lett.* **81**, 469 (2002).
35. C. Q. Chen, M. E. Gaevski, W. H. Sun, E. Kuokstis, J. P. Zhang, R. S. Q. Fareed, H. M. Wang, J. W. Yang, G. Simin, M. A. Khan, H. P. Maruska, D. W. Hill, M. M. C. Chou, and B. H. Chai, *Appl. Phys. Lett.* **81**, 3194 (2002).
36. A. Chakraborty, B. A. Haskell, S. Keller, J. S. Speck, S. P. DenBaars, S. Nakamura, and U. K. Mishra, *Appl. Phys. Lett.* **85**, 5143 (2004).
37. T. Onuma, A. Chakraborty, B. A. Haskell, S. Keller, S. P. DenBaars, J. S. Speck, S. Nakamura, U. K. Mishra, T. Sota, and S. F. Chichibu, *Appl. Phys. Lett.* **86**, 151918 (2005).

38. N. F. Gardner, J. C. Kim, J. J. Wierer, Y. C. Shen, and M. R. Krames, *Appl. Phys. Lett.* **86**, 111101 (2005).

39. S. L. Chuang, *Physics of Optoelectronic Devices*, Wiley, New York, 1995.

40. J. M. Hinckley and J. Singh, *Phys. Rev. B* **42**, 3546 (1990).

41. J. F. Nye, F. R. S., *Physical Properties of Crystals*, Chaps. 5 and 6, Clarendon, Oxford, 1989.

42. D. L. Smith and C. Mailhiot, *J. Appl. Phys.* **63**, 2717 (1988).

43. A. Bykhovski, B. Gelmont, and S. Shur, *Appl. Phys. Lett.* **63**, 2243 (1993).

44. D. Ahn, *Prog. Quantum Electron.* **21**, 249 (1997).

45. S. H. Park, S. L. Chuang, and D Ahn, *Semicond. Sci. Technol.* **15**, 203 (2000).

46. W. W. Chow, S. W. Koch, and M. Sergent III, *Semiconductor-Laser Physics*, Chapter 4, Springer, Berli, 1994.

47. H. P. Maruska and J. J. Tietjen, *Appl. Phys. Lett.* **15**, 327 (1969).

48. W. M. Yim, E. J. Stofko, P. J. Zanzucchi, J. I. Pankove, M. Ettenberg, and S. L. Gilbert, *J. Appl. Phys.* **44**, 292(1973).

49. K. Kim, W. R. L. Lambrecht, and B. Segall, *Phys. Rev. B* **53**, 16310 (1996).

50. K. H. Hellwege and O. Madelung, *Physics of Group IV Elements and III-V Compounds*, Landolt-Börnstein, New Series, Group III, Vol.17, Pt.a, Springer-Verlag, Berlin, 1982.

51. T. L. Tansley and C. P. Foley, *J. Appl. Phys.* **59**, 3241 (1986).

52. A. Shikanai, T. Azuhata, T. Sota, S. Chichibu, A. Kuramata, K. Horino, and S. Nakamura, *J. Appl. Phys.* **81**, 417 (1997).

53. M. Suzuki, T. Uenoyama, and A. Yanase, *Phys. Rev. B* **52**, 8132 (1995).

54. S. Nakamura, *IEEE J. Select. Topics Quantum Electron.* **4**, 483 (1998).

55. W. W. Chow, M. Hagerott, A. Gimdt, and S. W. Koch, *IEEE J. Select. Topics Quantum Electron.* **4**, 514 (1998).

56. M. Kumagai, S. L. Chuang, and H. Ando, *Phys. Rev. B* **57**, 15303 (1998).

57. S. H. Park and S. L. Chuang, *J. Appl. Phys.* **87**, 353 (2000).

58. T. Ohtoshi, A. Niwa, and T. Kuroda, *Jpn. J. Appl. Phys.* **35**, L1566 (1996).

59. V. W. L. Chin, T. L. Tansley, and Osotchan, *J. Appl. Phys.* **75**, 7365 (1994).

60. A. Polian, M. Grimsditch, and I. Grzegory, *J. Appl. Phys.* **79**, 3343 (1996).

61. O. Ambacher, J. Smart, J. R. Shealy, N. G. Weimann, K. Chu, M. Murphy, W. J. Schaff, L. F. Eastman, R. Dimitrov, L. Wittmer, M. Stutzmann, W. Rieger and J. Hilsenbeck, *J. Appl. Phys.* **85**, 3222 (1999).

9
Carrier Scattering in Quantum-Dot Systems
Frank Jahnke

9.1
Introduction

For the use of quantum dots (QDs) as the active material in future semiconductor laser devices, several advantages are expected over the currently employed quantum wells. Among these are lower pump power, better temperature stability, lower linewidth-enhancement factor, and larger differential gain. For a realistic estimation of the underlying physical effects, a detailed understanding of carrier interaction and scattering processes is necessary.

The purpose of this chapter is to present an overview of the microscopic description of scattering processes in nitride-based QD systems. The starting point of our theory will be the electronic states in group-III nitride nanostructures that are also an ongoing subject of investigations in the literature [1,2]. Complicated band-mixing effects result in a valence-band structure with strongly nonparabolic dispersion and a pronounced mass anisotropy, as discussed on the level of k·p calculations in [3,4]. Furthermore, nitride-based heterostructures with wurtzite crystal structure are known to have strong built-in electrostatic fields due to spontaneous polarization and piezoelectric effects that have been analyzed in ab-initio electronic structure calculations [5,6] and in comparison with photoluminescence experiments [7]. Recently, atomistic tight-binding models have been used for a systematic description of these effects on single-particle and optical properties [8].

While most of the aforementioned theoretical investigations are devoted to the *single-particle* states and transitions, it is known from InGaAs-based QD systems, that the emission properties are strongly influenced by many-body effects. In the subsequent paragraphs we address the combined influence of intrinsic nitride effects and many-body effects on carrier scattering. The light-emission properties of QD-based devices critically depend on the scattering processes that populate the states involved in optical transitions. The efficiency of carrier scattering also influences the dynamical emission properties. On a more general level, carrier scattering is the source of dephasing, that

Nitride Semiconductor Devices: Principles and Simulation. Joachim Piprek (Ed.)
Copyright © 2007 WILEY-VCH Verlag GmbH & Co. KGaA, Weinheim
ISBN: 978-3-527-40667-8

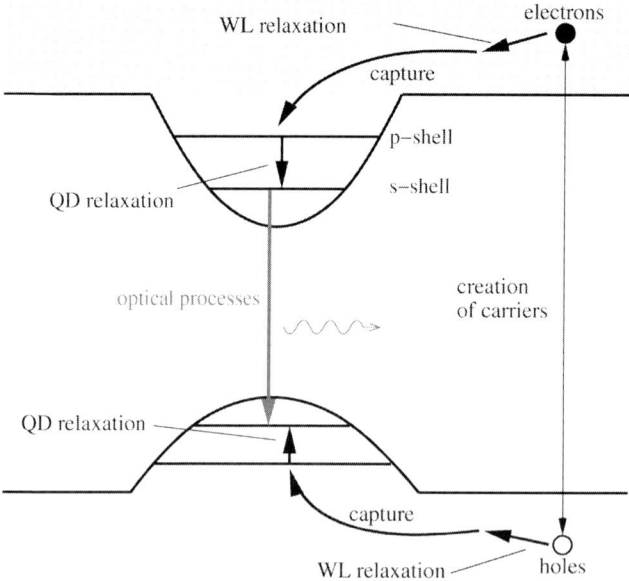

Fig. 9.1 Schematic picture of the states in a quantum-dot wetting-layer system. Two sets of localized states with discrete energies are labeled s- and p-shell. The wetting-layer states are represented by shaded areas. Scattering processes are indicated by arrows.

leads to broadening and shifts in the transition energies, as discussed for QD systems in [9].

Self-organized QDs grown in the Stranski–Krastanov mode typically consist of localized states with discrete energies and also a quasi-continuum of delocalized wetting-layer (WL) states at higher energies. In Fig. 9.1, the corresponding energy spectrum is schematically shown (along the vertical direction) together with the important intrinsic scattering processes. For applications in light emitters, electrons and holes are typically generated in the delocalized WL states or at even higher energies in the bulk continuum. The requirements for efficient light emission are fast scattering processes from the delocalized into the localized states (carrier capture) as well as fast transitions between localized states (carrier relaxation). The two important intrinsic scattering mechanisms, that contribute here, are the carrier–carrier Coulomb interaction, which can be assisted by carriers in bound (QD) or extended (WL) states, as well as the carrier–phonon interaction.

9.2
Scattering Due to Carrier–Carrier Coulomb Interaction

9.2.1
Formulation of the Problem and Previous Developments

In the recent past, theoretical tools for the description of carrier–carrier scattering in QD systems have been developed to various levels of refinement. Thereby the dependence of the scattering efficiency on material composition, geometrical parameters, and on the excitation conditions has been studied. Most of the subsequently discussed references focus on the InGaAs-based material system. They are, however, also the predecessors for properly extended models applicable to the nitride-based material system.

Capture and relaxation processes have been analyzed in [10, 11] under the assumption that in Auger-like transitions the excess energy is transferred to other carriers in the WL. A detailed inspection of the kinetic equation for Coulomb scattering, which is discussed below, shows that these are only some of the possible processes. A relaxation process starting with two carriers in an excited QD state where one is scattered into the QD ground state and the other one into the WL, studied in [12], is only one among several examples. Furthermore, often the WL states are described by plane waves that are not orthogonal to the QD states. Other approximations concern the omission of exchange Coulomb interaction or the screening of the Coulomb potential. The large number of applied approximations made it difficult to systematically compare the results.

To overcome these problems, a microscopic model for the carrier–carrier Coulomb scattering has been developed in [13], that avoids the discussed limitations: (i) calculations are not restricted to Fermi's golden rule but include population effects; (ii) both in- and out-scattering processes are considered for the calculation of capture and relaxation times; (iii) *all possible types* of scattering processes and the role of Coulomb exchange contributions to the scattering integrals are examined; (iv) properly orthogonalized states are systematically used and the influence of the wavefunction model on the scattering rates is analyzed; and (v) a theoretical model of screening in the coupled QD-WL system is provided.

The application of this theory to nitride-based QD systems has been given in [14]. The influence of built-in electrostatic fields in the wurtzite structure on the single-particle states and energies as well as screening effects due to excited carriers are described using coupled Schrödinger and Poisson equations. The resulting modified single-particle states are integrated in the microscopic model discussed above. From the calculated single-particle states the Coulomb interaction matrix elements are determined, that are used in the evaluation of scattering integrals.

Another new ingredient of [14], whose application is not restricted to the nitride-based material system, is the use of single-particle energies in the scattering integrals, which are renormalized by the many-body Coulomb interaction. These modifications are divided into Hartree, exchange and Coulomb correlation contributions. Thereby, also effects like charging of the QDs due to different numbers of electrons and holes in the QDs are considered. It turns out that the resulting electrostatic interaction (described by Hartree contributions) competes with the Coulomb exchange interaction and correlation effects. Considering the electrostatic Hartree contributions alone, or neglecting them in the presence of the other Coulomb effects can strongly modify the results as discussed below.

The application of these models to nitride-based QDs allows to study the efficiency of carrier–carrier scattering processes in the presence of the self-consistently determined built-in electrostatic fields. The electrostatic interaction causes a concentration of electrons and holes in different spatial regions along the growth direction, i.e., localization and charge separation, according to the quantum-confined Stark effect. In turn, this reduces the electron–hole interaction and increases the electron–electron and hole–hole interaction. Our results show that the usually discussed importance of this effect on the interaction matrix elements only weakly influences the scattering rates. It turns out that the change of the self-consistently renormalized energies due to charge separation leads to a much stronger modification of the scattering efficiencies. Specifically, for the influence on the interaction matrix elements, the reduction of electron–hole scattering is partly compensated by the increase of electron–electron and hole–hole scattering. In contrast, for the energy renormalization, the charge separation leads to increased repulsion and decreased attraction in the Hartree terms, both effects working in the same direction of shallower confined levels. This in turn causes an enhancement of the scattering efficiency.

In the calculation of scattering rates for the capture and relaxation processes, one can assume different scenarios. Considering optoelectronic devices, the carriers are typically generated in the delocalized states. Assuming empty QD states, the "initial" in-scattering rates can be determined. More realistically, for an operating light-emitter, one can also assume a quasi-equilibrium distribution of electrons and holes over the *combined* system of localized and delocalized states, that is justified by the fast scattering itself. Then the scattering times for capture and relaxation processes can be determined in *relaxation time approximation*, where the action against a small deviation from the equilibrium situation is considered [13].

The rates for carrier capture and relaxation in the combined QD-WL system strongly depend on the density of excited carriers in the localized and delocalized states. This is expected from the dependence of the rates on the availability of scattering partners and on the requirement of populated (unpopulated) initial (final) states. For intermediate densities, the scattering efficiency increases with carrier density. For large densities, Pauli-blocking and screening of the interaction matrix elements slow down a further increase of the scattering rates. Considering typical InGaN QD parameters, a QD density of 10^{10} cm^{-2} and a carrier density of 10^{11} cm^{-2} at room temperature, we find for the direct capture of electrons (holes) to excited QD states, scattering times on the order of 100 (10) ps. Relaxation times for scattering between the QD hole states are more than one order of magnitude shorter than capture, while at elevated densities the electron relaxation of the higher confined states is of the same order of magnitude as capture.

9.2.2
Kinetic Equation and Scattering Rates

The dynamics of the carrier population $f_\nu(t)$ as a function of time t for an arbitrary state ν of the combined QD-WL system can be determined from the solution of kinetic equations. The simplest form can be obtained in the Markov approximation with Boltzmann scattering rates. The sum of in- and out-scattering processes leads to

$$\frac{\partial}{\partial t} f_\nu = (1 - f_\nu) S_\nu^{in} - f_\nu S_\nu^{out} \tag{9.1}$$

For the Coulomb interaction, where two carriers are scattered from the states ν_1 and ν_3 into ν and ν_2 as well as vice versa, the in-scattering rate S_ν^{in} is given by

$$S_\nu^{in} = \frac{2\pi}{\hbar} \sum_{\nu_1,\nu_2,\nu_3} W_{\nu\nu_2\nu_3\nu_1} [W^*_{\nu\nu_2\nu_3\nu_1} - W^*_{\nu\nu_2\nu_1\nu_3}]$$
$$\times f_{\nu_1}(1 - f_{\nu_2}) f_{\nu_3} \, \delta(\tilde{\varepsilon}_\nu - \tilde{\varepsilon}_{\nu_1} + \tilde{\varepsilon}_{\nu_2} - \tilde{\varepsilon}_{\nu_3}) \tag{9.2}$$

A similar expression for the out-scattering rate S_ν^{out} is obtained by replacing $f \to 1 - f$. The population factors account for the availability of scattering partners. The scattering rates are proportional to the population f of the initial states and to the nonoccupation $1 - f$ of the final states. The sum involves all available initial states for in-scattering processes as well as all possible scattering partners, that can be provided in QD or WL states.

In the Markov approximation, the delta function accounts for strict energy conservation of the scattering processes. This approximation is valid in the long-time limit, where scattering events are considered for times that are larger than the inverse scattering rates. Even in this case, it is necessary to account for renormalizations of the single-particle energies $\tilde{\varepsilon}_\nu$ by the Coulomb interaction instead of using free-carrier energies ε_ν on the basis of perturbation theory.

The scattering cross-section is also determined by the matrix elements of the screened Coulomb interaction, $W_{\nu\nu_2\nu_3\nu_1}$. The first and second term in the square brackets correspond to direct and exchange Coulomb scattering, respectively.

A treatment beyond the Markov approximation is possible within a quantum-kinetic description. Then the delta-function is replaced by the spectral functions of the involved carriers and time-integrals over the past of the system, account for memory effects [15]. The simplest form of a quantum-kinetic equation requires two further approximations. i) Within the generalized Kadanoff–Baym ansatz (GKBA) closed equations for single-time population functions are obtained; see [15] for detailed discussions. ii) The quasi-particle properties of the involved carriers are described by renormalized energies and damping (finite lifetime of the quasi-particle state) that can be summarized in complex energies $\hat{\varepsilon}_\nu$. The latter approximation is related to the Fermi-liquid theory and is valid for large carrier densities, where excitonic effects can be neglected.

The quantum-kinetic equation with the discussed approximations has the form

$$\frac{\partial}{\partial t} f_\nu(t) = \int_{-\infty}^{t} dt' \left\{ [1 - f_\nu(t')] S_\nu^{\text{in}}(t,t') - f_\nu(t') S_\nu^{\text{out}}(t,t') \right\} \tag{9.3}$$

where the in-scattering rate is given by

$$S_\nu^{\text{in}}(t,t') = \frac{1}{\hbar} \text{Re} \sum_{\nu_1,\nu_2,\nu_3} W_{\nu\nu_2\nu_3\nu_1}(t) \left[W^*_{\nu\nu_2\nu_3\nu_1}(t') - W^*_{\nu\nu_2\nu_1\nu_3}(t') \right]$$

$$\times f_{\nu_1}(t')[1 - f_{\nu_2}(t')] f_{\nu_3}(t') \exp\left(-\frac{i}{\hbar}(\hat{\varepsilon}_\nu - \hat{\varepsilon}_{\nu_1} + \hat{\varepsilon}_{\nu_2} - \hat{\varepsilon}_{\nu_3})(t - t')\right) \tag{9.4}$$

and the out-scattering rate is obtained again from $f \to 1 - f$. It turns out that, especially for the calculation of dephasing and its influence on optical absorption and gain spectra, the quantum-kinetic description leads to more realistic results. As discussed in [9], for the description of dephasing in optical spectra, equations similar to (9.3) and (9.4) are used for polarization dynamics.

The determination of Coulomb interaction matrix elements, that enter in Eqs (9.2) and (9.4), involves two steps. The bare (unscreened) interaction matrix elements can be calculated from the wavefunctions $\Phi_\nu(\mathbf{r})$ of the involved electronic states according to

$$V_{\nu\nu_2\nu_3\nu_1} = \int d^3r\, d^3r'\, \Phi_\nu^*(\mathbf{r}) \Phi_{\nu_2}^*(\mathbf{r}') v(\mathbf{r} - \mathbf{r}')\, \Phi_{\nu_3}(\mathbf{r}') \Phi_{\nu_1}(\mathbf{r}) \tag{9.5}$$

where $v(\mathbf{r} - \mathbf{r}') = e^2/(4\pi\varepsilon_0\epsilon|\mathbf{r} - \mathbf{r}'|)$ is the Coulomb potential with the background dielectric function ϵ. The electron charge and the vacuum dielectric constant are given by e and ε_0, respectively, and the space integrals involve the three-dimensional vectors $\mathbf{r}, \mathbf{r}'$. Various levels of refinement can be employed at this point. A simple approximation consists in a separation of the wavefunction into in-plane and growth-direction parts [13,14]. Assuming a harmonic in-plane confinement potential, most of the resulting in-plane integrals can be evaluated analytically, which greatly simplifies the calculation of interaction-matrix elements. On the other end, atomistic tight-binding models, that allow for a detailed inclusion of structural properties and the resulting band-mixing effects, provide numerical wavefunctions. In [8] the Coulomb matrix elements have been determined for the localized QD states on this level. An approximate treatment of delocalized states is possible with the scheme of orthogonalized plane waves (OPW) as discussed in detail in [13]. A second step involves the inclusion of screening effects due to excited carriers. From the nature of screening, as well as from the excitation conditions at elevated temperatures (considered for optoelectronic applications), it appears natural to account only for screening due to the carriers in the delocalized WL states. A consistent theory of screening for the QD-WL system is developed in [14]. On these grounds, a simplified treatment can be justified, that consists in the replacement of $v(\mathbf{r}-\mathbf{r}')$ in (9.5) by the Fourier transform of a two-dimensionally screened Coulomb potential.

The renormalized single-particle energies that enter in the scattering rates (9.2) and (9.4), can be obtained from the free-carrier energies by adding Hartree, exchange, and correlation self-energy contributions. The Hartree self-energy involves electrostatic interaction with all other excited electrons and holes. Only in a system with local charge neutrality, do the Hartree terms cancel. In quantum dots, local charging appears as a result of different envelope wavefunctions for electrons and holes, as well as due to the charge separation of electrons and holes in the built-in electrostatic fields. An extended discussion in [14] shows how the calculation of Hartree-terms can be simplified. The calculation of Hartree terms for QD levels can be restricted to charge carriers from the same QD since, for randomly distributed QDs on a WL, the Hartree contributions from other QDs cancel the WL contribution due to global charge neutrality. On the same grounds, the Hartree shifts of the WL states can be neglected. For a uniform system, the Hartree terms are unscreened. However, to account for all possible interaction processes in our system composed of QDs and WL, it is a meaningful approximation to screen the QD Hartree terms by carriers in the WL, as discussed in detail in [14]. There it has also been used, that the exchange and correlation self-energies can be approximately described by screened-exchange and Coulomb-hole contributions, for which examples are subsequently shown.

9.2.3
Results for Carrier–Carrier Scattering

For the calculation of the following results, we consider InGaN QDs with two confined shells for electrons and three confined shells for holes. The energies of these shells and the other structural parameters that enter our calculations, are listed in Table 9.1. Otherwise, we use standard material parameters given in Chapter 2 of this book. A quasi-equilibrium situation is assumed, where the carrier population functions f_ν are described by Fermi–Dirac functions with renormalized energies. This requires a self-consistent calculation of the energy renormalization for the QD and WL states. An example for the p-shell energy of holes is shown in Fig. 9.2 with and without the influence of the built-in field.

Tab. 9.1 Quantum dot parameters used in the calculations. ε are the energies relative to the wetting-layer band edge. Further entries refer to the density of quantum dots on the wetting layer, wetting-layer thickness, and built-in electrostatic fields.

Parameter	Electrons	Holes
	s, p	s, p, d
Level spacing (meV)	90.0	30.0
ε_s (meV)	−160.0	−80.0
ε_p (meV)	−70.0	−50.0
ε_d (meV)		−20.0
QD density (cm^{-2})		10^{10}
WL thickness (nm)		3.0
F_{QD-WL} (MV cm^{-1})		1.5
$F_{barrier}$ (MV cm^{-1})		−0.75

Generally, a substantial lowering of the hole energy is found, which increases for larger carrier densities and leads to a shallower confinement. The energy shift in the presence of the built-in field is smaller than the zero-field case. The origin of this difference lies in the Hartree term, which reflects the electrostatic interaction of a given carrier with all the others. The field-induced change of the z-confinement functions tends to separate the electrons from the holes and, as a consequence for both, the repulsive part of the Hartree term is increased and the attractive part is decreased. For electrons the Hartree shift is repulsive both in the presence and in the absence of the built-in field, but more so in the former case. For holes, the built-in field makes the Hartree term less attractive. In both cases, the net result is a set of shallower bound states. This turns out to be the most important effect of the built-in field on the scattering rates, which are very sensitive to the energy spacing between the states involved (see below). The screened exchange and Coulomb hole terms are not significantly changed by the built-in field, see Figs 9.2 (c) and (d).

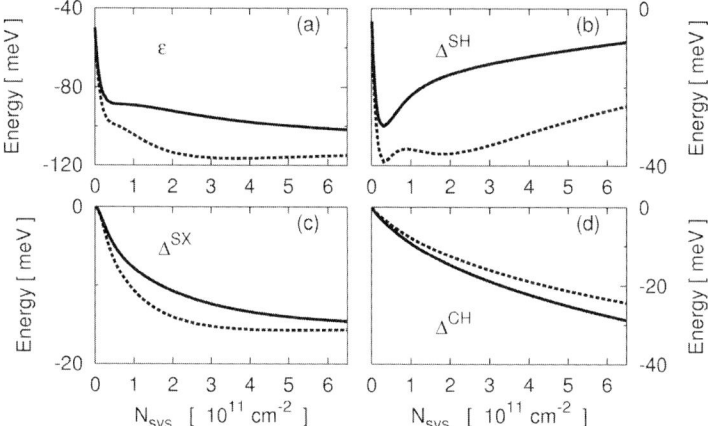

Fig. 9.2 Total renormalized energy for p-shell holes (a) as well as its contributions due to screened Hartree shift (b), screened exchange shift (c), and Coulomb hole shift (d) with (solid line) and without (dotted line) built-in field. All curves are plotted as a function of the total carrier density for a temperature of 300 K.

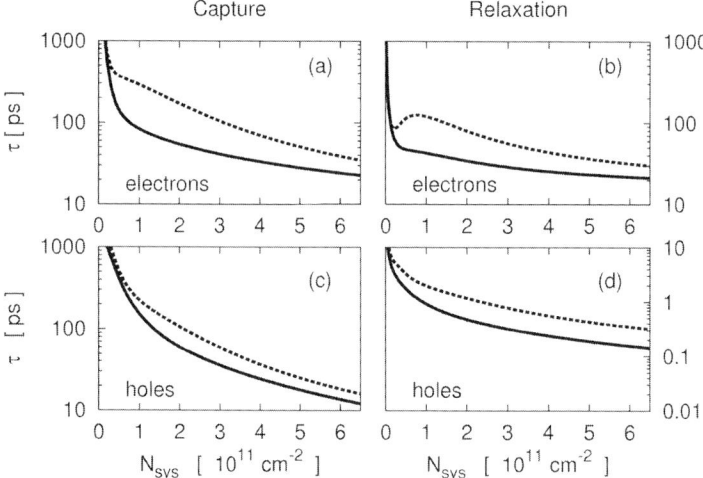

Fig. 9.3 Capture times for the p-shell electrons (a) and holes (c) and relaxation times for the p-shell electrons (b) and holes (d) with (solid-line) and without (dotted line) built-in field.

As the extended WL states are only renormalized by the screened exchange and Coulomb hole term, we find an overall negative energy shift that is almost momentum-independent (not shown).

For the calculation of the capture times, various possible capture processes [13] based on the renormalized energies are included in the scattering integrals $S_\nu^{in,out}$. As an example, capture times for the p-shell, calculated in scattering-time approximation [13], are shown in Fig. 9.3 (a) and (c) for electrons and holes, respectively. The capture times become substantially shorter in the presence of the built-in field compared to the zero-field case. This can be explained as follows. The energy released in the capture event is transferred to a WL carrier. For such a process, the matrix element is larger if the energy transfer is smaller [14]. The charge separation induced by the built-in field reduces this energy and therefore enhances the capture rates. This effect is stronger than the reduction of the electron–hole scattering efficiency. The WL-assisted QD relaxation times, cf. [13], for the p-shell are shown in Fig. 9.3 (b) and (d) for electrons and holes, respectively. Generally, the relaxation times for holes are one or two orders of magnitude shorter than for electrons, since the QD energy level spacing is larger for the latter. With built-in field, the relaxation times become shorter compared to the zero-field case, where the QD energy level spacing is larger. Relaxation times for scattering between QD states are in general more than one order of magnitude shorter than capture. For the p-shell electron relaxation, a saturation effect due to Pauli blocking is observed at higher densities, which leads to comparable capture and relaxation times.

9.3
Scattering Due to Carrier–Phonon Interaction

9.3.1
Perturbation Theory Versus Polaron Picture

For low carrier densities, when carrier–carrier Coulomb scattering is less important, the interaction of carriers with LO-phonons provides the only remaining efficient scattering processes. At higher carrier densities, carrier–phonon interaction is important as well, since the Coulomb scattering is unable to dissipate kinetic energy from the carrier system. A thermalization of the carriers with respect to the crystal lattice is essential for efficient optical gain, since otherwise the inversion is reduced.

The early work on carrier–phonon interaction in semiconductor QDs has been based on the application of time-dependent perturbation theory. This method has been previously applied to bulk semiconductors or quantum wells. In this case, its success is based on the application to carriers within a continuum of states. For the discussion of carrier–phonon interaction involving localized QD states with discrete energies, one has to consider the fact that the transition matrix elements provide only efficient coupling to phonons with small momenta [16, 17]. This strongly reduces the efficiency of carrier interaction with LA-phonons, at least for elevated temperatures. Regarding the coupling to LO-phonons, which provides the other potential process, it has been concluded on the basis of Fermi's golden rule that, except for the unlikely coincidence of transition energies between the QD states with the nearly constant LO-phonon energy, the energy-conservation requirement should prevent efficient carrier scattering [18]. In this case of the so-called *phonon bottleneck*, only much less efficient higher-order processes, like those involving combinations of LA- and LO-phonons can contribute [19, 20].

For a more advanced description of the interaction with LO-phonons, one has to take into account that freely moving charge carriers in ionic crystals are surrounded by a cloud of lattice distortions. The corresponding quasi-particle is called a polaron. In the past, polarons have been mostly studied for semiconductor structures with a quasi-continuum of electronic states, like bulk materials or quantum wells. Then electronic transitions are already possible within a free-carrier picture. For the usually considered materials with weak polar coupling, this explains why the application of perturbation theory provides, in this case, the leading contributions to polaron effects. They consist in a lowering of the carrier energy (polaron shift), in a renormalization of the carrier mass and dielectric constant for their Coulomb interaction, and in a small broadening of the electronic states [21].

Contrariwise, it has been pointed out in [17, 22, 23] that even in materials with weak polar coupling, the interaction of carriers in localized QD states with LO-phonons leads to strong modifications of the electronic properties. In this case, the phonon satellites of one state hybridize with other discrete states, even if the level spacing does not match the LO-phonon energy. The resulting new quasi-particles of the strongly interacting system are called QD-polarons. They cannot be treated within perturbation theory. Therefore, a simple discussion of carrier–phonon interaction based on Fermi's golden rule is not applicable to QDs. Related to this is an ongoing controversy in the QD literature regarding experimental evidence for [24–26] and against [27–29] the phonon bottleneck.

Early investigations of QD-polarons have been based on the "random phase approximation" (RPA) [17, 22] or on the direct diagonalization of a restricted state space [23], where only localized states are considered. For a single discrete electronic state, an analytic treatment of the interacting with LO-phonons is possible within the "independent Boson model" [30]. The numerical extension to several discrete electronic states is discussed in [31]. However, there the influence of delocalized WL states with a quasi-continuum of energies, which plays a key role for carrier generation in optoelectronic devices, has been neglected. When investigations are restricted to localized states with discrete energies, RPA results deviate from a more accurate restricted-direct diagonalization. In turn, the inclusion of WL states provides a natural source of broadening of the electronic states that justifies the application of the RPA.

The next step in the analysis of carrier–phonon scattering in QD systems was the development of a theory that accounts for the scattering of renormalized quasi-particles (QD-polarons). This became possible within a quantum-kinetic theory that has been applied to the problem of carrier capture and relaxation in [32]. Such a quantum-kinetic description naturally incorporates both non-Markovian effects on ultrafast timescales and a nonperturbative treatment of the carrier–phonon interaction in the polaron picture. Polaronic renormalization effects have been found as the main origin for fast carrier capture and relaxation processes as well as efficient carrier thermalization on a picosecond timescale. The predictions strongly deviate from a perturbative treatment using Fermi's golden rule. It has also been shown that quantum-kinetic models provide proper carrier relaxation towards the equilibrium state of the interacting system [33].

9.3.2
Polaron States and Kinetics

In the following, the retarded Green's function $G_\alpha^r(\tau)$ with the time τ and the state index α is used to describe the quasi-particle renormalizations of QD electrons and holes in the single-particle states α due to the nonperturbative interaction with LO-phonons. In the absence of interaction, this function oscillates with the free-carrier energy ϵ_α according to

$$G_\alpha^r(\tau) = \frac{i}{\hbar} \Theta(\tau) e^{-\frac{i}{\hbar} \epsilon_\alpha \tau} \tag{9.6}$$

and the Fourier transform

$$G_\alpha^r(\omega) = \frac{1}{\hbar\omega - \epsilon_\alpha + i\varepsilon} \tag{9.7}$$

with the frequency ω has a pole at the free-carrier energy with $\varepsilon > 0$, $\varepsilon \to 0$. Quasi-particle renormalizations are obtained from the solution of the Dyson equation [21]

$$\left[i\hbar \frac{\partial}{\partial \tau} - \epsilon_\alpha\right] G_\alpha^r(\tau) = \delta(\tau) + \int_0^\tau d\tau' \, G_\alpha^r(\tau') \sum_\beta G_\beta^r(\tau - \tau') D_{\alpha\beta}^>(\tau - \tau') \tag{9.8}$$

where the sum involves all available states β and a retarded self-energy in random-phase approximation has been used to formulate the interaction term with the phonon propagator

$$D_{\alpha\beta}^>(\tau) = \sum_{\vec{q}} |M_{\alpha\beta}(\vec{q})|^2 \times \left[n_{LO} \, e^{\pm i\omega_{LO}\tau} + (1 + n_{LO}) \, e^{\mp i\omega_{LO}\tau} \right] \tag{9.9}$$

Here n_{LO} is a Bose–Einstein function for the population of the phonon modes (assumed to act as a bath in thermal equilibrium), ω_{LO} is the LO-phonon frequency, and the sum over the three-dimensional phonon wave vector $\vec{q}$ involves all phonon modes.

The Dyson equation takes into account all possible virtual transitions due to carrier–phonon interaction for a carrier in the state α. The single-particle energies for both localized and delocalized states are considered and the corresponding wavefunctions (which incorporate effects of the built-in field) enter via the interaction matrix elements $M_{\alpha\beta}$.

The Fourier transform of the retarded Green's function directly provides the spectral function $\hat{G}_\alpha(\omega) = -2\,\mathrm{Im}\,G_\alpha^r(\omega)$ which reflects the density of states (DOS) for a carrier in state α. Considering noninteracting carriers, the spectral function contains a δ-function at the free-particle energies, which shows that each state is associated with a single energy. This changes due to the quasi-particle renormalizations as shown below.

The dynamics of the carrier population due to interaction with LO-phonons follows on the level of perturbation theory from a kinetic equation with Boltzmann scattering integrals,

$$\frac{\partial}{\partial t} f_\alpha = \frac{2\pi}{\hbar} \sum_{\beta,\vec{q}} |M_{\beta\alpha}(\vec{q})|^2$$

$$\times \Big\{ (1-f_\alpha)f_\beta \big[(1+n_{LO})\,\delta(\varepsilon_\alpha - \varepsilon_\beta + \hbar\omega_{LO}) $$
$$+ n_{LO}\,\delta(\varepsilon_\alpha - \varepsilon_\beta - \hbar\omega_{LO})\big] \qquad (9.10)$$
$$- f_\alpha(1-f_\beta)\big[n_{LO}\,\delta(\varepsilon_\alpha - \varepsilon_\beta + \hbar\omega_{LO}) $$
$$+ (1+n_{LO})\,\delta(\varepsilon_\alpha - \varepsilon_\beta - \hbar\omega_{LO})\big] \Big\}$$

This equation describes the temporal changes of the carrier population f in the state α due to scattering processes from other states β into this state α, $\sim (1-f_\alpha)f_\beta$, and vice versa involving phonon emission $\sim (1+n_{LO})$ and absorption $\sim n_{LO}$. Similar to the corresponding equation for carrier–carrier scattering, the in- and out-scattering rates contain delta-functions with free-carrier energies. However, since perturbation theory is not applicable, this equation needs to be generalized within a quantum-kinetic theory. The simplest version of such a generalization is again obtained using the GKBA and has the form

$$\frac{\partial}{\partial t_1} f_\alpha(t_1) = 2\,\mathrm{Re} \sum_\beta \int_{-\infty}^{t_1} dt_2\; G_\beta^r(t_1-t_2)\,[G_\alpha^r(t_1-t_2)]^*$$
$$\times \Big\{ f_\alpha(t_2)[1-f_\beta(t_2)]\,D_{\alpha\beta}^<(t_2-t_1) \qquad (9.11)$$
$$- [1-f_\alpha(t_2)]f_\beta(t_2)\,D_{\alpha\beta}^>(t_2-t_1) \Big\}$$

In comparison to the Boltzmann scattering rates, the delta-functions are replaced by convolutions of retarded Green's functions, which account for the polaronic renormalizations of the initial and final carrier states, with the phonon propagator. Furthermore, memory effects are included by the time dependence of the population factors on the past evolution of the system. Note that the Markov approximation *in the renormalized quasi-particle picture* corresponds to the assumption of a slow time-dependence of the population $f_a(t_2)$ in comparison with the retarded Green's functions, such that the population can be taken at the external time t_1. The Boltzmann scattering integral of Eq. (9.10) follows if one additionally neglects quasi-particle renormalizations and uses free-carrier retarded Green's functions (9.6).

9.3.3
Results for Carrier Scattering Due to LO-phonons

In the following we discuss results for nitride-based QDs that are characterized by intermediate polar coupling. In comparison to the InGaAs material system with weak polar coupling, this leads to enhanced polaron effects. We consider QDs with two confined shells that, due to their assumed angular momentum properties, are referred to as the s-shell (ground state) and the p-shell (excited states). To study the dependence of the electron scattering on the level spacing between s- and p-states and between p-states and WL band-edge, corresponding values between 1.0 and 1.4 $\hbar\omega_{LO}$ are used in the calculations. For the intrinsic electrostatic field, we include a typical field strength of $2(-0.5)$ MV cm^{-1} inside (outside) the QD and WL region. For the polar coupling constant and the LO-phonon energy, $\alpha = 0.49$ and $\hbar\omega_{LO} = 92$ meV are taken. All calculations are performed at room temperature.

In a first step, results for the spectral functions of QD and WL carriers under the influence of interaction with LO-phonons are discussed. They are obtained from a direct numerical solution of Eqs (9.8) and (9.9). For a conduction-band electron in the WL with zero momentum, the spectral function in the absence of carrier–phonon interaction is a delta function peaked at the band-gap energy. The position of this peak is indicated by the vertical line in Fig. 9.4. This delta function expresses the fact that, in a free-carrier picture, every state is associated with a single energy. Including the interaction with LO-phonons, the spectral function is broadened and exhibits a polaron shift. Furthermore, sidebands with a spacing of the LO-phonon energy appear. The picture shows, that in the interacting case a state is connected with a certain range of energies. The relative weight of these energies is given by the discussed spectral function. When the corresponding retarded function is used in a kinetic equation, the presence of sidebands leads to the inclusion of multi-phonon processes. (The appearance of sidebands is itself a consequence of higher-order processes.)

For simplicity, the influence of the QD states on the WL spectral function has been neglected in Fig. 9.4. As can be seen in [32] for the weak polar cou-

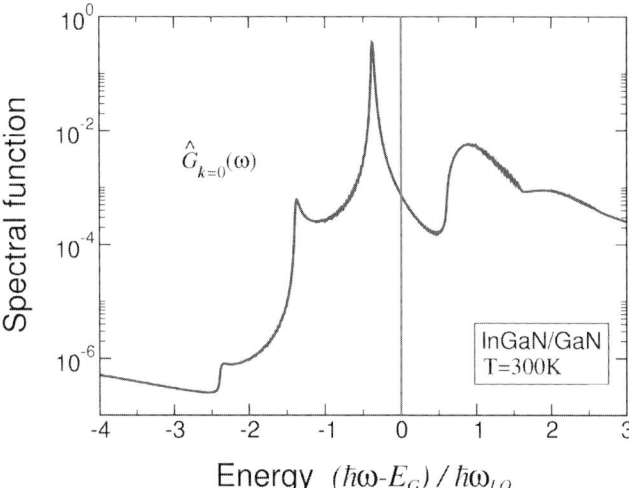

Fig. 9.4 Spectral function for a WL electron with zero momentum showing the increased range of energies $\hbar\omega$ associated with a single carrier state due to interaction with LO-phonons. The energies are given relative to the band-gap E_G in units of the LO-phonon energy $\hbar\omega_{LO}$. The vertical line indicates the position of the free-carrier energy.

pling InGaAs material system, and in [34] for the intermediate polar coupling InGaN material system, this leads only to small corrections.

A more complicated structure is obtained for the spectral function of the QD carriers in Fig. 9.5. Vertical lines indicate the free-carrier energies of the two considered QD shells, that appear below the continuum edge of the WL with the energy E_G. In the figure, the energy spacing of the QD states relative to the LO-phonon energy is increased from top to bottom. We obtain a strong departure from the picture, where single energies are associated with the QD states. The series of strong phonon satellites is modified by hybridization effects, in which the satellites from different shells are mixed.

Due to the broad range of energies, available for carriers in different shells, efficient carrier-scattering processes are expected, which is in contrast to the simple predictions of time-dependent perturbation theory. In the Markovian limit, the transition rate between two states α and β is proportional to the overlap of the spectral functions, shown in Figs 9.4 and 9.5, according to $\Lambda_{\alpha\beta} = \int d\omega \, \hat{G}_\alpha(\omega) \, \hat{G}_\beta(\omega \pm \omega_{LO})$. Note that both carrier capture and relaxation processes are described according to the choice of QD or WL states for α and β.

From a numerical solution of quantum-kinetic equations, one can directly access the efficiency of the scattering processes. In the following, we consider the simplest situation where Eq. (9.11) is used to describe the relaxation of an initial nonequilibrium carrier distribution. More complicated and for inter-

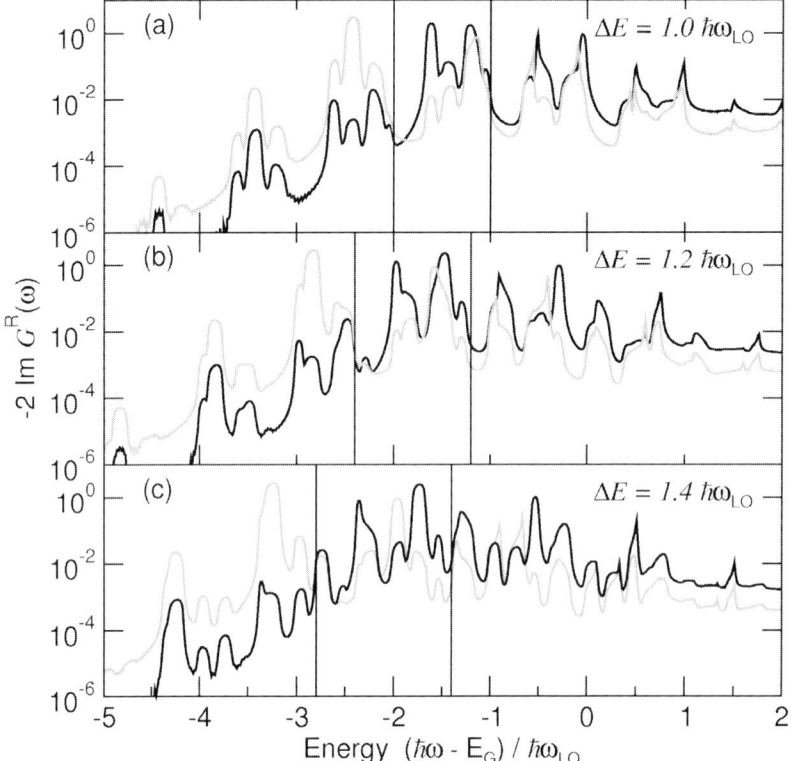

Fig. 9.5 Spectral function for quantum-dot electrons in the p-shell (black lines) and in the s-shell (gray lines). Increasing values for the quantum-dot level spacing ΔE relative to the LO-phonon energy $\hbar\omega_{LO}$ are used from top to bottom. Vertical lines indicate the positions of the free-carrier energies.

mediate polar coupling more accurate models use correlation functions depending on two independent times, that are discussed and evaluated in [33]. Furthermore, the dynamics of carrier generation with short optical pulses and the dephasing of the induced coherent polarization due to carrier–phonon interaction can be incorporated in such a description.

Results for the dynamical evolution of the carrier population are shown in Fig. 9.6 for a situation, where initially only the p-shell contains a small carrier population while the s-shell and WL states are unpopulated. Even for a large detuning of 40% between the transition energy and the LO-phonon energy, an ultrafast redistribution of carriers from the p-shell into the s-shell on a sub-ps timescale is obtained. This is due to the, above discussed, quasi-particle renormalizations, which strongly modify the picture of scattering between free-carrier states, as well as due to memory effects. Furthermore, the time evolution reveals oscillations that reflect the hybridization of electronic states

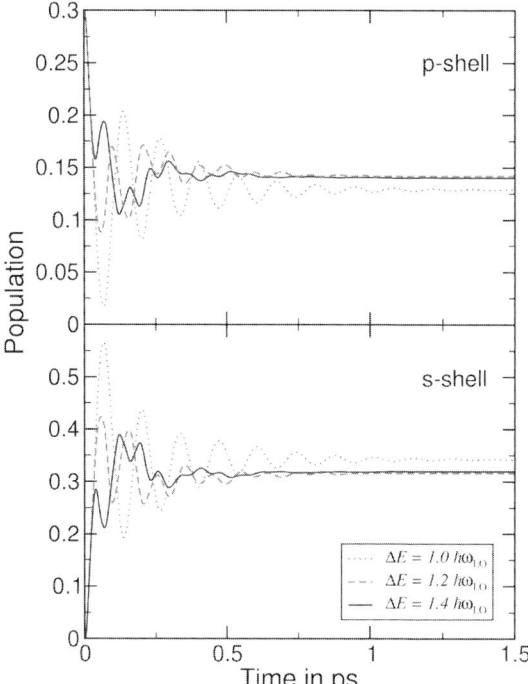

Fig. 9.6 Time evolution of the carrier occupation function $f_\alpha(t)$ for the initially populated p-shell and initially empty s-shell.

in the time domain. In comparison to InGaAs QDs, these effects are more pronounced due to the stronger polar coupling. The equilibrium population, obtained in Fig. 9.6, does not simply correspond to a Fermi distribution, but is connected to the density of states of the interacting system by the Kubo–Martin–Schwinger relation [33].

9.4
Summary and Outlook

In this chapter, the application of a microscopic theory for carrier–carrier and carrier–phonon interaction to scattering processes in nitride-based QD systems, has been presented. For both interaction processes, the inclusion of renormalization effects of the single-particle properties for the involved carriers leads to substantial modifications of the results. In the case of Coulomb interaction between the carriers, energy shifts of the confined levels relative to each other and to the WL, directly influence the excitation-density dependence of the scattering rates. For the carrier–phonon interaction, polaron effects lead to the breakdown of Fermi's golden rule and provide efficient scattering channels due to the interaction with LO-phonons.

Another application of the microscopic theory of scattering processes involves calculations of optical gain, where dephasing and carrier correlations account for the proper gain saturation and gain dynamics. Such a theory, that requires a self-consistent treatment of both carrier–carrier and carrier–phonon interaction, has recently been presented in [9] for InGaAs QD systems. The inclusion of nitride-specific effects like band-mixing and built-in fields is in preparation and will provide a better understanding of this interesting material system.

Acknowledgments

This work was supported by the Deutsche Forschungsgemeinschaft with a grant for CPU time at the NIC, Forschungszentrum Jülich. The author thanks Jan Seebeck and Torben Nielsen for preparing the figures, as well as Paul Gartner and Michael Lorke for valuable discussions.

References

1. I. Vurgaftman and J. Meyer, J. Appl. Phys. **94**, 3675 (2003).
2. A. Ranjan, G. Allen, C. Priester, and C. Delerue, Phys. Rev. B **68**, 115305 (2003).
3. S. Chuang and C. Chang, Phys. Rev. B **54**, 2491 (1996).
4. W. W. Chow and S. W. Koch, *Semiconductor-Laser Fundamentals*, Springer, Berlin, 1999.
5. F. Bernardini, V. Fiorentini, and D. Vanderbilt, Phys. Rev. Lett. **79**, 3958 (1997).
6. F. Bernardini, V. Fiorentini, and D. Vanderbilt, Phys. Rev. B **57**, R9427 (1998).
7. R. Cingolani, A. Botchkarev, H. Tang, H. Morko, G. Traetta, G. Coli, M. Lomascolo, A. Carlo, F. Sala, and P. Lugli, Phys. Rev. B **61**, 2711 (2000).
8. N. Baer, S. Schulz, S. Schumacher, P. Gartner, G. Czycholl, and F. Jahnke, Appl. Phys. Lett. **87**, 231114 (2005).
9. M. Lorke, T. R. Nielsen, J. Seebeck, P. Gartner, and F. Jahnke, Phys. Rev. B **73**, 085324 (2006).
10. U. Bockelmann and T. Egeler, Phys. Rev. B **46**, 15574 (1992).
11. A. V. Uskov, F. Adler, H. Schweizer, and M. H. Pilkuhn, J. Appl. Phys. **81**, 7895 (1997).
12. R. Ferreira and G. Bastard, Appl. Phys. Lett. **74**, 2818 (1999).
13. T. R. Nielsen, P. Gartner, and F. Jahnke, Phys. Rev. B **69**, 235314 (2004).
14. T. Nielsen, P. Gartner, M. Lorke, J. Seebeck, and F. Jahnke, Phys. Rev. B **72**, 235311 (2005).
15. H. Haug and A.-P. Jauho, *Quantum Kinetics in Transport & Optics of Semiconductors*, Springer-Verlag, Berlin, 1. edition, 1996.
16. U. Bockelmann and G. Bastard, Phys. Rev. B **42**, 8947 (1990).
17. T. Inoshita and H. Sakaki, Phys. Rev. B **56**, 4355 (1997).
18. H. Benisty, C. M. Sotomayor-Torres, and C. Weisbuch, Phys. Rev. B **44**, 10945 (1991).
19. T. Inoshita and H. Sakaki, Phys. Rev. B **46**, 7260 (1992).
20. H. Jiang and J. Singh, IEEE J. Quantum Electron. **34**, 1188 (1998).
21. W. Schäfer and M. Wegener, *Semiconductor Optics and Transport Phenomena*, Springer-Verlag, Berlin, 1. edition, 2002.

22. K. Kral and Z. Khas, Phys. Rev. B **57**, 2061 (1998).

23. O. Verzelen, R. Ferreira, G. Bastard, T. Inoshita, and H. Sakaki, phys. stat. sol. (a) **190**, 213 (2002).

24. J. Urayama, T. B. Norris, J. Singh, and P. Bhattacharya, Phys. Rev. Lett. **86**, 4930 (2001).

25. A. W. E. Minnaert, A. Y. Silov, W. van der Vleuten, J. E. M. Haverkort, and J. H. Wolter, Phys. Rev. B **63**, 075303 (2001).

26. S. Xu, A. A. Mikhailovsky, J. A. Hollingsworth, and V. I. Klimov, Phys. Rev. B **65**, 045319 (2002).

27. E. Tsitsishvili, R. v. Baltz, and H. Kalt, Phys. Rev. B **66**, 161405 (2002).

28. E. Peronne, F. Fossard, F. H. Julien, J. Brault, M. Gendry, B. Salem, G. Bremond, and A. Alexandrou, Phys. Rev. B **67**, 205329 (2003).

29. F. Quochi, M. Dinu, L. N. Pfeiffer, K. W. West, C. Kerbage, R. S. Windeler, and B. J. Eggleton, Phys. Rev. B **67**, 235323 (2003).

30. G. D. Mahan, *Many-Particle Physics*, Plenum, New York, 2. edition, 1990.

31. T. Stauber, R. Zimmermann, and H. Castella, Phys. Rev. B **62**, 7336 (2000).

32. J. Seebeck, T. R. Nielsen, P. Gartner, and F. Jahnke, Phys. Rev. B **71**, 125327 (2005).

33. P. Gartner, J. Seebeck, and F. Jahnke, Phys. Rev. B **73**, 115307 (2006).

34. J. Seebeck, T. R. Nielsen, P. Gartner, and F. Jahnke, Eur. Phys. J. B **49**, 167 (2006).

Part 2 Devices

10
AlGaN/GaN High Electron Mobility Transistors

Tomás Palacios and Umesh K. Mishra

10.1
Introduction

GaN-based transistors have greatly evolved since their first demonstration in 1993 [1]. In less than 15 years, they have developed from devices with less than 20 mA mm^{-1} of output current and virtually no high-frequency performance, to world-wide commercialization as power amplifiers at high frequencies [2].

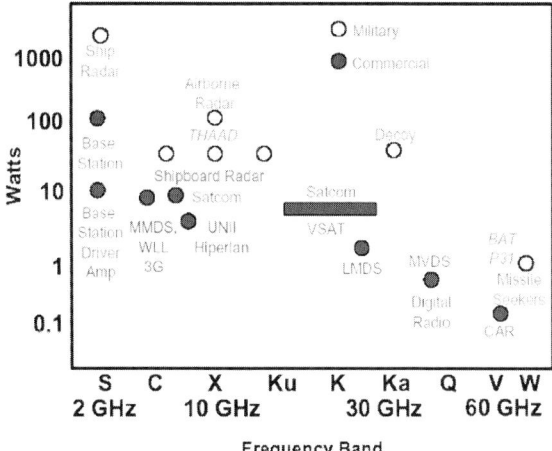

Fig. 10.1 Some of the most important potential applications for GaN-based power transistors

AlGaN/GaN structures have outstanding electronic properties. The high electron mobility (typically in the 1500–2200 cm^{2}V^{-1}s^{-1} range at room temperature) and carrier concentration (normally 0.7–1.4 × 10^{13} cm^{-2}) allow the fabrication of devices with maximum drain current densities in excess of 2 A mm^{-1} [3]. On the other hand, their very large critical breakdown electric field (> 3 MV cm^{-1}) renders very large breakdown voltages. Both high

Nitride Semiconductor Devices: Principles and Simulation. Joachim Piprek (Ed.)
Copyright © 2007 WILEY-VCH Verlag GmbH & Co. KGaA, Weinheim
ISBN: 978-3-527-40667-8

current density and large breakdown voltages are important requirements for power amplifiers, where the maximum linear output power ($P_{max,lin}$) can be calculated from:

$$P_{max,lin} = I_{max}(V_{BD} - V_{knee})/8$$

where I_{max} is the maximum drain current density, V_{BD} is the breakdown voltage of the device and V_{knee} is the knee voltage. It can be shown [4] that if some distortion is allowed in the output signal, the power performance can be increased by a factor of $16/\pi^2$. On the other hand, the high electron velocity ($> 2 \times 10^7$ cm s^{-1} [5]) makes this material system one of the best candidates for very high-frequency operation.

Figure 10.1 summarizes some of the multiple civil and military applications that could benefit from the use of AlGaN/GaN high electron mobility transistors (HEMTs) as power devices. As shown in the figure, these transistors can be used in many applications requiring high frequency and power operation. At lower frequencies, AlGaN/GaN HEMTs are being pursued for power-switching applications [6] and even for biological [7] and chemico/physical sensors [8].

The increase in performance of AlGaN/GaN HEMTs has been very fast over the last 10 years. At the time of writing this chapter, AlGaN/GaN HEMTs are already commercially available for power amplification in cell-phone base stations at 2 GHz [2]. Power amplifiers with a total output power in excess of 250 W have been reported at this frequency [9]. At a frequency of 4 GHz, AlGaN/GaN HEMTs have demonstrated more than 32 W mm^{-1} of output power with a power added efficiency (PAE) of 54.8% at $V_{DS} = 120$ V [10]. This performance is more than one order of magnitude better than in any other competing semiconductor technology. By correctly matching the output harmonic frequencies, it is also possible to get a combination of very high output power and PAE. For example, an output power of 7.8 W mm^{-1} and a PAE of 65% have been simultaneously achieved at 15 GHz when the device was biased at $V_{DS} = 30$ V [11].

The performance at mm-wave frequencies is also impressive. A maximum current gain cut-off frequency (f_T) of 163 GHz has been reported in devices with a gate length in the 60–90 nm range [12, 13]. These devices, when biased for maximum power gain, have shown 230 GHz of maximum power gain cut-off frequency (f_{max}) [14]. The very high gain in combination with the large breakdown voltage allow more than 10.5 W mm^{-1} of output power at 40 GHz with $V_{DS} = 30$ V [15]. This excellent performance has motivated the use of GaN devices in mm-wave power amplifiers. A record output power of 8 W and 31% PAE have been reported out of a single-stage power amplifier at 30 GHz [16].

Although the performance achieved in AlGaN/GaN HEMTs is remarkable, the understanding of the fundamental material and device properties of AlGaN/GaN HEMTs is still a pending issue which makes the simulation of these devices more challenging than in other material systems. Not only is the value of some of the basic material parameters unknown, but also important device properties, such as polarization, source of gate leakage, quasi-saturation of the electron velocity or breakdown and degradation mechanisms, are still highly controversial issues. In spite of all these challenges, the performance of AlGaN/GaN HEMTs can be accurately modeled by simulations at several levels of abstraction.

In a first level, physics-based simulations are fundamental for the basic modeling of the device operation. The knowledge of the electric field, electron density and currents in the different regions of the transistor is critical for the design of new devices with improved functionality as well as for the identification of weak points in the device performance.

In a second level of abstraction, the use of small-signal equivalent circuit models to study AlGaN/GaN HEMTs is also extremely useful when studying complex devices like multichannel structures [11,17,18]. This approach, more empirical than the physics-based simulations, does not require the knowledge of many of the material parameters needed in the physics-based simulations. However, it requires some measurements of actual devices to calculate the different fitting parameters.

To study the large signal performance of nitride HEMTs, a third level of abstraction is needed. By careful measurements and fitting of the current-voltage curves and intrinsic elements of the transistor under large signal operation, it is possible to design large-signal equivalent circuit models which allow accurate predictions of the expected output power, efficiency and linearity performance of these devices, once inserted in a circuit [19,20].

The forth level of abstraction involves the modeling of the noise performance. Numerous attempts have been made to model and predict the noise of these devices, and to use these simulations in actual circuit-level simulators [21,22].

This chapter will only cover the modeling and simulation of AlGaN/GaN transistors at the first level of abstraction (i.e., physics-based simulations). The reader interested in the other approaches will find some of the references given above useful.

This chapter does not try to be either exhaustive or complete. Many different research groups have performed excellent work on the simulation and modeling of AlGaN/GaN HEMTs. However, in an attempt to provide a more consistent view of the modeling of these transistors, most of the results provided in this chapter come from the simulation and modeling work performed over the years at the University of California, Santa Barbara.

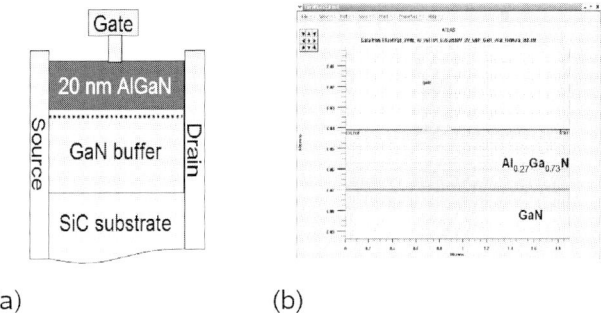

Fig. 10.2 (a) Typical device structure of an AlGaN/GaN HEMT, and (b) mesh structure used in the simulation of the DC performance of this device.

10.2
Physics-based Simulations

Under the general term of "physics-based simulators", we will consider a large variety of device simulators which have in common the use of physical equations to simulate the performance of AlGaN/GaN HEMTs. We refer to these simulators as "physics-based" in contrast with circuit-based simulators where the physical phenomena that happen in the device are somehow hidden from the user by the equivalent circuit model.

Most of the physics-based simulators can predict the electrical characteristics that are associated with specific device structures and bias conditions. In this simulation, the device structure is discretized in one, two or three dimensions (Fig. 10.2) and the physical equations governing the physics of the device are numerically solved. Each software has different built-in physical models with different degrees of functionality. For example, the simplest simulator would solve the Poisson equation to get the band diagram of the device in equilibrium. More accurate results of the electron distribution can be obtained when the Poisson solver is coupled to a Schrödinger solver, as in the case of the BandEng software [23]. The addition of the equations for drift-diffusion and thermionic emission allow the calculation of current densities and other important device parameters in nonequilibrium situations. Finally, the most complete simulation packages, like ATLAS by Silvaco [24], use energy balance models (or even Monte Carlo techniques) to account for hot carriers and overshoot effects. Other models like heat transfer are also frequently available.

Many different reasons justify the use of physics-based simulations to complement experimental measurements. First, the simulation of a device is normally much more time efficient and economical than the growth, fabrication and measurement of the actual device. Therefore, these simulators are widely used to get an initial idea of the design space and expected performance of new devices. Second, the use of this kind of simulation allows access to electrical variables in regions of the device which may be inaccessible in actual devices. Finally, it also provides insight and captures theoretical knowledge in a way that makes this knowledge available to nonexperts, which allows a quick understanding of the device behavior.

Most of the physics-based simulation programs were originally written for Si and GaAs devices. Slowly, they are being upgraded to nitride materials. However, in most of the cases, the user still has to manually introduce many of the particularities of nitride materials, such as polarization, surface states and traps in order to get accurate results.

In most of the examples shown in this section, we will use two different simulation packages. The software BandEng developed by Michael Grundmann at the University of California-Santa Barbara is a powerful band-diagram simulator device [23]. It solves the Schrödinger and Poisson equations self-consistently in a one-dimensional device structure. This solver has been specifically designed for nitride-based devices and it incorporates all the basic material parameters of nitrides as well as polarization effects. To calculate the performance of these devices in nonequilibrium conditions we have used the simulation package ATLAS by the company Silvaco [24]. This software provides general capabilities for numerical, physics-based, two-dimensional simulation of semiconductor devices. It allows the simulation of electrical characteristics like DC, small-signal AC, and time-domain transient behavior. It also has a library of physical models to accurately model tunneling, nonlinear mobility or thermionic currents.

10.2.1
Basic Material Properties

The great majority of the available commercial drift-diffusion simulators have not been originally designed for AlGaN/GaN devices and they do not include nitrides in their material database. Therefore, it is necessary to manually define the properties for these semiconductors. Table 10.1 summarizes the main material parameters used in the simulations presented in this chapter. An excellent compilation of material properties in different semiconductor families can be found in [25] and [26].

The bandgap of $Al_xGa_{1-x}N$ as a function of the Al composition (x) is given by [27]:

$$E_g(x) = 6.13x + 3.42(1-x) - x(1-x) \text{ eV}$$

and a linear interpolation of the dielectric constant is given by [28]:

$$\epsilon_r(x) = 10.4 - 0.3x$$

Tab. 10.1 Main material parameters used in the simulations shown in this chapter.

Parameter	Description	Units	GaN	Al$_{0.27}$Ga$_{0.73}$N
E_g	Bandgap @ 300K	eV	3.42	3.95
ΔE_c	Conduction band offset between AlGaN and GaN	Fraction of E_g	0.68	0.68
A_n^*	Richardson constant for electrons	A cm^{-2}K^{-2}	24	24
A_p^*	Richardson constant for holes	A cm^{-2}K^{-2}	96	96
μ_n	Low-field electron mobility	cm^{2}V^{-1}s^{-1}	425	10
μ_p	Low-field hole mobility	cm^{2}V^{-1}s^{-1}	5	5
E_d	Donor energy level	eV	0.025	0.025
E_a	Acceptor energy level	eV	0.160	0.160
ϵ_r	Relative permittivity		10.4	10.32
N_c	Conduction band effective density of states @ 300 K	$\times 10^{18}$cm^{-3}	2.24	2.24
N_v	Valence band effective density of states @ 300 K	$\times 10^{19}$cm^{-3}	1.80	1.80
X	Electron affinity	eV	3.5	3.14
$v_{sat,n}$	Electron saturation velocity	cm s^{-1}	1.3×10^7	10^4

In our simulations, the electron affinity is calculated assuming a conduction band offset of 68%. Using the measured electron affinity values of 3.5 ± 0.1 eV and 1.9 ± 0.2 eV for GaN and AlN, respectively [29, 30], 68% is within experimental error.

The electron velocity as a function of electric field in AlGaN/GaN structures is a very important parameter for the understanding of the performance of these transistors. It has been calculated by Monte Carlo techniques [5] (Fig. 10.3). This profile has also been measured by multiple methods as described in [31] and [32]. Due to the strong nonlinear behavior of the electron velocity with electric field, it is normally challenging, from the computational point of view, to incorporate the full electron velocity profile into a drift-diffusion simulator. In Section 10.2.4 we will discuss the effect of different simplifications of the electron velocity model in the final simulated performance of the devices.

10.2.2
Polarization

The combination of the lack of inversion symmetry of the wurtzite structure with the high electronegativity of the nitrogen atom make nitride semiconductors very polar materials [33]. Two different types of polarization can be identified in these semiconductors: spontaneous and piezoelectric. The spontaneous polarization in c-plane AlGaN as a function of Al composition is given

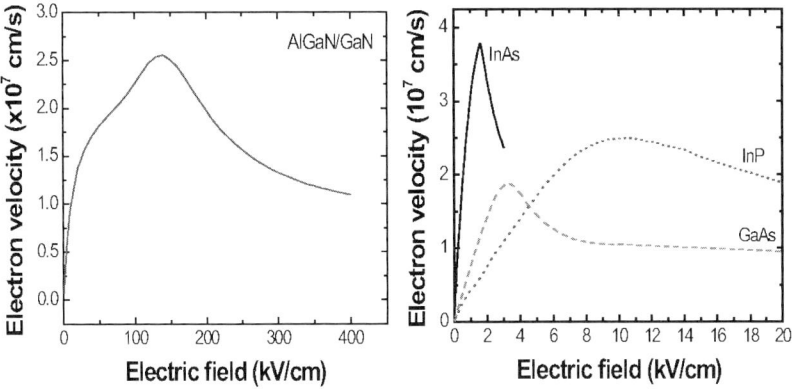

Fig. 10.3 Electron velocity versus electric field in different semiconductor materials.

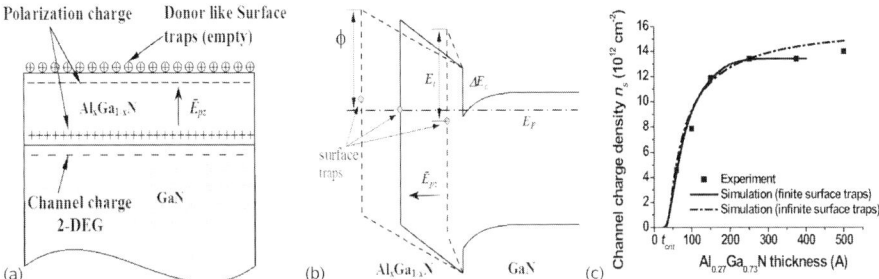

Fig. 10.4 (a) Diagram of the position of the different sheets of polarization-induced fixed charge in an AlGaN/GaN transistor. (b) Position of the surface traps energy level with respect to the Fermi level for different AlGaN thicknesses. In very thin AlGaN barriers, the trap level is below the Fermi level and the donor-like traps are neutral. For thicker AlGaN layers, the Fermi level touches the trap level and the traps become positively charged as they transfer their electrons to the channel. (c) Simulated and experimental dependence of the channel charge density, n_s, with the AlGaN thickness t_{AlGaN} [35].

by [33]:

$$P^{sp}_{Al_xGa_{1-x}N}(x) = [-0.090x - 0.034(1-x) + 0.019x(1-x)]\,\mathrm{C\,m^{-2}}$$

while the piezoelectric polarization of AlGaN pseudomorphically grown on a GaN template can be calculated from [34]:

$$P^{pz}_{AlGaN/GaN}(x) = [-0.0525x - 0.0282x(1-x)]\,\mathrm{C\,m^{-2}}$$

If a heterojunction made of these materials is synthesized and the two semiconductors forming the heterojunction have different polarization values, there will be a net fixed polar charge at the interface as shown in Figures 10.4(a) and 10.5, which will cause an electric field in the top material.

10 AlGaN/GaN High Electron Mobility Transistors

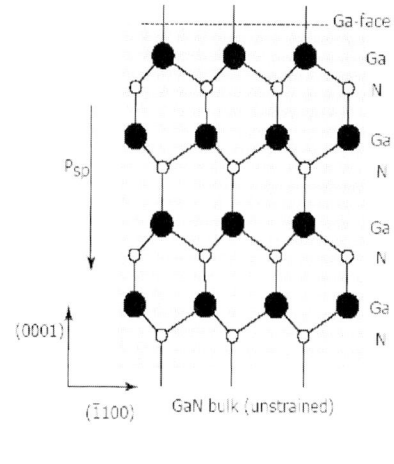

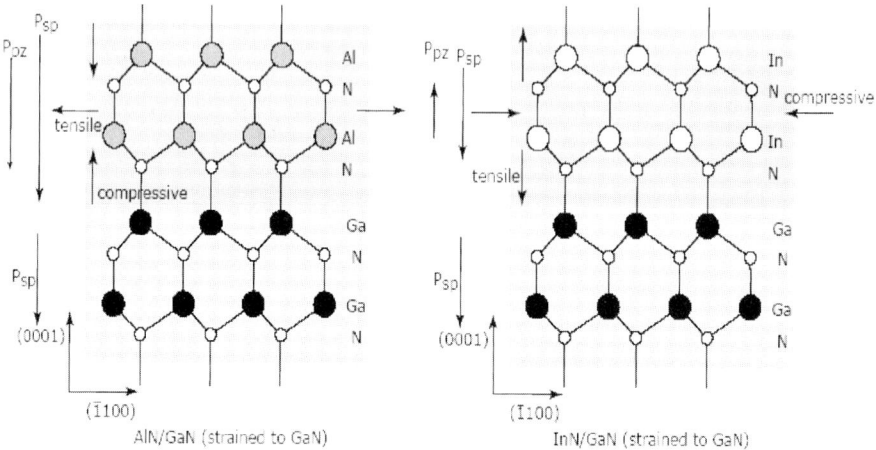

Fig. 10.5 Crystal structures and polarization fields in GaN, AlN on GaN and InN on GaN. GaN is relaxed, and AlN and InN are coherently strained leading to a piezoelectric component of the polarization [36].

Very few device simulators automatically take into account the polarization fields that exist in the vast majority of nitride-based devices[1]. Therefore, in most of the cases, polarization needs to be added manually to the structure. The easiest way to do this is by introducing sheets of fixed charge at each interface to account for the difference in polarization between the two semiconductor layers.

1) A remarkable exception is the program BandEng written by Michael Grundmann in the University of California-Santa Barbara. This program can be obtained from http://my.ece.ucsb.edu/mgrundmann/bandeng.htm

As an example, we will analyze the band diagram of a standard AlGaN/GaN transistor with the structure shown in Figure 10.2. These transistors are normally grown by molecular beam epitaxy or metalorganic chemical vapor deposition on SiC, Si or sapphire substrates. A relatively thick Ga-face GaN buffer layer (typically 0.5–2 μm thick) is grown on top of the substrate, with the help of an initial nucleation layer, to compensate for the significant lattice mismatch between GaN and the substrate. Then a thin ($<$ 40 nm) layer of Al$_x$Ga$_{1-x}$N (typically, $x < 0.4$) is grown as a barrier. A two-dimensional electron gas (2DEG) is formed at the interface between the GaN buffer and the AlGaN barrier layer. An important difference between nitride-based HEMTs and HEMTs in other material systems is that in nitrides it is not necessary to dope the barrier to populate the channel with electrons. The carrier density, n_s, under the gate as a function of the AlGaN composition and gate voltage can be calculated by using the expression [37]:

$$n_s(x) = \frac{\sigma_{AlGaN/GaN}(x)}{q} - \frac{\epsilon_0 E_F}{q^2}\left[\frac{\epsilon_{AlGaN}(x)}{t_{AlGaN}} + \frac{\epsilon_{GaN}}{t_{GaN}}\right] - \frac{\epsilon_0 \epsilon_{AlGaN}(x)}{q^2 t_{AlGaN}}[q(\phi_{AlGaN}(x) - V_{GS}) + \Delta(x) - \Delta E^c_{AlGaN}(x)]$$

where q is the electron charge, $\sigma_{AlGaN/GaN}$ is the net polarization charge at the AlGaN/GaN heterointerface:

$$\sigma_{AlGaN/GaN} = (P^{sp}_{GaN} + P^{pz}_{GaN}) - (P^{sp}_{AlGaN} + P^{pz}_{AlGaN})$$

t_{AlGaN} and t_{GaN} are the AlGaN and GaN thicknesses, ϵ_{AlGaN} and ϵ_{GaN} are the dielectric constants of AlGaN and GaN, $q\phi_{AlGaN}$ is the Schottky barrier height of the gate contact on top of the AlGaN barrier, V_{GS} is the applied gate voltage, ΔE^c_{AlGaN} is the conduction band discontinuity between AlGaN and GaN, $\Delta(x)$ is the penetration of the conduction band-edge below the Fermi level at the AlGaN/GaN interface, and E_F is the position of the Fermi level with respect to the GaN conduction-band-edge close to the GaN/substrate interface. $\Delta(x)$ can be approximated by:

$$\Delta(x) = E_0(x) + \frac{\pi \hbar^2}{m^*_{GaN}} n_s(x)$$

where m^*_{GaN} is the effective mass of electrons in GaN, and $E_0(x)$ is the lowest subband level of the 2DEG, which is given by

$$E_0(x) = \left[\frac{9\pi \hbar q^2}{8\epsilon_0 \sqrt{8m^*_{GaN}}} \frac{n_s(x)}{\epsilon_{GaN}}\right]$$

Smorchkova [38] and Ibbetson [39] proposed that, in equilibrium, surface donor-like traps are the source of the donors in the channel. Electrons from the surface are driven into the channel by the very strong polarization-induced electric fields in the AlGaN cap layer. By using this *polarization doping*, carrier densities as high as 1.5×10^{13} cm^{-2} have been reported for a 35% AlGaN barrier.

Two important features are needed to accurately simulate polarization doping. First of all, a surface trap layer needs to be added to the top surface of the semiconductor, as shown in Figure 10.4(a). Then, the polarization-induced electric field in the AlGaN layer is simulated by a positive charge layer at the AlGaN/GaN interface and an equal amount of negative charge at the surface of the AlGaN layer. The electric field in the GaN buffer will be assumed to be negligible, as impurities, defects and traps added during the early stages of the growth of this layer are expected to screen the polarization charge at the bottom interface. This assumption has been proven correct by numerous experimental measurements.

Polarization-induced doping is not the only useful result of the strong polarization fields in nitrides. Many other advanced structures are possible by using the extra degree of freedom given by polarization. Some of them are AlN interlayers between the GaN buffer and the AlGaN barrier to increase the electron mobility [40], GaN spacer HEMTs with 20% higher electron velocity [41] and InGaN back-barriers to increase the electron confinement and reduce short-channel effects in deeply scaled devices [14]. To properly model all these devices it is therefore critical to take into account polarization.

10.2.3
Surface States

As mentioned in Section 10.2.2, it is widely accepted that the electrons that populate the 2DEG channel in AlGaN/GaN HEMTs under equilibrium originate from surface states. The concentration of surface states in AlGaN/GaN material is difficult to estimate, but its density should be higher than 1.5×10^{13} cm^{-2}. It is very challenging to design experiments to study the surface trap energy level and density of these surface states. However, important insight can be obtained from simulations [35].

Figure 10.6(a) shows the simulated band diagram of an AlGaN/GaN HEMT where the surface charges and polarization effects have been taken into account. For comparison, Fig. 10.6(b) shows the band diagram of the same heterostructure without the polarization model or surface traps. Figure 10.6(c) shows the band diagram of the heterojunction with surface traps but without the polarization model and Figure 10.6(d) shows the band diagram of the heterojunction with the polarization model but no surface traps.

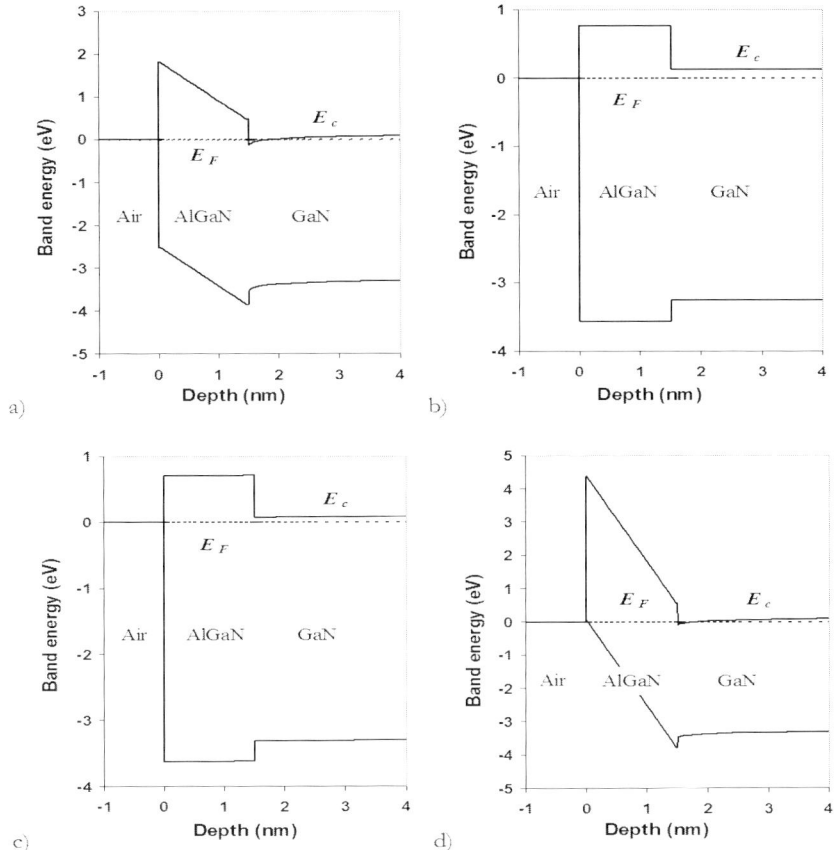

Fig. 10.6 Comparison of band diagrams of $Al_{0.27}Ga_{0.73}N/GaN$ heterojunctions with and without polarization field or surface traps. The AlGaN thickness is 150 Å. (a) With polarization model and surface traps, $n_s = 1.34 \times 10^{13}$ cm^{-2}. (b) No polarization model or surface trap, $n_s = 1.1 \times 10^{10}$ cm^{-2}. (c) No polarization model but with surface traps, $n_s = 8.01 \times 10^{10}$ cm^{-2}. (d) With polarization model but no surface trap, $n_s = 3.48 \times 10^{12}$ cm^{-2} [35].

For case (a), the 2DEG charge density (n_s) in the channel is 1.34×10^{13} cm^{-2}, but for case (b) n_s is only slightly larger than the background doping. n_s is 8.01×10^{10} cm^{-2} and 3.48×10^{12} cm^{-2} for case (c) and (d), respectively. In case (d), if we intentionally dope the AlGaN layer to 5×10^{18} cm^{-3}, n_s only increases to 9.87×10^{12} cm^{-2}, a value much lower than the experimentally determined n_s of doped $Al_{0.27}Ga_{0.73}N/GaN$ HEMTs. These simulations show that surface traps, rather than intentional doping, are the leading source of channel electrons, and the polarization electric field is the driving force to move the trapped electrons into the channel to form 2-DEG.

To obtain the polarization electric field and the surface trap parameters, the trap energy level E_t, trap density D_t, and polarization charge density σ_{pz} (i.e., polarization field, E_{pz}), were adjusted until simulation data matched experimental n_s vs $Al_{0.27}Ga_{0.73}N$ cap-layer thickness t_{AlGaN} (Fig. 10.4c) [38]. Although three parameters were adjusted, there was only one unique combination of all three parameters that led to the fitted n_s vs t_{AlGaN} curve for a single trap level. As shown in Figure 10.4(b); for very thin AlGaN thickness, the surface potential ϕ_s is smaller than the depth of the surface trap level E_t. Therefore, the Fermi-level lies above the surface trap level, and the traps are filled with electrons and are neutral. As the AlGaN layer becomes thicker, the polarization field increases the surface potential. Once the Fermi-level at the surface equals the trap level, the electrons in these traps begin to transfer to the channel, and t_{crit}, defined as the minimum thickness of the AlGaN cap layer to produce a 2DEG, is reached. The critical thickness is determined by both the surface-trap depth E_t, and the magnitude of E_{pz}. As the AlGaN layer becomes thicker than t_{crit}, more and more electrons move to the channel. Finally, when the trap level goes above the Fermi-level ($\phi_s > E_t$), all the electrons in the surface traps have moved to the channel. At this point, no more electrons can be brought to the channel from the surface states despite further increase in the AlGaN thickness. Therefore the channel charge density saturates at a certain AlGaN thickness and the saturation level of n_s is equal to the surface-trap density. The rate of n_s saturation with AlGaN thickness is dependent on how fast the surface-trap energy rises, and therefore dependent on the magnitude of E_{pz}.

The n_s vs thickness curve fitted to n_s measured by the Hall effect is shown in Figure 10.4(c). A good fit was achieved for a polarization charge density $\sigma_{pz} = 1.75 \times 10^{13}$ cm^{-2} and a surface trap density of 1.36×10^{13} cm^{-2} at an energy of 1.85 eV below the conduction band. Figure 10.4(c) also shows the simulated channel charge density when the trap density is infinite. From the almost negligible difference between simulations and experiments, it can be deduced that the trap density is very high but the actual value does not play an important role in the channel charge density. Also, the fitted net polarization charge in AlGaN is a little higher than the theoretical value of 1.56×10^{13} cm^{-2}. This small difference may be the result of a slightly lower-than-expected piezoelectric polarization, due to a lower stress in the AlGaN barrier layer.

10.2.4
Electron Mobility

In drift-diffusion simulations of AlGaN/GaN HEMTs, the use of the proper electron transport model is critical. In many early simulations of AlGaN/GaN devices, the main focus was on the low-field mobility of electrons and the high-field transport model was often neglected. However, AlGaN/GaN devices usually operate under very high electric field regimes. Traditionally, the high-field transport model of electrons was assumed to follow the Caughey and Thomas model [42]:

$$\mu_n(E) = \mu_{n0} \left[\frac{1}{1 + \left(\frac{\mu_{n0} E}{v_{sat,n}} \right)^{\beta_n}} \right]^{1/\beta_n}$$

where $\mu_n(E)$ is the electron mobility as a function of the longitudinal electric field in the channel, μ_{n0} is the low-field electron mobility as measured by Hall technique, $v_{sat,n}$ is the saturated value of the electron velocity and β_n is a fitting parameter. This model can approximate very well the electron velocity profile in GaAs devices. However, the shape of the electron velocity versus field profile in AlGaN and GaAs are very different from each other, as shown in Fig. 10.3.

The use of a GaAs-like high-field transport model significantly affects the simulated performance. Figure 10.7 shows the simulated DC performance of an AlGaN/GaN HEMT modeled using the Caughey and Thomas expression for the electron mobility with the values of μ_{n0} and $v_{sat,n}$ of AlGaN/GaN devices (i.e., $\mu_{n0} = 1700 \text{ cm}^2\text{V}^{-1}\text{s}^{-1}$ and $v_{sat,n} = 2.3 \times 10^7 \text{ cm s}^{-1}$). Figure 10.8 (squares) shows the actual performance measured in a typical device. There are two main differences between the simulations and the experimental results. First, in the experimental results, as V_{GS} becomes positive, the transconductance increases, it reaches a maximum and then it decreases again. However, the simulations are unable to reproduce the decrease in g_m at high current levels. Also, the simulations cannot reproduce the increase in the small signal source access resistance with drain current [43].

To increase the accuracy of the simulations, it is necessary to include the actual electron velocity vs the electric field profile in the model. This can be done in most commercial simulators by writing a small function or look-up table in a programming language like C, to approximate the electron velocity profile. However, when using the full profile, the convergence of the drift-diffusion simulation becomes difficult and special care must be taken to assure convergence. The result of such a simulation is shown in Figure 10.8 (dashed lines) and in this case the simulation reproduces the decrease in g_m with drain current very accurately. The increase of r_s with drain current is also captured by the simulation.

The reason for the different outcome of the two high-energy transport models lies in the peculiar shape of electron velocity vs electric field in GaN. Figure 10.3 shows that the electron velocity in GaN increases for low electric fields and then it *quasi-saturates* for electric fields close to 20 KV cm^{-1}. After this

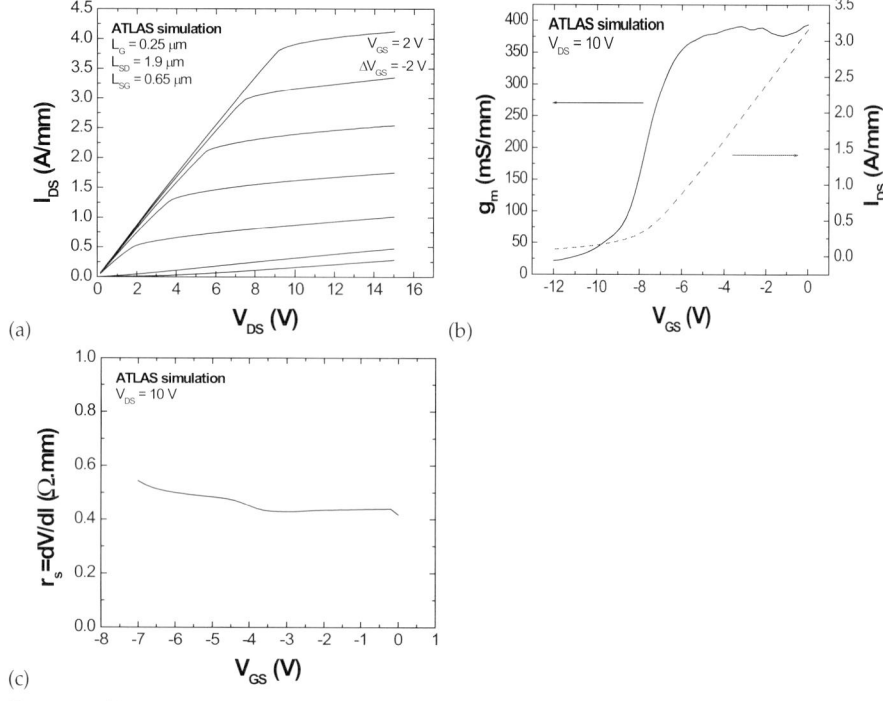

Fig. 10.7 Simulated performance of an AlGaN/GaN HEMT with the high-energy transport model of AlGaAs/GaAs HEMT. (a) Drain current (I_{DS}) as a function of drain voltage (V_{DS}) for different gate voltages (V_{GS}). (b) Transconductance (g_m) and I_{DS} as a function of V_{GS}. (c) Small signal source access resistance (r_s) as a function of V_{GS}.

quasi-saturation, the electron velocity increases again, reaching a maximum for an electric field of 100 KV cm^{-1} and then it decreases to its saturated value of 1.2×10^7 cm s^{-1}. A very important difference between the high-field transport model of GaN and that in other semiconductors is that in the latter, there is no quasi-saturation of the electron velocity. The quasi-saturation behavior is generally associated with the onset of optical phonon emission in nitride-based semiconductors [5, 44]. In other semiconductors like Si or GaAs, the energy for optical phonon emission is much lower and the quasi-saturation cannot be distinguished from the standard saturation. Some authors have also associated the early saturation of the electron velocity in AlGaN/GaN structures with hot phonon scattering [45, 46].

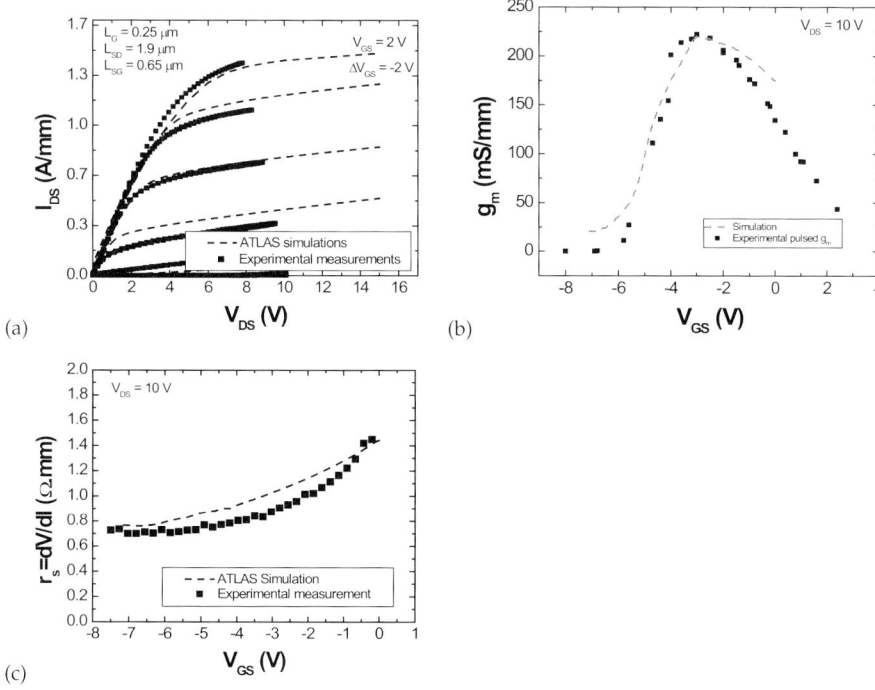

Fig. 10.8 Simulated performance of an AlGaN/GaN HEMT with the high-energy transport model as obtained from Monte Carlo simulations (dashed lines), cf. Figure 10.7. The figure also shows the experimental results obtained in a standard AlGaN/GaN HEMT (squares).

10.2.5
Breakdown Voltage

10.2.5.1 Impact ionization model

To accurately simulate the breakdown voltage and near-breakdown behavior of GaN-based transistors, it is necessary to use the correct model for impact ionization. The generation rate of electron–hole pairs due to impact ionization is given by:

$$G = \alpha_n \frac{J_n}{q} + \alpha_n \frac{J_p}{q}$$

where α_n and α_p are the electron and hole ionization rates and J_n and J_p are the current densities for electrons and holes, respectively. The value of these coefficients is a function of the electric field and lattice temperature. Different authors have proposed analytical expressions for these coefficients. Some of the most popular impact-ionization models are the Selberherr model [47], the Grant model [48] and the Crowell–Sze model [49]. Out of all of them, the simplest one, in terms of the required number of fitting parameters, is the

Grant model. Due to its simplicity, this is the model that is normally used to describe the field-dependency of the ionization rates in GaN-based devices during temperature-independent simulations. In this model, the ionization rates are given by:

$$\alpha_{n,p} = \alpha_{n,p}^{\infty} exp\left[-\left(\frac{E_{n,p}^{crit}}{E}\right)\right]$$

From Monte Carlo simulations of impact ionization in wurtzite GaN [50], the values of coefficients α_n^{∞} and E_n^{crit} can be calculated to be 2.60×10^8 cm^{-1} and 3.42×10^7 V cm^{-1} respectively for electrons and 4.98×10^6 cm^{-1} and 1.95×10^7 V cm^{-1} for holes. These simulated values show good agreement with experimental results of $\alpha_n^{\infty} = 2.9 \times 10^8$ cm^{-1} and $E_n^{crit} = 3.4 \times 10^7$ V cm^{-1} [51]. No experimental values exist for the impact ionization coefficients of holes. There are no enough studies about the ionization rate in AlGaN material, therefore as a first order approximation the above values for GaN are usually also applied to AlGaN.

Due to the approximations described above, the results of the simulations of the breakdown behavior of GaN devices have to be analyzed cautiously. Normally only the qualitative behavior of the device can be calculated.

10.2.5.2 Effect of field plates on breakdown voltage

The use of field plates in AlGaN/GaN HEMTs has been very useful in order to achieve high breakdown voltages and power efficiencies [52,53]. Due to its great importance, significant effort has recently been applied to simulate their effects and to understand their working principles.

The use of gate field plates in the drain access region reduces the peak electric field in this region. This reduction in the electric field has two very important beneficial effects in the performance of AlGaN/GaN HEMTs. First, the lower electric field increases the breakdown voltage as it reduces impact ionization. Figure 10.9 shows how the use of field plates allows a higher drain voltage for the same electric field peak. Second, the lower electric field also alleviates dispersion, which increases the PAE of the device and allows the operation of the transistor in a safer region from the reliability and breakdown voltage point of view.

10.2.6
Energy Balance Models

The traditional drift-diffusion model of charge transport does not take into account effects such as velocity overshoot and reduced energy dependent impact ionization. To model these effects more accurately it is convenient to use models that keep track of the total energy of the carriers instead of only the

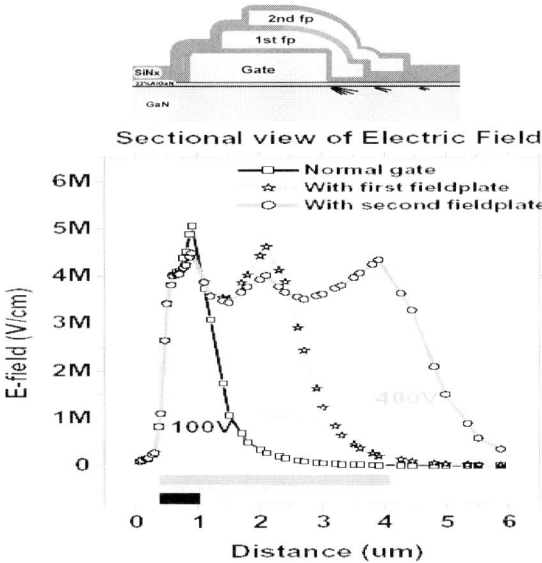

Fig. 10.9 For the same peak electric field value, the voltage that can be supported by the devices with field-plates is larger due to the increasing area under the E-field vs distance curve [54].

electric field in the structure, one example being the energy balance models. These models use a higher order approximation of the Boltzmann transport equation where the mobility and impact ionization are a function of the electron energy.

The drawback of using these more advanced models is that the convergence of the simulations is often compromised. In these cases, special care is needed to correctly mesh the device structure and to identify suitable initial conditions. Also, especially in the case of III-N semiconductors, many parameters needed in these simulations are still unknown or under investigation. For all these reasons, very little work on energy balance simulations has been done in GaN devices even though it is a topic of important interest.

Although most of the simulations of GaN transistors have been based on models that do not take into account energy balance, in first approximation their results are still valid due to the very short scattering time of electrons in AlGaN/GaN heterostructures (∼120 fs). As the simulated devices are much longer than the mean free path of the electrons in these materials (10–20 nm), no overshoot effects have been observed.

As an intermediate step between pure drift-diffusion simulations and the use of energy balance models, some authors have used drift-diffusion simulations coupled self-consistently to Monte Carlo simulations [43, 55]. In these simulations, the electric field in the structure is calculated by drift-diffusion

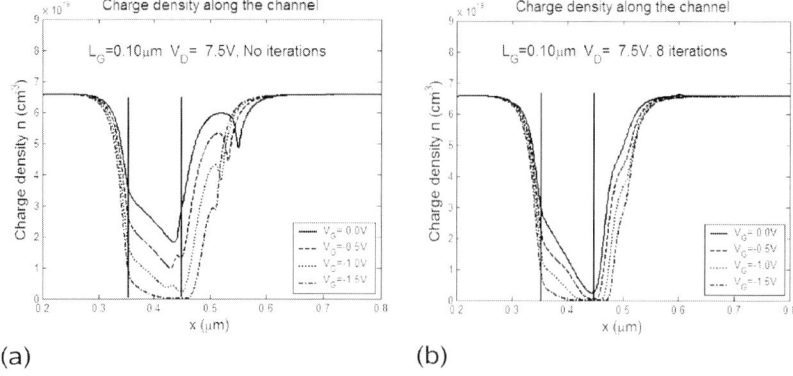

Fig. 10.10 Carrier density along the channel as calculated from a Poisson–Schrödinger simulation (a), and from a Poisson–Schrödinger simulation self-consistently coupled to a Monte Carlo simulator (b). The small notch on the carrier density in the drain side of the gate of simulation (a) is due to not taking into account velocity overshoot effects in the channel. This problem is solved when the simulation is coupled to the Monte Carlo simulator [55, 56].

simulations and then fed into a Monte Carlo simulator to calculate the electron velocity at each point of the structure. The electron velocity profile is then introduced into the drift diffusion simulator to calculate and update the electric field profile which is then fed into the Monte Carlo system again. After 2–3 iterations, the results between the different iterations are close enough (Fig. 10.10). The simulations obtained with this technique have shown excellent agreement with experimental measurements.

10.3
Conclusions

The simulation and modeling of AlGaN/GaN HEMTs constitutes an extremely important tool to enhance our understanding of these devices, as well as to speed up the design of new transistors and circuits with improved performance. The working principle of AlGaN/GaN devices has many different layers of complexity and to fully understand their performance it is necessary to run simulations at different levels. In this chapter we have studied how Poisson, drift-diffusion and Schrödinger simulations bring insight into the band diagram and carrier distribution of the transistors and they provide a first approximation to the IV performance of the device.

All these simulations need to include the many particularities of nitride semiconductors. Effects like polarization, quasi-saturation of the electron velocity and impact ionization have to be carefully considered for accurate results. The on going work in studying some of the fundamental effects and material parameters of these devices will allow the accuracy and power of these simulations to be increased even further in the near future.

Acknowledgments

The authors would like to thank all the different researchers at UCSB who have contributed through their work to the increased understanding and accuracy of the simulations of AlGaN/GaN HEMTs presented in this chapter: Likun Shen, Siddharth Rajan, Yuvaraj Dora and Naiqian Zhang. This work has been partially funded by the Office of Naval Research through the CANE and MINE MURI projects, monitored by Dr. H. Dietrich and Dr. P. Maki.

References

1. M. A. Khan A. Bhattarai, J. N. Kuznia, and D. T. Olson, Appl. Phys. Lett. 63, pp. 1214–15 (1993).
2. www.eudyna.com
3. A. Chini, R. Coffie, G. Meneghesso, E. Zanoni, D. Buttari, S. Heikman, S. Keller, and U. K. Mishra, Electron. Lett. 39, pp. 625–626 (2003).
4. D. M. Snider, IEEE Trans. Electron Dev. 14, no. 12, pp. 851–857 (1967).
5. M. Singh and J. Singh, J. Appl. Phys. 94, pp. 2498–2506 (2003).
6. W. Saito, Y. Takada, M. Kuraguchi, K. Tsuda, I. Omura, T. Ogura, and H. Ohashi, IEEE Trans. Electron Dev. 50, pp. 2528–2531 (2003).
7. G. Steinhoff, B. Baur, G. Wrobel, S. Ingebrandt, A. Offenhäusser, A. Dadgar, A. Krost, M. Stutzmann, and M. Eickhoff, Appl. Phys. Lett. 86, 033901 (2005).
8. M. Stutzmann, G. Steinhoff, M. Eickhoff, O. Ambacher, C. E. Nebel, J. Schalwig, R. Neuberger, and G. Müller, Diam. Relat. Mater. 11, pp. 886–891 (2002).
9. A. Wakejima, K. Matsunaga, Y. Okamoto, Y. Ando, T. Nakayama, K. Kasahara, and H. Miyamoto, Electronics Letters 41, no. 18 (2005).
10. Y.-F. Wu, A. Saxler, M. Moore, R. P. Smith, S. Sheppard, P. M. Chavarkar, T. Wisleder, U. K. Mishra, and P. Parikh, IEEE Electron Dev. Lett. 25, 3, pp. 117–9 (2004).
11. T. Palacios, A. Chakraborty, S. Keller, S. P. DenBaars and U. K. Mishra, Proc. of the International Symposium on Compound Semiconductors, Freiburg, Germany, 2005.
12. T. Palacios, E. Snow, Y. Pei, A. Chakraborty, S. Keller, S. P. DenBaars, and U. K. Mishra, Proc. of the IEEE Int. Electron Device Meeting, IEDM-2005, Washington, December 5–7, 2005.
13. M. Higashiwaki, T. Matsui, and T. Mimura, IEEE Electron Dev. Lett. 27, 1, pp. 16–18 (2006).
14. T. Palacios, A. Chakraborty, S. Heikman, S. Keller, S. P. DenBaars, and U. K. Mishra, IEEE Electron Dev. Lett. 27, pp. 13–15 (2006).
15. T. Palacios, A. Chakraborty, S. Rajan, C. Poblenz, S. Keller, S. P. DenBaars, J. S. Speck, and U. K. Mishra, IEEE Electron Dev. Lett. 26, pp. 781–783 (2005).
16. Y.-F. Wu, M. Moore, A. Saxler, T. Wisleder, U. K. Mishra, and P. Parikhm, Proc. of the IEEE International Electron Device Conference, paper 23–5, Washington DC, Dec. 2005.
17. B. Hughes and P. J. Tasker, IEEE Trans. Electron Dev. 36, 12, pp. 2267–2273 (1989).

18 T. Palacios, A. Chini, D. Buttari, S. Heikman, A. Chakraborty, S. Keller, S. P. DenBaars, and U. K. Mishra, IEEE Trans. Electron Devices 53, pp. 562–565 (2006).

19 V. Paidi, S. Xie, R. Coffie, B. Moran, S. Heikman, S. Keller, A. Chini, S. P. DenBaars, U. K. Mishra, S. Long, and M. J. W. Rodwell, IEEE Trans. Microwave Theory 51, 2, pp. 643–652 (2003).

20 V. Paidi, *MMIC Power Amplifiers in GaN HEMT and InP HBT Technologies*, Ph.D. Dissertation, University of California at Santa Barbara (2004).

21 S. Lee, K. J. Webb, V. Tilak, and L. Eastman, IEEE Trans. Microwave Theory 51, pp. 1567–1577 (2003).

22 C. Sanabria, H. Xu, T. Palacios, A. Chakraborty, S. Heikman, U. K. Mishra, and R. A. York, IEEE Trans. Microwave Theory 53, 2, pp. 762–769 (2005).

23 The software BandEng is a band diagram simulator developed by Michael Grundmann at the University of California at Santa Barbara. It is freely available in http://my.ece.ucsb.edu/mgrundmann/bandeng.htm

24 http://www.silvaco.com/products/device_simulation/atlas.html

25 http://www.ioffe.rssi.ru/SVA/NSM/Semicond/

26 O. Madelung, *Semiconductors: Data Handbook*, Springer Verlag, 3rd edition (2004).

27 D. Brunner, H. Angerer, E. Bustarret, F. Freudenberg, R. Hopler, R. Mimitrov, O. Ambacher, and M. Stutzmann, J. Appl. Phys. 82, pp. 5090–6 (1997).

28 O. Ambacher, B. Foutz, J. Smart, J. R. Shealy, N. G. Weimann, K. Chu, M. Murphy, A. J. Sierakowski, W. J. Schaff, L. F. Eastman, R. Dimitrov, A. Mitchell, and M. Stutzmann, J. Appl. Phys. 87, pp. 334–344 (2000).

29 C. I. Wu, A. Kahn, N. Taskar, D. Dorman, and D. Gallagher, J. Appl. Phys. 83, pp. 4349–4352 (1998).

30 V. M. Bermudes, C. I. Wu, and A. Kahn, J. Appl. Phys. 89, pp. 1991 (2001).

31 J. M. Baker, D. K. Ferry, D. D. Koleske, and R. J. Shul, J. Appl. Phys. 97, 063705 (2005).

32 T. Inoue, Y. Ando, K. Kasahara, Y. Okamoto, T. Nakayama, H. Miyamoto, and M. Kuzuhara, IEICE Trans. Electron., E86-C, 10, 2065 (2003).

33 F. Bernardini and V. Fiorentini, Phys. Rev. B 56, 16, pp. 24–27 (1997).

34 O. Ambacher, M. Eickhoff, A. Link, M. Hermann, M. Stutzmann, F. Bernardini, V. Fiorentini, Y. Smorchkova, J. Speck, U. Mishra, W. Schaff, V. Tilak, and L. F. Eastman, phys. status solidi (c) 6, pp. 1878–1907 (2003).

35 N. Zhang, *High voltage GaN HEMTs with low on-resistance for switching applications*, Ph.D. Dissertation, University of California, Santa Barbara (2002).

36 D. Jena, *Polarization induced electron populations in III-V nitride semiconductors. Transport, growth, and device applications*, Ph.D. Dissertation, University of California, Santa Barbara (2003).

37 E. T. Yu, G. J. Sullivan, P. M. Asbeck, C. D. Wang, D. Qiao, and S. S. Lau, Appl. Phys. Lett. 71, 19, pp. 2794–2796 (1997).

38 I. P. Smorchkova, C. R. Elsass, J. P. Ibbetson, R. Vetury, B. Heying, P. Fini, E. Haus, S. P. DenBaars, J. S. Speck, and U. K. Mishra, J. Appl. Phys. 86, pp. 4520–4526 (1999).

39 J. P. Ibbetson, P. T. Fini, K. D. Ness, S. P. DenBars, J. S. Speck, and U. K. Mishra, Appl. Phys. Lett. 77, pp. 250–2 (2000).

40 L. Shen, S. Heikman, B. Moran, R. Coffie, N.-Q. Zhang, D. Buttari, I. P. Smorchkova, S. Keller, S. P. DenBaars, and U. K. Mishra, IEEE Electron Dev. Lett. 22, no. 10, pp. 457–459 (2001).

41 T. Palacios, L. Shen, S. Keller, A. Chakraborty, S. Heikman, D. Buttari, S. P. DenBaars, and U. K. Mishra, phys. status solidi (a) 202, no. 5, pp. 837–840 (2005).

42 D. M. Caughey and R. E. Thomas, Proc. IEEE 55, pp. 2192–2193 (1967).

43 T. Palacios, S. Rajan, A. Chakraborty, S. Heikman, S. Keller, S. P. DenBaars, and U. K. Mishra, IEEE Trans. Electron Dev. 52, 10, pp. 2117–2123 (2005).

44 B. Gelmont, K. Kim, and M. Shur, J. Appl. Phys. 74, pp. 1818–1821 (1993).

45 L. Ardaravicius, A. Matulionis, J. Liberis, O. Kiprijanovic, M. Ramonas, L. F. Eastman, J. R. Shealy, and A. Vertiatchikh, Appl. Phys. Lett. 83, pp. 4038–4040 (2003).

46 B. K. Ridley, W. J. Schaff, and L. F. Eastman, J. Appl. Phys. 96, pp. 1499–1502 (2004).

47 S. Selberherr, *Analysis and Simulation of Semiconductor Devices*, Springer-Verlag, Wien-New York (1984).

48 W. N. Grant, Solid-State Elect. 16, pp. 1189–1203 (1973).

49 C. R. Crowell and S. M. Sze, Appl. Phys. Lett. 9, pp. 242–244 (1966).

50 I. H. Oguzman, E. Bellotti, K. F. Brennan, J. Kolnik, R. Wang, and P. P. Ruden, J. Appl. Phys. 81, pp. 7827–34 (1997).

51 K. Kunihiro, K. Kasahara, Y. Takahashi, and Y. Ohno, IEEE Electron Dev. Lett. 20, pp. 608–10 (1999).

52 Y. Ando, Y. Okamoto, H. Miyamoto, T. Nakayama, T. Inoue, and M. Kuzuhara, IEEE Electron Dev. Lett. 24, pp. 289–291 (2003).

53 A. Chini, D. Buttari, R. Coffie, L. Shen, S. Heikman, A. Chakraborty, S. Keller, and U. K. Mishra, IEEE Electron Dev. Lett. 25, 5, pp. 229–231 (2004).

54 Y. Dora, *Understanding materials and process limits for high breakdown voltage AlGaN/GaN HEMTs*, Ph.D. Dissertation, University of California at Santa Barbara (2006).

55 Y.-R. Wu, M. Singh, and J. Singh, IEEE Trans. On Electron Dev. 52, 6, pp. 1048–1054 (2005).

56 Y.-R. Wu and J. Singh, *private communication* (2006).

11
Intersubband Optical Switches for Optical Communications
Nobuo Suzuki

11.1
Introduction

Ultrafast all-optical switches will be crucially important in future photonic networks. Semiconductor all-optical switches are superior to those based on the optical nonlinearities in optical fibers or ferroelectric materials from the viewpoints of cost, size, integration, stability, and latency. However, pattern-effect-free operation above 100 Gbs^{-1} is extremely difficult to achieve in conventional semiconductor optical switches based on semiconductor optical amplifiers (SOAs) or electro-absorption modulators (EAMs), since the response time is limited by the carrier lifetime (>10 ps). On the other hand, the intersubband (ISB) relaxation time, which is governed by the electron scattering time due to the longitudinal-optical (LO) phonons, is less than 10 ps [1]. Therefore, pattern-effect-free Tbs^{-1} operation is expected in optical switches based on the saturation of ISB absorption. The near-infrared (1.3–1.55 μm) intersubband transitions (ISBT) have been realized in InGaAs-based coupled quantum wells (CQW) [2,3], CdS/ZnSe/BeTe multiple quantum wells (MQWs) [4], and GaN/Al(Ga)N MQWs [5–9]. The GaN-based ISBT switch is the fastest because of the strong Fröhlich interaction between electrons and LO phonons [10,11]. A 10-dB switching with a gate width of as short as 230 fs has been achieved in a waveguide-type GaN ISBT switch [12]. Ultrafast optical modulation corresponding to 1.5 Tbs^{-1} has also been demonstrated [13].

In this chapter, theoretical models concerning the ISBT in nitride MQWs are described. First, the basic physics of ISBT is summarized in Section 11.2. Section 11.3 concerns the calculation of absorption spectra in GaN/AlN MQWs, where the effects of strong built-in fields ($\sim$ MV cm^{-1}) due to the piezoelectric effect and the spontaneous polarization were taken into consideration. In Section 11.4, a finite-difference time-domain (FDTD) simulator of waveguide-type GaN/AlN ISBT switches is presented.

Nitride Semiconductor Devices: Principles and Simulation. Joachim Piprek (Ed.)
Copyright © 2007 WILEY-VCH Verlag GmbH & Co. KGaA, Weinheim
ISBN: 978-3-527-40667-8

11.2
Physics of ISBT in Nitride MQWs

11.2.1
Dipole Moment

A conduction band diagram of a quantum well is shown schematically in Fig. 11.1(a). The transition matrix element can be approximated by [14]

$$\langle u_{n'}\varphi_{j,n'}|H_I|u_n\varphi_{i,n}\rangle = \langle u_{n'}|H_I|u_n\rangle_{\text{cell}}\langle \varphi_{j,n'}|\varphi_{i,n}\rangle \\ + \langle u_{n'}|u_n\rangle_{\text{cell}}\langle \varphi_{j,n'}|H_I|\varphi_{i,n}\rangle \quad (11.1)$$

where u_n is the periodic part of the Bloch function for the band n, $\varphi_{j,n}(z)$ is the envelope function of the j-th subband of the band n, and H_I is the interaction Hamiltonian. For interband transition, the second term on the right side vanishes because $<u_{n'}|u_n> = 0$. In the case of ISBT in the conduction band ($n = n' = c$, and $j \neq i$), Eq. (11.1) reduces to $<\varphi_{j,c}|\mu_z|\varphi_{i,c}>E_z$, since $<\varphi_{j,c}|\varphi_{i,c}>= 0$, $<u_c|u_c>= 1$, and $H_I = \boldsymbol{\mu} \cdot \boldsymbol{E} = \mu_z E_z$, where μ_z is the z component of the dipole $\boldsymbol{\mu}$, and E_z is the z component of the electric field $\boldsymbol{E}$ of the light. Because of the symmetry, the ISBT occurs only for the light with the electric field normal to the well. The dipole moment of the transition between i-th and j-th subbands is given by [15]

$$\mu_{ji} = \langle \varphi_j|\mu_z|\varphi_i\rangle = e\int \varphi_j(z)^* z \varphi_i(z)\,dz \quad (11.2)$$

where e is the electron charge. Since μ_{ji} is an odd function of z, the transition between the subbands with the same parity is forbidden in a symmetric well. The transition between the adjacent subbands is much stronger than the other transitions. In an infinite square well, for example, $\mu_{21} = 0.18 e L_W$ and $\mu_{41} = 0.014 e L_W$, where L_W is the thickness of the well.

11.2.2
Rate Equations

The dynamics of carriers and polarization of a two-level system are described by the Bloch equations [16]:

$$\frac{d\rho_{11}}{dt} = -\frac{\mu_{21}E_z}{i\hbar}(\rho_{21} - \rho_{12}) - \frac{d\rho_{11}}{dt}\bigg|_{\text{rlx}} \quad (11.3\text{a})$$

$$\frac{d\rho_{22}}{dt} = \frac{\mu_{21}E_z}{i\hbar}(\rho_{21} - \rho_{12}) - \frac{d\rho_{22}}{dt}\bigg|_{\text{rlx}} \quad (11.3\text{b})$$

$$\frac{d\rho_{12}}{dt} = -\frac{\omega_{21}}{i}\rho_{12} - \frac{\mu_{21}E_z}{i\hbar}(\rho_{22} - \rho_{11}) - \frac{\rho_{12}}{\tau_{\text{ph}}} \quad (11.3\text{c})$$

11.2 Physics of ISBT in Nitride MQWs

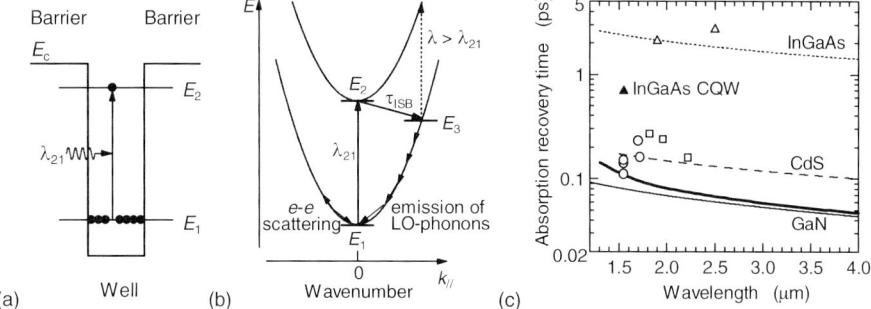

Fig. 11.1 Intersubband absorption and relaxation processes. (a) Conduction band structure of a quantum well. (b) Carrier relaxation processes. E_1, E_2 and E_3 are the levels assumed in Eq. (11.18). (c) Comparison of the measured absorption recovery time (symbols) [3, 12, 21–27] and the calculated intersubband relaxation time (lines) for InGaAs (dotted line and triangles), CdS (dashed line and squares) and GaN (solid lines and circles). The thin lines were estimated from Eq. (11.9), whereas the thick solid line was obtained by a numerical calculation [11].

where ρ_{ij} $(i,j = 1,2)$ is the density matrix element, $\hbar$ is Planck's constant divided by 2π, $\hbar\omega_{21}$ is the transition energy, and τ_{ph} is the dephasing time. The last term in Eqs (11.3a) and (11.3b) is the phenomenological relaxation term. Utilizing the relation, $\rho_{21} = \rho_{12}^* = \rho_R + i\rho_I$, Eqs (11.3a)–(11.3c) can be rewritten as [17]

$$\frac{d\rho_{11}}{dt} = -\frac{2\mu_{21}E_z}{\hbar\omega_{21}}\left(\frac{d\rho_R}{dt} + \frac{\rho_R}{\tau_{ph}}\right) - \frac{d\rho_{11}}{dt}\bigg|_{rlx} \quad (11.4a)$$

$$\frac{d\rho_{22}}{dt} = \frac{2\mu_{21}E_z}{\hbar\omega_{21}}\left(\frac{d\rho_R}{dt} + \frac{\rho_R}{\tau_{ph}}\right) - \frac{d\rho_{22}}{dt}\bigg|_{rlx} \quad (11.4b)$$

$$\frac{d^2\rho_R}{dt^2} + \frac{2}{\tau_{ph}}\frac{d\rho_R}{dt} + \omega_{21}^2\rho_R + \frac{\rho_R}{\tau_{ph}^2} = -\frac{\omega_{21}\mu_{21}E_z}{\hbar}(\rho_{22} - \rho_{11}) \quad (11.4c)$$

The terms in parentheses on the right side of Eqs (11.4a)–(11.4b) are introduced instead of $\omega_{21}\rho_I$ to remove the imaginary variables. The carrier densities in the two subbands, N_1 and N_2, and the macroscopic polarization P are associated with the density matrix elements as

$$N_i = \frac{1}{V}\sum_{\mathbf{k}_\|}\rho_{ii} \quad (i = 1, 2) \quad (11.5a)$$

$$P = \frac{1}{V}\text{Tr}(\mu_{21}\rho_R) \quad (11.5b)$$

respectively, where V is the volume and $\mathbf{k}_\|$ is the wave vector. Using these relations, the fundamental equations to describe the intersubband carrier dynamics can be derived [17]:

$$\frac{dN_1}{dt} = -\frac{2E_z}{\hbar\omega_{21}}\left(\frac{dP}{dt} + \frac{P}{\tau_{ph}}\right) + \frac{N_2}{\tau_{ISB}} \quad (11.6a)$$

$$\frac{dN_2}{dt} = \frac{2E_z}{\hbar\omega_{21}}\left(\frac{dP}{dt} + \frac{P}{\tau_{ph}}\right) - \frac{N_2}{\tau_{ISB}} \quad (11.6b)$$

$$\frac{d^2P}{dt^2} + \frac{2}{\tau_{ph}}\frac{dP}{dt} + \left(\omega_{21}^2 + \frac{1}{\tau_{ph}^2}\right)P = \frac{\omega_{21}\mu_{21}^2 E_z}{\hbar}(N_1 - N_2) \quad (11.6c)$$

where τ_{ISB} is the intersubband relaxation time. Note that all variables in Eqs (11.6a)–(11.6c) are real. If E_z and P are dealt with as complex variables, the real parts must be used in Eqs (11.6a) and (11.6b).

11.2.3
Absorption

When $\omega \sim \omega_{21} \gg \tau_{ph}^{-1}$, the absorption spectrum can be derived from the Fourier transform of Eq. (11.6c) as

$$\begin{aligned}\alpha(\omega) &= \frac{\omega\,\text{Im}[\chi(\omega)]}{c_0 n} = \frac{\omega\,\text{Im}[P(\omega)/E(\omega)]}{c_0\varepsilon_0 n} \\ &= \frac{\mu_{21}^2(N_1 - N_2)}{2c_0\varepsilon_0 n}\frac{\hbar\omega_{21}/\tau_{ph}}{(\hbar\omega_{21} - \hbar\omega)^2 + (\hbar/\tau_{ph})^2}\end{aligned} \quad (11.7)$$

where c_0 is the velocity of light in vacuum, ε_0 is the electric permittivity in vacuum, and n is the index of refraction. The full width at half maximum (FWHM) of the Lorentzian response is the homogeneous linewidth $\Delta\nu_{homo} = (\pi\tau_{ph})^{-1}$.

We utilize saturation of the intersubband absorption for ultrafast optical gate switches. We assume that a high density of the electrons locates in the ground subband due to the doping. Without a strong control optical pulse, weak signal light resonant to the ISBT is absorbed. When $(N_1 - N_2)$ approaches 0 due to the absorption of a strong control optical pulse, the absorption saturates, and hence the signal light can transmit. The saturation intensity at the absorption peak wavelength is given by [18]

$$I_S = \frac{c_0\varepsilon_0 n\hbar^2}{2\mu_{21}^2 \tau_{ISB}\tau_{ph}} \quad (11.8)$$

There is a trade-off between response time and switching energy. The switching energy of ultrafast ISBT switches tends to be high, but nitride semiconductors withstand such a high-power excitation. In the case of InGaAs, the saturation of absorption is partly canceled by two-photon absorption (TPA), which increases the switching energy and the insertion loss [19]. TPA is negligible in GaN since the bandgap is as large as 3.6 eV.

11.2.4
Relaxation Time

Figure 11.1(b) shows the relaxation process of the excited electron schematically. At room temperature, the dominant intersubband scattering process is the longitudinal optical (LO) phonon scattering [20]. Because of the band nonparabolicity, the energy separation of the two subbands is smaller for larger wave number $k_\parallel$. Hence, in the case of near-IR ISBT, electrons scattered from the upper subband to the high-energy state of the ground subband do not contribute to the absorption of the signal light until they lose the energy by emitting several LO-phonons. The carrier cooling time is generally comparable to or longer than τ_{ISB}. However, the thermalization time (electron–electron scattering time) is about one order of magnitude shorter than τ_{ISB}. Therefore, the absorption recovery time at a shorter wavelength (representing the transition for $k_\parallel \sim 0$) can be approximated by τ_{ISB}, whereas an additional slower time constant component appears in the absorption recovery at a longer wavelength (transition of electrons near the Fermi level) [21].

The intersubband scattering rate can be approximated by [20]

$$W_{21} = \frac{1}{2} W_0 \left(\frac{\hbar\omega_{LO}}{E_1}\right)^{1/2} \left[\frac{1}{4-(\hbar\omega_{LO}/E_1)} + \frac{1}{12-(\hbar\omega_{LO}/E_1)}\right] \quad (11.9)$$

where

$$W_0 = \frac{e^2}{4\pi\hbar}\left(\frac{2m^*\omega_{LO}}{\hbar}\right)^{1/2}\left[\frac{1}{\varepsilon_\infty} - \frac{1}{\varepsilon_s}\right] \quad (11.10)$$

is the basic rate, ω_{LO} is the angular frequency of the LO-phonon; ε_∞ and ε_s are the optical and the static dielectric permittivities, respectively; m^* is the effective mass; and E_1 is the energy of the ground subband. The relevant material constants for GaN, CdS, and InGaAs are compared in Table 11.1. The response of GaN is one order of magnitude faster than that in InGaAs because of strong interaction between electrons and LO-phonons [10].

In Fig. 11.1(c), absorption recovery times (τ_{rec}) measured in experiments are compared with the estimation ($\tau_{ISB} \sim 1/W_{21}$) by Eq. (11.9). The absorption recovery times at $\lambda \sim 1.55$ μm were reported to be 110–400 fs [12, 21–24] for GaN/Al(Ga)N MQWs, and 160–270 fs [25, 26] for II-VI MQWs, respectively. The absorption recovery time in InGaAs MQWs was about 2 ps [27], but it was shortened to 690 fs by utilizing a four-level system in a coupled double QW structure [3].

Tab. 11.1 Comparison of material constants related to the carrier relaxation processes in GaN, CdS, and InGaAs [29].

Material	m^*/m_0	$\varepsilon_s/\varepsilon_0$	$\varepsilon_\infty/\varepsilon_0$	$\hbar\omega_{LO}$	W_0
GaN	0.20	9.5	5.35	88 meV	1.2×10^{14} s^{-1}
CdS	0.19	10.3	5.2	38 meV	9.1×10^{13} s^{-1}
InGaAs	0.042	14.1	11.6	36 meV	6.7×10^{12} s^{-1}

Although Eq. (11.9) is simple, it tends to overestimate the scattering rate for near-IR ISBT. One of the reasons is that the barrier height is assumed to be infinite. Furthermore, the phonon model (a kind of guided mode model) assumed in the derivation of Eq. (11.9), cannot deal with the interface (IF) phonon modes [28], whose contributions are important in a narrow QW. The thick solid line in Fig. 11.1(c) is τ_{ISB} in an Al$_{0.8}$Ga$_{0.2}$N/GaN QW obtained by a numerical calculation [11, 29] based on the phonon model proposed by Huang and Zhu (a kind of modified dielectric continuum model) [30]. Material parameters of AlGaN were interpolated linearly from those of GaN and AlN [31]. The band nonparabolicity parameter was assumed to be $\alpha_{NP} = 0.187$ eV^{-1} [32]. Anisotropy and dispersion of the confined modes were ignored. Lattice and electrons were assumed to be in thermal equilibrium at $T = 300$ K. The intersubband scattering time obtained by this calculation fit the experimental results better than the estimation using Eq. (11.9). At $\lambda = 1.55$ μm, τ_{ISB} was calculated to be 109 fs, where the contribution of IF modes amounts to 60 %.

11.2.5
Dephasing Time and Spectral Linewidth

Spectral width of the intersubband absorption is typically 100–200 meV. It is generally dominated by the inhomogeneous broadening caused by structural fluctuation and the band nonparabolicity. However, ultrafast response is achievable only within the homogeneous linewidth ($\Delta \nu_{homo}$). Assuming that the pulse width is 100 fs, a bandwidth of about 100 meV is required to allocate different wavelengths to the signal and the control pulses. In InGaAs QWs, however, τ_{ph} is around 0.1 ps, corresponding to a homogeneous linewidth of about 10 meV.

The dephasing time is determined by various scattering mechanisms. The intrasubband LO-phonon scattering rate [28, 30], the electron–electron scattering rate [33] and the ionized impurity scattering rate [33] were calculated for an Al$_{0.8}$Ga$_{0.2}$N/GaN (6 MLs) QW at $T = 300$ K [11, 29]. The calculated electron–electron scattering rate $\Gamma_{11}^{(ee)}$ is shown in Fig. 11.2(a) for several carrier densities N as functions of the kinetic energy $E_\parallel$. For $N < 1 \times 10^{18}$ cm^{-3},

$\Gamma_{11}^{(ee)}$ increases with N because of an increasing probability of collision, but they decrease at a higher carrier density because of the screening. The exclusion principle also contributes to the reduction in scattering rates of low-energy electrons at high density. The effective electron–electron scattering rate $\Gamma_{av}^{(ee)}$ was calculated by averaging $\Gamma_{11}^{(ee)}(E_{\|})$ over the Fermi distribution. The effective scattering time $\tau_{ee} = 1/\Gamma_{av}^{(ee)}$ is about 10 fs, which is one order of magnitude shorter than that of GaAs ($\sim$ 100 fs). This is because the electron–electron scattering rate and the screened dielectric function are proportional to m^*/ε_∞^2 and m^*/ε_∞, respectively [33].

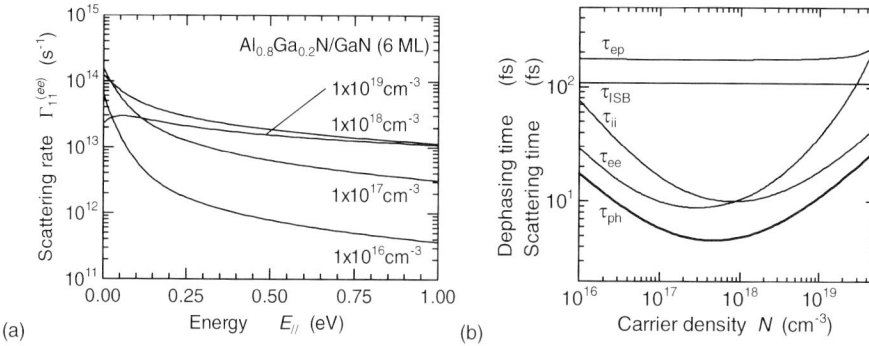

Fig. 11.2 Calculation of carrier scattering times and dephasing time in GaN/AlGaN MQWs [11, 29]. (a) Electron–electron scattering rate as functions of kinetic energy and carrier density. (b) Electron–electron scattering time τ_{ee}, ionized impurity scattering time τ_{ii}, intrasubband LO-phonon scattering time in the lowest subband τ_{ep}, intersubband relaxation time τ_{ISB}, and dephasing time τ_{ph} as functions of carrier density.

Similarly, the intrasubband LO-phonon scattering time τ_{ep} and the ionized impurity scattering time τ_{ii} were calculated. Then, the dephasing time was estimated from

$$\tau_{ph} = \left[\tau_{ee}^{-1} + \tau_{ii}^{-1} + \tau_{ep}^{-1} + (2\tau_{ISB})^{-1}\right]^{-1} \tag{11.11}$$

The results are shown in Fig. 11.2(b) [29]. The LO-phonon scattering time for an electron which can emit LO-phonons ($E_{\|} \geq \hbar\omega_{LO}$) is about 10 fs. Most of the electrons, however, have a kinetic energy of less than $\hbar\omega_{LO}$. The ionized impurity scattering shows a tendency similar to the electron–electron scattering. The dephasing time in GaN with a carrier density higher than 10^{19} cm^{-3} was estimated to be several tens of femtoseconds. Nonequilibrium carrier distribution under strong excitation may moderate the effect of the exclusion principle. Another scattering mechanism, such as interface roughness scattering may further shorten the dephasing time. The short dephasing time assures broadband (> 100 nm) operation of GaN ISBT switches, but leads to rather high switching pulse energy as suggested by Eq. (11.8). Modulation doping is considered to be effective for suppressing the ionized impurity scattering.

11.3
Calculation of Absorption Spectra

11.3.1
Transition Wavelength and Built-In Field

Wurtzite nitride semiconductors have a rather large piezoelectric effect and spontaneous polarization. A strong built-in field ($>$MV cm^{-1}) exists in GaN/Al(Ga)N MQWs grown along the c axis [34]. The built-in field drastically shortens the intersubband transition wavelengths (λ_{21}) in thick wells [35, 36]. The energy levels and the envelope functions of the subbands were calculated, assuming a uniform field F_w in the GaN well, utilizing the transfer matrix technique [37]. The conduction band discontinuity was assumed to be 1.76 eV, considering a deformation potential [38]. The calculated transition wavelengths were compared with some experimental results in Fig. 11.3(a) [36]. The solid and dotted lines denote GaN/AlN MQWs and GaN/Al$_{0.65}$Ga$_{0.35}$N MQWs, respectively. The circles and the triangles represent the data for AlN/GaN MQWs grown by molecular beam epitaxy (MBE) [7] and for GaN/AlGaN MQWs grown by metalorganic chemical deposition (MOCVD) [35, 39], respectively. Near-infrared ISBT has been achieved only in MBE-grown samples. From the figure, the effective field strengths in the well were estimated to be 5–8 MV cm^{-1} and 0–3 MV cm^{-1} for GaN/AlN MQWs and GaN/Al$_{0.65}$Ga$_{0.35}$N MQWs, respectively.

Next, the subbands and the potentials of the MBE-grown GaN/AlN samples were calculated in a self-consistent manner [36]. The thickness of the barriers (L_b) was 4.6 nm. The potential drop across one period of the quan-

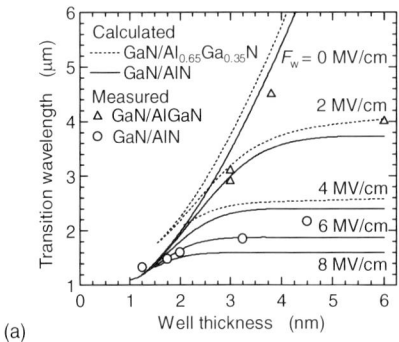
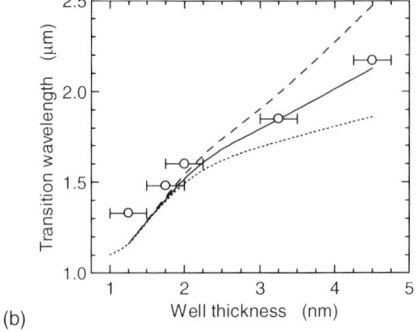

Fig. 11.3 Effect of built-in field on intersubband transition wavelength. (a) The transition wavelength of GaN/Al$_{0.65}$Ga$_{0.35}$N MQWs (dotted lines and triangles) [35, 39] and GaN/AlN MQWs (solid lines and circles) [7]. The lines were calculated assuming a constant built-in field in the well [35, 36]. (b) Self-consistent calculation of transition wavelength [36] for MBE-grown GaN/AlN samples (circles) [7]. The dashed, solid, and dotted lines were calculated for $N_D = 8 \times 10^{19}$ cm^{-3}, 4×10^{19} cm^{-3}, and 1×10^{16} cm^{-3}, respectively.

tum well (ΔV_{QW}) was assumed to be zero. This condition is satisfied when some amount of electrons is transferred across the MQW. First, the subband structure was calculated, assuming a certain potential. After the charge distribution was calculated, assuming the thermal equilibrium at 300 K, Poisson's equation was solved. The iteration was continued, updating the potential until the error in the potential became smaller than the predetermined value.

The calculated results are compared with the experimental data in Fig. 11.3(b). The dashed, solid and dotted lines represent the results for donor concentrations (N_D) of 8×10^{19} cm^{-3}, 4×10^{19} cm^{-3}, and 1×10^{16} cm^{-3}, respectively. The measured data (circles) [7] fit with the line for $N_D = 4 \times 10^{19}$ cm^{-3}, which is half of the Si-doping level. Some of electrons were considered to be transferred to the heterointerface of the underlying bulk GaN to flatten the Fermi level in the MQW. As discussed below, some of the electrons are considered to be trapped by edge dislocations. Therefore, the actual carrier density in the well is considered to be lower than the doping level. For thinner wells, the measured transition wavelengths were slightly longer than the calculated ones. This is attributable to interface gradient layers. Another possibility is a smaller conduction-band offset [40].

In waveguide samples with a few wells, the peak wavelength deviated from that predicted, because the built-in field in the MQW is not uniform. The peak wavelength also depends on the underlying structure. For example, the absorption wavelengths of GaN/AlN MQWs grown on AlN tended to be shorter than those grown on GaN.

11.3.2
Absorption Spectra

Absorption spectra of the MBE-grown samples were measured by Fourier-transform infrared spectroscopy (FTIR) in single pass geometry with Brewster angle incidence (Fig. 11.4(a)) [7]. Transmittance for this arrangement is given by [18]

$$T = \exp\left[-\alpha N_w L_w \frac{\sin^2 \theta_i}{\cos \theta_i}\right] \quad (11.12)$$

where N_w is the number of wells, and θ_i is the angle of incidence. The transmission spectra calculated utilizing Eqs (11.2), (11.7) and (11.12) were compared with the measured spectra [36]. The assumed MQW structures consist of 4.6-nm undoped AlN barriers and n-type GaN wells uniformly doped at 4×10^{19} cm^{-3}. As shown in Fig. 11.4(b), 2-monolayer step layers were assumed at the heterointerfaces to reduce the deviation from the measured peak wavelength for the samples with thinner wells. A thickness fluctuation of ± 1 ML (25 % each) was assumed to simulate the shoulders observed in the measured spectra. The FWHM of each spectral component was assumed to be 80 meV. The results (solid lines) are compared with the measured spectra (dashed lines) in Fig. 11.4(c), where the measured transmittance was normalized to unity in the longer wavelength region. Loss increase in the shorter wavelength region is attributable to surface scattering.

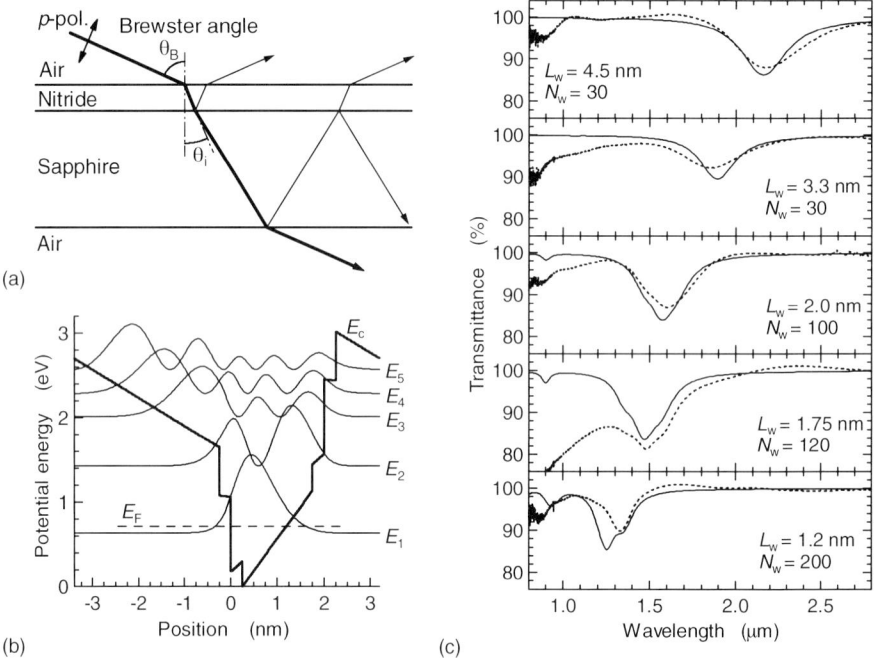

Fig. 11.4 Calculation of absorption spectra [36]. (a) Absorption spectra were measured in single pass geometry with Brewster angle incidence. (b) Subband structure for $L_w = 2$ nm. (c) Calculated (solid lines) and measured (dotted lines) spectra.

11.4
FDTD Simulator for GaN/AlGaN ISBT Switches

The FDTD [41–43] is suitable for calculating the time-dependent nonlinear response in waveguides, but it requires considerable computer resources. For personal computer simulation, a one-dimensional FDTD simulator, where the intersubband carrier dynamics were described by three-level rate equations, was developed [44, 45].

11.4.1
Model

Here, we suppose that the x direction is normal to the MQW and that the TM-mode optical pulses propagate in the z direction [45]. Then, the one-dimensional Maxwell's equations are

$$\frac{\partial H_y}{\partial t} = -\frac{1}{\mu_0}\frac{\partial E_x}{\partial z} \tag{11.13a}$$

$$\frac{\partial D_x}{\partial t} = -\frac{\partial H_y}{\partial z} \tag{11.13b}$$

where the relevant constitutive equation is

$$E_x = \frac{1}{\varepsilon_0}(D_x - P_x) \tag{11.14}$$

Here, E_x, H_y, D_x, and P_x are nonzero real components of the electric field, the magnetic field, the electric displacement, and the polarization, respectively; μ_0 is the permeability of the vacuum; and t is the time.

We consider here only the Lorentz-type linear polarization (P_L) and the polarizations due to the ISBT (P_{ISB}). In the time domain, P_L is given by [42]

$$\frac{d^2 P_L}{dt^2} + \gamma_L \frac{d P_L}{dt} + \omega_L^2 P_L = \varepsilon_0 \chi_s \omega_L^2 E_x \tag{11.15}$$

where ω_L is the resonant angular frequency; γ_L is the damping factor; and χ_s is the electric susceptibility. The background loss of the waveguide can be taken into consideration in terms of γ_L. In order to deal with the inhomogeneous broadening and the spatial mode profile, the polarizations due to the ISBT were treated as a sum of several independent components $P_{ISB}^{(i)}$. Then the total polarization is represented by [45]

$$P_x = P_L + \sum_i \Gamma_W^{(i)} P_{ISB}^{(i)} \tag{11.16}$$

where $\Gamma_W^{(i)}$ is the optical confinement factor of the wells relevant to the i-th ISBT.

To include the influence of the band nonparabolicity, the intersubband carrier dynamics were described by the three-level rate-equation model [44]. E_3 in Fig. 11.1(b) represents the states with a large kinetic energy in the lower subband, which cannot contribute to the absorption at λ_i. The time constants τ_{23} and τ_{31} are the intersubband relaxation time and the effective intrasubband relaxation time including the cooling time of the electrons, respectively. Then, Eqs (11.6a)–(11.6c) are modified as follows [45]:

$$\frac{d^2 P_{ISB}^{(i)}}{dt^2} + \frac{2}{\tau_{ph}}\frac{d P_{ISB}^{(i)}}{dt} + \left(\omega_i^2 + \frac{1}{\tau_{ph}^2}\right) P_{ISB}^{(i)} = -\frac{\omega_i \mu_i^2 \zeta_i E_x}{\hbar}\left(N_2^{(i)} - N_1^{(i)}\right) \tag{11.17}$$

$$\frac{dN_1^{(i)}}{dt} = -\frac{2\zeta_i E_x}{\hbar\omega_i}\left(\frac{dP_{ISB}^{(i)}}{dt} + \frac{P_{ISB}^{(i)}}{\tau_{ph}}\right) + \frac{N_3^{(i)}}{\tau_{31}} \qquad (11.18a)$$

$$\frac{dN_2^{(i)}}{dt} = \frac{2\zeta_i E_x}{\hbar\omega_i}\left(\frac{dP_{ISB}^{(i)}}{dt} + \frac{P_{ISB}^{(i)}}{\tau_{ph}}\right) - \frac{N_2^{(i)}}{\tau_{23}} \qquad (11.18b)$$

$$\frac{dN_3^{(i)}}{dt} = \frac{N_2^{(i)}}{\tau_{23}} - \frac{N_3^{(i)}}{\tau_{31}} \qquad (11.18c)$$

where ζ_i is the ratio of the electric field in the i-th well to that at the mode peak. In the case of inhomogeneous broadening, the absorption spectrum is composed of several homogeneous components with different ω_i and μ_i. The ratio of each spectral component is taken into consideration through $\Gamma_w^{(i)}$. In the case of the spatial mode profile, the wells are divided into several parts, of which $\Gamma_w^{(i)}$ and ζ_i were obtained through a mode calculation based on the finite-difference method (FDM).

Equations (11.13a)–(11.18c) were digitized utilizing the one-dimensional Yee's lattice with $\Delta t = 0.05$ fs and $\Delta z = 40$ nm. The absorbing boundary condition (ABC) was assumed. Power spectra of the optical pulses can be calculated by Fourier transformation of the temporal response of the waveguide.

11.4.2
Saturation of Absorption

The measured saturation characteristics were compared with the results of simulation. The first ISBT switch (SW-A) [24] consisted of a (0001) sapphire substrate, a 70-nm AlN buffer, a multiple intermediate layer (MIL) consisting of 10 pairs of 40-nm thick GaN and 10-nm thick AlN, a 0.5-μm thick GaN, an MQW, and a 1.0-μm thick GaN. Only the GaN wells were intentionally doped with Si at 5×10^{19} cm^{-3}. The upper GaN layer was etched to form mesas by Cl$_2$-based electron cyclotron resonance reactive ion beam etching (ECR-RIBE). The width of the mesa was 1 μm, but it broadened to 2 μm near the cleaved facets with 40-μm tapers. As shown in Fig. 11.5, the light is confined by high-index contrast between the nitride layers and the surroundings [45]. The facets were anti-reflection coated, and then the chip was housed in a module with fiber pigtails. The absorption peak wavelength and the linewidth of the sample were 1.75 μm and 120 meV, respectively. From the fitting, the MQW was assumed to be composed of the main component ($\lambda = 1.756$ μm, 50 %) and sub-components ($\lambda = 1.657$ μm and 1.838 μm, 25 % each) with a homogeneous linewidth of 80 meV [45].

The measured saturation characteristics of a 400-μm length waveguide for TM pulses are shown by solid circles in Fig. 11.6(a) [24]. The wavelength and

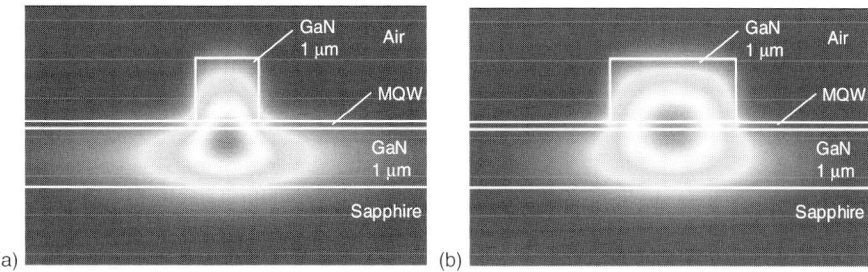

Fig. 11.5 Calculated field profile (E_y) of TM mode in GaN ISBT switch [45]. (a) W = 1 μm, (b) W = 2 μm.

the width of the pulses were 1.55 μm and 130 fs, respectively. The coupling loss and the background loss were estimated to be 3 dB/facet and 2 dB, respectively, from the device length dependency of the insertion loss for the TE mode. Polarization-dependent loss (PDL) was 26.5 dB, which was reduced to 21 dB when the fiber input pulse energy was 120 pJ. The change in transmittance was smaller than the expected. In nitride layers grown on a sapphire substrate by MBE, high-density ($> 10^{10}$ cm^{-2}) edge dislocations run perpendicular to the substrate. Electric charges trapped by the dislocations cause the excess loss for the TM mode. Actually, a PDL of 8–15 dB mm^{-1} was observed even in GaN waveguides without QWs [46]. The excess PDL (α_d) depended both on the dislocation density and the carrier density, but was almost independent of the optical pulse energy.

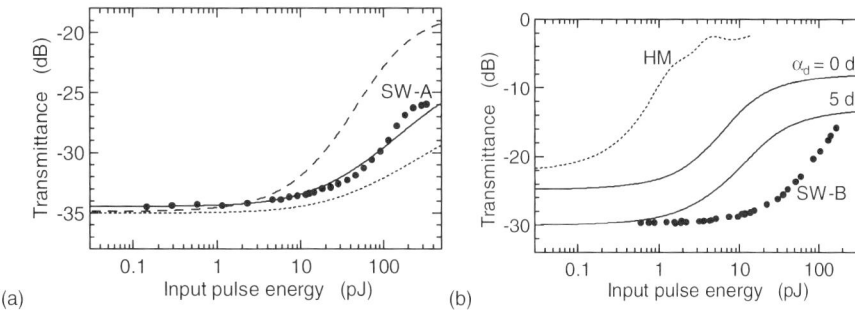

Fig. 11.6 Absorption saturation characteristics of GaN/AlN ISBT switches [45]. (a) Comparison of measured transmittance (solid circles) [24] and calculated transmittance (lines) for SW-A. The dashed, solid, and dotted lines are calculated for the combinations of $\xi_i = 0.75$ and $\alpha_d = 10$ dB, $\xi_i = 0.5$ and $\alpha_d = 15$ dB, and $\xi_i = 0.45$ and $\alpha_d = 18$ dB, respectively. (b) Improvement of the characteristics. The solid circles represent the measured transmittance of SW-B [12]. The solid lines denote the calculated results for an improved rib waveguide with $\alpha_d = 5$ dB and 0 dB. The dotted line shows the characteristic calculated for a 0.8×0.8-μm^2 high mesa waveguide.

The saturation characteristics were calculated for some combinations of ξ_i and α_d [45]. Loss before the saturation fit the measured data (34.5 dB) when (i) $\xi_i = 0.75$ and $\alpha_d = 10$ dB, (ii) $\xi_i = 0.5$ and $\alpha_d = 15$ dB, and (iii) $\xi_i = 0.45$ and $\alpha_d = 18$ dB, which are shown in Fig. 11.6(a) by the dashed, solid, and dotted lines, respectively. The best fit was obtained for (ii). For $W = 1$ µm, the mode peak position is in the lower GaN layer as shown in Fig. 11.5(a). The calculated ξ_i was 0.65. The fact that ξ_i was 25 % smaller than the calculated value is attributable to errors in the assumed values. For example, the actual carrier density in the wells is considered to be lower than the Si doping level, since some of electrons are trapped by the dislocations. As shown in Fig. 11.5(b), the mode peak shifted into the MQW for $W = 2$ µm, and hence the intersubband absorption will be stronger.

The solid circles in Fig. 11.6(b) show the saturation characteristics of another ISBT switch (SW-B), where the dislocation density was reduced to 1.7×10^9 cm^{-2} by means of a 0.8 µm-thick MOCVD-grown GaN template [12]. To suppress generation of edge dislocations in the MQW layer, the number of wells N_w was reduced to 2 and the thickness of the AlN barriers was reduced to 1.5 nm. To compensate for the reduction in N_w, the Si-doping level was increased to 2×10^{20} cm^{-3}. For 1.55 µm, 130 fs optical pulses, 10 dB saturation was achieved at a pulse energy of 101 pJ. The excess PDL was estimated to be 5–10 dB, and further improvement is required.

The solid lines in Fig. 11.6(b) show calculated saturation characteristics of an improved design [45], where the absorption peak wavelength, the number of wells, the carrier concentration, and the mesa width were assumed to be 1.66 µm, 3 wells, 4×10^{19} cm^{-3}, and 2 µm, respectively. The 10 dB switching energy was calculated to be 14 pJ and 8 pJ for $\alpha_d = 5$ dB and 0 dB, respectively. For further reduction of the switching pulse energy, waveguides with a smaller mode size are required. The dotted line in Fig. 11.6(b) shows the calculated result for a mesa waveguide with a cross-section of 0.8 µm $\times$ 0.8 µm. The wavelengths of the absorption peak, the pump and the probe were assumed to be 1.57 µm, 1.60 µm and 1.50 µm, respectively. The coupling loss was assumed to be 1 dB/facet, which will be achieved in future by means of spot-size converters. The 10 dB switching energy was calculated to be 0.76 pJ.

11.4.3
Temporal Response

Figures 11.7(a) and 11.7(b) show the pump and probe responses of SW-A [24] and SW-B [12], respectively. The wavelength and the pulse width of the probe were 1.55 µm and 130 fs, respectively. The modulation ratio of SW-A was about 2.4 dB for a 1.7 µm, 120 pJ, 230 fs control pulse. The FWHM of the response was 360 fs, which was limited by the convolution of the pulses.

The change in transmittance recovered within 1 ps. Pattern-effect free optical modulation corresponding to 1.0–1.5 Tbs^{-1} was demonstrated [13]. In SW-B, 11.5 dB switching was achieved at a control pulse energy of 150 pJ. The gate width was also improved to 230 fs by shortening the fiber pigtails to suppress the pulse broadening due to the nonlinearity and the dispersion. In addition to the fast recovery component ($\tau \sim$ 110 fs) due to the ISBT. However, a slow recovery component ($\tau \sim$ 2 ps) was observed in SW-B, which caused pattern effects at above 500 Gbs^{-1} [47]. The slow component was attributed to the tunneling of excited electrons to the adjacent bulk GaN layer. To suppress the pattern effect, the barrier thickness must be optimized.

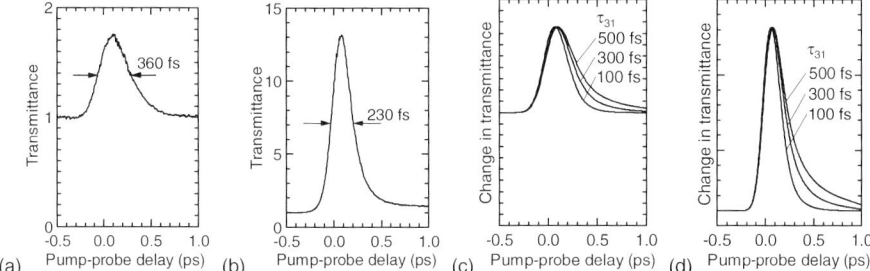

Fig. 11.7 Pump and probe response of GaN/AlN ISBT switches. (a) SW-A, measured [24]. (b) SW-B, measured [12]. (c) SW-A, calculated [45]. (d) SW-B, calculated. In (c) and (d), the scale of the ordinate was adjusted to fit the measured data.

The calculated responses [45] were compared in Figs 11.7(c) and (d), respectively. The intersubband relaxation time was assumed to be $\tau_{23} = 110$ fs [11]. In the present model, the intrasubband recovery time, τ_{31}, was treated as a phenomenological fitting parameter. The best fit was obtained for $\tau_{31} = 300$ fs.

11.4.4
Future Applications

From the viewpoint of switching energy, optical switches based on optical amplifiers are superior to saturable absorber-type optical switches. Troccoli et al. [48] realized InGaAs/InAlAs quantum cascade laser amplifiers (QCLAs) in the mid-IR region ($\sim$ 7.4 µm). However, realization of a near-IR GaN/AlN QCLA is considered to be very difficult, because a large voltage drop ($\sim$ 1 V/stage) and a large current density ($J \propto \tau_{ISB}^{-1}$) are required. As an alternative, optically pumped intersubband optical amplifiers (ISOAs) utilizing a four-level system in GaN/AlN coupled quantum wells were proposed [49]. The FDTD model combined with 4- or 5-level rate equations was applied to the feasibility study of some functional devices based on the ISOA: optical gate switches utilizing the cross-gain modulation (XGM) [49], 2 × 2 optical switches utilizing the cross-phase modulation (XPM) [50], and simultaneous multi-wavelength converters [51]. Although high-power (> 5 W) cw pumping is required to achieve a sufficient gain, the switching pulse energy was predicted to be reduced by one order of magnitude.

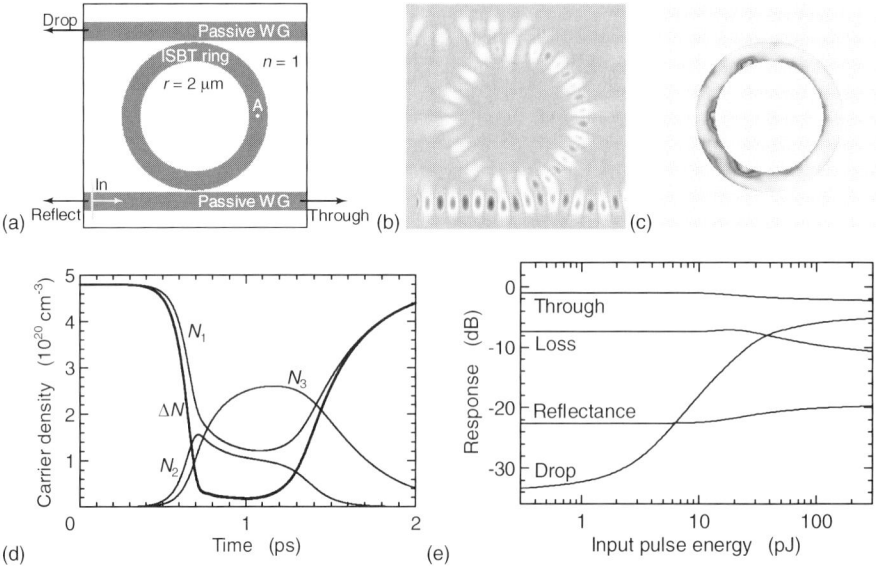

Fig. 11.8 An example of 2-D FDTD simulation of GaN/AlN ISBT ring resonator switches. (a) Plan view of the test structure; (b) and (c) show distributions of E_y and $\Delta N = N_1 - N_2$, respectively, at 0.2 ps after the pulse peak incidence. The input pulse energy was $I_{in} = 20$ pJ. (d) Response of carriers for $I_{in} = 20$ pJ at the point A on the ring. (e) Input pulse energy dependence of response.

The rate equations can be easily implemented in two-dimensional simulators. Figure 11.8(a) shows a test structure for 2-D simulations: a GaN ISBT switch consisting of a ring resonator with ISB absorption layers and two passive linear waveguides. The cross-section of the GaN ($n = 2.29$) high-index contrast (HIC) waveguide formed on a sapphire substrate ($n_{sub} = 1.72$) was assumed to be 0.6 μm (W) ×1 μm (H), which has only one TM mode ($n_{eff} = 2.01$). In 2-D calculations, it was emulated by a 0.6 μm waveguide ($n_{slab} = 2.18$) sandwiched by the air. The effective index calculated by a 1-D mode solver was 1.98. The radius of the ring and the gap width were assumed to be 2 μm and 40 nm, respectively. The carrier density, the optical confinement factor and the absorption peak wavelength of the wells were assumed to be 4.8×10^{20} cm^{-3}, 2.18 %, and 1.6 μm, respectively. Carrier diffusion and temperature rise were not taken into account. The wavelength and the width of the input Gaussian pulse were assumed to be 1.583 μm (resonant to the ring cavity) and 400 fs, respectively. The boundaries of the calculation region

(7 μm ×7 μm) were surrounded by perfectly-matched layers (PML) [52]. The lattice size and the time step were 20 nm and 0.04 fs, respectively. Since the structure was not optimized, the tracks of light in the output waveguides were not straight (Fig. 11.8(b)). The absorption (proportional to ΔN) in the ring was saturated by strong light (Figs 11.8(c) and (d)). At $I_{in} = 20$ pJ, the transmittance at the drop port was predicted to increase by 21.5 dB (Figs 11.8(e)). The width of the dropped pulse was about 500 fs. Three-dimensional simulations will be possible if a high performance computing system is utilized.

Acknowledgments

Part of this work was performed under the management of the Femtosecond Technology Research Association (FESTA, Dec., 1995-March, 2005), which was supported by the New Energy and Industrial Technology Development Organization (NEDO). The author would like to thank N. Iizuka and K. Kaneko for valuable discussion.

References

1 S. Noda, in T. Kamiya, F. Saito, O. Wada and H. Yajima (eds.), *Femtosecond Technology: From Basic Research to Application Prospects*, Springer, Berlin, **1999**, 222–233.

2 B. Sung, H. C. Chui, M. M. Fejer, J. S. Harris, Jr., *Electron. Lett.*, **1997**, 33, 818–820.

3 T. Akiyama, N. Georgiev, T. Mozume, H. Yoshida, A. V. Gopal, O. Wada, *Electron. Lett.*, **2001**, 37, 129–130.

4 R. Akimoto, Y. Kinpara, K. Akita, F. Sasaki, S. Kobayashi, *Appl. Phys. Lett.*, **2001**, 78, 580–582.

5 C. Gmachl, H. M. Ng, S.-N. G. Chu, A. Y. Cho, *Appl. Phys. Lett.*, **2000**, 77, 3722–3724.

6 K. Kishino, A. Kikuchi, H. Kanazawa, T. Tachibana, *Appl. Phys. Lett.*, **2002**, 81, 1234–1236.

7 N. Iizuka, K. Kaneko and N. Suzuki, *Appl. Phys. Lett.*, **2002**, 81, 1803–1805.

8 D. Hofstetter, S.-S. Schad, H. Wu, W. J. Schaff, L. F. Eastman, *Appl. Phys. Lett.*, **2003**, 83, 572–574.

9 A. Helman, M. Tchernycheva, A. Lusson, E. Warde, F. H. Julien, Kh. Moumanis, G. Fishman, E. Monroy, B. Daudin, Le Si Dang, E. Bellet-Amalric, D. Jalabert, *Appl. Phys. Lett.*, **2003**, 83, 5196–5198.

10 N. Suzuki, N. Iizuka, *Jpn. J. Appl. Phys.*, **1997**, 36, L1006-L1008.

11 N. Suzuki, N. Iizuka, *Jpn. J. Appl. Phys.*, **1998**, 37, L369-L371.

12 N. Iizuka, K. Kaneko, N. Suzuki, *Opt. Express*, **2005**, 13, 3835–3840.

13 N. Iizuka, K. Kaneko, N. Suzuki, *Technical Digest, LEOS 2004 Annual Meeting*, IEEE, **2004**, 665–666.

14 Y. Hirayama, J. H. Smet, L.-H. Peng, C. G. Fonstad, E. P. Ippen, *Jpn. J. Appl. Phys.*, **1994**, 33, 890–895.

15 L. C. West, S. J. Eglash, *Appl. Phys. Lett.*, **1985**, 46, 1156–1158.

16 A. Yariv, *Quantum Electronics, 3rd ed.*, John Wiley & Sons, USA, **1988**, Chap. 8.

17 N. Suzuki, N. Iizuka, K. Kaneko, *IEICE Trans. Electron.*, **2000**, E83-C, 981–988.

18 B. F. Levine, *J. Appl. Phys.*, **1993**, 74, R1-R81.

19 S. Sekiguchi, T. Shimoyama, H. Yoshida, J. Kasai, T, Mozume, H. Ishikawa, *Technical Digest, Opt. Fiber Commun. Conf.*, Vol. 5, OSA, **2005**, 92–94.

20. B. K. Ridley, in J. Shah (ed.) *Hot Carriers in Semiconductor Nanostructures: Physics and Applications*, Academic Press, Boston, **1992**, 17–51.
21. J. Hamazaki, H. Kunugita, K. Ema, A. Kikuchi, K. Kishino, *Phys. Rev. B*, **2005**, 71, 165334.
22. J. D. Heber, C. Gmachl, H. M. Ng, A. Y. Cho, *Appl. Phys. Lett.*, **2002**, 81, 1237–1239.
23. J. Hamazaki, S. Matsui, H. Kunugita, K. Ema, H. Kanazawa, T. Tachibana, A. Kikuchi, K. Kishino, *Appl. Phys. Lett.*, **2004**, 84, 1102–1104.
24. N. Iizuka, K. Kaneko, and N. Suzuki, *Electron. Lett.*, **2004**, 40, 962–963.
25. R. Akimoto, K. Akita, F. Sasaki, T. Hasama, *Appl. Phys. Lett.*, **2002**, 81, 2998–3000.
26. R. Akimoto, B. S. Li, K. Akita, T. Hasama, *Appl. Phys. Lett.*, **2005**, 87, 181104.
27. A. V. Gopal, T. Shimoyama, H. Yoshida, J. Kasai, T. Mozume, H. Ishikawa, *IEEE J. Quantum Electron.*, **2003**, 39, 1356–1361.
28. L. Wendler, R. Pechstedt, *phys. status solidi (b)*, **1987**, 141, 129–150.
29. N. Suzuki, N. Iizuka, in M. Osiński, P. Blood, and A. Ishibashi (eds.), *Physics and Simulation of Optoelectronic Devices VI, Proc. SPIE*, Vol. 3283, Pt.2, **1998**, 614–621.
30. K. Huang, B. Zhu, *Phys. Rev. B*, **1988**, 38, 13377–13386.
31. S. Strite, H. Morkoç, *J. Vac. Sci. & Technol. B*, **1992**, 10, 1237–1266.
32. B. E. Foutz, L. F. Eastman, U. V. Bhapkar, M. S. Shur, *Appl. Phys. Lett.*, **1997**, 70, 2849.
33. M. Dür, S. M. Goodnick, P. Lugli, *Phys. Rev. B*, **1996**, 54, 17794–17804.
34. F. Bernardini, V. Fiorentini, D. Vanderbilt, *Phys. Rev. B*, **1997**, 56, R10024-R10027.
35. N. Suzuki, N. Iizuka, *Jpn. J. Appl. Phys.*, **1999**, 38, L363-L365.
36. N. Suzuki, N. Iizuka, K. Kaneko, *Jpn. J. Appl. Phys.*, **2003**, 42, 132–139.
37. A. Harwit, J. S. Harris, Jr., A. Kapitulnik, *J. Appl. Phys.*, **1986**, 60, 3211–3213.
38. K. Kim, W. R. L. Lambrecht, B. Segall, *Phys. Rev. B*, **1996**, 53, 16310–16326.
39. N. Iizuka, K. Kaneko, N. Suzuki, T. Asano, S. Noda, *Appl. Phys. Lett.*, **2000**, 77, 648–650.
40. B. K. Ridley, W. J. Schaff, L. F. Eastman, *J. Appl. Phys.*, **2003**, 94, 3972–3978.
41. R. M. Joseph, A. Taflove, *IEEE Trans. Antennas & Propagat.*, **1997**, 45, 364–374.
42. R. W. Ziolkowski, *IEEE Trans. Antennas & Propagat.*, **1997**, 45, 375–391.
43. A. S. Nagra, R. A. York, *IEEE Trans. Antennas & Propagat.*, **1998**, 46, 334–340.
44. N. Suzuki, N. Iizuka, K. Kaneko, *IEICE Trans. Electron.*, **2000**, E83-C, 981–988.
45. N. Suzuki, N. Iizuka, K. Kaneko, *IEICE Trans. Electron.*, **2005**, E88-C, 342–348.
46. N. Iizuka, K. Kaneko, N. Suzuki, *J. Appl. Phys.*, **2006**, 99, 093107.
47. N. Iizuka, K. Kaneko, N. Suzuki, *Int. Quantum Electronics Conf. 2005 and Pacific Rim Conf. on Lasers and Electro-Optics 2005*, IEEE, **2005**, 1279–1280.
48. M. Troccoli, C. Gmachl, F. Capasso, D. L. Sivco, A. Y. Cho, *Appl. Phys. Lett.*, **2002**, 80, 4103–4105.
49. N. Suzuki, *Jpn. J. Appl. Phys.*, **2003**, 42, 5607–5612.
50. N. Suzuki, *J. Opt. Soc. Am. B*, **2004**, 21, 2017–2024.
51. N. Suzuki, *IEICE Trans. Electron.*, **2004**, E87-C, 1155–1160.
52. S. D. Gedney, *IEEE Trans. Antennas & Propagat.*, **1996**, 44, 1630–1639.

12
Intersubband Electroabsorption Modulator
Petter Holmström

12.1
Introduction

Optical modulators are important components in high-speed fiber-optical communication. The advantage over using a direct modulated laser is primarily that it is possible to generate optical pulses with lower frequency chirp, which considerably extends the transmission distance on the optical fiber. Further, it avoids the modulation speed limitations of interband semiconductor lasers. Currently, state of the art modulators at the communications wavelength 1.55 µm are typically based on interband transitions in semiconductor quantum wells (QWs), e.g., InGaAsP/InGaAsP QWs on InP, using the quantum confined Stark effect (QCSE) to achieve electroabsorption (EA) [1]. Modulators using electro-optic materials, notably $LiNbO_3$, in a Mach–Zehnder configuration to achieve intensity modulation, e.g., [2], have been constantly developed over more than four decades and are also much relied on. More recently promising results have been demonstrated in polymeric modulators [3], which can be produced at low cost, although the lifetime of the device is still a concern for these modulators.

This chapter describes the simulation of an optical modulator with a novel kind of EA based on intersubband (IS) transitions. It is not yet in practical use, but according to the simulation it has considerable potentials given that the quality of materials that exhibit IS transitions at the communications wavelength $\lambda = 1.55$ µm can be sufficiently developed. The EA in the simulated structure is due to four GaN/AlGaN/AlN step QWs. The core of the simulation is a self-consistent model of the active region including the MQW and adjoining layers, which, e.g., shows that it is possible to get the necessary near equal applied electric fields over all the QWs, despite the large built-in fields in nitrides. This is achieved by a proper choice of the composition in the adjoining AlGaN layers. In addition the IS transitions and corresponding absorption is modeled, and the optical mode is obtained by a standard transfer-matrix

Nitride Semiconductor Devices: Principles and Simulation. Joachim Piprek (Ed.)
Copyright © 2007 WILEY-VCH Verlag GmbH & Co. KGaA, Weinheim
ISBN: 978-3-527-40667-8

calculation. It is, e.g., shown that by high doping it is possible to achieve a very short modulator which is important for speed, and also that a small negative chirp parameter can be obtained which is nearly ideal for transmission on a standard single-mode optical fiber (SSMF). In order to obtain a low series resistance and simultaneously a strong confinement of the optical mode, the waveguide cladding layers employs the plasma effect in heavily doped GaN to achieve a low refractive index.

Typically an optical modulator is placed after a semiconductor laser that provides a continuous-wave light to inscribe the signal to be transmitted. A schematic picture of a fiber-optic communication link is shown in Fig. 12.1. Below the modulator are summarized its essential properties. A high-speed optical modulator should ideally have a low driving voltage V_{pp}, an extinction ratio of about 10 dB (the ratio of output optical power in transmitting and absorbing states), a low insertion loss (the optical loss sustained by the modulator in the transmitting state), a small, ideally slightly negative, chirp parameter α, a high optical saturation power P_s of about 10 mW or more, and of course a high f_{3dB} modulation frequency. Assuming an ideal detector, the maximum modulation frequency is defined as when the optical modulation depth has been reduced to $1/\sqrt{2}$ of the low-frequency value, i.e., the electrical $f_{3dB} = 1/2\pi RC$ in an RC-limited modulator. Here R is the impedance seen by the modulator (including the modulator series resistance) in the driver-modulator circuit and C is the modulator capacitance. The key point in developing high-speed modulators is really to reduce the driving voltage, i.e., the peak-to-peak voltage swing V_{pp}. By reducing the voltage, a separate amplifier can be avoided, although this would probably require a sub-1-V swing. It also allows driving the modulator at higher speed, since transistors in the driving electronics are restricted by power dissipation and by the breakdown voltage. Shrinking the size of transistors in order to achieve higher speed generally implies increased heating and reduced breakdown voltages.

High-speed modulators in use today, as presented above, each have their respective strong and weak points. Semiconductor modulators based on the QCSE can be made compact and with quite low voltage swings of typically 2–3 V. The current record for high-speed modulators was recently demonstrated at $f_{3dB} = 100$ GHz (2.1 V required voltage swing for 10 dB extinction ratio at DC) using QCSE in a travelling-wave configuration [1,4]. On the other hand the absorption modulation on the low-energy side of the exciton peak using QCSE, generally yields a substantial positive chirp parameter, especially at high transmission where it is most important. The optical saturation power is also moderate at typically a few mW due to the problem of extracting the photo-generated holes from the MQW. The electro-optic Mach–Zehnder modulator does not suffer from any of these drawbacks. In fact, it allows one to control the chirp by proper application of the microwave signal to the two

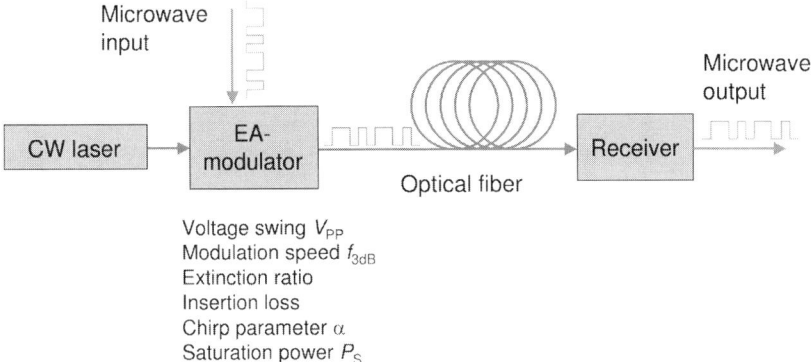

Fig. 12.1 Schematic figure of a fiber-optical communication system. Beneath the electroabsorption modulator are listed its essential properties.

arms, push-pull, so that zero chirp can be achieved. But the moderate electro-optical coefficients in LiNbO$_3$ necessitates high voltage swings of around 5 V and long modulator lengths of several cm. The high capacitance in such a large structure can be circumvented by using a travelling-wave solution, but the different velocities of the optical and microwave signals (walk-off) still limits the speed to maybe 40 GHz. And the large size in itself makes LiNbO$_3$ Mach–Zehnder modulators unsuitable for photonic integrated circuits.

Primarily with the target to reduce the driving voltage and increase bandwidth we have investigated IS transitions within the conduction band of quantum wells for high-speed EA modulators. Thus the optical modulation principle is still EA as in QCSE but the active layer is now unipolar, only involving electrons, which has several interesting consequences. Using IS transitions the subbands have a similar curvature giving a peak shaped joint density of states and thus the absorption can be stronger, enabling a shorter and thus faster modulator. In contrast interband transitions occur between subbands of opposite curvature. It is also interesting to note that the rapid LO-phonon mediated IS relaxation time on the order of 1 ps (only 0.14 ps in GaN/AlN QWs) makes IS based absorption insensitive to saturation. The short relaxation time means that excited electrons can rapidly relax within the QW and be "recycled". Thus the moderate optical saturation power afforded by interband QCSE, due to hole pileup, can be considerably improved.

However, in conventional III-V QW materials such as InGaAs/InAlAs and InGaAsP/InGaAsP the IS resonance energy is small, corresponding to wavelengths of 4 μm or longer. In order to reach the telecommunication wavelength of 1.55 μm (0.8 eV) a material combination with a large conduction band offset of about 1.5 eV or more is required. In recent years IS absorption at 1.55 μm have been demonstrated in several QW materials InGaAs/AlAsSb

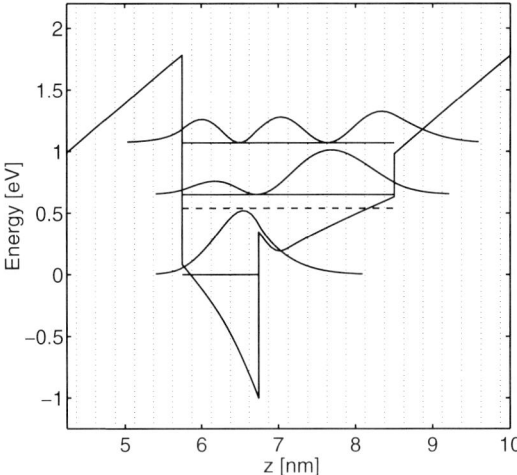

Fig. 12.2 Potential profile of the second GaN/Al$_{0.85}$Ga$_{0.15}$N/AlN step quantum well including moduli squared of the confined states at zero applied bias voltage. The dashed line indicates the Fermi level, while the vertical dotted lines indicate monolayers.

[5], GaN/Al(Ga)N [6–8], and ZnSe/BeTe [9]. We have performed simulations of RC-limited modulators at 1.55 μm based on IS transitions in GaN/AlGaN/AlN step QWs [10], and in InGaAs/InAlAs/AlAsSb coupled QWs [11]. The high-speed performance of such modulators will ultimately be determined by the IS absorption linewidth Γ that can be achieved. A small linewidth is essential. It can be shown that the capacitance of an IS modulator depends on the linewidth as $C \sim \Gamma^3$ [12]. At the small well-width required for IS transitions at 1.55 μm the absorption linewidth is generally limited by the material quality due to interface roughness and well-width fluctuations, and is also affected by doping.

12.2
Modulator Structure

12.2.1
Multiple-Quantum-Well Structure

We evaluate an IS modulator based on Stark shifting the IS resonance, i.e., the IS transition energy is shifted by applying an electric field. The Stark shift at any applied field is proportional to the dipole moment $e(z_{22} - z_{11})$, between the ground and first excited QW states. Here e is the elementary charge while $z_{ii} = <\phi_i|z|\phi_i>$, where $\phi_i(z)$, $i = 1, 2$, are the bound QW electron states. Ordinary single-layer QWs have a quadratic Stark shift vs the electric field, with a vanishing dipole moment at no applied field due to symmetry. However, in conventional III-V semiconductors, step QWs [12, 13] have been used to obtain the necessary asymmetry of the QWs. This achieves an efficient Stark shift, which is approximately linear to the applied electric field. In wurtzite nitride heterostructures there are very strong built-in fields, which also gives an asymmetry to single-layer QWs, see, e.g., [14]. But this is not quite enough for efficient modulation at 1.55 µm, due to the very thin QWs required. Thus we assess here a step QW structure. The built-in fields still imply though, that there is a significant difference in the potential profile of a step QW depending on whether the deep QW layer or the step layer of the QW is grown first, i.e., they are not only mirror images. We find by simulation that an IS resonance at 1.55 µm should be possible in both cases and, moreover, that two important properties for EA, namely the oscillator strength f_{12} and the dipole moment $e(z_{22} - z_{11})$ which can be achieved are comparable irrespective of the growth order. Although, if the AlGaN step layer is grown first (assuming Ga face growth, which is usually desired due to higher material quality), a significantly lower Al concentration is required in this layer. We choose here to evaluate a step QW structure with the deep layer grown prior to the step layer, see Fig. 12.2. Further, we assume that the barriers are pure AlN. The barriers should be thick enough to prevent leakage of carriers when the MQW structure is biased. Then using AlN rather than AlGaN the barriers can be thinner, which reduces the applied voltage that is needed to achieve the Stark shift. It may also be easier to grow, by molecular beam epitaxy (MBE), an MQW structure employing just one AlGaN composition (that of the QW step layer) since the need to change the temperature of the Al and/or Ga effusion cells is relaxed. However, MBE-systems employing migration-enhanced epitaxy (MEE) could achieve this by appropriate shutter sequences.

In order to obtain a strong IS absorption, the MQW structure must be heavily n-type doped. The IS absorption strength is proportional to the carrier density in the QWs. An interesting question, which however is not considered in any detail here, is the best placement of the dopants. Generally, modulation-doped structures with doping in the barriers, yield smaller IS absorption linewidths. This has been observed also in GaN/AlGaN single QWs [6]. But in order to effectively reduce broadening by barrier doping, barriers need to be relatively thick. In this paper a δ-doping layer in the well is assumed. Placing the dopants in the QW also allows a higher dopant concentration, due to the reduced band bending. However given the importance of the absorption linewidth for the performance (Section 12.4.7) it may still prove advantageous to use barrier doping. The sheet doping concentration is $n_D = 7 \times 10^{13}$ cm^{-2}. The position of the assumed δ-doping layer is clearly visible in Figs 12.2 and 12.3 as a dip in the potential profile.

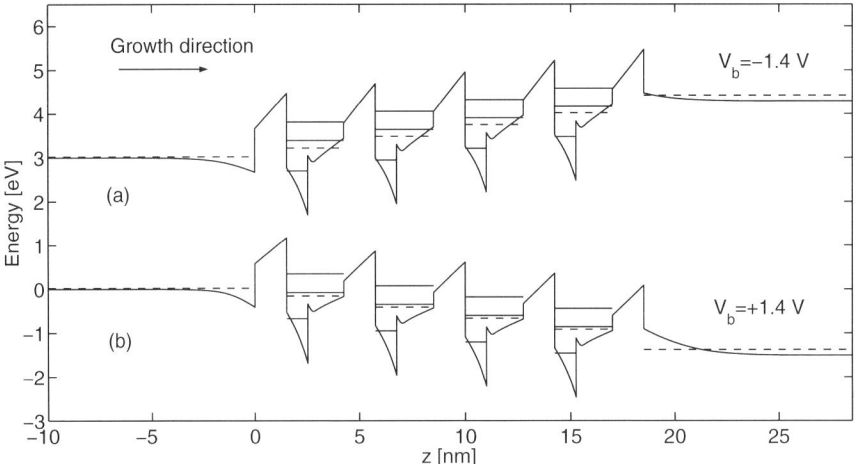

Fig. 12.3 Potential profile of the modulator structure including the energy levels of the confined states in the step quantum wells. The dashed lines indicate the Fermi levels. In (a) and (b) the applied bias voltages are $V_b = -1.4$ V and $V_b = 1.4$ V, and the modulator is opaque and transmitting, respectively, at $\lambda = 1.55$ µm. For clarity the potential was offset by 3 eV in (a).

On both sides of the MQW structure are AlGaN layers where electrons accumulate or are depleted depending on the applied bias. Hence, as in our previous paper on a mid-IR modulator [12], we call these layers accumulation–depletion layers (ADLs). Although it should be noted that, due to the strong built-in fields in the present structure, there is always accumulation in one of the layers and depletion in the other, as is evident in Fig. 12.3. As described in the previous paper [12] the ADL composition can be determined to minimize the band bending as seen over the whole MQW, and hence minimize undue spread in the IS transition energies amongst the four step QWs. This problem is seemingly a bit more complex in a nitride-based structure, since the Al content affects not only the conduction band-edge energy but also, at the same time, the built-in field. However, it is actually very straightforward to do by self-consistently simulating the MQW structure, with individual carrier densities n_1 in each of the step QWs, and the surrounding ADLs. Then by trying a range of AlGaN compositions in the ADLs, we can determine which Al content gives a flat band as seen over the whole MQW, i.e., similar electric field and IS transitions energies in all step QWs. In the present structure this procedure gives the ADL Al molar fraction $x = 0.52$. In order to reduce the depletion layer thickness of the topmost ADL, which as mentioned is always in depletion, a higher doping level, $n_D = 3 \times 10^{19}$ cm^{-3}, was assumed in that ADL than otherwise in the waveguide core. The complete layer sequence of the simulated structure is shown in Table 12.1.

Tab. 12.1 Description of the layers in the simulated device with n-type doping densities. The layers between the dotted lines constitute the active layer and are treated as an effective medium [12] when solving for the optical mode.

Material	Thickness [nm]	Doping [cm^{-3}]	
Top contact	-	-	
n++-GaN	≈ 1000	1×10^{20}	
n-GaN	400	5×10^{18}	
n-Al$_x$Ga$_{1-x}$N, graded	10	5×10^{18}	
n-Al$_{0.52}$Ga$_{0.48}$N	10	5×10^{18}	
....	....	....	
AlN	1.5	-	
GaN	1.0	-	
n-Al$_{0.85}$Ga$_{0.15}$N	1.75	7×10^{13} cm^{-2}	× 4 step QWs
AlN	1.5	-	
....	....	....	
n-Al$_{0.52}$Ga$_{0.48}$N	10	5×10^{18}	
n-Al$_x$Ga$_{1-x}$N, graded	10	5×10^{18}	
n-GaN	400	5×10^{18}	
n++-GaN	≈ 1000	1×10^{20}	

12.2.2
Waveguide and Contacting

A tight confinement of the optical mode is essential in a high-speed modulator, since the required modulator length is inversely proportional to the width of the optical mode. In other words the RC-limited speed can be increased in direct relation to the mode confinement. At least, this is true until the thus-increased intensity of the light leads to absorption saturation in the active layer. But, as discussed in Section 12.4.5, IS-based material is not prone to saturation, especially not using nitrides. Thus absorption saturation should not be an issue in this modulator.

A strong confinement requires a strong refractive index contrast between the core and cladding layers. That would suggest using AlN, with the refractive index $\eta = 2.05$, as cladding. However, the cladding must also have a good conductivity so that the applied voltage is indeed applied over the MQW, and in order not to compromise the modulation speed by a high series resistance. Unfortunately AlN is known to be all but impossible to dope, although in recent breakthroughs an electron density of up to 7×10^{17} cm^{-3} was achieved with a low mobility of 11 cm^{2}V^{-1}s^{-1} in a highly compensated material [15]. More adequate electron densities and mobilities have been achieved in AlGaN with a 20% Al content [16], but then of course the refractive index contrast compared to the GaN waveguide core is accordingly smaller.

Using the plasma effect (PE) the refractive index can be strongly reduced in a heavily doped material. The index reduction occurs when the plasma frequency ω_p approaches the light frequency ω, e.g., [17]. In a parabolic band $\omega_p^2 = ne^2/m_e m_0 \epsilon_\infty \epsilon_0$, where n is the electron density and m_0 is the free electron mass. In the simulation the PE is described by Eq. (12.16). Waveguides based on the PE, or at least enhanced by it, or on surface plasmons (SPs), have been used extensively in IS-based devices, notably quantum cascade lasers, for mid-IR and longer wavelengths. Tightly confining waveguides based on the PE at mid-IR wavelengths were proposed and compared to SPs [17]. Efficient waveguides based on the PE at wavelengths as short as $\lambda = 1.55$ µm have not to our knowledge been used, though weakly confining structures ($\Delta \eta \sim 0.01$) in conventional III-V materials were evaluated in the 1970s [18]. However, the reported high electron densities in GaN of up to 1×10^{20} cm^{-3} [19] with low compensation indicates that the PE can yield a substantial refractive index contrast without prohibitive absorption. Another important requirement is a large enough energetic separation between the Γ-valley containing the carriers and the satellite valleys. In GaN the $L - \Gamma$-valley separation is about 2 eV, which should preclude intervalley scattering and also indicates that electron densities $n \sim 5 \times 10^{20}$ cm^{-3} may be feasible to further reduce the refractive index. Although, at least using metal-organic chemical vapor deposition (MOCVD), the amphoteric nature of Si as a dopant in GaN becomes more apparent at such concentrations, resulting in highly compensated material [19]. Thus we here assume a waveguide based on the PE in GaN, as we have proposed earlier [10]. The GaN cladding and core layers are described in Table 12.1. The carrier density of $n = 1 \times 10^{20}$ cm^{-3} at the wavelength $\lambda = 1.55$ µm corresponds to a cladding complex refractive index of $\eta + i\kappa = 2.11 + i0.011$ (using Eq. (12.16)).

Using the PE rather than a compositional heterostructure has the advantage of achieving the combination of:

1. a high refractive index contrast ($\gtrsim 10\%$);

2. excellent conductivity of the cladding layer;

3. lattice matching of cladding and core layers.

On the downside is a higher waveguide absorption for a tightly confining waveguide, but it appears acceptable in a short modulator. In comparison, a metal-based SP waveguide at this short wavelength would give several times higher waveguide absorption.

The required modulator length for a 10 dB extinction ratio is only $L = 13$ µm in the considered structure with the assumed IS absorption linewidth $\Gamma = 100$ meV. In order to avoid stray light being transmitted outside such a short modulator, it would be practical to add passive waveguide sections. The modulator structure would then be etched down outside of the active length and, e.g., an AlN/GaN/AlN waveguide would be regrown for the passive sections. Such passive waveguides could also further facilitate coupling by allowing lateral tapering of the waveguide and somewhat increased waveguide thickness in the passive sections. Coupling losses to such passive sections can be low, as noted in Section 12.4.1. However, apart from this, only the active modulator part is considered in this chapter.

Due to the small size of the modulator a low resistivity contact on top of the mesa is necessary. Ohmic contacts with a low resistivity of $9 \times 10^{-8}\ \Omega\mathrm{cm}^2$ have been demonstrated using Ti/Al on n-GaN [20]. This corresponds to a clearly acceptable resistance of just under $1\ \Omega$ for the top contact.

12.3
Model

12.3.1
Conduction Band Potential and Active Layer Biasing

The assumed conduction band offset ($\Delta E_c = 1.7$ eV) is a conservative estimate for pseudomorphic GaN/AlN. In determining a suitable composition for the QW step layer material and the ADLs we use an AlGaN conduction band bowing parameter of 0.7 eV [21]. Thus the $\mathrm{Al}_x\mathrm{Ga}_{1-x}\mathrm{N}$ conduction band-edge energy relative to GaN is

$$V(x) = 1.7x - 0.7x(1-x)\ \mathrm{eV} \tag{12.1}$$

A prominent feature in wurtzite nitrides grown along the [0001] direction (the c-axis) is the strong material polarization. Spontaneous and strain-induced, i.e., piezoelectric, polarizations result in very strong built-in fields when two different III-N materials are connected in heterostructures such as QWs. The spontaneous polarization P_{sp} of AlGaN is interpolated linearly using data in Table 12.2. In a structure with biaxial strain the piezoelectric polarization in the growth direction is readily obtained as

$$P_{\mathrm{pz}} = e_{33}\epsilon_c + 2e_{31}\epsilon_a \tag{12.2}$$

where $\epsilon_a = (a_0 - a)/a$ and $\epsilon_c = -2C_{13}/C_{33}\epsilon_a$ are the strains in the plane and the growth direction, respectively. Here a_0 is the lateral lattice-parameter of the GaN buffer, while a is the native lattice-parameter of each layer material. The Al-containing active layer (i.e., the MQW) and the ADLs are thus assumed to be grown lattice-matched to the underlying GaN waveguide core layer. Experimentally GaN/AlGaN heterostructures support relatively thick strained layers [22,23], compared to the experience from conventional III-V semiconductors, e.g., in Ref. [22] a 600-nm $\mathrm{Al}_{0.104}\mathrm{Ga}_{0.896}\mathrm{N}$ layer was grown coherently on GaN.

Tab. 12.2 Parameter values [24–28] employed in the simulation.

Parameters		GaN	AlN
Conduction band offset	ΔE_c [eV]	—	1.7
Band gap	E_g [eV]	3.4	6.2
Electron mass	m_e	0.22	0.30
Static dielectric constant	$\epsilon_s (\parallel c)$	10.4	8.5
Refractive index	η	2.30	2.05
Spontaneous polarization	P_{sp} [C m^{-2}]	−0.034	−0.090
Piezoelectric coefficient	e_{33} [C m^{-2}]	0.67	1.50
"	e_{31} [C m^{-2}]	−0.34	−0.53
Elastic constant	C_{33} [GPa]	354	377
"	C_{31} [GPa]	68	94
Lattice parameter	a [Å]	3.819	3.112
"	c [Å]	5.185	4.982

In addition to the built-in fields there is also band bending resulting from the charge distribution in the structure, due to ionized donors and the electron densities in the QWs and outside of the MQW. The space charge (Hartree) potential $\Psi(z)$ is obtained for the whole structure using Poisson's equation

$$\frac{d}{dz}\epsilon_0 \epsilon_s(z) \frac{d\Psi}{dz} = e \left[n_{D\delta}(z) - \sum_{i=1}^{m} n_{1,i} |\phi_i(z)|^2 - n_a(z) + n_d(z) \right] \quad (12.3)$$

where $\epsilon_s(z)$ is the material-dependent static dielectric constant, e is the elementary charge, and the terms within the brackets are 3D densities: $n_{D\delta}$ is the positive charge of the ionized δ-doping in the QWs, the second term is the electron charge in the four ($m = 4$) step QWs, while $n_a(z)$ and $n_d(z)$ are charge densities adjacent to the active layer. These latter two densities change with the applied bias over the active region. In order to Stark shift the IS resonance and thus modulate the absorption, an electric field is applied over the active layer as was shown in Fig. 12.3. Depending on the strength of this electric field there will be a certain accumulation of carriers in the triangular potential well just below the MQW. This charge distribution is governed by the (ground) bound state (not shown in Fig. 12.3) of the triangular well

$$n_a(z) = n_a^{2D} |\phi_a(z)|^2 \quad (12.4)$$

where n_a^{2D} is the 2D electron concentration. The depletion of carriers in the AlGaN layer on top of the MQW is obtained by

$$n_d(z) = n_D - 2 \left(\frac{m_e m_0 k_B T}{2\pi \hbar^2} \right)^{3/2} \exp\left(-\frac{E_c(z) - E_F}{k_B T} \right) \quad (12.5)$$

where the first term is the doping density $n_D = 3 \times 10^{19}$ cm^{-3} and the second term is the expression for electron density in a bulk semiconductor, which is used to locally approximate the electron density at each position z in the depletion layer. Further, k_B is Boltzmann's constant, $T = 300$ K is the temperature and $E_F = -V_b$ is the Fermi energy in the upper ADL. Far from the MQW the two terms in Eq. (12.5) are equal and there is no net charge. Now the resulting conduction band-edge plotted in Fig. 12.3 is obtained by summing the contributions

$$E_c(z) = V(z) + V_p(z) + \Psi(z) \tag{12.6}$$

where $V_p(z)$ is the potential resulting from the material polarizations. The charge distributions in Eqs (12.4), (12.5) and of course the wavefunctions in the step QWs obtained from the Schrödinger Eq. (12.7) in turn depends on $E_c(z)$. Thus these must be solved self-consistently with the Poisson Eq. (12.3) in an iterative manner.

12.3.2
Intersubband Transitions

The envelope function approximation is employed to obtain the confined electron states in the step QWs. Since the IS transition energies are appreciable compared to the GaN bandgap a 2-band Kane model was used to account for the nonparabolicity of the conduction band. The Schrödinger equation then becomes

$$\frac{\hbar^2}{2} \frac{d}{dz} \frac{1}{m_e(E,z)m_0} \frac{d\phi(z)}{dz} + E_c(z)\phi(z) = E\phi(z) \tag{12.7}$$

The energy and position-dependent electron mass is given by

$$m_e(E,z) = m_e(z)\left(1 + \frac{E - E_c(z)}{E_g(z)}\right) \tag{12.8}$$

where $E_c(z)$ and $E_g(z)$ are the conduction band-edge and the unstrained band-gap, respectively. The 2-band Kane expression in Eq. (12.8) was recently found to describe the GaN nonparabolicity very well up to nearly 1 eV above the band-edge [29], which is what is required for 1.55 µm IS transitions. The in-plane mass, i.e., the subband mass, has been found to have an enhanced energy dependence, typically 2–3 times [30], compared to the bulk growth direction mass in Eq. (12.8). We assume parabolic subbands with masses $m_{e1} = 0.31$ and $m_{e2} = 0.35$ and room temperature (T=300 K), when calculating the subband populations in the step QWs.

The IS absorption is modeled by a wavelength-dependent permittivity in each step QW [31]

$$\epsilon_z = \epsilon_\infty - f_{12}\frac{\epsilon_\infty \omega_{p,z}^2}{\omega^2 - \omega_{12}^2 + i\gamma\omega} \tag{12.9}$$

$$\epsilon_x = \epsilon_\infty - \frac{\epsilon_\infty \omega_{p,x}^2}{\omega^2 + i\omega/\tau_w} \tag{12.10}$$

where f_{12} is the IS oscillator strength. IS transitions couple only to TM-polarized light, i.e., to the electric field in growth direction (z) as described by Eq. (12.9). This is clear from the dipole matrix element for the IS transition, which is $<\phi_1|z|\phi_2>$ in the *parabolic (single-band) case*. Note that here, however, we used a matrix element generalized to multiple bands [32], and obtained the oscillator strength f_{12} by Eq. (6) in Ref. [32], since a 2-band non-parabolic mass was used in the Schrödinger equation (12.7). Due to the optical mode confinement, the TM-polarized light will also have some component of the electric field in the propagation direction. This will be subject to absorption, due to excitation of in-plane plasma oscillations, as described by the Drude expression Eq. (12.10). In the plasma frequency

$$\omega_{p,z}^2 = \frac{n_1 e^2}{\epsilon_\infty \epsilon_0 m_0 L_w} \tag{12.11}$$

n_1 is the sheet electron density in the ground state and $L_w = 2.75$ nm is the QW width. The background dielectric constant $\epsilon_\infty = \eta_w^2$, where $\eta_w = 2.2$ is a refractive index relevant for the step QW material. The transition energy $\hbar\omega_{12} = E_2 - E_1$ and $\hbar\gamma = \Gamma = 100$ meV is a phenomenological IS resonance linewidth (FWHM) parameter. In the Drude expression (12.10) for ϵ_x, the plasma frequency is given by

$$\omega_{p,x}^2 = \frac{n_1 e^2}{\epsilon_\infty \epsilon_0 m_{e1} m_0 L_w} \tag{12.12}$$

and $\tau_w = 20$ fs is the electron dephasing time. The precise value of this dephasing time is not important here as it has a negligible influence on the final results.

The IS absorption, as determined from Eq. (12.9), corresponds to a Lorentzian lineshape, which is used for simplicity. Experimentally, IS resonances are subjected to a combination of finite-lifetime broadening and inhomogeneous broadening resulting in Lorentzian and Gaussian lineshapes, respectively. However, the absorption spectra of coupled QWs containing two IS transitions was recently well fitted to a sum of two Lorentzians [14]. The linewidth $\Gamma = 100$ meV appears possible in a high-quality step QW material. However, given the decisive importance of the absorption linewidth on (notably) the high-speed properties, extrapolations of the simulation result depending on the linewidth Γ are given in Section 12.4.7.

12.3.3
Optical Mode and the Plasma Effect

The optical mode and modal absorption were obtained by the transfer matrix method (TMM) [33]. Since the MQW period thickness is orders of magnitude smaller than the wavelength of the light we can simplify the TMM calculation by treating the active layer as an effective medium and thus obtain the complex permittivity of the whole MQW structure [34],

$$\frac{1}{\epsilon_{a,z}} = \frac{(m+1)L_b}{L_a}\frac{1}{\epsilon_b} + \frac{L_w}{L_a}\sum_{i=1}^{m}\frac{1}{\epsilon_{z,i}} \qquad (12.13)$$

$$\epsilon_{a,x} = \frac{(m+1)L_b}{L_a}\epsilon_b + \frac{L_w}{L_a}\sum_{i=1}^{m}\epsilon_{x,i} \qquad (12.14)$$

where $m = 4$ is the number of step QWs and $L_a = m(L_b + L_w) + L_b$ is the active layer thickness. The index i is included in $\epsilon_{z,i}$ and $\epsilon_{x,i}$ since f_{12}, ω_{12} and n_1 are modeled separately for each step QW. The dielectric constant in the barrier is $\epsilon_b = 2.05^2$.

As discussed in Section 12.2.2 we assume a waveguide employing the PE in heavily doped GaN. We are then faced with assessing the electron scattering time in order to estimate the free-carrier absorption (FCA) in the doped layers. In previous papers, e.g., [17], we emplyed a quantum mechanical calculation to obtain such scattering times relating to LO-phonons and ionized impurities in conventional III-V materials. However, the material quality of the III-nitrides is generally significantly lower, with, e.g., a high density of dislocations. Hence it would be appealing to obtain the scattering time from available experimental data for the electron mobility, as this would account for all scattering processes as well as for any compensation in the doping. We do this in the relaxation time approximation, which is applicable for elastic or nearly-elastic scattering events, such as impurity scattering which should dominate in a heavily doped material. The scattering time is then given by

$$\tau = \mu m_e m_0 (1 + 2E_F/E_g)/e \qquad (12.15)$$

where E_F is the Fermi energy and the term $2E_F/E_g$ accounts for the non-parabolicity. The mobility of n-doped GaN is $\mu = 100$ cm^{2}V^{-1}s^{-1} at $n = 1 \times 10^{20}$ cm^{-3} [19], implying the scattering time $\tau = 15$ fs using Eq. (12.15).

The PE is modeled by the classical Drude model[1]

$$(\eta + i\kappa)^2 = \epsilon = \epsilon_\infty \left[1 - \frac{\omega_p^2}{\omega^2}\left(1 - \frac{i}{\omega\tau}\right)\right] \qquad (12.16)$$

where ω_p is an effective plasma frequency [17], which accounts for the conduction band nonparabolicity of the heavily doped bulk GaN material, by assuming hyperbolic bands. Using a single scattering time τ as is done here, hence neglecting its dependence on the photon energy in the FCA process [35], gives the classical λ^2-dependence of the FCA, and is likely to overestimate the FCA. This is confirmed by a measured FCA of 300 cm^{-1} in MOCVD GaN at the photon energy $\hbar\omega = 0.8$ eV and electron concentration 5×10^{19} cm^{-3} [36], whereas the present model gives 1000 cm^{-1} for those parameters, assuming $\mu = 100$ cm^{2}V^{-1}s^{-1}.

12.4
Results

12.4.1
Electroabsorption

The absorption modulation around $\lambda = 1.55$ µm is achieved by Stark shifting the ground state to first excited state (1→2) IS resonance, Fig. 12.4. Since an IS resonance is peak shaped, in contrast to the step-like absorption spectrum of QW interband transitions, there is in principle a choice when employing the Stark effect as to using either the high or low-energy flank of the IS resonance for absorption modulation. However, several factors support using the high-energy flank as we do in this simulation:

1. it requires a lower IS transition energy, which is more easily realized;

2. it achieves a negative effective chirp parameter, as discussed in Section 12.4.2;

3. the oscillator strength is in general increased when the IS transition is Stark shifted to higher energy, thus yielding a stronger absorption modulation on the high-energy flank.

---

1) The positive sign of the imaginary part of the complex refractive index $\eta + i\kappa$ in Eq. (12.16) stems from assuming a negative sign in the time evolution of the optical field, i.e., $\exp(-i\omega t)$. The same time evolution is assumed in Eqs (12.9) and (12.10). Hence the sign difference in the imaginary part of the refractive index compared to our previous paper [17].

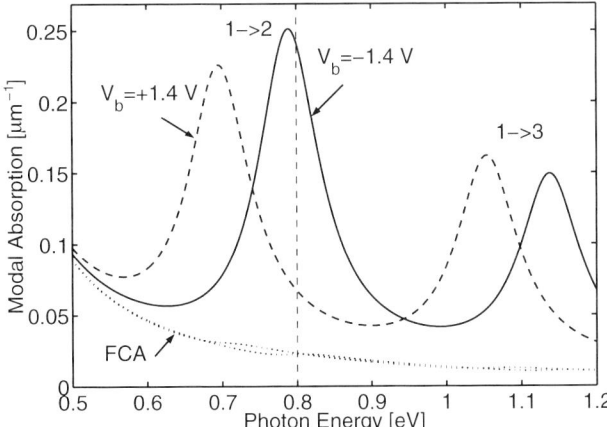

Fig. 12.4 Modal absorption spectrum in the modulator at the applied bias voltages $V_b = -1.4$ V (solid) and $V_b = 1.4$ V (dashed). The IS resonances due to the $1 \rightarrow 2$ and $1 \rightarrow 3$ transitions are visible. The dotted lines show the waveguide absorption due to FCA in the heavily doped n++-GaN cladding layers, at the same bias voltages. The vertical dashed line indicates the photon energy, $\hbar\omega = 0.8$ eV ($\lambda = 1.55$ μm), of the laser.

Under the application of the voltage swing the IS resonance energy is shifted from 698 meV at $V_b = 1.4$ V to 792 meV at $V_b = -1.4$ V. The IS resonance energy includes a sizeable depolarization blue shift (~ 90 meV), with respect to the single-particle IS transition energy $\hbar\omega_{12}$, due to the high electron concentration in the step QWs. A simple account (slab model) [37] of the depolarization is inherent in the model. Under the voltage swing the oscillator strength at the same time increases from $f_{12} = 1.99$ ($V_b = 1.4$ V) to $f_{12} = 2.23$ ($V_b = -1.4$ V) thus enhancing the absorption modulation. The carrier densities in the step QW ground subbands are on average $n_1 = 6.9 \times 10^{13}$ cm^{-2}. There is also a thermally excited carrier density in the first excited subbands, but it remains small $n_2 < 1 \times 10^{11}$ cm^{-2} for all applied biases. The absorption modulation is achieved for TM-polarized light, since conduction band IS transitions only couple to the growth direction component of the optical electric field, Eq. (12.9). Thus the modulator requires a highly linearly polarized optical input. This can be achieved by placing the modulator, as is often done, close to the laser. The waveguide absorption resulting from FCA in the heavily doped cladding layers is also indicated in Fig. 12.4. As mentioned above in Section 12.3 this is likely overestimated. Hence it appears that the PE in GaN can be used without a prohibitive FCA in the present modulator.

The active length of the modulator is determined by a required 10-dB extinction ratio, i.e., $\Delta\alpha_m L = \ln 10 = 2.3$. From Fig. 12.4 we have $\Delta\alpha_m = 0.17$ μm^{-1} for the applied voltage swing from $V_b = 1.4$ to -1.4 V, and thus an active

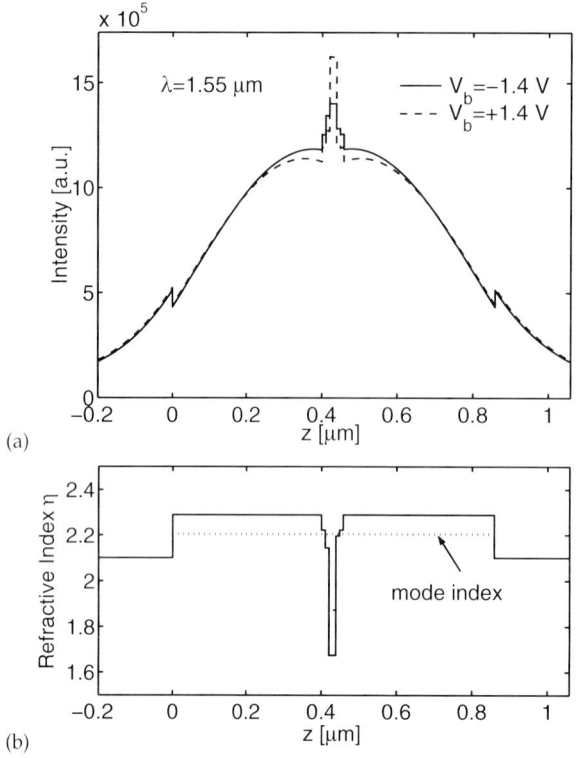

Fig. 12.5 (a) Mode intensity profiles and (b) profiles of the real refractive index in a direction perpendicular to the layers. The dotted line in the lower panel is the mode index. The results are for $V_b = -1.4$ V (solid) and $V_b = 1.4$ V (dashed).

length as short as $L = 13$ μm. For this active length the transmission loss 3.8 dB at $V_b = 1.4$ V, due to residual IS absorption and FCA in the cladding layers, is also readily obtained from Fig. 12.4. As discussed in Section 12.2.2 passive waveguide sections would probably be required to avoid transmission of stray light outside the short modulator. The additional coupling losses from such passive sections can be made very small, e.g., less than 1 dB in conventional III-V materials [38].

The transverse intensity profile of the optical mode is shown in Fig. 12.5. The mode overlap with the active layer is quite modest, e.g., 5.2% in the opaque state ($V_b = -1.4$ V). The lower refractive index in the cladding layers are due to the PE in heavily doped GaN ($n_D = 1 \times 10^{20}$ cm^{-3}) as discussed in Section 12.2.2. If higher electron concentrations can be achieved, this refractive index can be reduced considerably. This would enhance the mode confinement, though the confinement as judged by the mode width is already quite adequate.

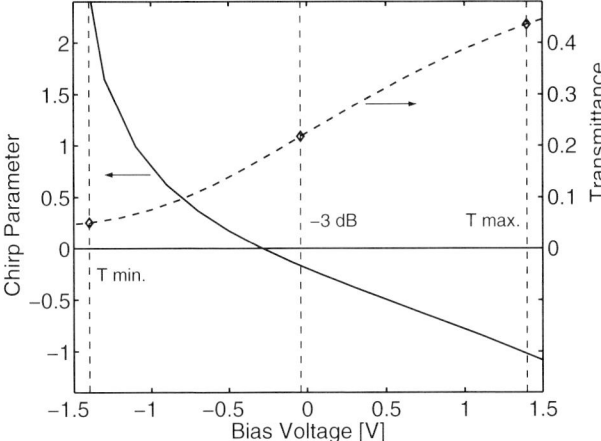

Fig. 12.6 Modulator chirp parameter α (solid line) and transmittance T (dashed line) versus the applied voltage V_b. The vertical dashed lines indicate the applied voltage swing with minimum and maximum transmittance T at $V_b = -1.4, +1.4$ V, respectively, and where T has dropped by 3 dB. The effective chirp parameter $\alpha_{\text{eff}} = -0.6$ is estimated by the average chirp during the top 3 dB of the transmittance.

12.4.2
Chirp Parameter

In current fiber-optic communication systems employing Er-doped fiber amplifiers the chromatic dispersion usually limits the performance. The dispersion penalty increases rapidly at high bit rates B, as the dispersion-limited transmission length $L_d \sim 1/B^2$. For example, at $B = 40$ Gb s^{-1} assuming zero chirp pulses on SSMF L_d is only about 4 km. However, this distance depends strongly on the phase modulation that in general accompanies the amplitude modulation in optical modulators [39, 40]. The amount of frequency chirp thus given to emitted optical pulses can be characterized by the modulator chirp parameter α [40]. A small negative chirp α is desirable on SSMF around 1.55 µm. Unfortunately direct modulated semiconductor lasers tend to have a large positive chirp parameter ($\alpha \sim 2$) which is indeed one of the motivations for using external modulators. However, it is interesting to note that IS-based lasers in general have close to zero chirp as a result of a symmetric IS gain spectrum [41].

The bias voltage dependent chirp parameter of an EA modulator can be obtained as

$$\alpha(V_b) = \frac{d\eta_m}{d\kappa_m} \qquad (12.17)$$

where η_m and κ_m are the real and imaginary parts, respectively, of the modal refractive index. The chirp parameter calculated according to Eq. (12.17) together with the transmittance of the modulator is shown in Fig. 12.6. The chirp parameter α is negative at high transmittance, while it becomes positive at low transmittance. This is interesting since it has been shown that the chirp parameter at high transmittance is more important, and the average of the chirp during the top 3 dB of transmittance was suggested as an effective chirp parameter [39]. Using that procedure we determine a small negative effective chirp parameter $α_{eff} = -0.6$, which is close to optimal on SSMF. We note that the negative chirp at high transmittance and hence a negative $α_{eff}$ is a direct consequence of using the high-energy flank of the IS resonance. In contrast, interband QCSE modulators, which naturally must use the low-energy side of the band-edge absorption step in that case, tend to have positive chirp at high transmittance and negative at low transmittance, e.g., [42]. Hence employing the interband QCSE there is a tradeoff that lower chirp can be achieved by accepting higher insertion loss, but a negative effective chirp is, in general, not achieved. Using the IS-based Stark effect such a tradeoff becomes unnecessary in the present modulator as the more negative chirp and the high transmittance is achieved in the same region of the voltage swing, namely, when the IS resonance is tuned away from the photon energy. We note that the modulator structure considered here, based on the Stark shift in a step QW, appears to have a better chirp performance than the modulator structure that we evaluated in [11]. It is also noteworthy that the transmittance curve in Fig. 12.6 has a desirable S-shape, i.e., it is steeper in the middle of the voltage swing. Thus, to some extent, noise in the input microwave signal is suppressed.

12.4.3
Electrical Properties

The results obtained thus far are steady state properties. The modulation speed of this modulator, which does not rely on carrier transport within the MQW structure, is limited by the RC time constant of the driver-modulator circuit.

We obtain the modulator capacitance as

$$C = eLw(\Delta n_a^{2D} + \Delta n_d^{2D})/2V_{pp} \tag{12.18}$$

where $w = 1.0$ μm is the modulator mesa width and Δn_a^{2D} (Δn_d^{2D}) is the change in sheet electron (ionized donor) density in the left (right) ADL under the application of the whole voltage swing $V_{pp} = 2.8$ V. The ionized donor density n_d^{2D} is obtained by integrating Eq. (12.5). For the present structure we obtain $C = 52$ fF. In order to fully benefit from this low capacitance the parasitic capacitances must of course be brought down well below this level. This does not appear to be completely impracticable. However, the parasitics are not considered further here.

The series resistance of the modulator will be mainly due to the contact resistance of the ohmic top contact and to the two GaN waveguide core layers with the total thickness $d_c = 0.8$ μm, i.e., $R_s \approx \rho_s/wL + d_c/wLen\mu = 1\,\Omega + 4\,\Omega = 5\,\Omega$. Here we assumed a specific contact resistance $\rho_s = 9 \times 10^{-8}\,\Omega\text{cm}^2$ [20], and the core layer doping $n = 5 \times 10^{18}$ cm^{-3} and mobility $\mu = 200$ cm^{2}V^{-1}s^{-1} [19].

The electrical properties of the driver-modulator circuit can be assessed in a simple lumped circuit model [12]. The electrical $f_{3dB} = 1/2\pi RC$ for small signals is defined as a 3 dB decrease in the received detector signal. With a standard driver impedance of $R_0 = 50\,\Omega$, and not considering a shunt (termination) resistance, we have $R = R_0 + R_s = 55\,\Omega$ and $f_{3dB} \approx 60$ GHz. We can consider f_{3dB} as a good estimate also for the large signal modulation speed, since the modal absorption is approximately linear vs the applied voltage V_b [12], while the capacitance, Eq. (12.18), is practically independent of the voltage swing. As mentioned in [12] the symmetric voltage swing of intersubband modulators, in contrast to interband-based EA modulators, implies that driving with transistors in a balanced configuration is more easily achieved. A balanced transistor amplifier output, in effect doubling the useful voltage swing, relaxes the requirements on the transistor breakdown voltage, thus allowing smaller higher-speed transistors.

12.4.4 Figure of Merit

It is now interesting to compare the present simulation results to current state-of-the-art *interband*-based EA modulators. The speed–power ratio f_{3dB}/P_{ac}, where the electrical signal power $P_{ac} = V_{pp}^2/8R_0$, is a relevant figure of merit in the context of an *RC*-limited EA modulator. This can be motivated by the fact that the required voltage swing V_{pp} can be chosen according to the number of QWs employed in the MQW structure, which in turn will affect the modulation speed according to $f_{3dB} \sim 1/V_{pp}^2$ [12]. Hence the speed–power ratio depends on the design of the QW-EA material and the waveguide, and is at least approximately invariant to the choice of the number of QWs.

In the present simulation the speed–power ratio $f_{3dB}/P_{ac} = 3.2$ GHz mW^{-1}. In interband QCSE lumped EA modulators there are a few demonstrations of f_{3dB} around 50 GHz in the last decade. E.g., Ido et al. [38] reported a modulator with $f_{3dB} \approx 50$ GHz at $V_{pp} = 2.5$ V yielding $f_{3dB}/P_{ac} = 3.2$ GHz mW^{-1}, the same as in the modulator simulated here. Their device had a higher extinction ratio of 15 dB, requiring a longer modulator length, thus increasing the capacitance. However, considering also that a 50 Ω termination resistor was used in [38] on their 10 Ω series resistance device, these two effects on the *RC*-limitation nearly cancel. Figures of merit for optical modulators should be used with some caution since there are generally trade-offs to be made, e.g., the speed can in general be increased at the expense of insertion loss. But the above comparison indicates that the speed–power ratio figure of merit of the simulated IS-based modulator is approximately on a par with high-performance lumped interband QCSE modulators. We should emphasize that this result was obtained at the assumed IS absorption linewidth $\Gamma = 100$ meV. As discussed in Section 12.4.7, the achievable speed of the simulated IS modulator depends decisively on the IS linewidth. It can also be noted that somewhat higher speed–power ratios have been demonstrated in interband QCSE modulators in a travelling-wave configuration, e.g., Ref. [1]. But such comparisons would be beyond the scope of this paper which considers a lumped RC-limited device. In a previous simulation [11] of an IS based EA modulator for $\lambda = 1.55$ μm using InGaAs/AlAsSb coupled QWs we obtained $f_{3dB} \approx 90$ GHz at $V_{pp} = 2.0$ V (and assuming a linewidth $\Gamma = 60$ meV), i.e., $f_{3dB}/P_{ac} = 9$ GHz mW^{-1}.

12.4.5
Absorption Saturation

One important advantage that has been predicted for IS-based modulators [12] compared to the interband-based QCSE type is the high optical powers tolerable without absorption saturation. The high absorption saturation power results from the rapid IS relaxation, allowing electrons excited from the first subband to the second to effectively relax and be "recycled" within the same QW. In contrast, in interband-based QCSE modulators with about four orders of magnitude longer relaxation times (over the bandgap in that case) electron and holes generated in the absorption process need to be efficiently swept out of the MQW region to avoid absorption saturation. Particularly for the holes, this is a problem leading to absorption saturation due to hole pile-up.

The absorption saturation intensity of a two-level system is given by (right-hand side equality)

$$\frac{\Gamma_{MQW} P_s}{d_{MQW} w} = I_s = \frac{\hbar \omega}{2\sigma T_1} \qquad (12.19)$$

where σ is the absorption cross-section and T_1 is the IS relaxation time. The left-hand side equality of Eq. (12.19) relates I_s to the saturation power P_s, where Γ_{MQW} is the optical mode overlap with the MQW, and d_{MQW} is the MQW thickness. If we then consider that the part of the modal absorption that is due to IS transitions (i.e., disregarding the waveguide absorption, which does not lead to IS absorption saturation) is $\alpha_{m,IS} = \Gamma_{MQW}(n_s/d_{MQW})\sigma$, where $n_s = 4n_1$ is the total sheet electron concentration in the four-step QWs, we can rearrange to get

$$P_s = \frac{wn_s\hbar\omega}{2\alpha_{m,IS}T_1} \tag{12.20}$$

From Fig. 12.4 we have the maximum $\alpha_{m,IS} = 0.22$ µm^{-1}. The IS scattering time in GaN/AlN QWs at telecommunications wavelengths was recently measured at $T_1 = 140$ fs [43]. This implies that the optical saturation power of the modulator is as huge as $P_s = 6$ W(!). This is about three orders of magnitude higher than the saturation power of interband QCSE modulators, which is typically several mW. The huge saturation power of the modulator analyzed here should, however, not be too surprising, since it is in accordance with the quite high optical pulse energies required to operate all-optical switches based on IS transitions in GaN/AlN QWs, as described in Chapter 11 on intersubband optical switches by N. Suzuki in this book. GaN being a highly polar semiconductor with large electron mass, $m_e = 0.22$, and LO-phonon energies, $\hbar\omega_{LO} \approx 90$ meV, exhibits very fast carrier dynamics. Indeed among the QW materials where IS resonances at 1.55 µm have been demonstrated GaN has the most rapid carrier dynamics, and notably the shortest IS relaxation time.

12.4.6
Thermal Properties and Current

Obviously it is thermally inconceivable to dissipate a multi-Watt power within the small modulator structure. The thermal conductivity of GaN is very good at $\sim 1-2$ W cm^{-1}K^{-1}. However, in sapphire, the most prevalent substrate for GaN, it is quite poor at 0.23 W cm^{-1}K^{-1} ($\parallel$ c-axis). The small footprint of the active area of the modulator (13×1 µm^2) then results in a thermal impedance of the simulated device of about 5 K mW^{-1}. Using instead SiC as a substrate, which is also attractive for growth of wurtzite GaN, but is about an order of magnitude more costly, its high thermal conductivity of 3.8 W cm^{-1}K^{-1} yields a device thermal impedance of about 1 K mW^{-1}. In recent experimental results the IS absorption even in a low-bandgap and hence high-nonparabolicity material, InAs/AlSb, is remarkably insensitive to temperatures up to 700 K [44]. Accepting say a heating of 100 K above room temperature we expect that optical and electrical powers of about 20 mW (100 mW) can be handled by the modulator when grown on a sapphire (SiC) substrate, without significant degrading of the performance, or of the structure itself, due to heating.

There is also heating due to the input microwave signal. During one cycle of the input signal the switching energy $E_{sw} = CV_{pp}^2 = 0.4$ pJ must be dissipated due to charge and discharge of C. If we as, a worst case, assume the frequency $f = f_{3dB} = 1/2\pi RC$, we obtain an upper limit for the dissipated electrical power $P_{sw} = V_{pp}^2/2\pi R = 23$ mW, which is independent of the capacitance.

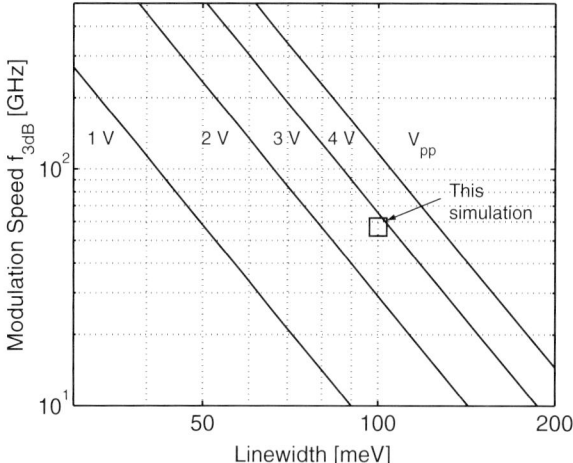

Fig. 12.7 RC-limited modulation frequency f_{3dB} plotted versus the IS absorption FWHM linewidth Γ for a few applied voltage swings V_{pp}. The square indicates the present simulation result, i.e., $f_{3dB} \approx 60$ GHz at $V_{pp} = 2.8$ V and $\Gamma = 100$ meV.

However, out of this only the fraction $R_s/R \approx 10\%$ is dissipated within the modulator and should have a negligible effect.

Thus the input power limitation in the present device is due to heating not, as for interband QCSE, absorption saturation. A power limitation of 20 mW as obtained for a sapphire substrate is significantly more than typically achieved in interband QCSE, and more than enough for fiber optical communication requirements.

Ideally there should not be any current through the device. What is important is to check that the leakage current through the MQW upon biasing is small enough not to cause undue heating of the modulator, and also small enough not to cause a voltage drop in the bulk layers and contacts outside the MQW, i.e., due to the series resistance R_s. In practice, in this very small device it is the first criterion that limits the acceptable leakage current. Accepting say 1 mW of heating yields the acceptable current 1 mW/1.4 V = 0.7 mA or equivalently the acceptable current density 5 kA cm^{-2}. The performance of unipolar nitride devices is often compromised by an excess leakage current, which has been attributed to threading dislocations with a screw component [45]. However, the excess leakage current density found due to dislocations in Ref. [45] was five orders of magnitude smaller than the acceptable current density in the present modulator. We also note that by incorporating AlN interlayers in the GaN buffer, screw component dislocations can be brought down below 10^6 cm^{-2} [46], so that usually none would be present in the modulator area.

12.4.7
Significance of the Linewidth

A small IS absorption linewidth is of vital importance in order to achieve high-speed modulation. As we have previously obtained [12] for IS-based Stark effect modulators the achievable device capacitance depends decisively on the absorption linewidth Γ. When the optical mode confinement is largely independent of the active layer, as in the present structure, we have [12]

$$C \sim \frac{\Gamma^3}{V_{pp}^2} \quad (12.21)$$

We note that Eq. (12.21) results from considering that the number of step QWs to use in the structure (under the constraint of an appropriate size Stark shift) is $m \sim V_{pp}/\Gamma$. Of course Eq. (12.21) should not be taken to mean that the capacitance is dependent on the linewidth or voltage swing in a given device. Hence, employing this simple relation, the RC-limited modulation speed can be extrapolated from the present simulation result as illustrated by Fig. 12.7, in the case of linewidths other than $\Gamma = 100$ meV, or when the number of step QWs is different from 4. Moving vertically up in this diagram means increasing the number of step QWs in the structure, hence increasing the required voltage swing V_{pp} at a given linewidth Γ. For example, we see that, according to this simulation, achieving $f_{3dB} = 40$ GHz at $V_{pp} = 1$ V would require an absorption linewidth Γ of just under 60 meV.

It follows directly from Eq. (12.21) that the speed–power ratio figure of merit depends as decisively on the IS resonance linewidth, i.e.,

$$f_{3dB}/P_{ac} \sim \Gamma^{-3} \quad (12.22)$$

As noted above in Section 12.4.4 the speed–power ratio is independent of the voltage swing V_{pp}. In Eq. (12.22) we neglected the Γ dependence of the series resistance R_s, since R_s has a marginal influence ($< 10\%$) on f_{3dB}.

Experimentally, the smallest observed IS linewidth in GaN/AlN QWs at, or at least overlapping with, $\lambda = 1.55$ μm is $\Gamma = 60$ meV, reported very recently [47]. At the small well-width required for IS resonances at 1.55 μm the linewidth is dominated by interface roughness scattering and inhomogeneous broadening due to well-width fluctuations. Thus there is still a significant potential for improvement of IS resonance linewidths in GaN-based QWs, if a suitable lattice-matched substrate material becomes available, and by further improving growth conditions. It can be expected, however, that the IS resonance linewidth is somewhat larger in step QWs, since they are inherently more sensitive to fluctuations in the transverse electric field. The high doping density required for strong IS absorption also generally increases broadening, and thus the dopant density and position are also important parameters for the linewidth. Optimizing these is the subject of future research.

12.5
Summary

This chapter has described the simulation of an electroabsorption modulator for $\lambda = 1.55$ μm based on intersubband transitions in GaN/AlGaN/AlN step QWs. The active region including the MQW and the adjoining layers were simulated self-consistently. This, e.g., allowed us to show that, despite large built-in fields, it is possible to achieve near equal applied electric fields over the step QWs. Concerning waveguide cladding, to overcome the problem of low electron density and mobility in AlGaN, the plasma effect in heavily doped GaN was employed to achieve cladding layers with high index contrast, low series resistance and lattice matching to the core layer.

The RC-limited speed is $f_{3dB} \approx 60$ GHz at an applied voltage swing of $V_{pp} = 2.8$ V. We also obtained that a desirable small negative effective chirp parameter (near perfect for standard single-mode fiber) results from Stark shifting the IS resonance. It is practically impossible to saturate the modulator. Instead the very small active section size of 13×1 μm^2 implies a (high) thermal limitation for the input power. The speed–power ratio of 3.2 GHz mW^{-1} is approximately on a par with present high-performance lumped interband QCSE modulators at the assumed IS linewidth of $\Gamma = 100$ meV, and increases strongly ($\sim \Gamma^{-3}$) with reduced IS linewidth. The main challenge is thus to achieve a narrow IS linewidth, in the high doping condition required for a strong IS absorption.

Acknowledgments

I am grateful to U. Westergren, L. Thylén, A. Kikuchi and K. Kishino for valuable discussions. This work was supported by the Swedish Research Council and the Japan Society for the Promotion of Science.

References

1 R. Lewén, S. Irmscher, U. Westergren, L. Thylén, U. Eriksson, *J. Lightwave Technol.* 22, (2004) p. 172

2 K. Kondo et al., *Electron. Lett.* 38, (2002) p. 472

3 W. H. Steier et al., *The 14th Annual Meeting of the IEEE Lasers and Electro-Optics Society*, vol. 1 (2001) p. 188

4 Y. Yu, R. Lewén, S. Irmscher, U. Westergren, L. Thylén, U. Eriksson, W. S. Lee, *Optical Fiber Communications Conference OFC2005*, vol. 3 (2005) p. 85

5 T. Mozume, H. Yoshida, A. Neogi, M. Kudo, *Jap. J. Appl. Phys.* 38 (1999) p. 1286

6 C. Gmachl, H. M. Ng, S.-N. G. Chu, A. Y. Cho, *Appl. Phys. Lett.* 77 (2000) p. 3722

7 K. Kishino, A. Kikuchi, H. Kanazawa, T. Tachibana, *phys. stat. sol. (a)* 192 (2002) p. 124

8 N. Iizuka, K. Kaneko, N. Suzuki, *Appl. Phys. Lett.* 81 (**2002**) p. 1803

9 R. Akimoto, Y. Kinpara, K. Akita, F. Sasaki, S. Kobayashi, *Appl. Phys. Lett.* 78 (**2001**) p. 580

10 P. Holmström, *IEEE J. Quantum Electron.* 42 (**2006**) p. 810

11 P. Jänes, P. Holmström, *Proc. of the 15th Int. Conf. on Indium Phosphide and Related Materials*, **2003**, p. 308

12 P. Holmström, *IEEE J. Quantum Electron.* 37 (**2001**) p. 1273

13 R. P. G. Karunasiri, Y. J. Mii, K. L. Wang, *IEEE Electron. Device Lett.* 11 (**1990**) p. 227

14 C. Gmachl, H. M. Ng, A. Y. Cho, *Appl. Phys. Lett.* 79 (**2001**) p. 1590

15 T. Ive, O. Brandt, H. Kostial, K. J. Friedland, L. Däweritz, K. H. Ploog, *Appl. Phys. Lett.* 86 (**2005**) p. 024106

16 M. Ahoujja, J. L. McFall, Y. K. Yeo, R. L. Hengehold, J. E. Van Nostrand, *Mater. Sci. Eng.* B91–92 (**2002**) p. 285

17 P. Holmström, *Physica E* 7 (**2000**) p. 40

18 T. Tamir, E. Garmire, Eds., *Integrated Optics*, vol. 7 of *Topics in Applied Physics*, Springer Verlag, Berlin, **1975**

19 J. H. Edgar, Ed., *Properties of Group III Nitrides*, no. 11 in *EMIS Datareviews Series*, INSPEC, London, UK, **1994**

20 Z. Fan, S. N. Mohammad, W. Kim, Ö. Aktas, A. E. Botchkarev, *Appl. Phys. Lett.* 68 (**1996**) p. 1672

21 S. R. Lee, A. F. Wright, M. H. Crawford, G. A. Petersen, J. Han, R. M. Biefeld, *Appl. Phys. Lett.* 74 (**1999**) p. 3344

22 T. Takeuchi, H. Takeuchi, S. Sota, H. Sakai, H. Amano, I. Akasaki, *Jap. J. Appl. Phys., Pt. 2* 36 (**1997**) p. L177

23 D. Korakakis, Jr. K. F. Ludwig, T. D. Moustakas, *Appl. Phys. Lett.* 72 (**1998**) p. 1004

24 I. Vurgaftman, J. R. Meyer, L. R. Ram-Mohan, *J. Appl. Phys.* 89 (**2001**) p. 5815

25 F. Bernardini, V. Fiorentini, D. Vanderbilt, *Phys. Rev. B* 63 (**2001**) p. 193201

26 S. Shokhovets et al., *J. Appl. Phys.* 94 (**2003**) p. 307

27 A. S. Barker Jr., M. Ilegems, *Phys. Rev. B* 7 (**1973**) p. 743

28 S. Strite, H. Morkoç, *J. Vac. Sci. Technol. B* 10 (**1992**) p. 1237

29 S. Shokovets, G. Gobsch, O. Ambacher, *Appl. Phys. Lett.* 86 (**2005**) p. 161908

30 U. Ekenberg, *Phys. Rev. B* 40 (**1989**) p. 7714

31 W. P. Chen, Y. J. Chen, E. Burstein, *Surf. Sci.* 58 (**1976**) p. 263

32 C. Sirtori, F. Capasso, J. Faist, S. Scandolo, *Phys. Rev. B* 50 (**1994**) p. 8663

33 P. Yeh, *Optical Waves in Layered Media*, Wiley Interscience, New York, **2005**

34 K. B. Ozanyan, O. Hunderi, B. O. Fimland, *J. Appl. Phys.* 75 (**1994**) p. 5347

35 B. Jensen, in *Handbook of Optical Constants of Solids*, E. D. Palik, Ed., p. 169, Academic Press, San Diego, **1985**

36 G. Bentoumi, A. Deneuville, B. Beaumont, P. Gibart, *Mater. Sci. Eng.* B50 (**1997**) p. 142

37 S. J. Allen Jr., D. C. Tsui, B. Vinter, *Solid State Commun.* 20 (**1976**) p. 425

38 T. Ido, S. Tanaka, M. Suzuki, M. Koizumi, H. Sano, H. Inoue, *J. Lightwave Technol.* 14 (**1995**) p. 2026

39 F. Dorgeuille, F. Devaux, *IEEE J. Quantum Electron.* 30 (**1994**) p. 2565

40 F. Koyama, K. Iga, *J. Lightwave Technol.* 6 (**1988**) p. 87

41 J. Kim et al., *IEEE J. Quantum Electron.* 40 (**2004**) p. 1663

42 Y. Miyazaki, H. Tada, S. Tokizaki, K. Takagi, T. Aoyagi, Y. Mitsui, *IEEE J. Quantum Electron.* 39 (**2003**) p. 813

43 J. Hamazaki et al., *Appl. Phys. Lett.* 84 (**2004**) p. 1102

44 R. J. Warburton, K. Weilhammer, C. Jabs, J. P. Kotthaus, M. Thomas, H. Kroemer, *Physica E* 7 (**2000**) p. 191

45 J. W. P. Hsu et al., *Appl. Phys. Lett.* 78 (**2001**) p. 1685

46 K. Kishino, A. Kikuchi, *phys. stat. sol. (a)* 190 (**2002**) p. 23

47 I. Friel, K. Driscoll, E. Kulenica, M. Dutta, R. Paiella, T. D. Moustakas, *J. Crystal Growth* 278 (**2005**) p. 387

13
Ultraviolet Light-Emitting Diodes

Yen-Kuang Kuo[1], Sheng-Horng Yen, and Jun-Rong Chen

13.1
Introduction

High-efficiency ultraviolet light-emitting diodes (UV LEDs) have attracted great attention in recent decades due to their promising applications in many fields. In particular, UV LEDs can be used to pump red-green-blue phosphors to generate white light. Such solid-state light sources are expected to replace the traditional fluorescent and incandescent lamps from the viewpoint of energy saving. Furthermore, UV LEDs are useful for application in the medical and biochemical fields, and the purification of the environment.

Since the nitride-based LEDs and laser diodes were demonstrated successfully in the late 1900s; InGaN, AlGaN, and AlInGaN alloys with the wurtzite crystal structure are now widely studied and developed due to their extremely wide range of emission wavelengths and direct energy bandgap. Because the bandgap energy of AlGaN-based materials can be adjusted from 6.2 eV (AlN) to 3.4 eV (GaN), the (In)AlGaN alloys have become promising candidate materials for UV LEDs. In 1992, an efficient AlGaN/GaN double-heterostructure UV LED was realized by Akasaki *et al.* [1]. Then, an AlGaN/GaN multiple quantum-well (MQW) UV LED at an emission wavelength of 353 nm was demonstrated by Han *et al.* in 1998 [2]. In 2004, UV LED with an emission wavelength of as short as 250 nm was realized by Adivarahan *et al.* [3].

In order to obtain LEDs in the UV spectral range, various material systems including InGaN, AlGaN and AlInGaN have been employed in quantum-well structures. High-power near-UV LEDs which used an InGaN active region at a wavelength of 365 nm were demonstrated by Nichia Inc. [4]. However, because of the limit of the InGaN energy bandgap, AlGaN and AlInGaN need to be used for the deep-UV spectral range. Starting from 2004, Khan *et al.* reported a series of AlGaN-based deep-UV LEDs in a spectral range down to 250 nm [3,5–9]. Specifically, AlGaN UV LEDs with 10 mW pulse operation at 265 nm and 1.2 mW continuous-wave (CW) operation at 280 nm were demon-

1) Corresponding author

Nitride Semiconductor Devices: Principles and Simulation. Joachim Piprek (Ed.)
Copyright © 2007 WILEY-VCH Verlag GmbH & Co. KGaA, Weinheim
ISBN: 978-3-527-40667-8

strated in 2005 [8, 9]. In order to improve the external efficiency, they also reported high-power deep-UV LEDs-based on a micro-pixel design [10, 11]. Razeghi et al. realized high-power 280 nm AlGaN LEDs which reached a peak external quantum efficiency of 0.24% at 40 mA in 2004 [12].

Various attempts to achieve high-efficiency UV LEDs were proposed in the past few years. High-quality AlN substrates, which are transparent to UV light and have high thermal conductivity, were used to reduce the defect density [13, 14]. Improved current spreading in the AlGaN UV LEDs-based on a microring geometry was presented by Choi and Dawson [15]. Moreover, photonic crystals have been utilized in UV LEDs to obtain high extraction efficiency of UV light [16, 17]. However, the performance of UV LEDs with an emission wavelength below 360 nm has not reached the requirement for practical applications to date.

In the case of using the quaternary AlInGaN material system as the emitting layer of UV LEDs, the efficiency can be enhanced due to the effect of the In-segregation in quantum wells [18]. On the other hand, the emission intensity of the quaternary AlInGaN active layer is not affected obviously by the threading dislocations as compared with the AlGaN-based UV LEDs [19]. Hirayama et al. have already reported efficient AlInGaN-based UV LEDs at an emission wavelength near 350 nm [20–22]. Based on the experimental results, they confirmed the advantage of the use of quaternary AlInGaN for 350-nm UV emitters in comparison with the use of AlGaN. Wang et al. studied the effect of strain relaxation and exciton localization on the performance of 350-nm AlInGaN quaternary LEDs [23]. They also indicated that the quaternary AlInGaN was better suited as the active region for UV LEDs. Therefore, further improvement and optimization of AlInGaN-based UV LEDs are important for various promising applications of UV light.

In order to better understand the physics and characteristics of UV LEDs, a sample UV-LED structure is numerically studied using a self-consistent simulation program APSYS (Advanced Physical Models of Semiconductor Devices) based on two-dimensional (2D) models in this chapter [24]. Moreover, an AlInGaN-based 370-nm UV LED, which was fabricated by Chang et al. [25], is used as a reference for numerical simulation. Specifically, optimization of the LED performance is attempted by adjusting the aluminum composition in the AlGaN electron blocking layer (EBL) and the number of quantum wells.

The device structure of the UV LED under study is described in Section 13.2. The physical models and parameters used in simulation are given in Section 13.3. Comparisons of the calculated and measured results are discussed in Section 13.4. In Section 13.5, the optical performance of the UV LEDs with different aluminum compositions in the AlGaN electron blocking layer and different numbers of quantum wells are investigated in an attempt to optimize the UV LED device performance. In addition, the feasibility of employing a lattice-matched quaternary AlInGaN electron blocking layer in the UV LED structure under study is numerically evaluated with the APSYS simulation program. Finally, conclusions are given in Section 13.6.

13.2 Device Structure

The AlInGaN UV LED used as a reference for subsequent simulation was grown on a (0001) sapphire substrate by low-pressure horizontal-flow metal organic chemical vapor deposition (MOCVD), which had an emission wavelength of approximately 370 nm and a maximum output power of 4 mW at 125 mA under room temperature CW operation. Figure 13.1 depicts the structure of the AlInGaN/AlInGaN UV LED under study. A 30 nm-thick low-temperature GaN nucleation layer was grown on a sapphire substrate, followed by a 2 μm-thick GaN buffer layer and a 1 μm-thick n-GaN layer with a doping concentration of 2×10^{18} cm^{-3}. Next, a 50 nm-thick graded n-Al$_x$Ga$_{1-x}$N (x ranges from 0.10 to 0.14) was deposited with a doping concentration of 1×10^{18} cm^{-3} for cladding. The active region consisted of three Al$_{0.06}$In$_{0.09}$Ga$_{0.85}$N quantum wells, each 2.5 nm thick, separated by 3 nm-thick Al$_{0.05}$In$_{0.01}$Ga$_{0.94}$N barriers. On top of the MQW active region was a 25 nm-thick p-Al$_{0.19}$Ga$_{0.81}$N electron blocking layer with a doping concentration of 8×10^{17} cm^{-3}. A 125 nm-thick p-Al$_{0.09}$Ga$_{0.91}$N layer and 10 nm-thick p-GaN

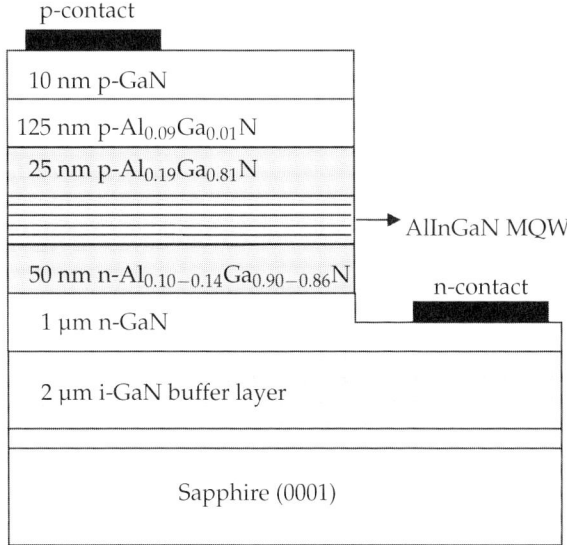

Fig. 13.1 Schematic drawing of AlInGaN/AlInGaN UV LED under study.

contact layer were grown with a doping concentration of 1×10^{18} cm^{-3} to complete the structure. The device geometry was designed with a rectangular shape of 300 μm × 300 μm. The detailed layer structure and room-temperature physical parameters of the AlInGaN/AlInGaN UV LED under study are tabulated in Table 13.1. The doped carrier densities considered in the simulation represent actual densities of free carriers.

Tab. 13.1 Layer structure and room-temperature physical parameters of the AlInGaN/AlInGaN UV LED under study (d, layer thickness; N_{dop}, doped carrier density; n, refractive index at wavelength 370 nm; κ, thermal conductivity). The doped carrier density, N_{dop}, represents actual density of free carriers.

Parameter (unit)	d (nm)	N_{dop} (cm^{-3})	n	κ (W cm^{-1}K^{-1})
p-GaN	10	1×10^{18}	2.647	1.30
p-Al$_{0.09}$Ga$_{0.91}$N	125	1×10^{18}	2.605	0.27
p-Al$_{0.19}$Ga$_{0.81}$N	25	8×10^{17}	2.548	0.27
i-Al$_{0.05}$In$_{0.01}$Ga$_{0.94}$N barrier	3		2.676	0.27
i-Al$_{0.06}$In$_{0.09}$Ga$_{0.85}$N well	2.5		3.048	0.27
i-Al$_{0.05}$In$_{0.01}$Ga$_{0.94}$N barrier	3		2.676	0.27
i-Al$_{0.06}$In$_{0.09}$Ga$_{0.85}$N well	2.5		3.048	0.27
i-Al$_{0.05}$In$_{0.01}$Ga$_{0.94}$N barrier	3		2.676	0.27
i-Al$_{0.06}$In$_{0.09}$Ga$_{0.85}$N well	2.5		3.048	0.27
i-Al$_{0.05}$In$_{0.01}$Ga$_{0.94}$N barrier	3		2.676	0.27
n-Al$_{0.10-0.14}$Ga$_{0.90-0.86}$N	50	1×10^{18}	2.585	0.27
n-GaN	1000	2×10^{18}	2.647	1.30
i-GaN	2000		2.647	1.30

13.3
Physical Models and Parameters

The self-consistent simulation program APSYS solves Poisson's equation, the current continuity equation, the carrier transport equations, the quantum mechanical wave equations, and the heat transfer equations. The transport model includes drift and diffusion of electrons and holes in devices. Built-in polarization induced by spontaneous and piezoelectric polarization is considered at hetero-interfaces of nitride related devices, which is vital because the polarization effect is of great importance for III-nitride devices. The contribution of spontaneous and defect related Shockley–Read–Hall (SRH) recombination of carriers can be analyzed respectively. Some of the key physical models used in this simulation are briefly described in the following subsections. More details can be found in the books of Piprek [26, 27].

13.3.1
Band Structure

In this work, the 6×6 $k \cdot p$ model, developed for strained wurtzite structures by Chuang and Chang [28, 29], is utilized to calculate the valence-subband structures of the UV-LED structure under study. It is assumed that the conduction bands can be characterized by a parabolic-band model. As for the valence subbands, the 6×6 Hamiltonian matrix including heavy-hole (HH), light-hole (LH), and crystal-field split-off hole (CH) bands can be block diagonalized into the following upper and lower 3×3 matrices to simplify the calculation [29, 30]

$$H_{6\times 6} = \begin{bmatrix} H^U & 0 \\ 0 & H^L \end{bmatrix} \tag{13.1}$$

where

$$H^U = \begin{bmatrix} F & K_t & -iH_t \\ K_t & G & \Delta - iH_t \\ iH_t & \Delta + iH_t & \lambda \end{bmatrix} \tag{13.2}$$

$$H^L = \begin{bmatrix} F & K_t & iH_t \\ K_t & G & \Delta + iH_t \\ -iH_t & \Delta - iH_t & \lambda \end{bmatrix} \tag{13.3}$$

The matrix elements are

$$\begin{aligned}
F &= \Delta_1 + \Delta_2 + \lambda + \theta \\
G &= \Delta_1 - \Delta_2 + \lambda + \theta \\
\lambda &= \frac{\hbar^2}{2m_0}(A_1 k_z^2 + A_2 k_t^2) + \lambda_\varepsilon \\
\lambda_\varepsilon &= D_1 \varepsilon_{zz} + D_2(\varepsilon_{xx} + \varepsilon_{yy}) \\
\theta &= \frac{\hbar^2}{2m_0}(A_3 k_z^2 + A_4 k_t^2) + \theta_\varepsilon \\
\theta_\varepsilon &= D_3 \varepsilon_{zz} + D_4(\varepsilon_{xx} + \varepsilon_{yy}) \\
K_t &= \frac{\hbar^2}{2m_0} A_5 k_t^2 \\
H_t &= \frac{\hbar^2}{2m_0} A_6 k_z k_t \\
\Delta &= \sqrt{2}\Delta_3
\end{aligned} \tag{13.4}$$

and

$$k_t^2 = k_x^2 + k_y^2 \tag{13.5}$$

The above Hamiltonian is considered under the assumption that the strained quantum well is grown along the (0001) (c axis) direction. Therefore, the strain tensor in the plane of the epitaxial growth is

$$\varepsilon_{xx} = \varepsilon_{yy} = \frac{a_0 - a}{a} \tag{13.6}$$

where a is the natural unstrained lattice constant of the quantum well and a_0 is the lattice constant of the GaN. The perpendicular strain tensor can be expressed as

$$\varepsilon_{zz} = -\frac{2C_{13}}{C_{33}} \varepsilon_{xx} \tag{13.7}$$

where C_{13} and C_{33} are the elastic stiffness constants. For quantum wells, k_z is replaced by $-i(d/dz)$. In order to obtain the numerical parameters of the quaternary AlInGaN materials, a linear interpolation between the parameters of the relevant binary semiconductors is utilized, except for the unstrained bandgap energies. For the physical parameter Q, the interpolation formula is

$$Q(Al_xIn_yGa_{1-x-y}N) = Q(AlN)x + Q(InN)y + Q(GaN)(1-x-y) \tag{13.8}$$

The unstrained AlInGaN bandgap energies can be expressed as a weighted sum of the bandgap energies of relevant ternary semiconductors with appropriate bandgap bowing parameters. Specifically, the unstrained AlInGaN bandgap energies are calculated by following expressions [31]

$$E_g(AlInGaN) = \frac{xyE_g^u(AlInN) + yzE_g^v(InGaN) + xzE_g^w(AlGaN)}{xy + yz + zx} \tag{13.9}$$

$$E_g^u(AlInN) = uE_g(InN) + (1-u)E_g(AlN) - u(1-u)B(AlInN) \tag{13.10}$$

$$E_g^v(InGaN) = vE_g(GaN) + (1-v)E_g(InN) - v(1-v)B(InGaN) \tag{13.11}$$

$$E_g^w(AlGaN) = wE_g(GaN) + (1-w)E_g(AlN) - w(1-w)B(AlGaN) \tag{13.12}$$

$$u = \frac{1-x+y}{2} \quad v = \frac{1-y+z}{2} \quad w = \frac{1-x+z}{2}, \tag{13.13}$$

where x, y, and $z = 1 - x - y$ represent the compositions of aluminum, indium, and gallium in the AlInGaN material system, respectively. $B(AlInN)$, $B(InGaN)$, and $B(AlGaN)$ are the bandgap bowing parameters of AlInN, InGaN, and AlGaN, respectively. Note that the bandgap bowing parameters of AlInN, InGaN, and AlGaN have been studied by several researchers in recent years and various different figures have been reported [32–35]. In this study, the values provided by Vurgaftman et al. [32] are adopted. Namely, the bandgap bowing parameters of AlInN, InGaN, and AlGaN are assumed to be 2.5 eV, 1.4 eV, and 0.7 eV, respectively. The material parameters of the relevant binary semiconductors used in the simulation can be found in Table 13.2 [32].

Tab. 13.2 Material parameters of the binary semiconductors GaN, AlN, and InN at room temperature. ($\Delta_{cr} = \Delta_1$, $\Delta_{so} = 3\Delta_2 = 3\Delta_3$.)

Parameter	Symbol (unit)	GaN	AlN	InN
Lattice constant	a_0 (Å)	3.189	3.112	3.545
Spin–orbit split energy	Δ_{so} (eV)	0.017	0.019	0.005
Crystal-field split energy	Δ_{cr} (eV)	0.010	−0.169	0.040
Hole effective mass parameter	A_1	−7.21	−3.86	−8.21
	A_2	−0.44	−0.25	−0.68
	A_3	6.68	3.58	7.57
	A_4	−3.46	−1.32	−5.23
	A_5	−3.40	−1.47	−5.11
	A_6	−4.90	−1.64	−5.96
Hydrost. deform. potential (c-axis)	a_z (eV)	−4.9	−3.4	−3.5
Hydrost. deform. potential (transverse)	a_t (eV)	−11.3	−11.8	−3.5
Shear deform. potential	D_1 (eV)	−3.7	−17.1	−3.7
	D_2 (eV)	4.5	7.9	4.5
	D_3 (eV)	8.2	8.8	8.2
	D_4 (eV)	−4.1	−3.9	−4.1
Elastic stiffness constant	C_{33} (GPa)	398	373	224
Elastic stiffness constant	C_{13} (GPa)	106	108	92
Electron effective mass (c-axis)	m_e^z/m_0	0.2	0.32	0.07
Electron effective mass (transverse)	m_e^t/m_0	0.2	0.30	0.07

The valence-subband structure for the upper Hamiltonian can be determined by solving the following expression

$$\sum_{j=1}^{3}\left[H_{ij}^{U}\left(k_z = -i\frac{\partial}{\partial z}\right) + \delta_{ij}E_v^0(z)\right] g_m^{(j)}(z;k_t) = E_m^U(k_t)g_m^{(i)}(z;k_t) \quad (13.14)$$

for $i = 1, 2, 3$ with different values of k_t. The valence band discontinuity is represented by a discontinuity in $E_v^0(z)$. The envelope functions of the mth valence subbands are denoted by $g_m^{(1)}$, $g_m^{(2)}$, and $g_m^{(3)}$. As for the lower Hamiltonian, we can get the same band structure when the quantum-well structure has reflection symmetry $E_v^0(z) = E_v^0(-z)$.

13.3.2
Polarization Effects

The piezoelectric and spontaneous polarization is a critical effect for nitride-based optoelectronic devices. The built-in polarization causes a strong deformation of the quantum wells accompanied by a strong electrostatic field.

Under this circumstance, electrons and holes are separated in the quantum well, which leads to a reduction in the photon emission rate.

In order to consider the built-in polarization within the interface, the method developed by Fiorentini et al. is employed to estimate the fixed interface charges [36]. The spontaneous polarization of ternary nitride alloys can be expressed by

$$P_{sp}(Al_xGa_{1-x}N) = -0.090x - 0.034(1-x) + 0.019x(1-x)$$
$$P_{sp}(In_xGa_{1-x}N) = -0.042x - 0.034(1-x) + 0.038x(1-x) \quad (13.15)$$
$$P_{sp}(Al_xIn_{1-x}N) = -0.090x - 0.042(1-x) + 0.071x(1-x)$$

The spontaneous polarization of quaternary AlInGaN can be calculated in a similar way to that shown in expression (13.9). On the other hand, the piezoelectric polarization of AlInGaN and AlGaN can be calculated by the following expression

$$P_{pz}(Al_xIn_yGa_{1-x-y}N)$$
$$= P_{pz}(AlN)x + P_{pz}(InN)y + P_{pz}(GaN)(1-x-y) \quad (13.16)$$

where

$$\begin{aligned} P_{pz}(AlN) &= -1.808\varepsilon + 5.624\varepsilon^2 \quad \text{for} \quad \varepsilon < 0 \\ P_{pz}(AlN) &= -1.808\varepsilon - 7.888\varepsilon^2 \quad \text{for} \quad \varepsilon > 0 \\ P_{pz}(GaN) &= -0.918\varepsilon + 9.541\varepsilon^2 \\ P_{pz}(InN) &= -1.373\varepsilon + 7.559\varepsilon^2 \end{aligned} \quad (13.17)$$

The total built-in polarization is the sum of spontaneous and piezoelectric polarization.

For the AlInGaN UV-LED structure under study, the net surface charges are calculated and listed in Table 13.3. Note that the amount of surface charges obtained from experimental measurement is usually smaller than that obtained by theoretical calculation. The surface charges could be screened due to the defects inside the device and the screening percentage commonly ranges from 20% to 80% for InGaN materials [37, 38]. For a better match of LED performance obtained numerically and experimentally, the screening percentages of AlInGaN/AlInGaN, AlInGaN/AlGaN, and AlGaN/AlGaN are set at 90%. Thus, only 10% of the calculated surface charge density is used in the following calculation. The surface charge densities shown in Table 13.3 are the values when the screening effect has been taken into account.

Tab. 13.3 Net surface charge density at each interface of the UV LED.

Interface	Surface charge density
GaN/Al$_{0.10}$Ga$_{0.90}$N	$+5.50 \times 10^{15}$ m^{-2}
Al$_{0.14}$Ga$_{0.86}$N/Al$_{0.05}$In$_{0.01}$Ga$_{0.94}$N	-6.60×10^{15} m^{-2}
Al$_{0.05}$In$_{0.01}$Ga$_{0.94}$N/Al$_{0.06}$In$_{0.09}$Ga$_{0.85}$N	-1.13×10^{16} m^{-2}
Al$_{0.06}$In$_{0.09}$Ga$_{0.85}$N/Al$_{0.05}$In$_{0.01}$Ga$_{0.94}$N	$+1.13 \times 10^{16}$ m^{-2}
Al$_{0.05}$In$_{0.01}$Ga$_{0.94}$N/Al$_{0.19}$Ga$_{0.81}$N	$+9.50 \times 10^{15}$ m^{-2}
Al$_{0.19}$Ga$_{0.81}$N/Al$_{0.09}$Ga$_{0.91}$N	-5.80×10^{15} m^{-2}
Al$_{0.09}$Ga$_{0.91}$N/GaN	-4.90×10^{15} m^{-2}

13.3.3
Carrier Transport Model

The governing model of carrier transport in a semiconductor is the drift-diffusion model. By employing the Poisson's equation, the electrostatic field is related to the charge distribution in following expression

$$\nabla \cdot (\varepsilon \mathbf{E}) = q \left(p - n + N_D^+ - N_A^- \pm N_f \right) \tag{13.18}$$

where $\mathbf{E}$ is the electric field, q is the magnitude of a unit charge, p is the hole concentration, n is the electron concentration, N_D^+ is the ionized donor concentration, N_A^- is the ionized acceptor concentration, and N_f represents other fixed charges such as the surface charges caused by spontaneous and piezoelectric polarization for nitride-based devices. The continuity equation describes a change in electrons and holes density over time. This equation can be separated into two parts for electrons and holes and expressed by

$$\frac{\partial n}{\partial t} = G_n - R_n + \frac{1}{q} \nabla \cdot \mathbf{J}_n \tag{13.19}$$

$$\frac{\partial p}{\partial t} = G_p - R_p - \frac{1}{q} \nabla \cdot \mathbf{J}_p \tag{13.20}$$

where G_n and R_n are the generation rates and recombination rates for electrons, G_p and R_p are the generation rates and recombination rates for holes, and $\mathbf{J}_n$ and $\mathbf{J}_p$ are the electron and hole current densities, respectively. In addition, the electron and hole transport equations can be written as

$$\mathbf{J}_n = q\mu_n n \mathbf{E} + q D_n \nabla n \tag{13.21}$$

$$\mathbf{J}_p = q\mu_p p \mathbf{E} - q D_p \nabla p \tag{13.22}$$

where μ_n and μ_p are the electron and hole mobilities and D_n and D_p are the electron and hole diffusion coefficients, respectively. The drift-diffusion model consists of Poisson's equation, the continuity equation, and the carrier transport equations. These equations govern the electrical behavior of a semiconductor device.

13.3.4
Thermal Model

The effect of self-heating has a major impact on the performance of LED devices. The increased internal temperature caused by current injection, limits the maximum output power due to the increase of nonradiative carrier recombination and the spread of the gain spectrum. One of the most important heat sources is the Joule heat caused by the electrical resistance of the material. The Joule heat density can be expressed by

$$H_J = \frac{J_n^2}{q\mu_n n} + \frac{J_p^2}{q\mu_p p} \tag{13.23}$$

On the other hand, the energy resulted from an electron–hole pair recombination is either transferred to a photon (radiative) or to phonons (nonradiative). The recombination heat density caused by nonradiative carrier recombination is given by

$$H_R = R_{SRH}(E_{Fn} - E_{Fp}) \tag{13.24}$$

with the quasi-Fermi levels E_{Fn} and E_{Fp} for electrons and holes, respectively. To calculate the internal device temperature $T(x,y,z)$, the steady state heat flux equation is introduced in the calculation model, which can be expressed by

$$H_{heat} = -\nabla(\kappa \nabla T) \tag{13.25}$$

where H_{heat} is the total heat power density and κ is the thermal conductivity. The thermal conductivity of 1.3 W cm^{-1}K^{-1} for bulk GaN at 300 K was first measured by Sichel and Pankove [39]. However, since the thermal conductivity is dependent on many factors, such as device temperature, alloy composition, crystal quality, and doping concentration [40,41], the thermal conductivities listed in Table 13.1 are roughly estimated, considering the effects of alloy composition and interface scattering of phonons on thermal conductivity.

For nitride-based LED devices, the temperature rise with injection current is mainly caused by the thermal resistance of the sapphire substrate and the mount. Therefore, an external thermal resistance of 83 K W^{-1} is added to the heat sink at the bottom of the UV LED.

13.3.5
Spontaneous Emission

The spontaneous recombination rate in each bulk layer of the AlInGaN UV LED can be characterized by the bimolecular recombination coefficient B and written approximately as [27]

$$R_{sp}^{bulk} = B(np - n_0 p_0) \tag{13.26}$$

where n_0 and p_0 are the electron and hole concentrations at thermal equilibrium. In this specific simulation, it is assumed that the bimolecular recombination coefficient B has a value of 2×10^{-10} cm^3 s^{-1} in each bulk layer. For the quantum wells, the spontaneous emission spectrum is calculated by [42]

$$r_{sp}^{qw}(E) = \frac{q^2 \hbar}{2m_0^2 \varepsilon E} D(E) \rho_{red}(E) |M|^2 f_c^n (1 - f_v^m) \tag{13.27}$$

where $|M|^2$ is the momentum matrix element in the strained quantum well, E is the photon energy, f_c^n and f_v^m are the Fermi functions for the conduction band states and the valence band states respectively, $D(E)$ is the optical mode density, and $\rho_{red}(E)$ is the reduced density of states in each subband. The optical mode density and reduced density of states can be expressed respectively as

$$D(E) = \frac{\varepsilon n E^2}{\pi^2 \hbar^3 c^3} \tag{13.28}$$

$$\rho_{red}(E) = \frac{m_r}{\pi \hbar^2 d_z} \tag{13.29}$$

where n is the index of refraction, c is the speed of light, d_z is the thickness of the quantum well, m_r is the reduced effective mass.

Based on the $k \cdot p$ theory described in Section 13.3.1, the momentum matrix elements for HH, LH, and CH can be written respectively as

$$\left|M_{HH}^{TE}\right|^2 = \frac{3}{2} O_{ij} \left(M_b^{TE}\right)^2 \tag{13.30}$$

$$\left|M_{LH}^{TE}\right|^2 = \frac{3}{2} \cos^2 \theta O_{ij} \left(M_b^{TE}\right)^2 \tag{13.31}$$

$$\left|M_{CH}^{TE}\right|^2 = 0 \tag{13.32}$$

for TE polarization and

$$\left|M_{HH}^{TM}\right|^2 = 0 \tag{13.33}$$

$$\left|M_{LH}^{TM}\right|^2 = 3\left(1 - \cos^2 \theta\right) O_{ij} \left(M_b^{TM}\right)^2 \tag{13.34}$$

$$\left|M_{CH}^{TM}\right|^2 = 3 O_{ij} \left(M_b^{TM}\right)^2 \tag{13.35}$$

for TM polarization. The angular factor $\cos^2 \theta$ can be related to the z-component of the electron **k** vector, k_z, by

$$k_z = |\mathbf{k}| \cos \theta \tag{13.36}$$

O_{ij} is the wavefunction overlap integral and M_b^{TE} and M_b^{TM} are the bulk dipole moments given by [43]

$$\left(M_b^{TE}\right)^2 = \frac{m_0}{6}\left(\frac{m_0}{m_e^z} - 1\right)\frac{(E_g + \Delta_1 + \Delta_2)(E_g + 2\Delta_2) - 2\Delta_3^2}{E_g + 2\Delta_2} \quad (13.37)$$

$$\left(M_b^{TM}\right)^2 = \frac{m_0}{6}\left(\frac{m_0}{m_e^t} - 1\right)\frac{E_g\left[(E_g + \Delta_1 + \Delta_2)(E_g + 2\Delta_2) - 2\Delta_3^2\right]}{(E_g + \Delta_1 + \Delta_2)(E_g + \Delta_2) - \Delta_3^2} \quad (13.38)$$

where m_e^z and m_e^t are the electron effective masses parallel and perpendicular to the growth direction, respectively. The material parameters used in the simulation can be found in Table 13.1. To account for the broadening induced by intra-band scattering, the Lorentzian lineshape function is used in the final expression of the spontaneous emission spectrum, which is given by

$$R_{sp}^{qw} = \frac{1}{\pi}\int r_{sp}^{qw}(E)\frac{\Gamma}{(E_{ij} - E)^2 + \Gamma^2}dE \quad (13.39)$$

where $\Gamma = \hbar/\tau$, which represents the broadening due to the intraband scattering relaxation time τ, and E_{ij} is the transition energy from the ith conduction band to the jth valence band. To account for the broadening due to scattering, it is assumed that $\tau = 0.1$ ps in the calculations.

13.3.6
Ray Tracing

Because of the large difference in refractive index between the semiconductor and the surrounding air, the percentage of photons that can be extracted from the LED is severely limited by total internal reflection. The ray-tracing model, which is based on simple geometrical optics, is employed in the simulation to calculate the extraction efficiency [27].

The refractive index of each layer is required for the ray-tracing model, since the reflection and transmission of the optical field at each layer are evaluated by Fresnel's formula. Because the photon energy emitted from the LED device is smaller than the GaN energy bandgap, Adachi's model can be utilized to calculate the reflective index of each layer in the LED structure. In nitride III-V compounds, since the valence band splitting is very small, Adachi's model for the transparency region can be expressed as

$$n^2(\hbar\omega) = A\left(\frac{\hbar\omega}{E_g}\right)^{-2}\left[2 - \sqrt{1 + \left(\frac{\hbar\omega}{E_g}\right)} - \sqrt{1 - \left(\frac{\hbar\omega}{E_g}\right)}\right] + B \quad (13.40)$$

The refractive indices of all the layers in the LED structure under study are listed in Table 13.1. A detailed description of the method for deriving the refractive indices of AlInGaN alloys may be found in the paper by Peng and Piprek [44].

13.4
Comparison Between Simulated and Experimental Results

In this section, the optical performance of the UV-LED structure with three AlInGaN/AlInGaN quantum wells, as depicted in Section 13.2, is numerically investigated. The results obtained by experimental measurements are used as a reference for the adjustment of the physical parameters used in the simulation. Note that, in the simulation, the internal loss, thermal conductivity, and SRH recombination lifetime in quantum wells are employed as fit parameters [45]. The simulated L–I curve at room temperature is in good agreement with that obtained from experimental measurements when the internal loss is 10 cm^{-1}, the thermal conductivity of active layers is 0.27 W cm^{-1}K^{-1}, and the SRH recombination lifetime in quantum wells is 1 ns.

Figure 13.2 shows the emission wavelength obtained by experimental measurements and simulation when the injection current increases from 10 to 100 mA. According to the electroluminescence spectra, it is observed that the peak emission wavelength is red-shifted from 368.68 nm to 373.88 nm as the injection current increases from 10 to 100 mA due to a thermal effect. The emission wavelength is designed in such a way that light absorption by bulk GaN, which has a characteristic wavelength of 365 nm, can be avoided.

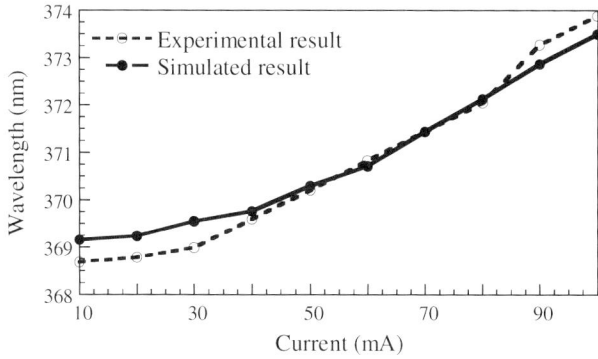

Fig. 13.2 Emission wavelength obtained by experimental measurements and simulation when the injection current increases from 10 to 100 mA.

The L–I–V curves obtained from the experimental and simulated results are shown in Fig. 13.3. The maximum LED output power is 3.97 mW at 300 K and 2.05 mW at 380 K. Both the experimental and simulated results indicate that the turn-on voltage is approximately 3.1 V. An average extraction efficiency of approximately 25% is determined using the ray-tracing model. Figures 13.2 and 13.3 show that the results obtained from simulation are in close agreement with those obtained by experimental measurements.

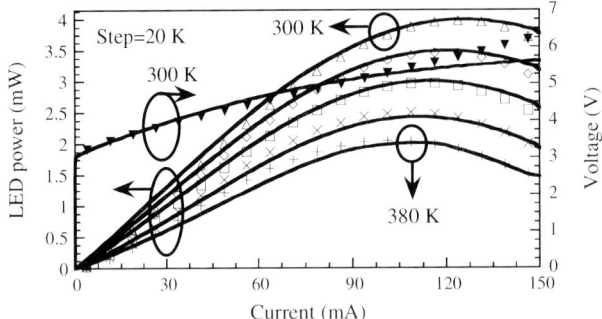

Fig. 13.3 LED power and voltage versus injection current in CW operation at different heat sink temperatures. The lines and dots represent simulated and experimental results, respectively.

The relative temperature rise in active region as a function of injection current at different heat sink temperatures are shown in Fig. 13.4. Note that, in the simulation it is assumed that a heat sink is located at the bottom of the LED. When the heat sink temperature increases, the relative temperature rise in the active region increases accordingly. Therefore, the increase of temperature-dependent nonradiative recombination and electronic overflow are the main causes for the power roll-off at the elevated heat sink temperature.

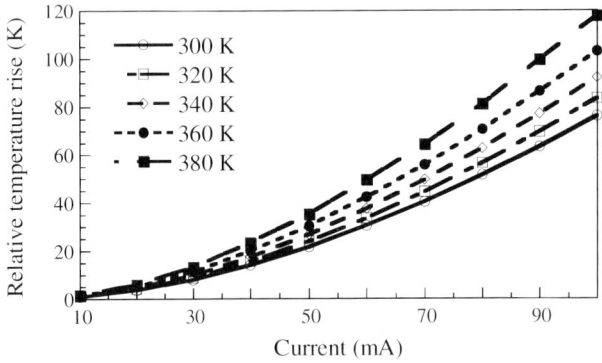

Fig. 13.4 Relative temperature rise in active region as a function of injection current at different heat sink temperatures.

The band-offset ratio, which is defined as the ratio between the conduction band offset ΔE_c and the valence band offset ΔE_v, of the III-nitride material system is an important issue for the study of LED device performance [46]. The band-offset ratio of III-nitride quaternary compounds has not been well documented in the literature [47–49]. In this work, a band-offset ratio of 0.7/0.3, i.e., $\Delta E_c/\Delta E_v = 0.7/0.3$, is assumed for the AlInGaN/AlInGaN material system.

13.5 Performance Optimization

In this section, the optical performance of the UV LED with different aluminum compositions in the AlGaN electron blocking layer and different numbers of quantum wells are investigated in an attempt to optimize the device performance.

13.5.1 Optimal Aluminum Composition in p-AlGaN Electron Blocking Layer

The problem of electronic overflow plays an important role for the optical performance of III-nitride light-emitting devices [46, 50, 51]. A p-AlGaN layer is commonly used in III-nitride LEDs to prevent electronic overflow from the active region to the p-side layer. Note that the aluminum composition in the AlGaN electron blocking layer needs to be properly adjusted since the high potential barrier created by the AlGaN electron blocking layer may also hinder holes from entering the active region. In this subsection, the 370 nm UV LED performance as a function of aluminum composition in the p-AlGaN electron blocking layer is studied.

Figure 13.5 shows the L–I performance curves of the UV LED structure for different aluminum compositions in p-AlGaN electron blocking layer at 300 K and 360 K, which indicates that increase in output power of the LED becomes less evident when the aluminum composition in p-AlGaN electron blocking layer is larger than 0.19.

The internal efficiencies for different aluminum compositions in p-AlGaN electron blocking layer at 300 K and 360 K are plotted in Fig. 13.6, which also indicates that the internal efficiency of the LED cannot be markedly improved when the aluminum composition is larger than 0.19. Therefore, the $Al_{0.19}Ga_{0.81}N$ electron blocking layer used in the original LED structure was an appropriate design.

When the device temperature increases, the internal efficiency decreases due to the increase of recombination loss and electron current leakage from the active region to the p-type layers. The simulated results shown in Fig. 13.6 suggest that the low internal efficiency may be limited by the increased current leakage and nonradiative recombination.

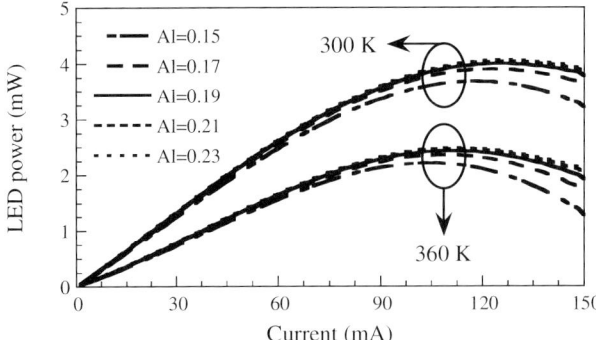

Fig. 13.5 L–I curves under different aluminum compositions in p-AlGaN EBL.

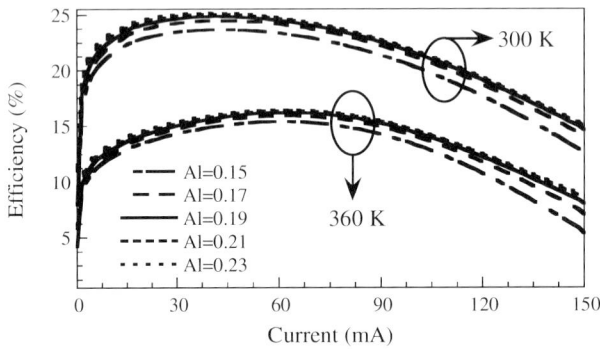

Fig. 13.6 Internal efficiencies for different aluminum compositions in p-AlGaN EBL.

13.5.2
Optimal Number of Quantum Wells

Figures 13.7(a) and (b) show the L–I curves for different numbers of quantum wells (ranging from 1 to 11) at 300 K and 360 K, respectively. It is found that the LED device demonstrates a good performance when the number of quantum wells is three as the temperature increases from 300 to 360 K. The lowest LED power is obtained when the number of the quantum well is one, which is due mainly to the large percentage of electron leakage current.

For convenient comparison between the number of quantum wells and electron leakage current, the percentage of electron leakage current, which is defined as the ratio of the current overflow to the p-type layer to that injected into the active region of the LED device are listed in Table 13.4, for different numbers of quantum wells and temperatures. It is apparent that the percentage of electron leakage current increases from 8.96% to 13.17% for the AlInGaN LED with a single quantum well when the temperature increases from 300 to 360 K. However, the percentage of electron leakage current only increases from 0.11% to 0.25% when the temperature varies from 300 to 360 K for the AlInGaN LED with eleven quantum wells. The simulated results indicate that the percentage of electron leakage current may be efficiently reduced when the number of quantum wells is larger than three.

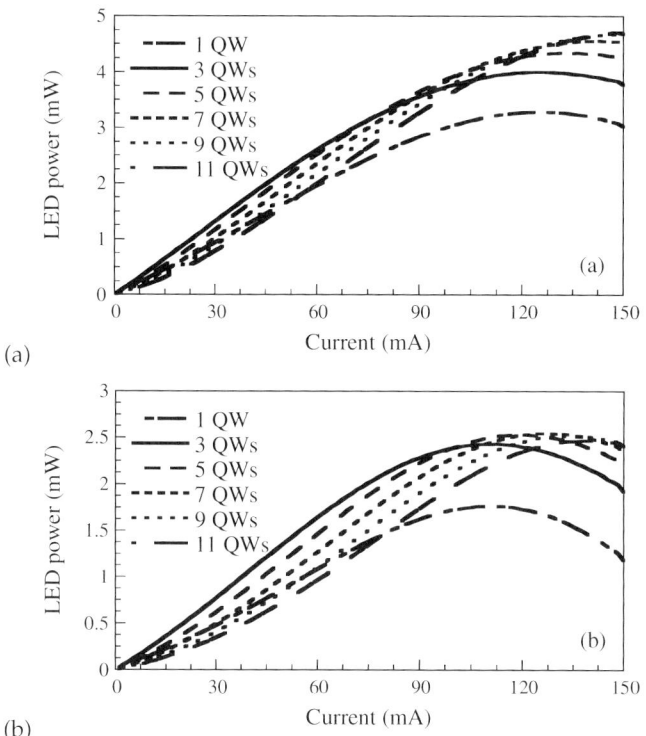

Fig. 13.7 L–I curves of the AlInGaN UV LEDs with different numbers of quantum wells at (a) 300 K and (b) 360 K.

Tab. 13.4 Percentage of electron leakage current for different numbers of quantum wells and temperatures.

Leakage percentage (%)	300 K	310 K	320 K	330 K	340 K	350 K	360 K
1 well	8.96	9.59	10.28	11.00	11.74	12.48	13.17
3 wells	1.56	1.73	1.93	2.17	2.46	2.77	3.22
5 wells	0.53	0.59	0.66	0.75	0.86	0.99	1.16
7 wells	0.29	0.32	0.36	0.41	0.48	0.55	0.66
9 wells	0.16	0.18	0.21	0.24	0.27	0.32	0.37
11 wells	0.11	0.12	0.14	0.16	0.18	0.21	0.25

The nonuniform carrier distribution within the quantum wells could be one of the major causes for the decrease in LED power when the number of quantum wells is large. Figure 13.8 shows the distribution of electrons and holes within the quantum wells at 300 K when the number of quantum wells is eleven. It can be found that the electrons accumulate in the quantum well close to the n-side and holes accumulate in the quantum well close to the p-side. In these circumstances, a poor LED performance can be expected.

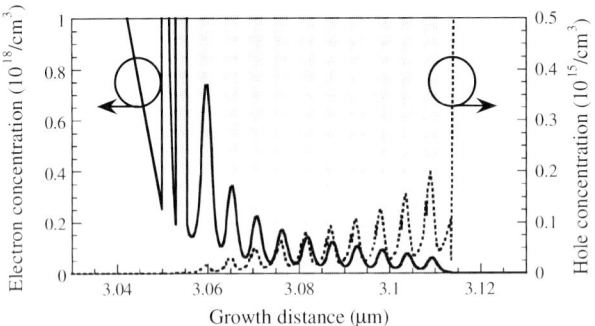

Fig. 13.8 Distribution of electrons and holes within the quantum wells at 300 K when the number of quantum wells is eleven. Quantum wells are marked by the gray areas.

Since the power roll-off of the AlInGaN LED is typically caused by the self-heating in high injection current and in continuous-wave operation, suppression of the leakage current can prevent an increase in the device temperature. When the number of quantum wells increases, the output power increases accordingly; however, the output power at low injection current decreases as the number of quantum wells is more than three. Thus, the simulated results suggest that the optimal number of quantum wells in the AlInGaN UV LED is three.

Note that, in addition to the electronic current overflow, the hole current overflow is also examined. It is found that the hole leakage current is very small, even when the number of quantum wells is one. Therefore, the problem of hole leakage current is not an important issue for the AlInGaN UV LED under study.

13.5.3
Lattice-matched AlInGaN Electron Blocking Layer

Asano *et al.* found that the strained AlGaN electron blocking layer might induce crystal defects due to the lattice mismatch to InGaN or GaN layers [52]. Therefore, in this subsection, the use of a lattice-matched AlInGaN electron blocking layer in the UV LED is investigated and compared with the AlGaN electron blocking layer. For this study, special attention is paid to the effect of built-in polarization on the LED properties, assuming that the crystal quality is the same for the LED with either AlGaN or AlInGaN electron blocking layer. For comparison purpose, $Al_{0.18}In_{0.039}Ga_{0.781}N$ is chosen to be the quaternary electron blocking layer because its energy bandgap is identical to that of the original $Al_{0.19}Ga_{0.81}N$ electron blocking layer. With the lattice-matched $Al_{0.18}In_{0.039}Ga_{0.781}N$ electron blocking layer, the polarization charge densities in the heterostructure interfaces can be reduced since the piezoelectric polarization is caused by strain. Specifically, the polarization charge density within the $Al_{0.18}In_{0.039}Ga_{0.781}N/Al_{0.09}Ga_{0.91}N$ interface is decreased from -5.80×10^{15} m^{-2} to -1.81×10^{15} m^{-2} when the $Al_{0.19}Ga_{0.81}N$ electron blocking layer is replaced by the $Al_{0.18}In_{0.039}Ga_{0.781}N$ electron blocking layer.

The room-temperature L–I performance curves of the AlInGaN UV LEDs with $Al_{0.18}In_{0.039}Ga_{0.781}N$ and $Al_{0.19}Ga_{0.81}N$ electron blocking layers are shown in Fig. 13.9. It is found that the maximum output power is improved at 300 K when the $Al_{0.19}Ga_{0.81}N$ electron blocking layer is replaced by the $Al_{0.18}In_{0.039}Ga_{0.781}N$ electron blocking layer.

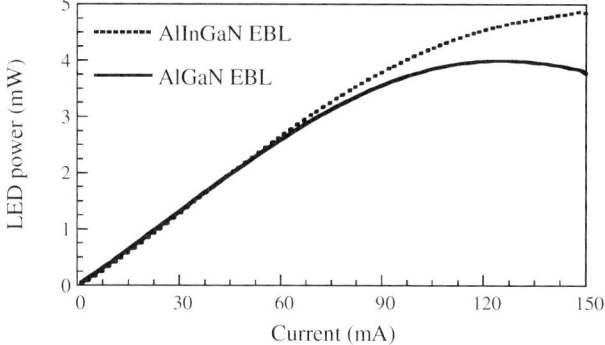

Fig. 13.9 L–I curves of the AlInGaN UV LEDs with AlInGaN and AlGaN EBL at 300 K.

Generally, the electron and hole carriers in nitride-based quantum wells are separated due to a strong built-in electrostatic field as a large amount of polarization charges exist in the interfaces of the quantum wells. Under these circumstances, several disadvantages, including less overlap between electron and hole wavefunctions, more nonradiative recombination rates and more electron leakage current result in III-nitride devices. In order to analyze the mechanisms between the UV LED with $Al_{0.18}In_{0.039}Ga_{0.781}N$ or $Al_{0.19}Ga_{0.81}N$ electron blocking layer, the conduction and valence band diagrams of these two devices are plotted in Fig. 13.10. For the device with an $Al_{0.19}Ga_{0.81}N$ electron blocking layer, the band diagrams in quantum wells have more bending than the device with an $Al_{0.18}In_{0.039}Ga_{0.781}N$ electron blocking layer. Thus,

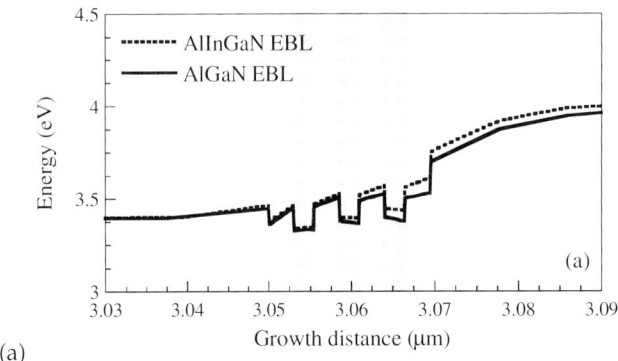

(a)

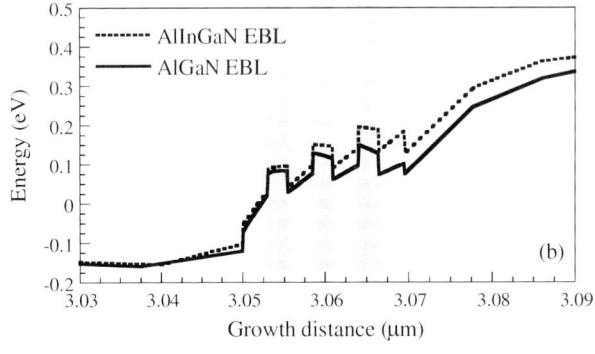

(b)

Fig. 13.10 (a) Conduction band diagram and (b) valence band diagram of the UV LED with $Al_{0.19}Ga_{0.81}N$ or $Al_{0.18}In_{0.039}Ga_{0.781}N$ EBL. Quantum wells are marked by the gray areas.

better radiative recombination and less electrostatic field within the active region can be expected when the UV LED is with the $Al_{0.18}In_{0.039}Ga_{0.781}N$ electron blocking layer.

Figure 13.11 shows the electrostatic field in the active region of the UV LED with $Al_{0.19}Ga_{0.81}N$ or $Al_{0.18}In_{0.039}Ga_{0.781}N$ electron blocking layer. Note that the electrostatic field in quantum wells for the situation when the $Al_{0.18}In_{0.039}Ga_{0.781}N$ electron blocking layer is used in the UV LED is less than that when the $Al_{0.19}Ga_{0.81}N$ electron blocking layer is used in the UV LED. Specifically, the amplitude of the electrostatic field in the quantum well close to the p-side is reduced from -0.096 MV cm^{-1} to -0.035 MV cm^{-1} when the $Al_{0.19}Ga_{0.81}N$ electron blocking layer is replaced by the $Al_{0.18}In_{0.039}Ga_{0.781}N$ electron blocking layer.

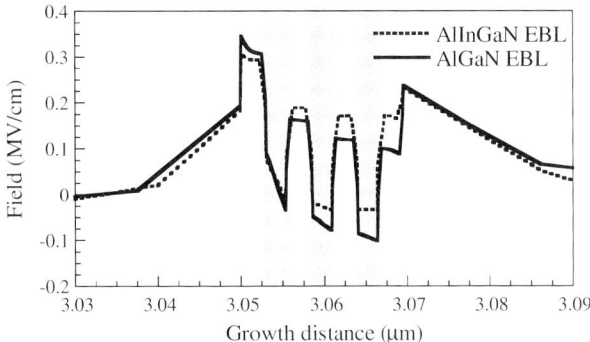

Fig. 13.11 Electrostatic field in the active region of the UV LED with $Al_{0.19}Ga_{0.81}N$ or $Al_{0.18}In_{0.039}Ga_{0.781}N$ EBL. Quantum wells are marked by the gray areas.

According to the simulated results, when compared to the 370 nm AlInGaN UV LED with an $Al_{0.19}Ga_{0.81}N$ electron blocking layer, it is evident that the electrostatic field within the active region can be reduced when the lattice-matched $Al_{0.18}In_{0.039}Ga_{0.781}N$ electron blocking layer is used in the AlInGaN UV LED. The output power of the LED with an $Al_{0.18}In_{0.039}Ga_{0.781}N$ electron blocking layer can be improved as well.

13.6 Conclusion

In this chapter, the development of UV LEDs with ternary AlGaN and the quaternary AlInGaN active region is briefly reviewed. The optical properties of a 370-nm UV LED is also investigated with a self-consistent simulation program APSYS. Moreover, the optical performance of the UV LEDs with different aluminum compositions in the AlGaN electron blocking layer and different numbers of quantum wells are investigated in an attempt to optimize the UV LED device performance. Since the built-in polarization is one of the most important factors for the deterioration of III-nitride LED performance, the feasibility of using a lattice-matched quaternary $Al_{0.18}In_{0.039}Ga_{0.781}N$ electron blocking layer, which has the same energy bandgap as that of the ternary $Al_{0.19}Ga_{0.81}N$ electron blocking layer, in the UV LED to improve the LED performance is numerically evaluated. The simulated results suggest that, with the use of a lattice-matched AlInGaN electron blocking layer, the polarization charge density in each heterostructure interface is reduced, the electrostatic field in quantum wells is reduced, and the maximum output power is sufficiently enhanced.

Acknowledgments

The authors would like to thank Yi-An Chang, Hao-Chung Kuo, Tien-Chang Lu, and Shing-Chung Wang, Institute of Electro-Optical Engineering, National Chiao-Tung University, Hsinchu, Taiwan, for providing the experimental data used in this study. This work is supported by the National Science Council of Taiwan under grant NSC-94-2112-M-018-009.

References

1. I. Akasaki and H. Amano, "Room temperature ultraviolet/blue light emitting devices based on AlGaN/GaN multi-layered structure," in Extended Abstracts of the 1992 International Conf. Solid State Devices and Materials, Aug. 1992, pp. 327–329.
2. J. Han, M. H. Crawford, R. J. Shul, J. J. Figiel, M. Banas, L. Zhang, Y. K. Song, H. Zhou, and A. V. Nurmikko, Appl. Phys. Lett., vol. 73, pp. 1688–1670, 1998.
3. V. Adivarahan, W. H. Sun, A. Chitnis, M. Shatalov, S. Wu, H. P. Maruska, and M. A. Khan, Appl. Phys. Lett., vol. 85, pp. 2175–2177, 2004.
4. D. Morita, M. Sano, M. Yamamoto, T. Murayama, S. Nagahama, and T. Mukai, Jpn. J. Appl. Phys., vol. 41, pp. L1434–L1436, 2002.
5. V. Adivarahan, S. Wu, J. P. Zhang, A. Chitnis, M. Shatalov, V. Mandavilli, R. Gaska, and M. A. Khan, Appl. Phys. Lett., vol. 84, pp. 4762–4764, 2004.
6. W. Sun, V. Adivarahan, M. Shatalov, Y. Lee, S. Wu, J. Yang, J. Zhang, and M. A. Khan, Jpn. J. Appl. Phys., vol. 43, pp. L1419–L1421, 2004.
7. W. H. Sun, J. P. Zhang, V. Adivarahan, A. Chitnis, M. Shatalov, S. Wu, V. Mandavilli, J. W. Yang, and M. A. Khan, Appl. Phys. Lett., vol. 85, pp. 531–533, 2004.
8. J. Zhang, X. Hu, A. Lunev, J. Deng, Y. Bilenko, T. M. Katona, M. S. Shur, R. Gaska, and M. A. Khan, Jpn. J. Appl. Phys., vol. 44, pp. 7250–7253, 2005.
9. Y. Bilenko, A. Lunev, X. Hu, J. Deng, T. M Katona, J. Zhang, R. Gaska, M. S Shur, W. Sun, V. Adivarahan, M. Shatalov, and M. A. Khan, Jpn. J. Appl. Phys., vol. 44, pp. L98–L100, 2005.
10. S. Wu, V. Adivarahan, M. Shatalov, A. Chitnis, W.-H. Sun, and M. A. Khan, Jpn. J. Appl. Phys., vol. 43, pp. L1035–L1037, 2004.
11. V. Adivarahan, S. Wu. W. H. Sun, V. Mandavilli, M. S. Shatalov, G. Simin, J. W. Yang, H. P. Maruska, and M. A. Khan, Appl. Phys. Lett., vol. 85, pp. 1838–1840, 2004.
12. K. Mayes, A. Yasan, R. McClintock, D. Shiell, S. R. Darvish, P. Kung, and M. Razeghi, Appl. Phys. Lett., vol. 84, pp. 1046–1048, 2004.
13. G. Tamulaitis, I. Yilmaz, M. S. Shur, Q. Fareed, R. Gaska, M. A. Khan, Appl. Phys. Lett., vol. 85, pp. 206–208, 2004.
14. T. Nishida, T. Makimoto, H. Saito, and T. Ban, Appl. Phys. Lett., vol. 84, pp. 1002–1003, 2004.
15. H. W. Choi and M. D. Dawson, Appl. Phys. Lett., vol. 86, pp. 053504–1–3, 2005.
16. J. Shakya, K. H. Kim, J. Y. Lin, and H. X. Jiang, Appl. Phys. Lett., vol. 85, pp. 142–144, 2004.
17. T. N. Oder, K. H. Kim, J. Y. Lin, and H. X. Jiang, Appl. Phys. Lett., vol. 84, pp. 466–468, 2004.
18. H. Hirayama, A. Kinoshita, T. Yamabi, Y. Enomoto, A. Hirata, T. Araki, Y. Nanishi, and Y. Aoyagi, Appl. Phys. Lett., vol. 80, pp. 207–209, 2002.
19. H. Hirayama, J. Appl. Phys., vol. 97, pp. 091101–1–19, 2005.
20. H. Hirayama, K. Akita, T. Kyono, T. Nakamura, and K. Ishibashi, Jpn. J. Appl. Phys., vol. 43, pp. L1241–L1243, 2004.

21. K. Akita, T. Nakamura, and H. Hirayama, Jpn. J. Appl. Phys., vol. 43, pp. 8030–8031, 2004.
22. H. Hirayama, T. Kyono, K. Akita, T. Nakamura, K. Ishibashi, phys. stat. sol. (c), vol. 2, pp. 2899–2902, 2005.
23. T. Wang, G. Raviprakash, F. Ranalli, C. N. Harrison, J. Bai, J. P. R. David, and P. J. Parbrook, J. Appl. Phys., vol. 97, pp. 083104–1–4, 2005.
24. APSYS Version 2005.03.11 by Crosslight Software Inc., Burnaby, Canada. (http://www.crosslight.com)
25. Y.-A. Chang, S.-H. Yen, T.-C. Wang, H.-C. Kuo, Y.-K. Kuo, T.-C. Lu, and S.-C. Wang, Semicond. Sci. Technol., vol. 21, pp. 598–603, 2006.
26. J. Piprek: *Semiconductor Optoelectronic Device: Introduction to Physics and Simulation* (Academic Press, San Diego, 2003).
27. J. Piprek and S. Li: GaN-based Light-Emitting Diodes. In: *Optoelectronic Devices: Advanced Simulation and Analysis*, ed. by J. Piprek (Springer Verlag, New York, 2005).
28. S. L. Chuang and C. S. Chang, Semicond. Sci. Technol., vol. 12, pp. 252–263, 1997.
29. S. L. Chuang and C. S. Chang, Phys. Rev. B, vol. 54, pp. 2491–2504, 1996.
30. S. L. Chuang and C. S. Chang, Appl. Phys. Lett., vol. 68, pp. 1657–1659, 1996.
31. I. Vurgaftman, J. R. Meyer, and L. R. Ram-Mohan, J. Appl. Phys., vol. 89, pp. 5815–5875, 2001.
32. I. Vurgaftman and J. R. Meyer, J. Appl. Phys., vol. 94, pp. 3675–3696, 2003.
33. B.-T. Liou, S.-H. Yen, and Y.-K. Kuo, Appl. Phys. A, vol. 81, pp. 651–655, 2005.
34. B.-T. Liou, C.-Y. Lin, S.-H. Yen, and Y.-K. Kuo, Opt. Communi., vol. 249, pp. 217–223, 2005.
35. B.-T. Liou, S.-H. Yen, and Y.-K. Kuo, Appl. Phys. A, vol. 81, pp. 1459–1463, 2005.
36. V. Fiorentini, F. Bernardini, and O. Ambacher, Appl. Phys. Lett., vol. 80, pp. 1204–1206, 2002.
37. H. Zhang, E. J. Miller, E. T. Yu, C. Poblenz, and J. S. Speck, Appl. Phys. Lett., vol. 84, pp. 4644–4646, 2004.
38. F. Renner, P. Kiesel, G. H. Döhler, M. Kneissl, C. G. Van de Walle, and N. M. Johnson, Appl. Phys. Lett., vol. 81, pp. 490–492, 2002.
39. E. K. Sichel and J. I. Pankove, J. Phys. Chem. Solids, vol. 38, pp. 330–330, 1977.
40. B. C. Daly, H. J. Maris, and A. V. Nurmikko, M. Kuball, and J. Han, J. Appl. Phys., vol. 92, pp. 3820–3824, 2002.
41. J. Zou, D. Kotchetkov, A. A. Balandin, D. I. Florescu, and F. H. Pollak, J. Appl. Phys., vol. 92, pp. 2534–2539, 2002.
42. R. H. Yan, S. W. Corzine, L. A. Coldren, and I. Suemune, IEEE J. Quantum Electron., vol. 26, pp. 213–216, 1990.
43. S. L. Chuang, IEEE J. Quantum Electron., vol. 32, pp. 1791–1800, 1996.
44. T. Peng and J. Piprek, Electron. Lett., vol. 32, pp. 2285–2286, 1996.
45. J. Piprek and S. Nakamura, Proc. IEE-Optoelectron., vol. 149, pp. 145–151, 2002.
46. Y.-K. Kuo, B.-T. Liou, M.-L. Chen, S.-H. Yen, and C.-Y. Lin, Opt. Communi., vol. 231, pp. 395–402, 2004.
47. T. H. Yu and K. F. Brennan, J. Appl. Phys., vol. 89, pp. 3827–3834, 2001.
48. K. A. Bulashevich, V. F. Mymrin, S. Y. Karpov, I. A. Zhmakin, and A. I. Zhmakin, J. Comput. Phys., vol. 213, pp. 214–238, 2006.
49. C. B. Soh, S. J. Chua, S. Tripathy, S. Y. Chow, D. Z. Chi, and W. Liu, J. Appl. Phys., vol. 98, pp. 103704–1–8, 2005.
50. J.-Y. Chang and Y.-K. Kuo, J. Appl. Phys., vol. 93, pp. 4992–4998, 2003.
51. Y.-K. Kuo and Y.-A. Chang, IEEE J. Quantum Electron., vol. 40, pp. 437–444, 2004.
52. T. Asano, T. Tojyo, T. Mizuno, M. Takeya, S. Ikeda, K. Shibuya, T. Hino, S. Uchida, and M. Ikeda, IEEE J. Quantum Electron., vol. 39, pp. 135–140, 2003.

14
Visible Light-Emitting Diodes

Sergey Yu. Karpov

14.1
Introduction

Since the first demonstration of high-brightness blue light-emitting diodes (LEDs) in the early 90s [1, 2], group-III nitride semiconductors have gone a long way from being just promising materials in the industrial world to the basis for modern visible and ultra-violet (UV) optoelectronics. This has happened, on the one hand, due to the unique physical properties of these semiconductors, namely, a direct bandgap that could be varied from 0.65 to 6.2 eV, depending on the III-nitride alloy composition [3–5]; and, on the other, to reliable and effective *p*-doping demonstrated in earlier studies [6,7]. Very quickly, blue III-nitride LEDs were capable of providing an emission efficiency suitable for application, in spite of an extremely high threading dislocation density (TDD), $\sim 10^8$–10^{10} cm^{-2}, inherent in epitaxial materials on mismatched sapphire or SiC substrates. This breakthrough stimulated an enormous amount of activity in III-nitride semiconductor technology, and soon the practical achievements outpaced the understanding of physical mechanisms underlying the LED operation. In part, this was caused by some novel physical factors, such as electrical spontaneous polarization and a strong piezoeffect [8,9], the significance of which for LED operation was recognized only after a remarkable delay.

The research and development stage has been completed with a world-wide mass production of III-nitride LEDs, establishing new targets aimed at improving the device performance; primarily brightness and efficiency. These targets require a deeper insight into the physics of LED operation, as further effort can no longer rely on the intuitive design of LED heterostructures and analogy with conventional III-V devices. As the potential of many experimental approaches to LED optimization is likely to be almost exhausted, the role of theoretical work in the further improvement of device performance has become much more important.

Nitride Semiconductor Devices: Principles and Simulation. Joachim Piprek (Ed.)
Copyright © 2007 WILEY-VCH Verlag GmbH & Co. KGaA, Weinheim
ISBN: 978-3-527-40667-8

This chapter presents the simulation of visible III-nitride LEDs with the focus on heterostructure design and nanoscale physical mechanisms critical to device performance. Special attention is given to highlighting the capabilities of modeling for the identification of physical mechanisms controlling the LED performance and optimization of the heterostructure.

14.2
Simulation Approach and Materials Properties

Group-III nitride LEDs were simulated by using the SiLENSe 2.0 package produced by STR, Inc. [10]. The package implements a 1D carrier transport and recombination model based on the drift-diffusion equations which accounts for such important features of III-nitride semiconductors as: spontaneous electric polarization, a strong piezoeffect, a low acceptor activation efficiency, and a high dislocation density inherent in epitaxial materials [11]. Within the model, a LED heterostructure is considered as a stack of epitaxial layers coherently grown on a template (normally, a buffer or an n-contact layer) in the z-direction along the hexagonal C-axis of the wurtzite crystal. As a result, all the layers have the in-plane lattice constants equal to the lattice constant of the template a_s. The misfit strain components in every biaxially stressed epitaxial layer are $\epsilon_1 = \epsilon_2 = -(a - a_s)/a$, $\epsilon_3 = -2\epsilon_1 C_{13}/C_{33}$, where ϵ_i and C_{ij} are the components of the strain and elastic stiffness tensors in the Voigt notation [12]. The unstrained lattice constant a of an arbitrary $Al_xIn_yGa_{1-x-y}N$ alloy obeys the Vegard rule:

$$a = xa_{AlN} + ya_{InN} + (1 - x - y)a_{GaN} \tag{14.1}$$

where a_{AlN}, a_{InN}, and a_{GaN} are the lattice constants of the binary constituents of the alloy. The electric polarization vector $\mathbf{P}^{tot}$ has only a z-component and includes both piezo- (P^{pz}) and spontaneous polarization (P^{sp}) allowable by the symmetry of wurtzite crystals:

$$\mathbf{P}^{tot} = \mathbf{e}_z \cdot (P^{sp} + P^{pz}) \quad , \quad P^{pz} = 2\epsilon_1 \Lambda \quad , \quad \Lambda = \left(e_{13} - e_{33}\frac{C_{13}}{C_{33}} \right) \tag{14.2}$$

where $\mathbf{e}_z$ is the unit vector along the C-axis and e_{ij} are the components of the piezoelectric tensor in the Voigt notation. The Vegard law is used to calculate the spontaneous polarization and Λ-constant of an $Al_xIn_yGa_{1-x-y}N$ alloy from the values corresponding to binary nitrides:

$$P^{sp} = xP^{sp}_{AlN} + yP^{sp}_{InN} + (1 - x - y)P^{sp}_{GaN} \tag{14.3}$$

$$\Lambda = x\Lambda_{AlN} + y\Lambda_{InN} + (1 - x - y)\Lambda_{GaN} \tag{14.4}$$

This procedure of estimating the electric polarization in the alloy, in combination with the measured lattice constants of AlN, GaN, and InN, is found to reproduce the experimental sheet electron concentrations in high-electron mobility transistors more accurately than the nonlinear approximation suggested in [13] and referred to the computed values of the lattice constants.

The governing equations of the drift-diffusion model include the Poisson equation for the electric potential φ,

$$\nabla \cdot (\mathbf{P}^{\text{tot}} - \hat{\varepsilon}^* \nabla \varphi) = q \, (N_D^+ - N_A^- + p - n) \tag{14.5}$$

and the continuity equations for electrons and holes:

$$\nabla \cdot \mathbf{j}_n = +qR \, , \quad \mathbf{j}_n = \mu_n n \, \nabla F_n \tag{14.6}$$

$$\nabla \cdot \mathbf{j}_p = -qR \, , \quad \mathbf{j}_p = \mu_p p \, \nabla F_p \tag{14.7}$$

Here q is the electron charge, n and p are the electron and hole concentrations, respectively, $\hat{\varepsilon}^*$ is the effective dielectric permittivity tensor with the components $\varepsilon_{11}^* = \varepsilon_{22}^* = \varepsilon_0 \varepsilon_{11} + e_{15}^2/C_{44}$, $\varepsilon_{33}^* = \varepsilon_0 \varepsilon_{33} + e_{33}^2/C_{33}$, ε_0 is the dielectric permittivity of vacuum, ε_{ij} is the diagonal dielectric constant tensor, μ_n is the electron mobility, μ_p is the hole mobility assumed to be the same for all the valence subbands, $\mathbf{j}_n$ and $\mathbf{j}_p$ are the vectors of electron and hole current densities, respectively, and F_n and F_p are the quasi-Fermi levels of electrons and holes. The effective dielectric permittivity tensor originates in Eq. (14.5) from the back electro-mechanical coupling of the electrostatic induction and a uniform biaxial mismatch strain existing in every epitaxial layer.

The carrier concentrations and the concentrations of the ionized donors and acceptors, N_D^+ and N_A^-, depend on the quasi-Fermi level positions with respect to the conduction and valence band-edges and on the total donor (N_D) and acceptor (N_A) concentrations as

$$n = N_c \cdot \mathcal{F}_{1/2}\left(\frac{F_n - E_c + q\varphi}{kT}\right) \, , \quad p = \sum_s N_s \cdot \mathcal{F}_{1/2}\left(\frac{E_s - F_p - q\varphi}{kT}\right) \tag{14.8}$$

$$N_D^+ = \frac{N_D}{1 + g_D \exp\left(\dfrac{F_n - E_c + q\varphi + E_D}{kT}\right)} \tag{14.9}$$

$$N_A^- = \frac{N_A}{1 + g_A \exp\left(\dfrac{E_v - q\varphi + E_A - F_p}{kT}\right)} \tag{14.10}$$

where $\mathcal{F}_{1/2}(x)$ is the Fermi integral, k is the Boltzmann constant, T is temperature, E_c denotes the conduction band bottom, E_s corresponds to the top of the s-th valence subband, $E_v = \max_s(E_s)$ denotes the valence band top, $g_D = 2$ and $g_A = 4$ are the degeneracy factors of the donor and acceptor states, respectively. The densities of states in the conduction band, N_c, and in the s-th valence subband, N_s, are given by the expressions

$$N_c = 2m_n^\perp (m_n^z)^{1/2} \left(\frac{kT}{2\pi\hbar^2}\right)^{3/2}, \quad N_s = 2m_s^\perp (m_s^z)^{1/2} \left(\frac{kT}{2\pi\hbar^2}\right)^{3/2} \quad (14.11)$$

where $\hbar$ is the Plank constant, $m_n^\perp$ and m_n^z are the in-plane and normal (along the C-axis) electron effective masses, $m_s^\perp$ and m_s^z are the in-plane and normal hole effective masses in the respective subband.

The recombination rate R in Eqs (14.6)–(14.7) accounts for both nonradiative and radiative channels, $R = R^{nr} + R^{rad}$. The nonradiative carrier recombination is assumed to proceed primarily on threading dislocation cores and can be accounted for within the Shockley–Read approach, i.e.

$$R^{nr} = \frac{np}{\tau_p(n+n_d) + \tau_n(p+p_d)} \cdot \left[1 - \exp\left(-\frac{F_n - F_p}{kT}\right)\right] \quad (14.12)$$

where

$$n_d = n \cdot \exp\left(\frac{E_d - F_n}{kT}\right), \quad p_d = p \cdot \exp\left(\frac{F_p - E_d}{kT}\right) \quad (14.13)$$

and the electron (hole) lifetime, $\tau_{n(p)}$, is [14]

$$\tau_{n(p)} = \frac{1}{4\pi D_{n(p)} N_d} \left[\ln\left(\frac{1}{\pi a^2 N_d}\right) - \frac{3}{2} + \frac{2D_{n(p)}}{aV_{n(p)}S}\right] \quad (14.14)$$

Here, N_d is the dislocation density, $D_{n(p)} = (kT/q)\mu_{n(p)}$ is the diffusivity of electrons (holes), $V_{n(p)} = (3kT/m_{n(p)}^{av})^{1/2}$ is the thermal velocity of electrons (holes), $(m_{n(p)}^{av})^{3/2} = m_{n(p)}^\perp (m_{n(p)}^z)^{1/2}$, $S = 0.5$ is the fraction of electrically active sites on a dislocation core, and E_d is the energy level associated with the dislocation traps (see [14] for more detail). The bimolecular radiative recombination rate is defined as

$$R^{rad} = Bnp \cdot \left[1 - \exp\left(-\frac{F_n - F_p}{kT}\right)\right] \quad (14.15)$$

where B is the radiative recombination rate constant.

Figure 14.1(a) shows the electron and hole lifetimes as a function of TDD calculated for the typical electron and hole mobilities of 100 and 10 cm^{2}V^{-1}s^{-1}, respectively. The radiative lifetime, $\tau_{rad} = 1/B\Delta n$, where Δn is a nonequilibrium carrier concentration in the active region, is also plotted in the figure for comparison, assuming that $\Delta n = 10^{18}$ cm^{-3}. One can see that τ_n becomes less than τ_{rad} at the TDD greater than $\sim 10^7$ cm^{-2}. Actually, from this density and onwards, the threading dislocations start to affect the internal quantum efficiency (IQE) of the material.

According to Eq. (14.14), the lower the carrier mobility the longer is its non-radiative lifetime (see also Fig. 14.1(b)). So, any factors lowering the carrier mobility in the LED active region, like composition fluctuations or band-edge modulation due to donor/acceptor co-doping, are favorable for IQE improvement.

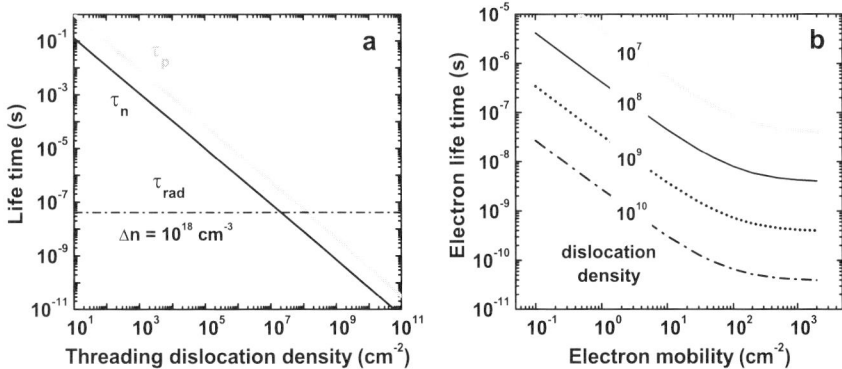

Fig. 14.1 Nonradiative electron (τ_n) and hole (τ_p) lifetimes as a function of TDD (a). The horizontal line shows the radiative lifetime (τ_{rad}) for $\Delta n = 10^{18}$ cm^{-3}. Electron lifetime as a function of mobility (b) at different TDDs.

The drift-diffusion approach in the form of Eqs (14.5)–(14.7) is capable of considering not only the polarization charges accumulated at abrupt heterostructure interfaces but also the distributed polarization charges in III-nitride graded-composition alloys [15]. As the TDD effect is rather crucial to the LED performance, the dislocation-mediated nonradiative recombination model has been carefully validated in [14] by comparison with available observations. Other nonradiative recombination channels are ignored in this study because of the uncertainty in their recombination rates which frequently depend on the heterostructure growth conditions.

As soon as the conduction and valence band profiles, along with the electric potential distribution in a LED heterostructure are found from Eqs (14.5)–(14.15), the emission spectrum can be computed by solving the Schrödinger equations for electrons and holes with the account of the complex valence-band structure of III-nitride materials within the 8×8 Kane Hamiltonian [16] (see *Ultraviolet Light-Emitting Diodes* by Y.-K. Kuo et al. for details of the Kane approach). Actually, such a procedure considers the electric field screening by free carriers within the quasi-classical approximation underlying the drift-diffusion model, but rigorously computes the separation of the electron and hole wavefunctions in a quantum-well active region, which is important for spectrum prediction. In order to make simulations more operative, the quantum-confinement effects on the radiative recombination rate were ignored, and the radiative recombination rate constant B was considered as that of the bulk material.

A parameter characterizing the overall performance of a LED heterostructure is its internal quantum efficiency, η_{int}, defined within the 1D approach as

$$\eta_{int} = \frac{q}{j}\int_0^L dz \cdot R^{rad}(z) \qquad (14.16)$$

Here $j = j_n + j_p$ is the total current density in the diode; the points $z = 0$ and $z = L$ correspond to the bottom and the top of the LED structure, respectively. The IQE, η_{int}, accounts not only for the competition between the radiative and nonradiative recombination in the structure but also for the losses of electrons and holes at the contact electrodes produced by carrier leakage (see Section 14.3.2 for more details).

An important issue of simulation largely affecting its predictability is the choice of reliable material properties. Omitting here a detailed analysis of experimental and theoretical works underlying this choice, we just summarize the properties of binary nitrides and ZnO in Table 14.1 and Table 14.2. The properties of $Al_xIn_yGa_{1-x-y}N$ alloys, are found by using the Vegard law, i.e., relationships similar to Eq. (14.1). The only exception is the bandgap of the alloy E_G, for which the quadratic approximation is employed:

$$E_G = xE_G^{AlN} + yE_G^{InN} + (1 - x - y)E_G^{GaN}$$
$$- b_{AlGaN}x(1 - x - y) - b_{InGaN}y(1 - x - y) - b_{AlInN}xy \qquad (14.17)$$

where E_G^{AlN}, E_G^{InN}, and E_G^{GaN} are the bandgaps of the binary nitrides and $b_{AlGaN} = 0.5$ eV, $b_{InGaN} = 2.5$ eV, and $b_{AlInN} = 4.8$ eV are the bowing parameters derived from the luminescence spectra of ternary alloys reported in

Tab. 14.1 Electro-mechanical properties of wurtzite semiconductors at room temperature.

Parameter	Symbol (unit)	GaN	AlN	InN	ZnO
In-plane lattice constant	a (Å)	3.188	3.112	3.540	3.250
Stiffness constant	C_{11} (GPa)	375	395	225	207
Stiffness constant	C_{12} (GPa)	140	140	110	118
Stiffness constant	C_{13} (GPa)	115	105	95	106
Stiffness constant	C_{33} (GPa)	385	395	200	210
Stiffness constant	C_{44} (GPa)	120	100	45	45
Piezoelectric constant	e_{33} (C m^{-2})	0.65	1.55	0.43	0.96
Piezoelectric constant	e_{31} (C m^{-2})	-0.33	-0.58	-0.22	-0.62
Piezoelectric constant	e_{15} (C m^{-2})	-0.33	-0.48	-0.22	-0.37
Spontaneous polarization	P^{sp} (C m^{-2})	-0.029	-0.081	-0.032	-0.050
Dielectric constant	ε_{33}	8.9	8.5	15.3	8.75

Tab. 14.2 Band parameters and impurity properties at room temperature.

Parameter	Symbol (unit)	GaN	AlN	InN	ZnO
Bandgap	E_G (eV)	3.40	6.20	0.65	3.37
Crystal-field splitting	Δ_{cr} (meV)	160	−170	35	38
Spin–orbital splitting	Δ_{so} (meV)	12	17	9	10
In-plane electron mass	$m_n^\perp / m_0$	0.20	0.26	0.10	0.27
Normal electron mass	m_n^z / m_0	0.20	0.26	0.10	0.27
In-plane heavy hole mass	$m_{hh}^\perp / m_0$	1.65	2.58	1.45	0.64
Normal heavy hole mass	m_{hh}^z / m_0	1.10	1.95	1.35	0.64
In-plane light hole mass	$m_{lh}^\perp / m_0$	0.15	0.25	0.10	0.62
Normal light hole mass	m_{lh}^z / m_0	1.10	1.95	1.35	0.62
In-plane split-off hole mass	$m_{so}^\perp / m_0$	1.10	1.93	1.54	0.59
Normal split-off hole mass	m_{so}^z / m_0	0.15	0.23	0.09	0.32
Donor ionization energy	E_D (meV)	13	13	13	50
Acceptor ionization energy	E_A (meV)	170	350	170	135
Recombination constant $\times 10^{-11}$	B (cm^{3}s^{-1})	2.4	2.0	0.66	5.5

numerous publications. The valence band offset $\Delta E_V \simeq 0.3 \Delta E_G$ is assumed for all the alloys, in accordance with recommendations given in [11, 17]. The parameters of wurtzite ZnO given in Table 14.1 and Table 14.2 were mainly borrowed from [18–20].

14.3
Device Analysis

In this section, we consider typical blue single-quantum-well (SQW) and multiple-quantum-well (MQW) LED heterostructures. A representative SQW structure, hereafter referred to as structure A, consists of an n-GaN contact layer ([Si] = 5×10^{18} cm^{-3}) 3–4 μm thick, an unintentionally doped 2 nm In$_{0.2}$Ga$_{0.8}$N SQW followed by a 100 nm p-Al$_{0.2}$Ga$_{0.8}$N electron blocking layer (EBL) ([Mg] = 5×10^{19} cm^{-3}) and a 200 nm p-GaN contact layer ([Mg] = 7×10^{19} cm^{-3}). The structure is very similar to those reported in [21] and [22].

An MQW LED structure, hereafter referred to as structure B, contains an n-GaN contact layer ([Si] = 2×10^{18} cm^{-3}) 4–5 μm thick and four unintentionally doped 3 nm In$_{0.13}$Ga$_{0.87}$N quantum wells (QWs) separated by 12 nm n-GaN barriers ([Si] = 2×10^{18} cm^{-3}). The MQW active region is followed by a 60 nm p-Al$_{0.15}$Ga$_{0.85}$N EBL ([Mg] = 1.5×10^{19} cm^{-3}) and a 500 nm p-GaN contact layer with [Mg] = 2×10^{19} cm^{-3}. Such a heterostructure has been studied in [23, 24].

The electron and hole mobilities of 100 and 10 cm^{2}V^{-1}s^{-1} were typically chosen for every epitaxial layer. The mobility variation was found to provide a minor effect on the simulation results. The dislocation density $N_d = 10^8$ cm^{-2} was assumed in all the computations. This value is an order of magnitude lower than the typical TDD in LED heterostructures grown by metalorganic vapor-phase epitaxy (MOVPE) on mismatched substrates. The reduced N_d used in the simulations accounts, at least qualitatively, for the effect of indium composition fluctuations responsible for electron and hole capture in the In-rich regions away from threading dislocations, thus suppressing the nonradiative carrier recombination [15].

14.3.1
Band Diagrams, Carrier Concentrations, and Partial Currents

Figure 14.2 compares the simulation results on the above SQW and MQW LED heterostructures operating at nearly the same current density. It is seen that both heterostructures provide a good carrier confinement in the active region, i.e., there is no marked electron leakage into the *p*-region or the hole leakage into the *n*-contact layer (see the partial electron and hole current densities plotted in Fig. 14.2(a,b)). However, there is a certain asymmetry in the behavior of electrons and holes in the MQW structure. More mobile electrons, whose concentration in the *n*-GaN contact layer exceeds that of holes in the *p*-GaN contact layer, penetrate readily through all the QWs and stop only at the *p*-AlGaN EBL interface. In contrast, holes primarily fill the QW adjacent to the *p*-AlGaN EBL (Fig. 14.2(d)). As a result, the radiative recombination largely occurs in that QW (Fig. 14.2(f)), making other wells ineffective for light emission; actually less than ∼10% of all photons are emitted from the latter QWs. This effect is typical of MQW LED heterostructures, and special efforts are required to make the light emission more uniform.

Figure 14.2(d) illustrates that almost all electrons leave the barriers for QWs in the MQW structure. In the case of both structures A and B, the nonequilibrium carrier concentrations in the QWs are 1.5–2 orders of magnitude higher than in the contact layers and *p*-AlGaN EBLs, which is due to the well-known superinjection effect occurring in double-heterostructures (DHS) with a QW active region. A high carrier concentration favors the bimolecular radiative recombination, providing enhancement of the LED IQE, as compared to the case of a wide active region.

In addition to a strong nonradiative carrier recombination in the MQWs, we have predicted a noticeable recombination in the barriers and the *n*-GaN contact layer (Fig. 14.2(f)). However, the total recombination rate in these regions is much lower than in the QWs, despite the very different layer thicknesses.

The electric field distribution in structure A is shown in Fig. 14.3(a) for different values of the current density. One can see that the field inside the InGaN QW is negative, while outside the well it is positive. This is due to the fact that

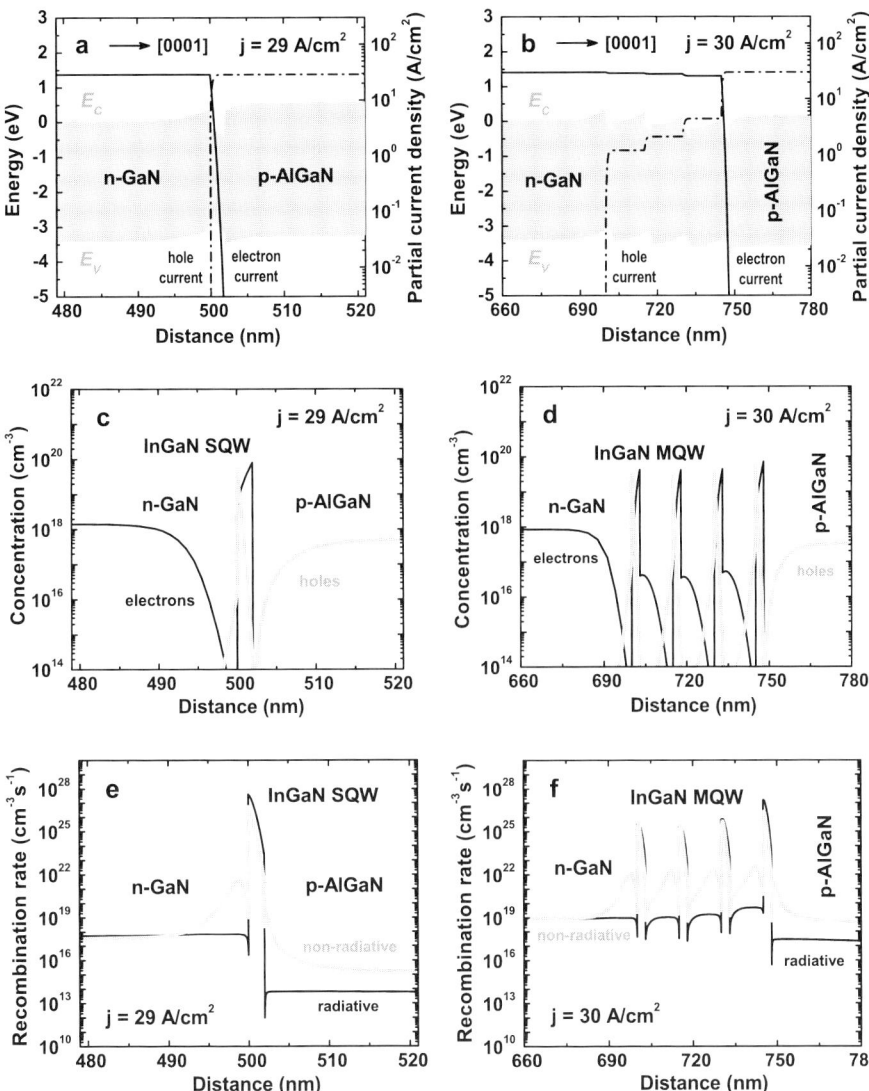

Fig. 14.2 Room temperature band diagrams and partial current densities (a,b), carrier concentrations (c,d), and recombination rates (e,f) in SQW and MQW blue LED structures. The bandgap is shown in (a,b) by gray shadow.

the p-n junction field in the diode is oppositely directed to the polarization field induced in the QW. The current density/bias variation changes the magnitude of the p-n junction field. As a result, the total electric field in the well also varies with current, as shown in Fig. 14.3(a). Generally, the field magni-

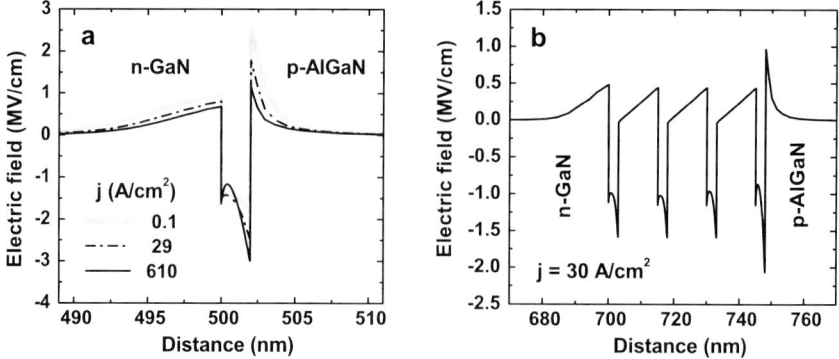

Fig. 14.3 Electric field distributions in SQW (a) and MQW (b) LED structures. Redistribution of the electric field upon the current density variation is shown in (a) for the SQW LED.

tude approaches the value of ~2–3 MV cm^{-1} in the In$_{0.2}$Ga$_{0.8}$N QW and of ~1–2 MV cm^{-1} in the cladding layers. The p-n junction field vanishes when a high bias is applied. A nonmonotonic electric field distribution inside the QW is due to its partial screening with nonequilibrium carriers.

Figure 14.3(b) presents the electric field distribution in structure B. The field magnitude, ~1–2 MV cm^{-1}, is here smaller than in structure A, which is due to a lower InN fraction in the In$_{0.13}$Ga$_{0.87}$N MQW. The field in the QW adjacent to the p-AlGaN EBL is higher than in the other wells. The reason for this is a greater polarization charge formed at the InGaN/AlGaN interface compared to the charges at the rest InGaN/GaN interfaces.

14.3.2
Internal Quantum Efficiency and Carrier Leakage

Of special interest are factors and physical mechanisms controlling the IQE of LED heterostructures. One of the factors is the nonradiative electron–hole recombination at threading dislocations affecting the efficiency at a TDD higher than ~10^6–10^7 cm^{-2} (see [14] and references therein). Owing to the optimization of MOVPE of III-nitride LED structures, the dislocation density N_d has been gradually reduced from ~10^9–10^{10} cm^{-2} to 2–5×10^8 cm^{-2}. This, however, is still insufficient to avoid the dislocation impact on the emission efficiency (Fig. 14.1(a)). In LEDs utilizing an InGaN SQW or MQW active region, the dislocation effect is partly suppressed by the composition fluctuations in the ternary compound, localizing carriers away from the dislocations and, hence, enhancing the IQE [25]. Our previous results have shown that the effect of composition fluctuations may be roughly accounted for by using an effective TDD approximately an order of magnitude lower than its actual value [26].

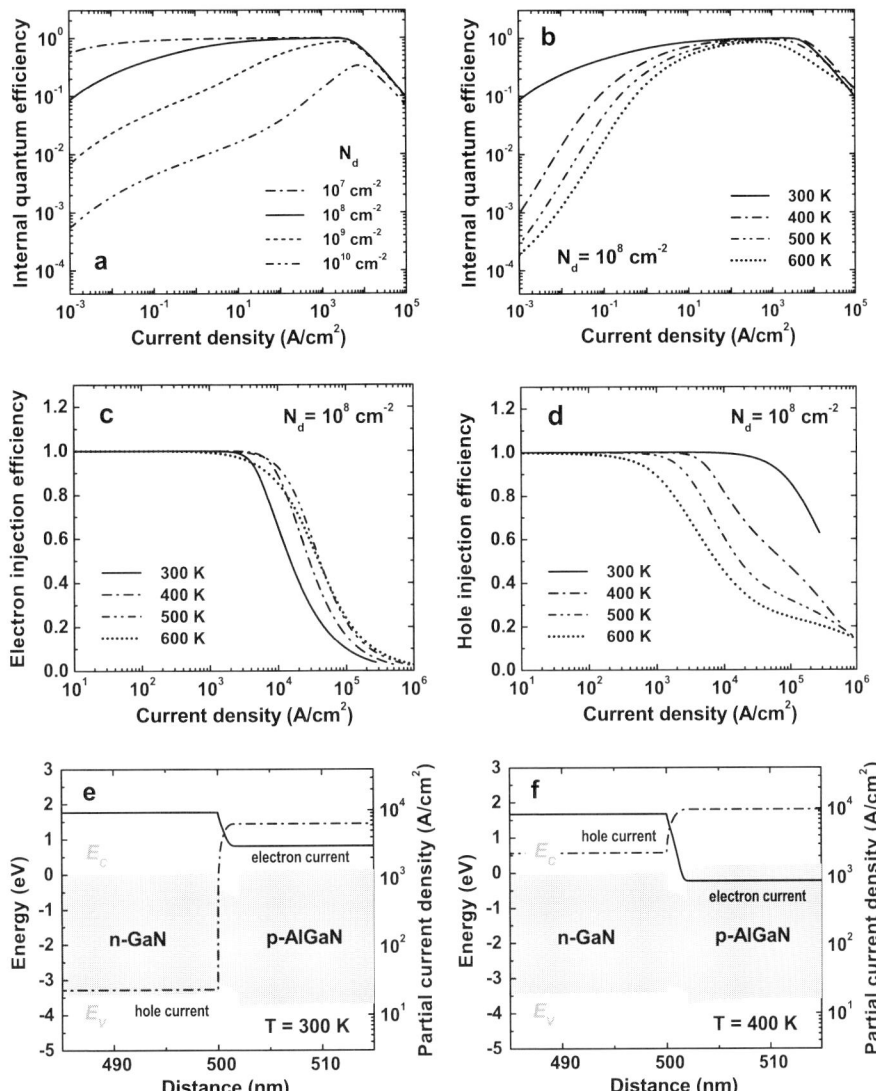

Fig. 14.4 IQE of structure A as a function of current density at different values of TDD (a) and operating temperature (b). The injection efficiency of electrons (c) and holes (d) in this structure at different temperatures. Band diagrams and partial electron/hole current densities at $j = 10$ kA cm^{-2} computed for 300 K (e) and 400 K (f).

Figure 14.4(a) plots the IQE of structure A vs current density for different values of the TDD. One can see that in the practically important range of the current density variation, 1–10^3 A cm^{-2}, the TDD dramatically affects the LED efficiency. Generally, the IQE first increases with current, then reaches a maximum (or exhibits a plateau), and finally falls. The rise in the predicted IQE, followed by its saturation, is due to the fact that the bimolecular radiative recombination rate increases with the injected carrier concentration faster than the nonradiative recombination rate, which is nearly linearly dependent on the concentration. The further IQE rollover occurring at a higher current density requires, however, a more detailed explanation.

Let us consider the electron (η_e) and hole (η_h) injection efficiency defined by the expressions: $\eta_e = \left(j_n - j_n^{leak}\right)/j_n$ and $\eta_h = \left(j_p - j_p^{leak}\right)/j_p$. Here, j_n and j_p are the total current densities of electrons and holes at the n- and p-electrodes, respectively, and j_n^{leak} and j_p^{leak} are the leakage current densities of the respective carriers at the p- and n-electrodes. Actually, the injection efficiencies defined in this way correspond to the fractions of carriers recombining inside the heterostructure, primarily in the active region of the diode. The other carriers become lost at the contact electrodes.

Figure 14.4(c,d) shows the electron and hole injection efficiencies as a function of the current density computed for different operating temperatures. One can see that the electron injection efficiency starts to decrease at a current density of $\sim$ 2–3 kA cm^{-2}, which is practically independent of temperature (Fig. 14.4(c)). A slight variation in η_e, nonmonotonic with temperature, is due to the band diagram transformation, as illustrated in Fig. 14.4(e,f). Indeed, the temperature rise from 300 to 400 K produces a wider space-charge region in the p-AlGaN EBL where a potential barrier is formed in addition to the band offset, reducing the electron leakage current. The further temperature rise leads to the opposite effect, a leakage increase, which is caused by enhancement of the electron spillover from the InGaN QW. The reduction in the hole injection efficiency starts, however, between 100 and 1000 A cm^{-2} at a temperature higher than $\sim$500 K (Fig. 14.4(d)), i.e., it can easily be observed under practical operation conditions. It follows from the comparison of the computed IQE and injection efficiencies that it is the carrier leakage which is responsible for the high-current IQE rollover.

The IQE temperature dependence of structure A is given in Fig. 14.4(b). A general trend clearly seen from the figure is the IQE reduction with temperature. This is in agreement with numerous observations of the LED efficiency reduction caused by the device self-heating.

Qualitatively, IQE of the MQW LED (structure B) behaves in a very similar way to that of the SQW LED. We found that an MQW active region provided a slightly higher hole injection efficiency, but a slightly lower electron injection efficiency. As a result, both structure A and structure B have a comparable IQE. Among other factors affecting the IQE of MQW structures is the doping of the barriers separating the individual QWs [27].

14.3.3
Emission Spectra

Normally, an InGaN SQW or MQW serves as the active region of a visible III-nitride LED. The thickness of the InGaN layers should be small enough to avoid stress relaxation accompanied by generation of numerous misfit dislocations. This means, however, that a huge electric field, $\sim$1–3 MV cm^{-1}, induced by the polarization charges at the QW interfaces is typically formed in the device heterostructure. Therefore, a built-in electric field is one of the critical factors controlling the LED emission spectra.

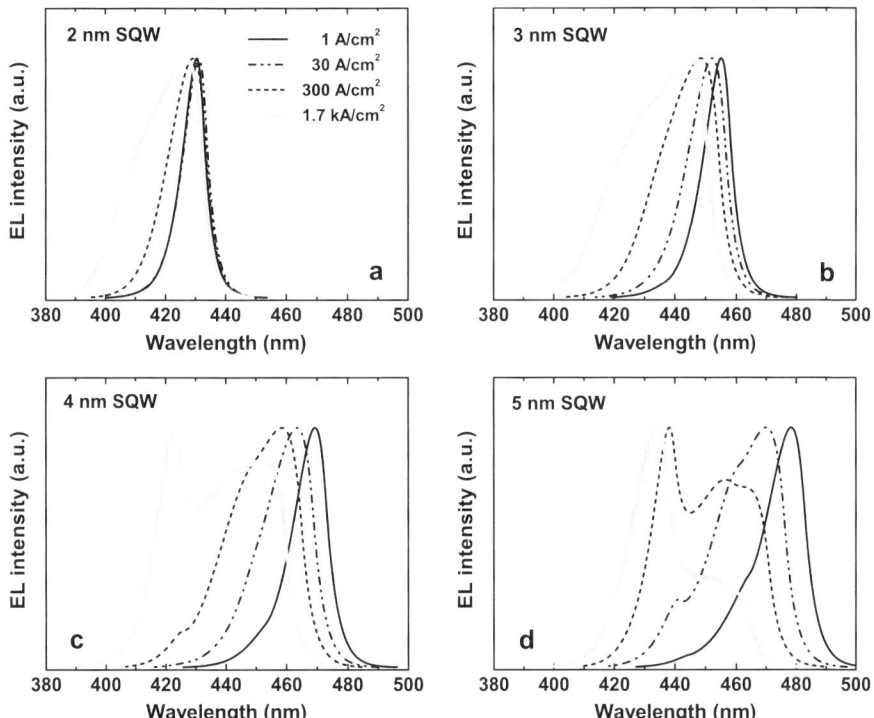

Fig. 14.5 Normalized emission spectra of GaN/In$_{0.2}$Ga$_{0.8}$N/AlGaN SQW LEDs with different quantum well thicknesses. In each plot, the spectra shown correspond to the same set of current densities indicated in (a). The uniform broadening parameter $\Gamma = 20$ meV was used in the spectra computations.

The role of the polarization field is clearly seen from Fig. 14.5. Here, the normalized emission spectra from GaN/InGaN/AlGaN SQW LEDs with various QW widths are plotted for different current densities. The spectra from a 3 nm InGaN SQW peak at about 450 nm with a distinct blue shift of the peak wavelength with current (Fig. 14.5(b)). The spectral broadening at high currents is caused by the contribution of the excited electron state in the SQW. The emission spectrum from a 2 nm SQW LED is shifted to ~430 nm (Fig. 14.5(a)) because of the pushing up of the electronic levels in the QW. The spectral broadening with current is here much less pronounced due to a larger separation of the electronic states in a narrow well. In turn, a huge red shift of the emission spectra is predicted for 4 nm and 5 nm QWs (Fig. 14.5(c,d)). This is just the manifestation of the quantum-confined Stark effect (QCSE) induced by the built-in polarization field. In "thick" QWs, the contribution of the excited electron states to the emission spectra becomes essential, leading first to the broadening and then to the splitting of the spectra with current.

To understand better the current-induced spectral transformation, we consider the SQW structure with a 5 nm InGaN quantum well in more detail. The band diagrams at different current densities are shown in Fig. 14.6. Additionally, the figure gives the wavefunctions of electron (marked by symbol "e") and hole states providing the major contribution to the emission spectra. The pedestal of every wavefunction corresponds here to the energy level of the respective state. Because of the chosen band parameters, the energy levels and wave functions of the heavy (marked as "hh") and light (marked as "lh") holes are indistinguishable with a graphical accuracy.

It is seen from Fig. 14.6 that the increasing current density results in a partial screening of the electric field in the QW, but even at $j = 1.7$ kA cm^{-2} it is still insufficient to avoid spacial separation of the ground-state electron and hole wavefunctions. At the current density of 1 A cm^{-2}, the emission spectrum is primarily due to the $e_0 \to hh_0$ transitions with a minor contribution of the excited hole states. The $e_0 \to hh_0$ and $e_0 \to hh_1$ transitions dominate at $j = 30$ A cm^{-2}. But if the current density is increased to 300 A cm^{-2} and higher, the $e_1 \to hh_0$ and $e_1 \to hh_1$ transitions, involving the excited electron state e_1, provide the major rise to the emission. This occurs because of both the filling of the excited state with electrons at a high nonequilibrium carrier concentration in the well and the good overlap of the e_1 and hh_0 wavefunctions. The excitation of the $e_1 \to hh_0$ transitions gives rise to an extra peak at ~420–430 nm, as shown in Fig. 14.5. Finally, the $e_0 \to sh_0$ transitions involving the ground-state split-off holes (marked as "sh") becomes noticeable at $j = 1.7$ kA cm^{-2}.

The spectra presented in Fig. 14.5 were computed on the assumption of a uniform broadening parameter $\Gamma = 20$ meV typical of conventional III-V compounds. The analysis of the long-wavelength tails in the experimental spectra of blue and green LEDs reported in literature shows that the broadening parameter Γ may be as large as ~50–70 meV. The computations, using the latter value, predict broad featureless spectra instead of split ones, like those plotted in Fig. 14.5(c,d).

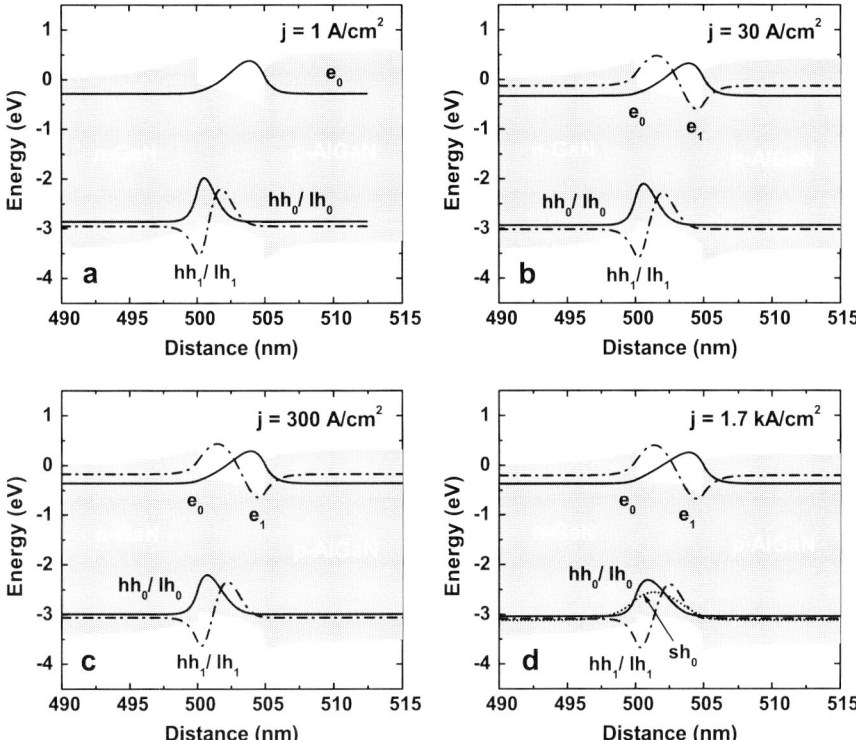

Fig. 14.6 Band diagrams of the SQW LED with a 5 nm QW at different current densities. Each plot shows the wavefunctions of electrons and holes providing the major contribution to the emission spectra. The pedestals of every wave function corresponds to the energy level of the respective state.

14.3.4
Polarity Effects

The crystal polarity largely determines the charge distribution in a III-nitride LED heterostructure. Therefore, the same structure exhibits quite different behavior, being grown with either Ga- or N-polarity [26]. This means that the growth direction corresponds to the positive direction of the hexagonal axis[1] of a wurtzite crystal in the former case and is opposite to this direction in the latter case. We consider here two aspects of the polarity impact on the LED operation: (i) its effect on the carrier confinement in an SQW active region and (ii) its influence on the emission line stability with current.

1) Conventionally, the positive direction of the hexagonal [0001] axis in a III-nitride crystal corresponds to the direction from a group-III metal atom to the nearest nitrogen atom bound to the metal atom covalently along the hexagonal axis.

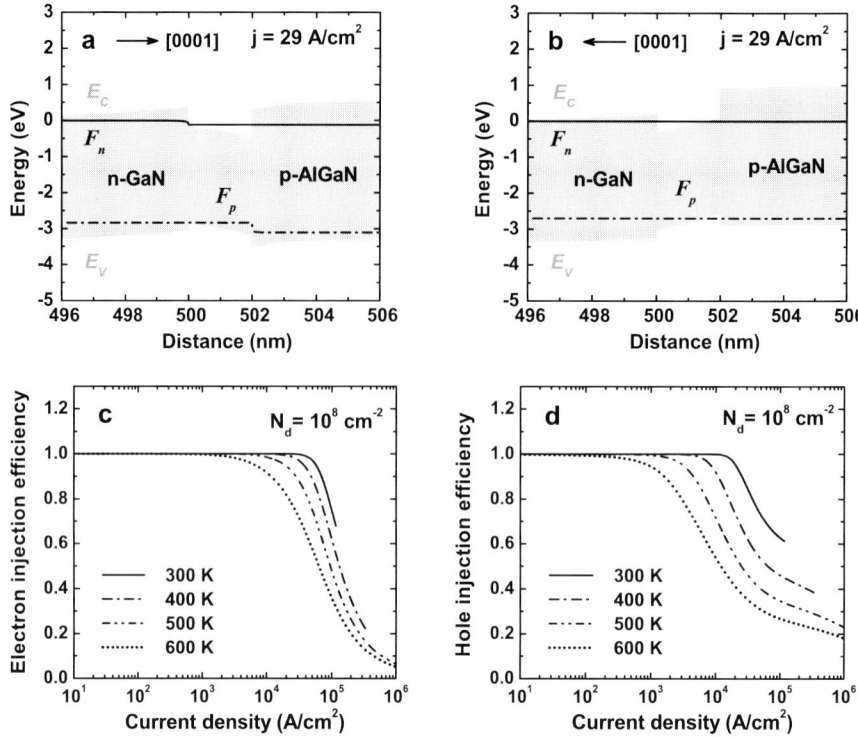

Fig. 14.7 Band diagrams of Ga-polar (a) and N-polar (b) SQW LED. Electron (c) and hole (d) injection efficiency of the N-polar LED as a function of the current density at different temperatures. The electron and hole quasi-Fermi levels are denoted as F_n and F_p, respectively.

Figure 14.7(a,b) compares the band diagrams of structure A with Ga- and N-polarity. The N-polar structure differs from the Ga-polar one primarily in higher potential barriers formed to electrons and holes by the p-AlGaN stopper layer and the n-GaN contact layer, respectively. These barriers favor the electron/hole confinement in the SQW active region, suppressing the carrier leakage from the heterostructure. To illustrate this conclusion, we have plotted the electron and hole injection efficiencies of the N-polar heterostructure in Fig. 14.7(c) and Fig. 14.7(d), respectively. One can see from the comparison of these plots with those given in Figs 14.4(c,d) for the Ga-polar heterostructure that the electron injection efficiency is systematically higher in the N-polar LED. The hole injection efficiency of the N-polar structure also exceeds that of the Ga-polar structure at operation temperatures as high as ~400–500 K. At 600 K, however, the difference between the electron/hole injection efficiencies of the Ga-polar and N-polar LEDs becomes negligible because of the enhanced thermal carrier ejection from the InGaN quantum well.

Let us discuss now the effect of the crystal polarity on light emission spectra. Generally, the electric field in SQW or MQW LEDs is controlled not only by the polarization charges induced at the QW interfaces but also by the *p-n* junction field in the depletion regions (see Section 14.3.1). The direction of the polarization field is generally determined by the crystal polarity, while the *p-n* junction field is always directed from the *n* to the *p* region. This results in a remarkably different behavior of the emission spectra from Ga-polar and N-polar LED structures.

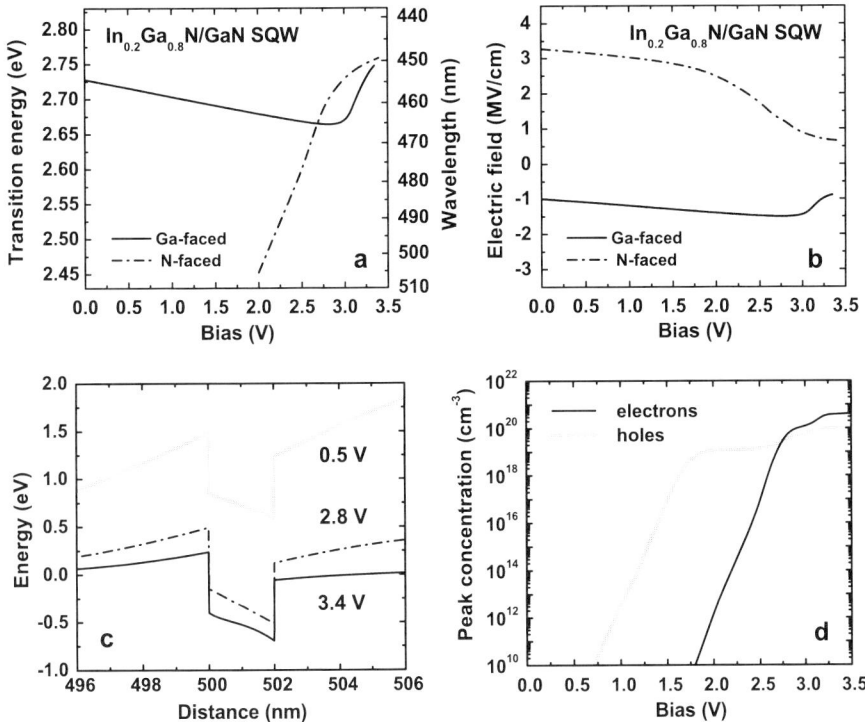

Fig. 14.8 Transition energy (a) and built-in electric field (b) in a GaN/InGaN/GaN SQW of either Ga- or N-polarity as a function of applied bias. The conduction band profile near the quantum well (c) and the peak carrier concentrations in the well as a function of bias (d).

Following [28], consider a simple *n*-GaN/InGaN/*p*-GaN SQW heterostructure with either Ga- or N-polarity. If the diode *n*-region is followed by the *p*-region in the [0001] direction (Ga-polarity), the built-in polarization field is opposite to the *p-n* junction field. For a typical $In_{0.2}Ga_{0.8}N$/GaN structure with the doping level of $1-5\times10^{18}$ cm^{-3}, the polarization field in a well exceeds the *p-n* junction field, so the total electric field in the well is directed from the *p*- to the *n*-region (Fig. 14.8(b)). A forward bias leads to a narrowing of the depletion regions in the barriers, reducing the *p-n* junction field. As a result, the total electric field first increases, producing a red shift of the emission wavelength. At a higher bias, however, the carriers injected into the well partly screen the electric field, causing a spectral blue shift (see Fig. 14.8(c) that compares the conduction band profiles near the InGaN QW region computed for different values of bias). Therefore, the emission wavelength is predicted to vary nonmonotonically with bias (Fig. 14.8(a)). The transition point corresponds well to a dramatic increase in the electron concentration in the quantum well (Fig. 14.8(d)). In contrast, the polarization field in the N-polar structure is always aligned with the *p-n* junction field, and the total electric field in the well monotonically decreases with bias, resulting in a considerable blue shift of the emission spectrum (Fig. 14.8(a)).

It is clear from the above consideration that Ga-polar structures exhibit a much better wavelength stability, since the polarization and *p-n* junction fields partly compensate for each other. In nonpolar heterostructures with the in-plane C-axis, it is only the *p-n* junction field that controls the QCSE in the QW active region. Hence, an intermediate wavelength stability can be expected for such structures.

14.4
Novel LED Structures

Simulations are especially helpful in the research and development of novel types of LED, as they enable a comprehensive insight into basic mechanisms involved and are capable to reveal some obscure features of the device operation. In this section, two special cases will be considered to illustrate the capabilities of the modeling approach.

14.4.1
LED with Indium-free Active Region

UV LEDs emitting light with a wavelength less than 360 nm are normally made from GaN and AlGaN materials having a relatively small lattice mismatch. This allows growth of a rather thick active region, in contrast to an InGaN one that has to be extremely thin to prevent strain relaxation via the mismatch dislocation formation.

Consider, for instance, the indium-free DHS LED structure suggested in [29]. It consists of an *n*-GaN active region ([Si] = 1.7×10^{18} cm^{-3}) 200 nm thick sandwiched between 300 nm *n*- and *p*-Al$_x$Ga$_{1-x}$N stopper layers (x = 0.21–0.24, [Si] = 1.7×10^{18} cm^{-3}, [Mg] = 1×10^{20} cm^{-3}), all grown on an *n*-GaN contact layer ([Si] = 6×10^{18} cm^{-3}) 4–6 μm thick. A 200 nm *p*-GaN contact layer ([Mg] = 1.5×10^{20} cm^{-3}) is deposited on the *p*-AlGaN EBL. The above structure with a thick active layer was grown by hydride vapor-phase epitaxy (HVPE), a technique which is much cheaper than conventional MOVPE. Due to a larger range of the growth rate variation, HVPE enables remarkably thicker GaN buffer layers, providing TDD control in the heterostructure.

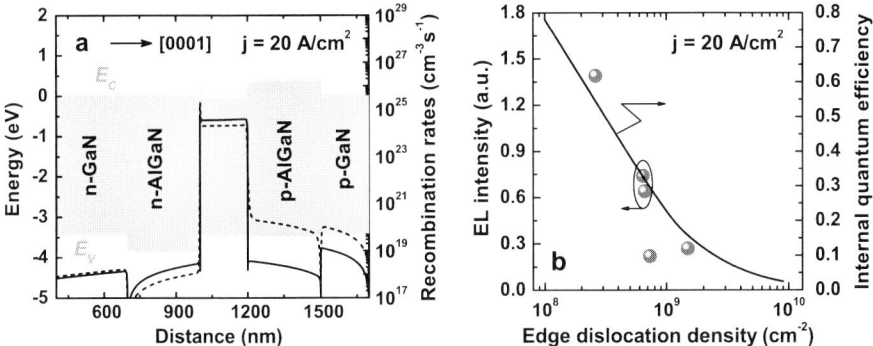

Fig. 14.9 Band diagram, radiative (solid line), and nonradiative (dotted line) recombination rates (a) and the correlation between the measured EL intensity and the computed IQE as a function of the dislocation density (b) for the DHS LED structure with a thick active region. Balls are experimental data borrowed from [29]; lines are computations.

Our simulations show that the carrier recombination in the InGaN DHS largely occurs in the n-GaN active layer due to a good carrier confinement provided by the AlGaN EBLs (see Fig. 14.9(a)). However, because of the absence of compositional fluctuations in the indium-free active region, the EL intensity in the DHS LED structure depends strongly on the TDD (Fig. 14.9(b)). This observation is in excellent agreement with the predicted IQE variation vs the dislocation density. The use of a thick HVPE-grown buffer layer has allowed reduction in the TDD in the LED structure and, hence, improvement of its efficiency.

As the DHS LED operates at a relatively low nonequilibrium carrier concentration (see discussion of this issue in Section 14.3.1), the reduction in its IQE caused by the electron leakage starts only at the current density as high as ~ 500 A cm^{-2}. At a lower current density, no IQE reduction is predicted by modeling, in close agreement with observation [29].

14.4.2
Hybrid ZnO/AlGaN LED

Another interesting example of a novel device is the hybrid ZnO/AlGaN single-heterostructure (SHS) LED reported in [30]. The interest in ZnO-based LEDs has been stimulated by a direct bandgap of 3.37 eV in this material and a high, ~60 meV, exciton binding energy, which are believed to open up the opportunity for the fabrication of highly efficient UV light emitters. However, the poorly reproducible and unstable p-doping of ZnO impedes the development of such devices. The use of hybrid n-ZnO/p-AlGaN heterostructures is one of the ways to overcome this problem.

The hybrid SHS LED contains a 0.8 μm p-$Al_{0.12}Ga_{0.88}N$ contact layer ($p = 5\times10^{17}$ cm^{-3}) grown on a 6H-SiC substrate followed by a 1 μm n-ZnO active/contact layer ($n = 7\times10^{17}$ cm^{-3}). As both ZnO and AlGaN layers are sufficiently thick, a complete stress relaxation was assumed in the simulation of the hybrid heterostructure.

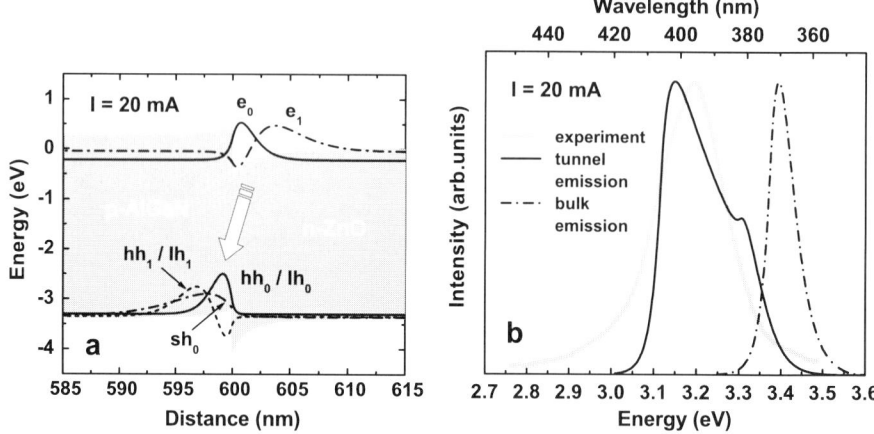

Fig. 14.10 Band diagram with electron/hole wavefunctions (a) and EL spectra (b) of the ZnO/AlGaN hybrid LED structure from [30](see text for detail).

Generally, a ZnO/AlGaN heterojunction exhibits a type-II band alignment [31]. Due to the band lineup and the charge at the ZnO/AlGaN interface induced by spontaneous polarization, the local potential wells adjacent to the interface are formed for electrons and holes in n-ZnO and p-AlGaN, respectively (Fig. 14.10(a)). The carrier concentration in these wells may be 1–2 orders of magnitude higher than in the semiconductor bulk. In this case, the tunnel radiative recombination of electrons and holes confined in the wells becomes possible, despite a small overlap of their wave functions [32]. Figure 14.10(b) compares the experimental emission spectrum of the hybrid SHS LED with those predicted for bulk ZnO (dash-dotted line) and for the tunnel recombination (solid line). It is seen that the computed tunnel emission spectrum fits the experimental peak position well. The larger width of the experimental spectrum may arise from the heterojunction nonuniformity, i.e., from local fluctuations of the composition, doping, and strain.

14.5 Conclusion

The state-of-the-art modeling of III-nitride LEDs has now reached a predictability level sufficient to explain the basic features of their operation and to identify a number of factors that control the device performance. Among these, the most important ones seem to be the crystal polarity and the related polarization charge distribution in an LED heterostructure, the threading dislocation density that determines the nonradiative carrier recombination, and the operation temperature that may affect the carrier leakage from the active region. The models accounting for these factors are capable of predicting general trends of III-nitride LED operation to form the basis for optimization of the heterostructure and chip design. Nevertheless, many aspects of the physical mechanisms and the approaches used to take them into account still remain open to question.

Our experience shows that standard drift-diffusion models of carrier transport similar to that used in this study, frequently overestimate the room-temperature turn-on voltage of LEDs with a high In concentration in the QWs (e.g., in green LEDs), especially in the case of MQW heterostructures. This is not related directly to the existence of polarization charges at the structure interfaces but is rather a specific feature of wide-bandgap materials exhibiting a negligible intrinsic carrier concentration. At least two mechanisms may be considered to improve the model – a tunneling current in the potential barriers surrounding the QWs and hot carrier transport. The former mechanism requires a nonlocal description of the electron/hole transport, while the latter needs consideration of the energy transfer coupled with the drift-diffusion equations.

Another problem is associated with the InGaN composition fluctuation impact on the nonradiative recombination rate, the emission line position, and the spectral broadening. Currently, there is no quantitative model capable of an adequate description of these effects. The difficulties of building up such a model arise from the fact that the composition fluctuations result not only in a microscopic modulation of the conduction and valence band edges, but are also accompanied by a variation in the QW width and strain, inducing local piezoelectric fields. Besides, dislocations threading through the active region may additionally generate regions with an enriched/depleted composition. In this situation, a phenomenological approach may be more effective than a rigorous microscopic model accounting for the above factors.

These and other issues should be addressed in future work aimed at further improvement of the simulation predictability, reliability and effectiveness.

Acknowledgments

Many thanks to my colleagues, K. A. Bulashevich, V. F. Mymrin, I. Yu. Evstratov, and A. I. Zhmakin, who contributed much to the development of physical models and numerical approaches to LED simulations. I am also grateful to J. Piprek for a critical reading of the manuscript and valuable suggestions on its improvement.

References

1 Nakamura, S., *J. Cryst. Growth* 145 (**1994**), p. 911.

2 Nakamura, S., Mukai, T., Senoh, M., *Appl. Phys. Lett.* 64 (**1994**), p. 1687.

3 Vurgaftman, I., Meyer, J.R., Ram-Mohan, L.R., *J. Appl. Phys.* 89 (**2001**), p. 5815.

4 Davydov, V.Yu., Klochikhin, A.A., Emtsev, V.V., Ivanov, S.V., Vekshin, V.A., Bechstedt, F., Furthmüller, J., Harima, H., Mudryi, A. V., Hashimoto, A., Yamamoto, A., Aderhold, J., Graul, J., Haller, E. E., *phys. stat. sol. (b)* 230 (**2002**) p. R4.

5 Wu, J., Walukiewicz, W., Yu, K.M., Ager III, J.W., Haller, E.E., Lu, H., Schaff, W.J., *Appl. Phys. Lett.* 80 (**2002**), p. 4741.

6 Amano, H., Kito, M., Hiramatsu, K., Akasaki, I., *Jpn. J. Appl. Phys.* 28 (**1989**), p. L2112.

7 Nakamura, S., Mukai, T., Senoh, M., Iwasa, N., *Jpn. J. Appl. Phys.* 31, (**1992**), p. L139.

8 Bernardini, F., Fiorentini, V., Vanderbilt, D., *Phys. Rev.* B56 (**1997**), p. R10024.

9 Bernardini, F., Fiorentini, V., *Phys. Rev.* B64 (**2001**), p. 085207.

10 http://www.semitech.us/products/

11 Ambacher, O., *J. Phys. D - Applied Physics* 31 (**1998**), p. 2653.

12 Nye, J. F. *Physical Properties of Crystals. The Representation by Tensors and Martices*, Oxford at the Clarendon Press (**1964**).

13 Fiorentini, V., Bernardini, F., and Ambacher, O., *Appl. Phys. Lett.* 80 (**2002**) p. 1204.

14 Karpov, S.Yu., Makarov, Yu. N., *Appl. Phys. Lett.* 81 (**2002**), p. 4721.

15 Bulashevich, K.A., Mymrin, V.F., Karpov, S.Yu., Zhmakin, I.A., Zhmakin, A.I., *J. Comput. Phys.* 213 (**2006**), p. 214.

16 Chuang, S.L., Chang, C.S., *Phys. Rev.* B54 (**1996**), p. 2491.

17 Yu, T.-H. and Brennan, K.F., *J. Appl. Phys.* 89 (**2001**) p. 3827.

18 Özgür, Ü., Alivov, Ya.I., Liu, C., Teke, A., Reshchikov, M.A., Dogan, S., Avrutin, V., Cho, S.-J., and Morkoc, H., *J. Appl. Phys.* 98, (**2005**) p. 041301.

19 Osinsky, A., Dong, J.W., Kauser, M.Z., Hertog, B., Dabiran, A.M., Plaut, C., Chow, P.P., Pearton, S.J., Dong, J.W., and Palmström, C.J. in *Electrochem. Soc. Proceedings* 2004-6, (**2004**) p. 70.

20 O. Madelung, editor, *Numerical Data and Functional Relationships in Science and Technology. Group III: Crystal and Solid State Physics* Vol.17, Subvolume b, Springer Verlag, New York (**1982**).

21 Nakamura, S., *Mat. Res. Soc. Symp. Proc.* 395 (**1996**), p. 879.

22 Kaufmann, U., Kunzer, M., Köhler, K., Obloh, H., Pletschen, W., Schlotter, P., Wagner, J., Ellens, A., Rossner, W., Kobusch, M., *phys. stat. sol. (a)* 192 (**2002**), p. 246.

23 Kim, A.Y., Götz, W., Steigerwald, D.A., Wierer, J.J., Gardner, N.F., Sun, J., Stockman, S.A., Martin, P.S., Kramers, M.R., Kern, R.S., Steranka, F.M., *phys. stat. sol. (a)* 188 (**2001**), p. 15.

24 Mamakin, S.S., Yunovich, A.E., Wattana, A.B., Manyakhin, F.I., *Semiconductors* 37 (**2003**), p. 1107.

25 Nakamura, S., *Semicond. Sci. Technol.* 14 (**1999**), p. R27.

26 Karpov, S.Yu., Bulashevich, K.A., Zhmakin, I.A., Nestoklon, M.O., Mymrin, V.F., Makarov, Yu.N., *phys. stat. sol. (b)* 241 (**2004**), p. 2668.

27 Mymrin, V.F., Bulashevich, K.A., Podolskaya, N.I., Zhmakin, I.A., Karpov, S.Yu., Makarov, Yu.N., *phys. stat. sol. (c)* 2 (**2005**), p. 2928.

28 Bulashevich, K.A., Karpov, S.Yu., Suris, R.A., *phys. stat. sol. (b)* 243 (**2006**), p. 1625.

29 Usikov, A.S., Tsvetkov, D.V., Mastro, M.A., Pechnikov, A.I., Soukhoveev, V.A., Shapovalova, Y. V., Kovalenkov, O.V., Gainer, G.H., Karpov, S.Yu., Dmitriev, V.A., O'Meara, B.S., Gurevich, S.A., Arakcheeva, E.M., Zakhgeim, A.L. Helava, H., *phys. stat. sol. (c)* 0 (**2005**), p. 2265.

30 Alivov, Ya.I., Kalinina, E.V., Cherenkov, A.E., Look, D.C., Ataev, B.M., Omaev, A.K., Chukichev, M.V., Bagnall, D.M., *Appl. Phys. Lett.* 83 (**2003**), p. 4719.

31 Hong, S.-K., Hanada, T., Makino, H., Chen, Y., Ko, H.-J., Yao, T., Tanaka, A., Sasaki, H., Sato, S., *Appl. Phys. Lett.* 78 (**2001**), p. 3349.

32 Bulashevich, K.A., Evstratov, I.Yu., Nabokov, V.N., Karpov, S.Yu., *Appl. Phys. Lett* 87 (**2005**), p. 243502.

15
Simulation of LEDs with Phosphorescent Media for the Generation of White Light

Norbert Linder, Dominik Eisert, Frank Jermann, and Dirk Berben

15.1
Introduction

The development of white-light-emitting diodes has been the latest step towards a revolution in lighting technology. Because of their advantages in size, versatility, energy consumption, and reliability they have conquered a wide range of applications ranging from LCD backlighting in mobile phones and other hand-held devices, to dashboard lighting and even the first prototypes of headlights in automotive use. Huge efforts are made worldwide to increase the efficiency and the output power of these devices and it is believed that they will ultimately replace conventional light sources in many general lighting applications as soon as the device costs drop below a critical level.

Most of the "white" LEDs today consist of a blue-emitting InGaN LED chip and a conversion phosphor with emission in the yellow-green spectral range. By a suitable choice of the LED wavelength and the material composition of the phosphor, various color impressions between "cold" bluish and "warm" yellowish whites can be achieved. This device concept allows for highly efficient, inexpensive white-light generation, but its color-rendering properties, i.e., the color perception from illuminated objects, can be poor because of missing parts in the visible spectrum, in particular in the red wavelength range. Mixtures of phosphors with different emission spectra allow us to improve this situation and offer a way to other nonsaturated colors in LEDs ("Color on demand"), but usually at the expense of device efficiency. Nevertheless, single-chip phosphor conversion LEDs have been so successful that RGB concepts with three different LED colors are used only where color control is necessary. More recently, even conversion from UV emitters to spectrally pure colors have been investigated in order to fill, e.g., the green gap around 550 nm, where no efficient semiconductor light sources exist.

Converter materials usually consist of a few-micrometer-diameter particles immersed in an epoxy resin or silicone matrix. Designing a conversion LED can be a highly sophisticated problem because the scattering of exciting and converted photons, the absorption and reemission properties of the phosphor and the interaction between chip, phosphor and package have to be taken into account. Since all the constituents are in close neighborhood and the device geometries are complex, simple theoretical approaches are of limited usefulness and numerical methods involving ray-tracing algorithms and advanced material models have to be used to optimize the device properties. In these models the LED chip enters through its optical properties only, i.e., its emission, absorption and reflection characteristics, which can be determined by simulation or measurement. In this chapter emphasis will be given to the phosphor and the full device model, rather than a detailed description of the LED chip.

15.2
Requirements for a Conversion LED Model

A cross-sectional view of a typical geometry for a conversion LED is shown in Fig. 15.1. The package is usually filled with converter material immersed in a silicone matrix. For most of the white LEDs Cerium-doped Yttrium Aluminum Garnet (YAG) is used as the luminescence converter. The LED spectrum is then composed of the blue emission line of the LED die, usually having a peak wavelength between 450 and 460 nm, and the broad yellow-green band of the YAG:Ce converter (cf. Fig. 15.8). The color impression is determined by the balance between the two spectral components. Although a wide range of variation between blue and yellow is perceived as white, the human eye is quite sensitive to hues within this range [1]. Therefore, if multiple LEDs are used for area lighting in, e.g., backlights of LCD panels or automobile dashboards, precise color control is necessary to avoid inhomogeneities of the color impression.

Color control is usually achieved by choosing an appropriate concentration of phosphor in the resin. For many applications, however, it is not sufficient to adjust the overall spectral emission of the device only. In particular, if imaging optics is involved, the angular emission pattern of the LED has to be considered as well. Generally, light emitted in the vertical direction has a larger blue component than light emitted under oblique angles, because the path through the converter is shorter and less primary emission is absorbed. These color variations are visible in many devices with simple design. Advanced devices use layers of highly concentrated converter on top of the die, which are able to reduce, but not completely suppress, the color gradients.

Since phosphor particle sizes are in the order of microns, both scattering and conversion properties have to be taken into account. Simple models for scattering media like, e.g., the Kubelka–Munk theory [2], which is essentially a one-dimensional description, fail to work, because device geometries are

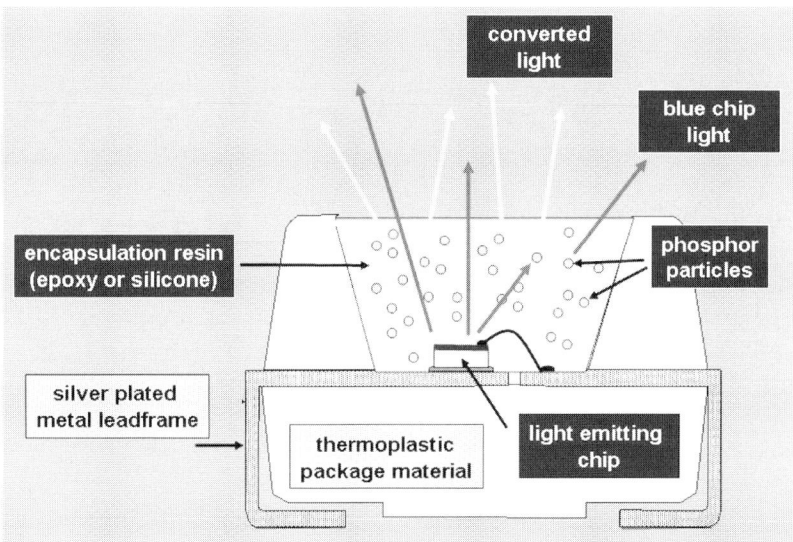

Fig. 15.1 Schematic setup of a luminescence conversion LED.

complex and the free path lengths of the propagating light are in the order of the device dimensions. Moreover, the interaction of primary and secondary emission with the reflector, the chip and the remaining parts of the package are also important for the device characteristics. If it is desired, for example, to homogenize the color gradients in the angular emission pattern of an LED, the use of fine rather than coarse grain converter may appear favorable. The scattering from small particles is more isotropic and leads to a uniform angular distribution of blue and yellow rays. On the other hand, more light is scattered back onto the chip and the cavity parts of the package, where it is partly lost by absorption. Also, scattering and conversion efficiencies of the phosphor are below 100%, so that optimizing a device for color homogeneity may be possible but with a significant penalty in overall efficiency.

Therefore, an optical model that allows us to sufficiently describe the properties of real LEDs has to be based on a precise description of the converter and it has to take into account the optical properties of the package as well as those of the LED die. Moreover, it requires spectral resolution and the capability of performing color calculations that allow to derive *physiological* quantities from the results of the *physical* calculations.

15.3
Color Metrics for Conversion LEDs

A model for human color perception based on physiological experiments was standardized by the CIE (Commission Internationale de l'Éclairage) in 1931. Although there have been extensions and refinements in recent years (see, e.g., in [3]) it is still the basis of most color calculations. It resorts to the fact that color vision is based on the excitation of certain cells in the retina of the human eye, the cones. There are three types of cones that differ in spectral sensitivity. Therefore, the color impression of every spectral power distribution $S(\lambda)$ can be represented by three signals X, Y, and Z, the tristimulus values, via [1]

$$X = k \int_{380\,\text{nm}}^{780\,\text{nm}} S(\lambda) \bar{x}(\lambda)\, d\lambda \tag{15.1a}$$

$$Y = k \int_{380\,\text{nm}}^{780\,\text{nm}} S(\lambda) \bar{y}(\lambda)\, d\lambda \tag{15.1b}$$

$$Z = k \int_{380\,\text{nm}}^{780\,\text{nm}} S(\lambda) \bar{z}(\lambda)\, d\lambda \tag{15.1c}$$

where $\bar{x}(\lambda)$, $\bar{y}(\lambda)$, and $\bar{z}(\lambda)$ are the color matching functions of the CIE 1931 Standard Colorimetric Observer (Fig. 15.2) [1]. These functions are related but not identical to the spectral sensitivity of the cones. $k = 683$ lm W^{-1} is the maximum luminous efficacy of the human eye and determines the brightness impression of $S(\lambda)$. If only color is considered, the tristimulus values can be represented by their normalized values

$$c_x = \frac{X}{X+Y+Z}, \quad c_y = \frac{Y}{X+Y+Z} \tag{15.2}$$

in the two-dimensional chromaticity diagram of Fig. 15.3, where the fact is used that $c_z = Z/(X+Y+Z) = 1 - c_x - c_y$ is redundant.

In the chromaticity diagram, spectrally pure, i.e., monochromatic, colors are located at the perimeter of the CIE chromaticity space, while equi-energetic "white" radiation is represented by locus E in the center of the diagram.

If two color stimuli are given, i.e., the blue radiation from the LED die and the yellow-green emission from the phosphor, the locus of the resulting color impression can be calculated by additive mixture of the two components. It can be seen from elementary calculations that it is located on a straight line between the loci of the two stimuli. Similarly, if more than two stimuli are used, the locus of the resulting color signal is always located on the surface of the polygon, whose corner points are defined by the multiple stimuli.

If one wants to determine the physiological "brightness" impression of an LED, the spectral sensitivity $V(\lambda)$ of the human eye has to be taken into account. For photopic (daylight) vision the color matching function $\bar{y}(\lambda)$ of the CIE Standard Observer has been chosen such that it is identical with this function. Hence, Eq. (15.1b) allows us to calculate, e.g., the luminous flux of a light source if its spectral power distribution is known. Since the sensitivity $V(\lambda)$ of the human eye is weak in the blue spectral range (cf. Fig. 15.2), it is obvious that the luminous flux of a white LED is almost completely determined by the luminescent power of the phosphor, although the optical power of the primary emission from the die is in the same range. Therefore, the effective conversion efficiency of the device, which is determined not only by the quantum efficiency of the phosphor, but also by absorption, scattering and reflection within the cavity, is essential for the performance of the LED.

The color locus in the chromaticity diagram determines the color perception of a light source if directly viewed by the eye or under the illumination of an ideal white surface. For many applications, e.g., street and area lighting or monochromatic backlights, this is sufficient and attention can be paid to maximizing its efficiency. If, however, the light source is used to illuminate colored surfaces, its color-rendering properties are important. Colors are generally perceived as most natural if illuminated by solar radiation, which is close to blackbody radiation, and it is obvious from the spectra in Fig. 15.8 that the color-rendering quality of standard high-luminance white LEDs is poor, in particular in the red wavelength range. Here the addition of a red-emitting phosphor in a "warm-white" LED helps to improve the situation. The efficiency of such devices, however, drops because of reabsorption of the yellow-green emission and increased light scattering.

Generally, each phosphor adds another degree of freedom to the device design and optimization is only possible if the properties of the phosphors are precisely known and well represented in the simulation models.

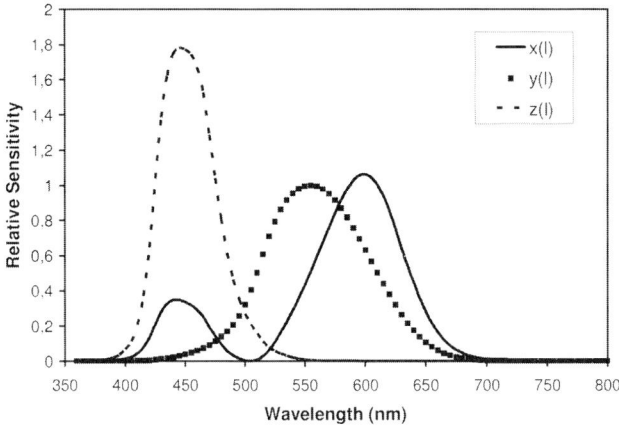

Fig. 15.2 $\bar{x}(\lambda)$, $\bar{y}(\lambda)$, and $\bar{z}(\lambda)$ color matching functions of the CIE 1931 Standard Colorimetric Observer.

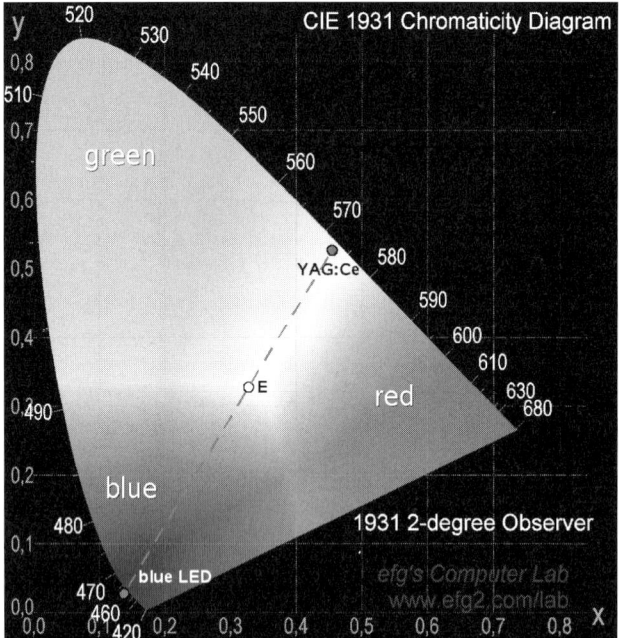

Fig. 15.3 CIE 1931 chromaticity diagram. E is the equi-energetic excitation point. Also included are the color loci of a blue LED, of YAG:Ce phosphor and the loci of conversion LEDs. Full color plots can be found at http://www.efg2.com/Lab/Graphics/Colors/Chromaticity.htm

15.4 Phosphor Model

15.4.1 Phosphor Materials

The first white phosphor conversion LEDs were based on a cerium activated, yellow-emitting garnet phosphor powder $(Y_{1-a}Gd_a)_3(Al_{1-b}Ga_b)_5O_{12}:Ce^{3+}$ (YAG:Ce, typically $a, b \leq 0.2$). Yttrium aluminum garnet phosphors had been known before from other applications. For LED manufacturers they were a Godsend, since they "accidentally" fulfilled all the requirements of the, at that time, new application.

The basic requirements of a conversion phosphor for single-chip white LEDs based on blue InGaN dies are a high quantum efficiency, a high absorption at blue LED wavelengths (typically 450–470 nm), and an emission spectrum which is well matched to the desired color loci. All these items describe optical phosphor properties which have to be considered in a simulation model. Besides optical, there are also nonoptical requirements like

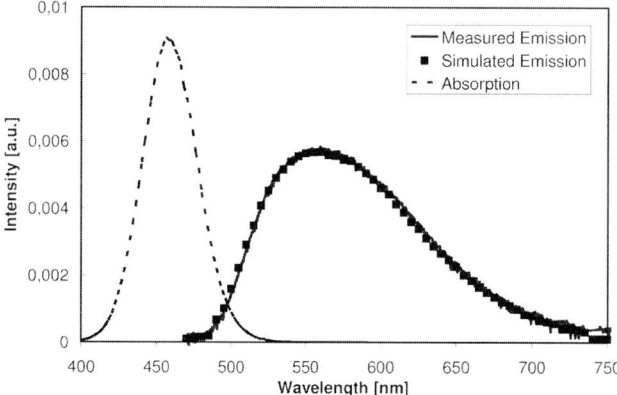

Fig. 15.4 Absorption and luminescence Spectrum of YAG:Ce phosphor. Dots: simulation.

chemical and thermal stability, toxicity and production costs, which further restrict the choice of a suitable LED phosphor.

Despite significant research on new phosphor systems, yellow garnet phosphors are still the most important phosphors today for white phosphor conversion LEDs [4]. They have quantum efficiencies mostly well above 90% and a very strong absorption at blue LED wavelengths due to an allowed 4f→5d transition of the Ce^{3+} activator ion [5]. Garnet phosphors typically show a broad yellow-green emission spectrum (Fig. 15.4) with a dominant wavelength of about 570 ± 7 nm and outstanding chemical stability.

A commercially used alternative to garnet-type phosphors are yellow-emitting orthosilicate-type compounds like $(Sr_{1-a}Ba_a)_2SiO_4:Eu^{2+}$ [6]. However, quantum efficiency and optical absorption at typical blue LED wavelengths cannot usually compete with garnet phosphors [7]. Chemical stability problems and a strong drop in quantum efficiency at temperatures above 100 °C (thermal quenching) are further drawbacks of these phosphors.

Nevertheless, future requirements of solid state lighting on color rendering and color temperature of white LEDs cannot be met using garnet phosphors solely. Thus, many companies and institutes are searching for novel blue-excitable phosphors, especially green and red-emitting ones. The first phosphors proposed were sulfides like green-emitting thiogallates, for example $SrGa_2S_4:Eu^{2+}$, and red-emitting $(Sr,Ca)S:Eu^{2+}$ [8]. Despite showing a relatively low chemical stability against humidity and a pronounced thermal quenching, sulfide phosphors are used in warm white LEDs for lack of alternatives [9]. A novel class of phosphors coming up are novel nitride-based phosphors like the red-emitting nitridosilicates $(Ba,Sr,Ca)_2Si_5N_8:Eu^{2+}$ [10] and the blue-green to yellow-emitting oxynitrides $(Ba,Sr,Ca)Si_2O_2N_2:Eu^{2+}$ [11]. Red nitride phosphors are used in highly efficient and long-lived warm white LEDs. An overview of phosphor materials and their characteristic properties is given in Table 15.1.

Tab. 15.1 Phosphor materials used for white and colored conversion LEDs.

Phosphor name	Chemical formula	Emission colors
Yttrium Aluminum Garnets	$Y_3Al_{5-x}Ga_xO_{12}$:Ce ($x = 0$: YAG:Ce)	Yellow – yellow-green
Orthosilicates	$(Ba,Sr,Ca)_2SiO_4$: Eu^{2+}	Blue-green – yellow
CaMg-Chlorosilicate	$Ca_8Mg(SiO_4)_4Cl_2$:Eu^{2+}	Blue-green – green
Thiogallates	$(Sr,Ca,Mg)Ga_2S_4$:Eu	Green
Sulfides	$(Sr,Ca)S$:Eu^2	Red – deep-red
Oxynitrides	$(Ba,Sr,Ca)Si_2O_2N_2$:Eu^{2+}	Blue-green – yellow
Nitridosilicates	$(Ba,Sr,Ca)_2Si_5N_8$: Eu^{2+}	Orange-yellow – deep red

15.4.2
Luminescence and Absorption of Phosphor Particles

LED phosphors are typically activated by rare-earth elements. In the majority of cases the activators are Ce^{3+} or Eu^{2+}. The absorption and luminescence of these ions is determined by the energy levels of their 5d electrons. The position of these levels strongly depends on the host lattice structure and the resulting local coordination sphere of the activator site [12]. Due to inhomogeneities of the host lattice and coupling between the activator atom and phonon states, the transitions are strongly broadened. Moreover, a significant Stokes shift is obtained because there is a considerable difference in chemical bonding, and thus in lattice distortion, between the excited and the ground state of the Ce^{3+} ions (Franck–Condon principle).

The absorption peak occurs at 460 nm, which almost perfectly agrees with the observed wavelength of maximum efficiency for blue nitride-LEDs. The phosphor emission has a dominant wavelength (for a definition see, e.g., [1]) of about 570 nm, which gives access to a wide range of white hues in the chromaticity diagram (see Fig. 15.3). There is only a small overlap between the absorption and luminescence spectrum, so little losses are expected by reabsorption.

The external quantum efficiency η_q of the phosphor, which is defined by the fraction of re-emitted over absorbed photons, depends on the wavelength λ_{exc} of the exciting radiation. Thus, the so-called excitability of a given phosphor is defined by the absorption spectrum $K(\lambda)$ multiplied by $\eta_q(\lambda)$. As it is difficult to determine the absolute values of $K(\lambda)$ and $\eta_q(\lambda)$ for a phosphor powder, usually a normalized spectrum $K(\lambda) * \eta_q(\lambda)$ is measured which is called the excitation spectrum. The Stokes shift is then defined by the difference in average energy of the excitation and the emission spectrum $E_{ss} = \langle E_{exc} \rangle - \langle E_{em} \rangle$.

Typically, commercially available garnet phosphors have quantum efficiencies of about 95% when excited in the blue spectral region [9]. Thus, the main energy loss in the conversion process (about 20%) results from the Stokes shift E_{ss}.

15.4.3
Scattering of Phosphor Particles

For the propagation of light in a granular medium the knowledge of the scattering properties of this medium is essential. To treat the most general case, the Maxwell equations should be solved for the exact spatial configuration of particles in the granular medium. But even with the most advanced computers, this can be done only for a very small volume. Besides, in most cases relevant for LEDs the microscopic arrangement of particles is random and only macroscopic parameters like particle size and density distributions are known.

Therefore, in order to reduce the computational complexity, the scattering problem is broken up into two parts. The first part is to solve the single-particle scattering problem more or less rigorously in the wave optics regime. The second part is then to propagate the light incoherently, applying the scattering properties of the single particles. This approximation should deliver good results as long as the density of the scattering particles is not too high (so there is no optical interaction between the particles) or large-scale ordering effects occur.

So the first task is to calculate the scattering cross-sections of a particle. We need the extinction scattering cross-section σ_{ext}, which describes the amount of energy ΔE taken from an electromagnetic wave with intensity I_0 impinging on the particle

$$\sigma_{ext} = \frac{\Delta E}{I_0} \tag{15.3}$$

The unit of the cross-section is that of an area. Because the phosphor particles can absorb energy, the total scattering cross-section can be separated into the scattering cross-section σ_{scatt} for elastic energy conserving scattering, and the absorption cross-section σ_{abs} for light absorbed by the particle. The total cross-section is then the sum of both fractions

$$\sigma_{ext} = \sigma_{scatt} + \sigma_{abs} \tag{15.4}$$

It is convenient to introduce the dimensionless scattering efficiency parameter Q by dividing the scattering cross-section by the geometrical particle cross-section a

$$Q = \frac{\sigma}{a} \tag{15.5}$$

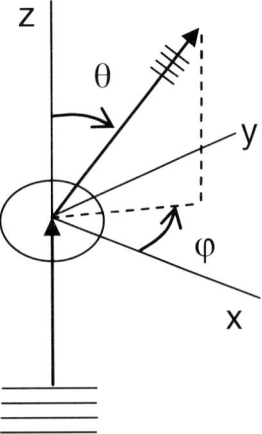

Fig. 15.5 Geometry definition of the scattering directions.

By elastically scattering from the particle, light is redistributed into the entire hemisphere. The scattering function

$$S(\theta, \varphi) = \frac{d\sigma}{d\Omega} \qquad (15.6)$$

describes the cross-section for scattering into a solid angle element $d\Omega$. The geometry definitions are shown in Fig. 15.5. The direction of propagation of the incident wave is in the z-direction. The directions of the incident light and the scattered light define the scattering plane, so that polarizations in the plane and perpendicular to that plane have to be distinguished. In our cases we consider only unpolarized light, so for the scattering function we can use an average over the two polarization directions. If the medium is isotropic, we can assume rotational symmetry about the z-axis so that the scattering amplitude is only dependent on the polar angle $S(\theta)$.

Even for a single particle an analytical solution of the electromagnetic scattering problem can be obtained only for the most simple cases. For dielectric spherical particles with diameters very small compared to the wavelength, the Rayleigh theory applies (for details, see, e.g., [13]).

Scattering properties scale with the ratio of particle size and wavelength, so it is convenient to express the scattering formula in terms of the dimensionless particle size parameter

$$\alpha = \frac{2\pi r}{\lambda} \qquad (15.7)$$

where r is the radius of the spherical particle wavelength, and λ the wavelength in the medium with refractive index m_0 in which the particle is embedded

$$\lambda = \frac{\lambda_0}{m_0} \tag{15.8}$$

The refractive index $m = n + ik$ of the particle itself can be complex to include absorption in the theory. Then the scattering coefficients in the Rayleigh limit are given by the following equations:

$$\sigma_{\text{scatt}} = \frac{2\lambda^2}{3\pi} \alpha^6 \left| \frac{m^2 - 1}{m^2 + 2} \right|^2 \tag{15.9a}$$

$$\sigma_{\text{abs}} = -\frac{\lambda^2}{\pi} \alpha^3 \, \text{Im} \left\{ \frac{m^2 - 1}{m^2 + 2} \right\} \tag{15.9b}$$

It is worth noting that the scattering cross-section scales with the sixth power of particle size, while the absorption cross-section scales only with the third power. Therefore, with increasing particle diameter, the scattered part will increase much more rapidly than the absorbed portion. In fluorescent media where absorption is needed to convert light, this has to be taken into account when selecting particle sizes. So particles with nanometer dimensions should be able to absorb and re-emit light without much additional scattering.

The scattering functions in the Rayleigh regime are given by

$$S_{\text{vert}} = \frac{3}{8\pi} \sigma_{\text{scatt}} \tag{15.10a}$$

$$S_{\text{hor}}(\theta) = \frac{3}{8\pi} \sigma_{\text{scatt}} \cdot \cos^2 \theta \tag{15.10b}$$

For vertical polarization with the field vector perpendicular to the scattering plane, the scattering function S_{vert} is isotropic. For horizontal polarization there cannot be radiation perpendicular to the direction of incidence, and S_{hor} becomes a function of the cosine of the polar angle.

Rayleigh scattering is only applicable if the electric field does not vary too much within the particle. Apparently, interference effects will set in if the particle dimension reaches that of the light wavelength. Then a rigorous solution of the Maxwell equations is needed. For a spherical particle this solution has already been given by Mie and Debye in terms of Legendre polynomial expansions.

$$S_{\text{vert}}(\theta) = \frac{\lambda^2}{4\pi^2} \left| \sum_{1}^{\infty} \frac{2n+1}{n(n+1)} [a_n \pi_n (\cos \theta) + b_n \tau_n (\cos \theta)] \right|^2 \tag{15.11a}$$

$$S_{\text{hor}}(\theta) = \frac{\lambda^2}{4\pi^2} \left| \sum_{1}^{\infty} \frac{2n+1}{n(n+1)} [a_n \tau_n (\cos \theta) + b_n \pi_n (\cos \theta)] \right|^2 \tag{15.11b}$$

where π_n and τ_n are functions of the n-th Legendre polynomial and its derivative, respectively. The coefficients a_n and b_n are composed from Ricatti–Bessel functions, which take the particle size parameter as argument. The details of Mie theory are described in the literature [14].

Mie theory is the exact solution to the Maxwell equations. Raleigh scattering is an approximate theory for small particles, contained in the first terms of the series expansion of Mie theory. With increasing particle size parameter, an increasing number of terms has to be summed up to achieve sufficient accuracy. This is why Mie theory requires high computational effort and generates results in tabulated numerical form.

Yet it is often desirable to write down the scattering function in a closed analytical form. This was even more important before powerful desktop computers became available. The astronomers Henyey and Greenstein [15] developed an empirical scattering function:

$$p_{H-G}(\theta) = \frac{1}{4\pi} \cdot \frac{1-g^2}{(1+g^2 - 2g\cos(\theta))^{3/2}} \tag{15.12}$$

As a probability distribution function it is normalized to 1:

$$\int_0^\pi p_{H-G}(\theta) \cdot 2\pi \sin(\theta) \, d\theta = 1 \tag{15.13}$$

The parameter g is chosen such that it is the asymmetry factor of the distribution, that is the expectation value of $\cos(\theta)$:

$$\int_0^\pi p_{H-G}(\theta) \cdot \cos(\theta) \cdot 2\pi \sin(\theta) \, d\theta = g \tag{15.14}$$

Therefore, it is quite straightforward to adjust the Henyey–Greenstein function to measured distributions by simply inserting the asymmetry factor. Due to its simple form the Henyey–Greenstein scattering function p_{H-G} is implemented in many optical simulation tools to describe particle scattering.

To illustrate the light propagation conditions that can be encountered in granular phosphorescent media, exemplary Mie scattering calculations are shown in Figs 15.6 and 15.7. In Fig. 15.6 the scattering efficiency of a dielectric particle with absorption is shown in dependence on the size parameter α. The scattering efficiency is a measure of how much the scattering cross-section is enhanced compared to the geometrical cross-section. For small size parameters, i.e., for example, nano-particles, the scattering efficiency approaches zero (if the assumption of a homogeneous refractive index still holds). In this Raleigh limit the efficiency for scattering increases with the fourth power of the size parameter, while the absorption increases linearly. So for nano particles, absorption dominates over scattering.

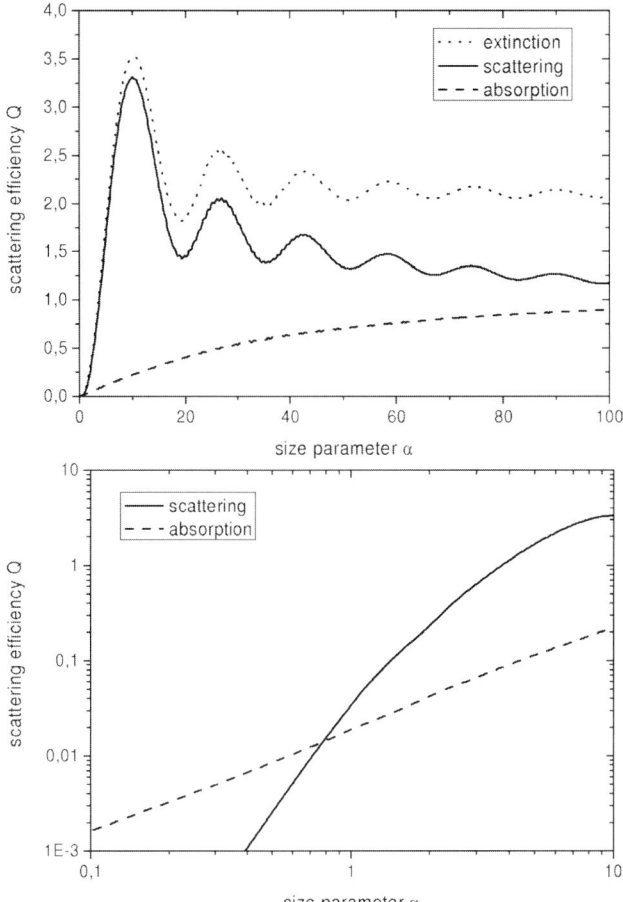

Fig. 15.6 Scattering efficiency of dielectric absorbing spherical particles as a function of the size parameter calculated by Mie theory. The Raleigh regime for small particles can be derived from the logarithmic plot on the right.

When the particle size comes into the range of the wavelength, standing waves can build up within the particle and scattering is strongly enhanced by interference effects. These scattering maxima repeat periodically, but decrease in peak height. In contrast, the absorption efficiency increases more or less monotonically. For large size parameters the extinction coefficient (the sum of scattering and absorption) approaches $Q_{\text{ext}} \rightarrow 2$, not 1 as might be expected from the geometrical cross-section. This fact is known as optical paradoxon.

Of similar importance for propagation is deflection of the scattered light. The scattering functions shown in Fig. 15.7(a) were averaged over the two polarization directions and normalized to unit scattering probability. Small particles scatter isotropically. The minimum at 90° occurs because the inten-

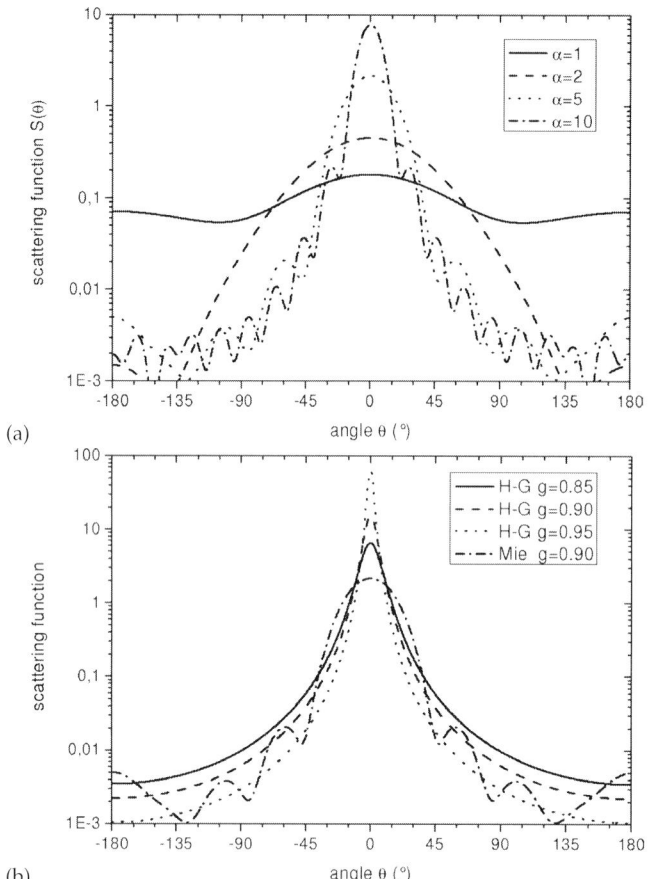

Fig. 15.7 (a) Averaged and normalized Mie scattering functions of spheres with increasing size parameter. The scattering changes from nearly isotropic for small particles, to pronounced forward scattering. (b) Henyey–Greenstein scattering functions for a few asymmetry parameters. For comparison, a Mie scattering function is overlaid.

sity of the in-plane polarization must vanish. With increasing size parameter the scattering function becomes "wriggly", but these features will average out for a particle size ensemble. Of more importance is the ratio of forward scattering to back scattered light – note the logarithmic scale in Fig. 15.7(a). This means that most of the scattered light will fall into a narrow cone about the axis of propagation. Therefore, incoming light keeps its general direction of propagation over many interaction lengths. The scattered light will not become diffused completely within the dimensions of an LED. In contrast the converted light is diffused due to the isotropic reemission process. This is the reason for different spatial and angular distributions of the wavelengths used for excitation and generated by conversion.

The asymmetry parameter g in the Henyey–Greenstein scattering function can be used to describe a wide range of scattering behavior from forward scattering ($g = +1$) over isotropic scattering ($g = 0$) to complete back scattering ($g = -1$). For micron-size particles, typical values for g are in the range of about 0.9. Some examples of the Henyey–Greenstein scattering function are shown in Fig. 15.7(b). For comparison, a Mie scattering function is also included. It can be seen that the Henyey–Greenstein function reproduces the scaling of forward scattering to back scattering, but omits the fine structure. It is a smooth function and therefore suited to approximate measured scattering functions or to enter analytical calculations where derivatives are needed.

For LED simulations it is necessary to account for the fact that neither the particle size nor the particle shape have uniform distributions. It is reasonable to assume that the influence of varying particle shapes is averaged if multiple scattering in an ensemble of arbitrarily oriented particles occurs. Hence Mie theory should be a valid approach for particle scattering in an LED when the particle volumes are known. Varying particle sizes in the converter powder, however, have to be taken into account by averaging the individual scattering functions. Knowledge of the particle size distribution, therefore, is an essential input parameter for device simulations.

15.4.4 Determination of Material Parameters

The quality of computer simulations depends strongly on the validity of the material parameters. While most of the optical properties of the package materials like the resin, the reflector plastics, the involved coatings, and the LED-chip itself are accessible to standard measurement techniques, the properties of the converter material are less easy to determine. As described in the previous section, one needs knowledge about the absorption and scattering characteristics of the phosphor, determined mainly by the complex refractive indices of the phosphor particles and the resin and by the particle size distribution of the phosphor material. The particle morphology, on the other hand, is hardly of importance since the statistic orientation of large amounts of nonspherical powder particles makes a description in terms of an "effective" particle size of spherical particles possible.

Particle size distribution

Various methods exist for the measurement of the particle size distributions. The most widely used method is laser diffraction on particle-liquid emulsions. A laser beam (sometimes two laser beams with different wavelengths to enhance the accessible particle size range) is transmitted through a cuvette with an emulsion of the powder in water (sometimes with detergent). To avoid sedimentation and agglomeration, the liquid is pumped through this optical cell via a liquid loop which often incorporates ultrasonic de-agglomeration transducers. The particle size distribution is determined by measuring the angular intensity distribution of the scattered laser light by means of a number of carefully positioned photo-detectors at specifically defined angles. Using certain additional assumptions, the particle size distribution is then determined from matching the calculated scattering pattern with the observed pattern. The particle size distribution obtained in this way is not unambiguous, since several particle size distributions can produce the same scattering pattern. Precise measurements additionally require detailed input data on the complex refractive index of the particles.

An alternative method utilizes sedimentation in a viscous gradient. The particles are detected as they pass a photo sensor. Usually the gradient is built inside a centrifuge to achieve a higher sedimentation speed. The measurement of the particle size is thus performed by measuring the time required to sink by a defined distance. The relative amount of particles at the respective particle size is determined by the amount of extinction seen by the photo sensor. For this technique the density of the particles has to be known, or a reference sample of specific particle size of the same material has to be available for calibration. The sensitivity of this method on the refractive index of the material is much smaller than for the laser diffraction methods, since only the extinction has to be considered.

Refractive index

Measuring the complex refractive index of a powder with particle sizes in the range of the wavelength of visible light, is challenging. Conventional methods require sample sizes on a millimeter scale and are therefore only suited for bulk materials. However, most phosphors are only available as powders.

Standard technologies for measuring the refractive index of a powder use their optical diffraction properties in certain environments. Similar to these methods is the immersion of small particles in a range of liquids with exactly defined refractive indices. With increasing refractive index of the immersion liquid a characteristic optical effect can be observed, thus indicating that the refractive index n of the powder and the refractive index of the respective immersion liquid have been matched. These elegant methods, however, suffer from a significant catch: typical LED phosphors have a refractive index around 2. Liquids with large refractive indices contain large amounts of arsenic which makes their handling rather uncomfortable. Apart from the handling issues the problem remains that only the real part of the refractive index can be measured in this way. While absorption measurements in bulk materials are relatively easy, the scattering properties of particles with dimensions in the order of the wavelength make the determination of the absorbance tricky.

Absorption

A feasible approach to determining the absorption of a powder is by measuring the diffuse reflectance of a powder plaque [16]. Special care must be taken to ensure a truly Lambertian characteristic of the plaque surface (avoiding glancing angles). For slightly absorbing powders the Kubelka–Munk equation

$$\left(\frac{K}{S}\right) = \frac{(1 - R_\infty)^2}{2 R_\infty} \tag{15.15}$$

can be used to determine the ratio of the absorption factor K and the scattering factor S from the diffuse reflectance R_∞ of an "infinitesimally thick" powder plaque, i.e., with negligible or no optical transmittance. The single values of K and S can be determined, e.g., by measuring the diffuse reflectance and transmittance of an optically thinner system with a clearly defined thickness d.

With increasing absorbance of the powders the Kubelka–Munk theory, although often used, becomes more and more incorrect. In this situation, results can be improved by diluting the powder plaque with absorption-free scattering particles until the remission of the plaque is sufficiently high again.

For powders with dimensions on the wavelength scale, absorbance and scattering are closely related. In the case of a powder plaque the penetration depth of light incident on a plaque is not only dependent on the absorbance of the powder, but also on the scattering properties. Strong scattering prevents light from entering the plaque by reducing the total propagation length inside the plaque, thus the plaque shows a higher reflectance. In the case of weaker scattering the penetration depth and therefore the optical path inside the plaque is much larger and therefore more light can be absorbed inside the plaque.

Phosphor resin model

An alternative approach to a direct experimental determination of the material parameters is to perform a well-designed experiment, yielding some physical output quantities, and then to determine the desired material parameters from a computer simulation of the experiment. If the measured quantities depend sensitively on the required material parameters, the correct values can be found by matching the experimental and the simulated results. This procedure bears an additional advantage in that inherent deficiencies of the simulation procedure such as, in our case, unknown properties of surfaces, statistical deviations in particle concentrations, particle morphologies, etc., can be compensated for by using effective material parameters determined from the same type of simulation that is used later to make predictions on device properties.

The key word here is *self-consistency*. The material parameters which the simulation uses must be matched to the inherent inaccuracies of the simulation itself. Thus, fine-tuning of the material parameters by simulating measurements on powder systems, i.e., powder plaques, and then adapting the material parameters until simulation and experiment match, has proved to be a reasonable approach. Extra care must be taken, as it is not clear from the start that this fine tuning will eventually converge towards a "correct" set of parameters or whether this tuning yields runaway parameters which eventually render the simulation useless.

As an example, the determination of the complex refractive index of YAG:Ce can be conducted as follows. First, the remission of a powder plaque with an experimentally determined volume fill factor (typically from 40 to 50 %) is measured. The measurement setup is then schematically rebuilt as a computer model and the simulation is fed with literature values for the refractive index of YAG and the specific particle size distribution. The simulation software then calculates the scattering properties of YAG:Ce powder by means of Mie theory and simulates the behavior of the powder plaque. As a result, the simulation will give a first remission spectrum which will most likely differ from the measured result. By introducing physically reasonable absorption bands of various kinds, the complex refractive index of the material is altered until finally the simulation matches the measurement. The correctness of this set of parameters can then be tested by simulating additional well-defined geometrical setups, for example, thin phosphor-resin layers. Within this approach the simulated emission spectrum of Fig. 15.4 has been obtained. The result is a physical material model which is specifically tailored to the utilized simulation procedure.

15.4.5
LED Ray Tracing Model

The problem of light propagation in a diffuse medium can be solved in analogy to a random walk on a three-dimensional domain. Although originally optical ray tracing was developed for optical systems with distinct surfaces (for a description of the method see, e.g., [17]), nonsequential ray tracing provides all the features necessary to implement a random walk theory. So most commercial ray-tracing programs support a volume scattering model in some way. To simulate a phosphor with scattering and fluorescence, the following parameters have to be implemented:

Extinction cross-section (free path length)	
absorption cross-section	scattering cross-section
– quantum efficiency – emission spectrum	– scattering function

The extinction coefficient or its equivalent, the free path length, determines the range of unperturbed propagation. The propagation distance is calculated according to the Lambert–Beer law

$$I(x) = I_0 \exp(-\mu_{tot} x) \qquad (15.16)$$

where

$$\mu_{tot} = \sigma_{tot} \cdot C \qquad (15.17)$$

Here C is the particle concentration (number of particles per volume) of the phosphor.

In the case of a scattering event within the boundaries of the medium, a decision has to be made of whether the scattering is elastic or inelastic. The probabilities are calculated from the absorption and scattering cross-sections. In the case of elastic scattering, the ray retains its original wavelength. The new direction of propagation, relative to the incoming direction, must be derived from the scattering function.

If the ray is absorbed, the quantum efficiency determines the probability of re-emission. The energy of the ray is distributed over the emission spectrum. Due to the Stokes shift, energy is lost and the ray is assigned a longer wavelength. The direction of propagation of the converted ray should be independent of the incoming ray and therefore isotropic in a homogeneous medium.

Not all commercial ray tracing programs allow a wavelength change for a scattered ray. In this case the energy of the absorbed ray is dumped to a temporary file. In a second program run this file is used as a new source with the desired wavelength.

Because a single ray can be scattered many times within the converter medium, all scattering events should be implemented in a probabilistic fashion without ray splitting. Otherwise memory consumption and computational effort can become enormous. Even for a single ray, a high number of scattering events need to be computed if the medium is large compared to the free path length. This should be considered when setting cut-off limits, to keep artificial energy loss negligible.

Another important topic is wavelength discretization. As a general rule, with ray tracing simulations, the trade-off between calculation time and accuracy governed by Poisson statistics has to be considered. At least the different properties of the exciting blue light and the converted yellow light need to be regarded. For many purposes it is sufficient to digitize the spectrum just into a blue and a yellow wavelength. Of course then the resulting color coordinates will always lie on a straight line. If more subtle effects like spectral narrowing due to wavelength dependent absorption or reflection are to be investigated, the wavelength discretization needs to be refined.

15.5
Simulation Examples

In the following section the simulation methods developed in the last chapters will be applied to model a realistic white LED device. The general layout of the device is shown in Fig. 15.1. A large number of device families have been derived, varying in size and shape, by utilizing this simple principle. A blue-emitting GaN-chip is mounted into a preformed cavity (also called a reflector because the sides should be reflective), and then casted with a transparent resin. The phosphor is dispersed in the resin, so light can be converted at any location within the cavity. This means that light emitted by the chip can be converted after multiple reflections from the cavity walls. An integral simulation approach is therefore essential to account for the optical details of the device.

The dimensions and parameters used in the following simulations have been chosen to represent OSRAM's basic TOPLED™ device, see Fig. 15.9. The cavity is a truncated cone of 2.4 mm diameter and a height of 0.8 mm. The semiconductor chip has a size of about a quarter millimeter and emits blue light with a dominant wavelength of 460 nm and a half-width of ca. 20 nm. The cavity is filled completely with the resin and phosphor compound. The emission of the phosphor has a yellow hue with a dominant wavelength of ca. 570 nm and a width of ca. 100 nm. Hence the resulting spectrum has the two-peak composition that is typical for LED lighting (Fig. 15.8).

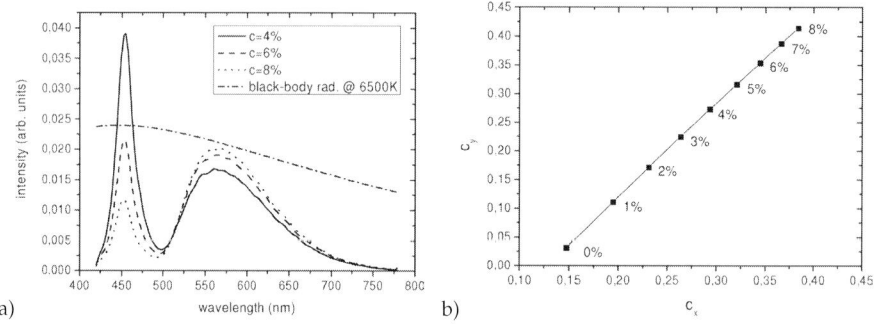

Fig. 15.8 Spectrum and CIE-color coordinates of LEDs with increasing converter concentration. For comparison, the spectrum of a black-body radiator at T=6500K is included.

One advantage of LED lighting technology is the fairly easy adjustability of color temperature by using the appropriate ratio of blue to yellow light. This can be done by increasing the phosphor concentration. Of course simulation should be able to give a quantitative answer as to how much phosphor is needed to obtain a specific color coordinate. The results of the ray tracing simulations are shown in Fig. 15.8. As expected, the blue peak decreases with

15.5 Simulation Examples

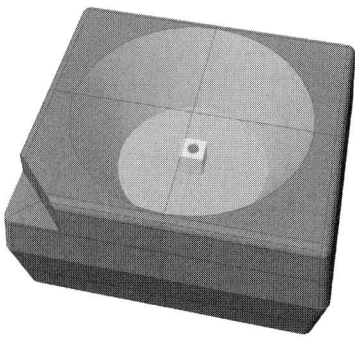
Low reflectivity (white reflector)

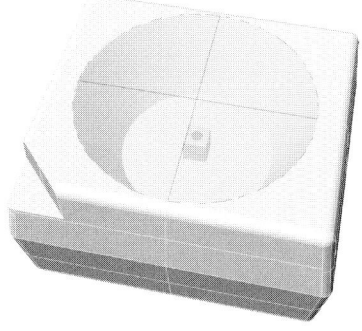
High diffuse reflectivity (black reflector)

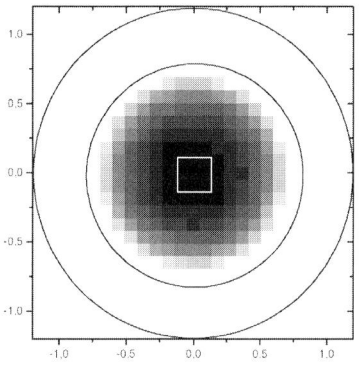
without converter, black reflector

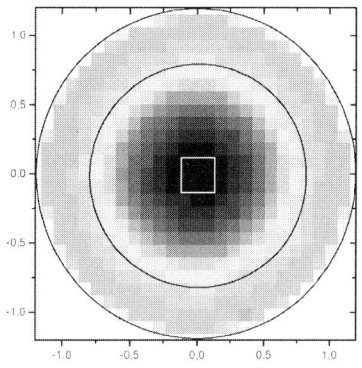
without converter, white reflector

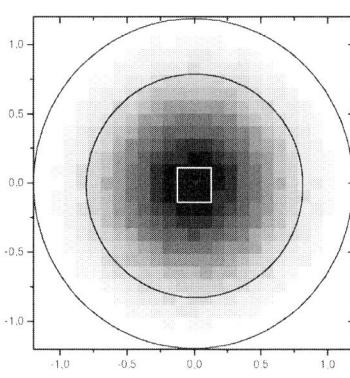
with converter, black reflector

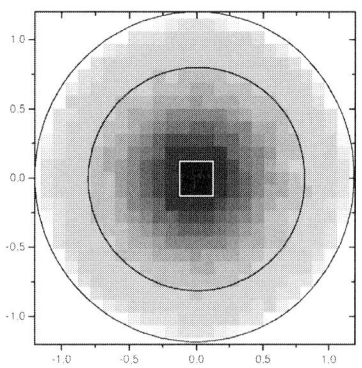
with converter, white reflector

Fig. 15.9 Simulation of light exiting from a package with planar surface. Plotted is the radiant flux through a surface element on a logarithmic scale. The outlines of the semiconductor chip and the reflector are included. To illustrate the versatility of the simulation two LEDs with black, respectively white, reflector have been computed. The simulation was performed for a casting without converter ("blue" LED) and with converter ("white" LED).

phosphor concentration, while the converted portion increases. A plot of the color coordinates results in a straight line in the color diagram. From this diagram we can read that, for white color with $c_x = 0.33/c_y = 0.33$, a phosphor concentration of 5.5% is needed.

In principle, to calculate the color coordinates, a simulation with only one blue and one yellow wavelength would have been satisfactory. Yet the spectral resolution used here was 5 nm to resolve spectral details. On a closer look the line in the c_x/c_y-diagram is not straight but slightly curved. This is due to the fact that the absorption and emission band of the phosphor overlap in the green spectral range, and therefore re-absorption of converted light occurs with increasing concentration. The depletion in the green spectral region leads to the observed lowering of the c_y-coordinate. For this phosphor the effect is small, but it can become a significant issue, e.g., for a mixture of different phosphors with shifted emission to cover a broader spectral range.

While simplified models could be used to calculate integral quantities, the full power of ray tracing comes into play when detailed geometrical problems have to be answered. So the designer of illumination applications wants to know the distribution of light on the exit facet of the LED light source.

The example from above will now be elaborated. To gain more insight, a simulative matrix of numerical experiments is set up (shown in Fig. 15.9). To study the effect of the reflector, in one set of simulations zero reflectivity (a black reflector) is assigned to the package. The other set of simulations is performed with highly diffuse reflectivity, as encountered in realistic LED packages. The package is then "filled up" with a resin without converter and a 5% converter dispersion, respectively. Results for the four combinations are shown in Fig. 15.9. Plotted is the radiative flux per area emerging from the top surface into the hemisphere. The scale is logarithmic because about two orders of magnitude have to be covered.

In the case of a black reflector with clear resin, the extraction is limited to a well-defined circle about the chip. The diameter of the circle is defined by the angle of total internal reflection between resin and air. Light emitted by the chip beyond this angle is reflected from the interface, and then absorbed by the package. If light is reflected by the white package as in the plot to the right, an additional ring of light is observed over the reflector walls.

Now the converter is added, shown in the last row of Fig. 15.9. In the case of the black reflector, the extraction area is no longer sharply defined, but gradually decreasing to the outside. Conversion is responsible for an exponential decay of the blue light for larger distances from the source, while part of the isotropically emitted yellow light can escape through the surface. In combination with the reflector, a more homogeneous light extraction from the LED can be achieved. But it should be noted that the major part of the light is still extracted from the center of the LED.

While Fig. 15.9 reports only on the energetic distribution of light over the package, Fig. 15.10 gives an account of the uniformity of color. Since, for conversion LEDs, the color locus is a straight line between blue and yellow, the c_x coordinate is a valid indicator for the color. A $c_x < 0.3$ corresponds to a bluish white, a $c_x > 0.33$ to a yellowish hue. So it can be seen from Fig. 15.10 that the color changes from bluish in the center of the LED to yellow in the exterior. It should be noted that, particularly for the absorbing reflector, the diameter of the region with high blue content corresponds to the region of high overall extraction. Outside this region the yellow light dominates rapidly, simply because here blue light can be extracted only after a scattering event. When the reflector is "switched on" again, the background level of yellow light and therefore the color coordinate, are increased. In a converter, LED color is determined by many feedback loops.

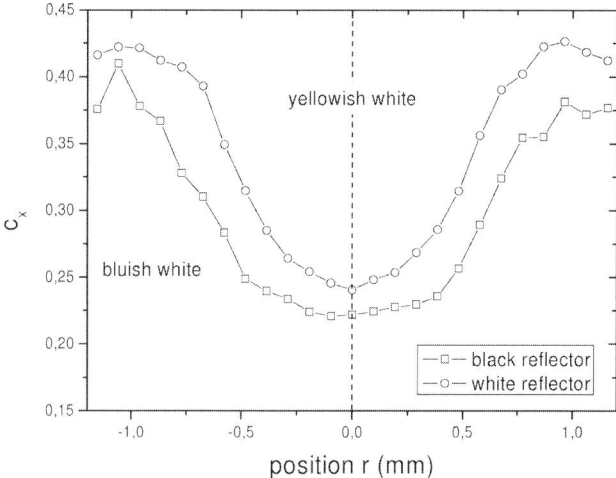

Fig. 15.10 Color coordinate c_x over the radial package position for converter-filled packages shown in Fig. 15.9.

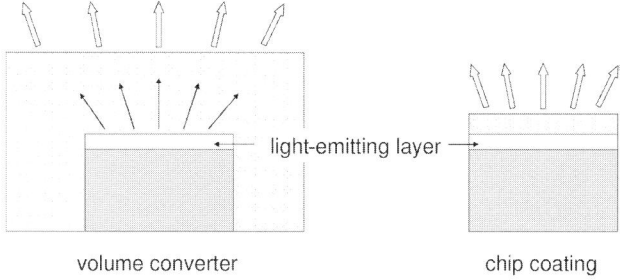

Fig. 15.11 Volume conversion concept vs chip coating.

The simulation analysis shows that most of the light is extracted from the center of the LED with a more or less constant chromaticity coordinate. Yet the ratio of yellow light increases when moving from the center. This behavior does not constitute a restriction for most illumination or backlighting applications, where the distance from observer to the light source is large or some light-mixing elements are provided. Yet it can lead to perturbing effects like color fringes when the LED is used in combination with imaging optics. Advanced white LED technologies attempt to avoid color separation by reducing the thickness of the converter layer (Fig. 15.11). This concept is called conversion by chip coating, in comparison with the conventional volume conversion. Since the light-emitting layer and the conversion layer now have the same geometrical dimension, the blue and the yellow converted beams cannot separate spatially. Thus chip coating conversion is well suited to projection applications because the luminance is not diminished. Of course the converter concentration has to be increased accordingly to reach the same chromaticity coordinates.

15.6
Conclusions

While there were only moderate requirements for the first generation of white conversion LEDs, advanced devices for use in general lighting applications need highly uniform color distributions, in particular, if imaging optics is involved. Therefore, empirical design approaches are being replaced by advanced methods involving optical simulations. We have shown in the previous sections that the optical simulation of white-emitting conversion LEDs is a complex task involving a precise knowledge of the material properties as well as sophisticated and efficient simulation techniques. There has been progress in both areas over the last years, but a lot of issues have yet to be solved. In multiple phosphor blends, for example, which are about to be used for high-color-rendering and color-on-demand applications, the interactions between different phosphor materials have to be taken into account. Moreover, as more and complex optical functionality is being integrated into the LED light source, more sophisticated optimization methods become necessary to allow for rapid design cycles. The software tool manufacturers have only recently started to incorporate the complex algorithms for particle scattering, luminescence conversion and a comprehensive and efficient treatment of colors into their software and a lot of work remains to be done to make these tools powerful and fast enough to meet the increasing demands of the rapidly growing LED industry. Nevertheless, full device simulations have become possible and the success of the new design methods are about to make them indispensable for the LED industry.

References

1. C. DeCusatis (ed.), *Handbook of Applied Photometry*, Springer, New York, 1998
2. P. Kubelka, J. Opt. Soc. Am. **44**, 330 (1954)
3. M.D. Fairchild, *Color Appearance Models*, 2nd edition, John Wiley & Sons, Chichester, England, 2005
4. M. Yamada et al., IEICE Trans. Electron., Vol. E88-C, No. 9, 1860 (2005)
5. G. Blasse, B.C. Grabmaier, *Luminescent Materials*, 45f, Springer, Berlin, Heidelberg, New York
6. J. K. Park et al., Appl. Phys. Lett. **87**, 031108 (2005)
7. M. Zachau et al., *Phosphors, Patents, and Products for LEDs*, Intertech Phosphor Global Summit 2006, San Diego, Conference Proceedings, 2006
8. G. O. Mueller, R. Mueller-Mach, Solid State Lighting II, Proceedings of SPIE, Vol. 4776, 122–130 (2002)
9. G. O. Mueller, R. Mueller-Mach, *Set The Pace in White Space – White LEDs for Illumination and Backlighting*, Intertech Phosphor Global Summit 2005, San Diego, Conference Proceedings, 2005
10. G. Bogner et al., *Light Source Using a Yellow-To-Red-Emitting Phosphor*, Osram Opto Semiconductors GmbH, Patent Application US 6,649,946 B2 (1999)
11. A. C. A. Delsing et al., *Luminescent Material, Especially for LED Application*, Osram Opto Semiconductors GmbH, Patent Application WO 2004/030109 A1 (2002)
12. S. Shionoya and Q. M. Yen (eds.), *Luminescence Centers of Rare-earth Ions*, Phosphor Handbook, pp. 178–201, CRC Press, 1999
13. J.D. Jackson, *Classical Electrodynamics*, 3rd edition, John Wiley & Sons, New York, 1999
14. C. F. Bohren, D. R. Huffmann, *Absorption and Scattering of Light by Small Particles*, Wiley-Interscience, New York 1983
15. L. Henyey, J. Greenstein, J. of Astrophysics **93,** 70–77 (1941)
16. G. Kortüm, *Reflexionsspektroskopie*, Springer Verlag 1969
17. W.J. Smith, *Modern Optical Engineering*, 3rd edition, McGraw-Hill New York 2000

16
Fundamental Characteristics of Edge-Emitting Lasers
Gen-ichi Hatakoshi

16.1
Introduction

Device design of GaN-based blue-violet semiconductor lasers is very important for practical applications such as high-density optical recording. In GaN lasers, the design of layer structures including MQW active region is indispensable in order to realize low-threshold characteristics, and to realize a high-quality output beam. Due to the practical restriction in the device fabrication, it is difficult to obtain a sufficient bandgap difference between the active layer and the cladding layers, and it is also difficult to obtain high ionized-acceptor concentrations of p-cladding layers. Therefore, the design for reducing carrier overflow is very important. Also, the design for improving beam quality is essential, because the GaN lasers have a peculiar waveguide structure compared with that of conventional lasers. In addition to the electrical and optical characteristics, thermal properties should be considered for realizing high-reliability devices.

In this chapter, device simulations for electrical, optical and thermal characteristics of GaN-based lasers are described. The basic laser structures treated here are Fabry–Perot type InGaN MQW-SCH structures on a sapphire substrate. Both facets were assumed to be cleaved; i.e., without coating. One-dimensional and two-dimensional models were applied for the electrical and optical simulation, and a three-dimensional model was adopted for the thermal simulation.

Nitride Semiconductor Devices: Principles and Simulation. Joachim Piprek (Ed.)
Copyright © 2007 WILEY-VCH Verlag GmbH & Co. KGaA, Weinheim
ISBN: 978-3-527-40667-8

16.2
Basic Equations for the Device Simulation

16.2.1
Electrical and Optical Simulation

Electrical characteristics of GaN-based lasers were analyzed by a conventional model described by Poisson's equation and current continuity equations for electrons and holes, as represented by the following equations [1].

$$\nabla \cdot (\varepsilon \nabla \psi) = -q(p - n + N_{D+} - N_{A-}) \tag{16.1}$$

$$\frac{dn}{dt} = \frac{1}{q} \nabla \cdot J_e - R \tag{16.2}$$

$$\frac{dp}{dt} = -\frac{1}{q} \nabla \cdot J_h - R \tag{16.3}$$

where, ψ is the electric potential, n is the electron density, p is the hole density, J_e and J_h are the current densities for the electron and hole, respectively, and R is a recombination term. Recombination by stimulated emission for lasers is included in R. In Eq. (16.1), N_{D+} and N_{A-} represent ionized donor concentration and ionized acceptor concentration, respectively. Note that these values are different, in general, from donor and acceptor concentrations. Especially for GaN materials, N_{A-} is significantly smaller than the acceptor concentration due to the deep Mg acceptor level. Fermi statistics were adopted for the description of electron and hole concentrations. Wave equations and the rate equation for photons are simultaneously solved [2–5], if necessary. In the simulation described in this chapter, the piezoelectric effect, which has an important bearing on the characteristics of GaN-based lasers, is not included, under the assumption that well layers are sufficiently thin.

Table 16.1 shows parameters used in the electrical simulation. These parameters were extracted from the reported data for nitride materials [6–14]. Note that the values in Table 16.1 are not necessarily consistent with various

Tab. 16.1 Parameters used in the electrical simulation.

Parameter	Symbol	Unit	$In_xGa_{1-x}N$	$Ga_{1-x}Al_xN$
Bandgap energy	E_g^Γ	eV	$3.39 - 2.5x + x^2$	$3.39 + 1.81x + x^2$
Electron affinity	χ	eV	$4.26 + 1.67x - 0.67x^2$	$4.26 - 1.29x - 0.71x^2$
Effective mass				
(electron)	m_e	m_0	$0.2 - 0.08x$	$0.2 + 0.13x$
(hole)	m_h	m_0	$0.8 - 0.3x$	$0.8 + 0.1x$
Dielectric constant	ε	ε_0	$9.8 + 5.2x$	$9.8 - 1.3x$

reports of experimental and theoretical results. In the simulation, anisotropic structures for the conduction and valence bands were not considered, and single values of effective mass for electron and hole, respectively, were used as shown in the table. In general, the mobility values μ_e and μ_h for electron and hole, respectively, vary depending on the composition and impurity concentrations. In the simulation, these values were fixed as $\mu_e = 200 \text{ cm}^2\text{V}^{-1}\text{s}^{-1}$ and $\mu_h = 10 \text{ cm}^2\text{V}^{-1}\text{s}^{-1}$. Also, the spontaneous emission coefficient was fixed as $B = 1 \times 10^{-10} \text{ cm}^3\text{s}^{-1}$. Nonradiative lifetimes for Shockley–Read–Hall type recombination were assumed as $\tau_n = \tau_p = 10$ ns for nondoped MWQ and $\tau_n = \tau_p = 1$ ns for the other layers.

Recombination by the stimulated emission is represented by the following equation.

$$R_{stim} = \frac{c}{n_{\text{eff}}} sg(n,p) \tag{16.4}$$

where, c is the velocity of light in a vacuum, n_{eff} is the effective refractive index of the laser cavity for the waveguide mode, s is the photon density, and $g(n,p)$ is the gain as a function of carrier density. In the calculation, the gain was given by the following simple approximation:

$$g(n,p) = g_0 + g_1 \min(n,p) + g_2 np \tag{16.5}$$

Parameters for InGaN MQW were set as follows: $g_0 = -1000 \text{ cm}^{-1}$, $g_1 = 1.43 \times 10^{-16} \text{ cm}^2$, $g_2 = 0$. In this case, the value $-g_0/g_1 (= 7 \times 10^{18} \text{ cm}^{-3})$ corresponds to the transparency carrier density. These parameters are roughly consistent with the reported gain characteristics [12–14]. Also, the value of g_1 is nearly consistent with the bandgap dependence of the gain parameter reported in Ref. [15].

Figure 16.1 shows a simulation example of current distribution [16] for an inner-strip-structure InGaN laser [17] with an n-GaN current blocking layer. Inner-stripe lasers have the advantage of low contact resistance at the p-electrode and p-GaN contact layer region, because large contact-area structure can be used. Injected current is confined in the stripe region defined by the n-GaN blocking layer, as shown in Fig. 16.1. In the simulator, current-flow lines were represented by utilizing the continuity of total current density $J (= J_e + J_h)$ [4].

Two-dimensional analysis, as shown in Fig. 16.1, is important for designing semiconductor lasers with superior electrical characteristics and with high-quality optical characteristics of the output light. Threshold current density J_{th}, which is a major concern in GaN lasers, is also affected by the two-dimensional structure. However, a dominant factor determining J_{th} is the one-dimensional layer structure for the MQW active region and cladding layers. Therefore, results of one-dimensional simulation will mainly be described for electrical characteristics analysis in Section 16.3.

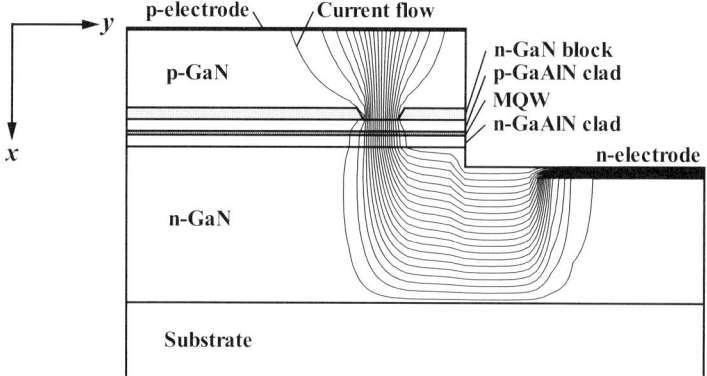

Fig. 16.1 Simulation example of two-dimensional current distribution for an inner-stripe-structure GaN laser.

The basic wave equations for the one-dimensional case are represented by the following equations.

$$\frac{d^2 E_y}{dx^2} + (k_0^2 n_r^2 - \beta^2) E_y = 0 \qquad \text{(for TE mode)} \tag{16.6}$$

$$n_r^2 \frac{d}{dx}\left(\frac{1}{n_r^2}\frac{dH_y}{dx}\right) + (k_0^2 n_r^2 - \beta^2) H_y = 0 \quad \text{(for TM mode)} \tag{16.7}$$

where, x is a coordinate perpendicular to the layers, E_y and H_y are the y components of the electric field and the magnetic field, respectively, n_r is the refractive index and β is the effective propagation constant defined by the following equation.

$$E_y, H_y = F(x) \exp(-i\beta z) \tag{16.8}$$

Both n_r and β in Eqs (16.6) and (16.7) are generally complex numbers. The wave equations (16.6) and (16.7) were extended to two-dimensional equations by the effective refractive index method when solving for two-dimensional structures.

The rate equation for photons is given as follows:

$$\frac{dS_m}{dt} = \left(\frac{c}{n_{\text{eff}}} G_m - \frac{1}{\tau_{ph}}\right) S_m + C_m \bar{R}_{sp} \tag{16.9}$$

where, S_m is the integral of the photon density $s_m(x, y)$ for waveguide mode m in the xy plane, G_m is the gain for mode m, C_m is the spontaneous emission factor, and $\bar{R}_{sp}$ is the integral of the spontaneous emission rate weighted by the light intensity of the waveguide mode. The photon lifetime τ_{ph} in Eq. (16.9) is given by the following equation.

$$\frac{1}{\tau_{ph}} = \frac{c}{n_{eff}} \left\{ \alpha_i + \frac{1}{2L} \ln\left(\frac{1}{R_1 R_2}\right) \right\} \quad (16.10)$$

where, α_i is the internal optical loss including free carrier absorption, R_1 and R_2 are the reflectivities of the front and rear facets, respectively, and L is the cavity length. Introduction of the rate equation is very convenient for analyzing semiconductor lasers, because Eq. (16.9) always holds independent of the current density: i.e., regardless of the laser oscillation. In the self-consistent simulation, the wave equation and the rate equation for the photon are coupled with the electronic equations (16.1)–(16.3) through the gain distribution and photon density distribution.

16.2.2
Simulation Model for Thermal Analysis

Analysis of the thermal characteristics for GaN lasers has been carried out by using a three-dimensional thermal conduction model [18], in order to take the effect of a three-dimensional structure into consideration. The following heat conduction equation was numerically solved.

$$C_p \rho \frac{\partial T}{\partial t} = \nabla \cdot (k \nabla T) + S \quad (16.11)$$

where T is the temperature, C_p is the specific heat, ρ is the density, k is the thermal conductivity and S is the heat source density. Boundary conditions were given by the following equations [19].

$$[k \nabla T]_n = q_c + q_r \quad (16.12)$$

$$q_c = \alpha (T - T_f) \quad (16.13)$$

$$q_r = \epsilon \sigma (T^4 - T_f^4) \quad (16.14)$$

where $[k \nabla T]_n$ is the heat flux component normal to the boundary, q_c and q_r are heat fluxes caused by natural convection heat transfer and by heat radiation, respectively, α is the heat transfer coefficient, ϵ is the emissivity of the surface, σ is the Stefan–Boltzmann constant [20] and T_f is the ambient temperature. In the numerical calculation, Eq. (16.11) was solved in three dimensions by the finite-difference method with controlled volume [19,21]. Boundary conditions were given as follows: $\alpha = 0$ and $\epsilon = 0.1$ for boundaries not in contact with a heat sink, and $\alpha = \infty$ (sufficiently large value in the actual simulation) for the boundaries of the heat sink side. The former condition is almost equivalent to the case of an adiabatic boundary, and the latter corresponds to the condition of constant temperature $T = T_f$.

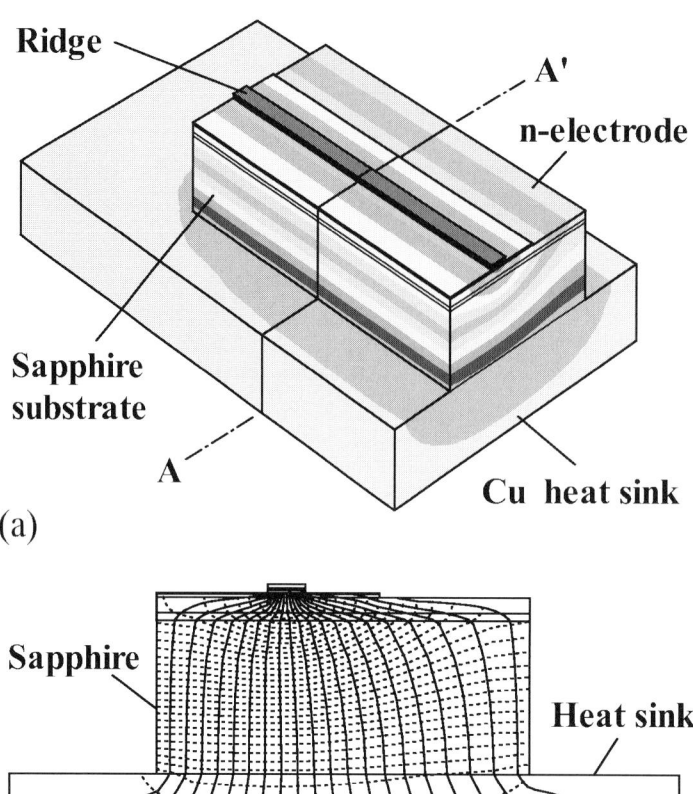

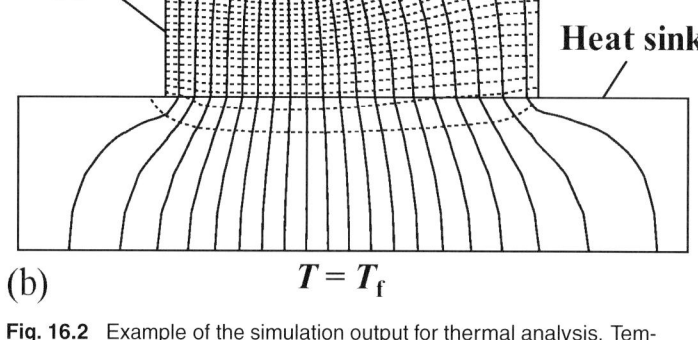

Fig. 16.2 Example of the simulation output for thermal analysis. Temperature distribution on the surface of the structure (a) and isotherms (dashed) and heat flow distribution (solid) (b) at cross-section A–A′ are represented. In the actual simulator, the temperature distribution in the upper figure is displayed by color gradations.

Figure 16.2 shows an example of the simulation output. Temperature distribution on the surface of the structure is represented in the upper figure. In the lower figure, temperature distribution (dashed lines) and heat flow distribution (solid lines) are represented for the cross-section indicated by A–A′ in the upper figure. The method of heat-flow display [4, 18] is similar to that described for current-flow lines. Note that the heat-flow line is not continuous due to the heat generation given by S in Eq. (16.11); i.e., discontinuity of the heat-flow line occurs at such a heat generation area.

16.3 Simulation for Electrical Characteristics and Carrier Overflow Analysis

In general, the transparency carrier density, which is the minimum carrier density required for laser oscillation, increases with increasing bandgap energy [15]. Therefore, a high injected carrier density will be required for the wide-gap GaN lasers. It is desirable to form cladding layers with large bandgaps and high carrier densities in order to realize effective carrier confinement in the active region. However, it is difficult to obtain a p-type GaAlN cladding layer with high carrier density and high Al composition, which corresponds to a high-bandgap-energy material. For this reason, carrier overflow from the active region to the p-type cladding layer can occur even in the case of a structure with a large bandgap difference between the active layer and cladding layers, resulting in a remarkable increase in the threshold current density [16, 22, 23].

Figure 16.3 shows the calculated examples of band diagram, carrier density distributions and current density distributions for the case in which significant carrier overflow occurs [16]. The large voltage drop in the p-GaAlN cladding layer, as shown in Fig. 16.3(a), is due to the low concentration of ionized acceptor in this layer. Also, it causes a reduction in the conduction-band heterobarrier at the interface of the p-type cladding layer and the active region, resulting in a significant electron overflow from the active region to the p-cladding layer. Figure 16.3(c) shows current density distributions for electrons and holes. The hole current injected from the p-type layer side decreases in the MQW region due to carrier recombination, and becomes almost zero in the n-type cladding layer. In the same way, the electron current injected from the n-type layer side decreases in the MQW region by the same amount as the decrease in the hole current. However, a large electron current remains in the p-type cladding layer, which corresponds to the overflow current. This overflow current is an extra current which does not contribute to the stimulated emission recombination, and should therefore be reduced.

An increase in the ionized-acceptor concentration in the p-type cladding layer has a remarkable effect on the reduction of electron overflow. Development of crystal growth technologies enabled high concentrations of acceptor and donor impurities for wide-gap GaN and GaAlN [24].

Another effective method for decreasing the overflow current is to reduce the active layer volume. In general, the threshold current density is nearly proportional to the active layer thickness, assuming a small optical loss. In the initial stage of the GaN laser development, MQW structures with rather large numbers of wells, as shown in Fig. 16.3, were used. Such large numbers of wells cause a nonuniform distribution of injected carriers in the MQW-active region. For this reason and the large active layer volume itself, the threshold current density and carrier overflow increase with increasing well number.

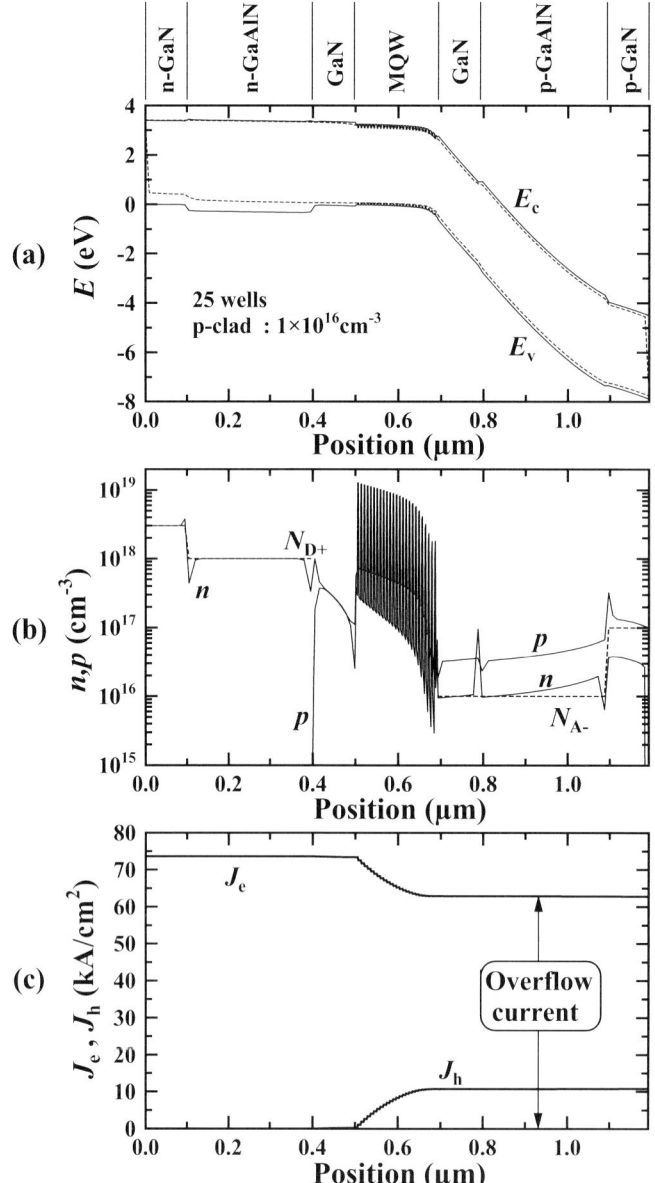

Fig. 16.3 Calculated examples of band diagram, carrier distributions, and current density distributions for the case in which significant carrier overflow occurs. A large electron current (J_e) is present in the p-type cladding layer, which corresponds to the overflow current. (G. Hatakoshi et al., "Analysis of device characteristics for InGaN laser diodes", Jpn. J. Appl. Phys., 38, pp. 1780–1785, **1999**, Copyright Institute of Pure and Applied Physics. Reproduced with permission.)

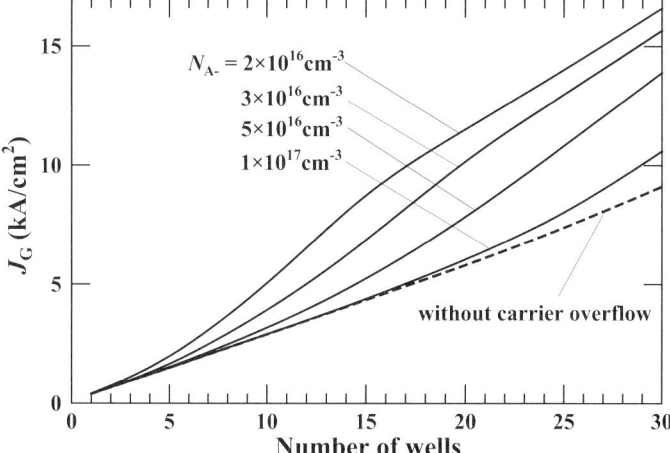

Fig. 16.4 Dependence of transparency current density on the acceptor concentration and the number of wells. J_G denotes the current density required for obtaining gain.

Figure 16.4 shows the calculated results of the dependence of "transparency" current density J_G on the ionized acceptor concentration and the number of wells [4, 22]. Note that J_G, in this figure, is the current density required for obtaining gain. The dependence of J_G on the well number can

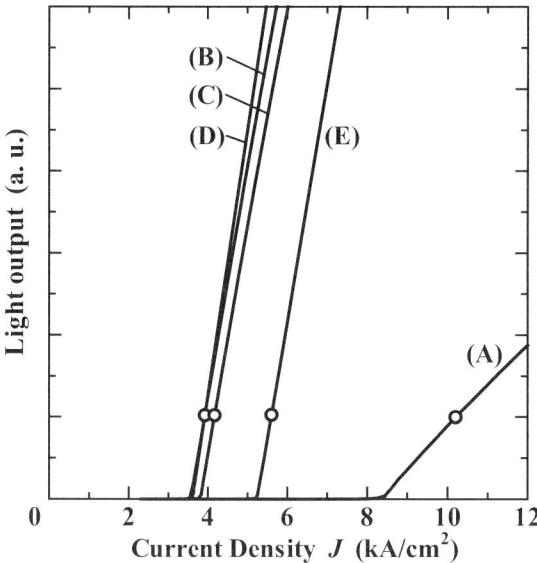

Fig. 16.5 Calculated examples for light output characteristics as a function of injected current density. Parameters for the layer structure used for the simulation are listed in Tables 16.2 and 16.3.

Tab. 16.2 Layer structures used for the simulation.

Layer			(A)	(B)	(C)	(D)	(E)
			\multicolumn{5}{c}{N_{D^+}, N_{A^-} (cm^{-3}), d (μm)}
p-GaN (1.0 μm)	$N_{A^-} =$		1×10^{18}	1×10^{18}	1×10^{18}	1×10^{18}	1×10^{18}
p-Ga$_{0.85}$Al$_{0.15}$N clad	$N_{A^-} =$		2×10^{16}	1×10^{17}	2×10^{16}	1×10^{17}	1×10^{17}
	$d =$		0.8	0.8	0.8	0.8	0.2
p-GaN guide (0.1 μm)	$N_{A^-} =$		1×10^{16}	1×10^{16}	1×10^{16}	1×10^{16}	1×10^{16}
p-Ga$_{0.85}$Al$_{0.15}$N barrier (20 nm)	$N_{A^-} =$		None	None	1×10^{16}	1×10^{17}	1×10^{17}
MQW (5 wells)							
In$_{0.15}$Ga$_{0.85}$N well	(2.5 nm)						
In$_{0.05}$Ga$_{0.95}$N barrier	(5.0 nm)						
n-GaN guide (0.1 μm)	$N_{D^+} =$		1×10^{16}	1×10^{16}	1×10^{16}	1×10^{16}	1×10^{16}
n-Ga$_{0.85}$Al$_{0.15}$N clad	$N_{D^+} =$		2×10^{18}	2×10^{18}	2×10^{18}	2×10^{18}	2×10^{18}
	$d =$		0.8	0.8	0.8	0.8	0.2
n-GaN (3.0 μm)	$N_{D^+} =$		2×10^{18}	2×10^{18}	2×10^{18}	2×10^{18}	2×10^{18}

Tab. 16.3 Refractive indices used in the optical simulation.

Material	Refractive index
p-electrode	0.9 − 2.5 i
GaN	2.537
Ga$_{0.85}$Al$_{0.15}$N	2.456
In$_{0.05}$Ga$_{0.95}$N	2.553
In$_{0.15}$Ga$_{0.85}$N	2.625
Sapphire	1.785

be understood from Eqs (16.2)–(16.4) and Fig. 16.3. As shown in Fig. 16.3(c), current density changes at a well by an amount determined by the carrier recombination as represented by Eqs (16.2)–(16.4). Under the assumption that the transparency carrier density is constant, the total change in the current density, which is a dominant part of J_G for the case of small overflow current, is nearly proportional to the number of wells. The threshold current density J_{th} is larger than J_G because the factor caused by the internal loss and cavity loss should be added. However, the dependence on the acceptor concentration and well number is thought to be similar to Fig. 16.4, except for the region of the very small well number.

Figure 16.5 shows calculated example for light output vs current density ($L - J$ characteristics). Parameters for the layer structure used for the analysis are shown in Table 16.2 and Table 16.3. As shown in Fig. 16.5, the layer structure has a significant effect on the threshold current density. Figures 16.6 and 16.7 show calculated band diagrams, carrier density distributions and current density distributions for the structures (A)–(D) at operation points indicated by circles in Fig. 16.5. Note that limited ranges near the active regions are represented for a clearer understanding.

Figure 16.6(a) is the result for the structure (A) where carrier overflow occurs due to the low acceptor concentration in the p-cladding layer. The structure (B) has the same parameters as (A) except for the acceptor concentration of the p-cladding layer. An increase in the acceptor concentration in the p-type cladding layer has a remarkable effect on the reduction of the threshold current density due to the reduction of the overflow electron current, as shown in Fig. 16.6(b). Electron overflow can also be reduced by introducing the electron barrier layer [22], which is widely used in InGaN lasers [25]. The technology

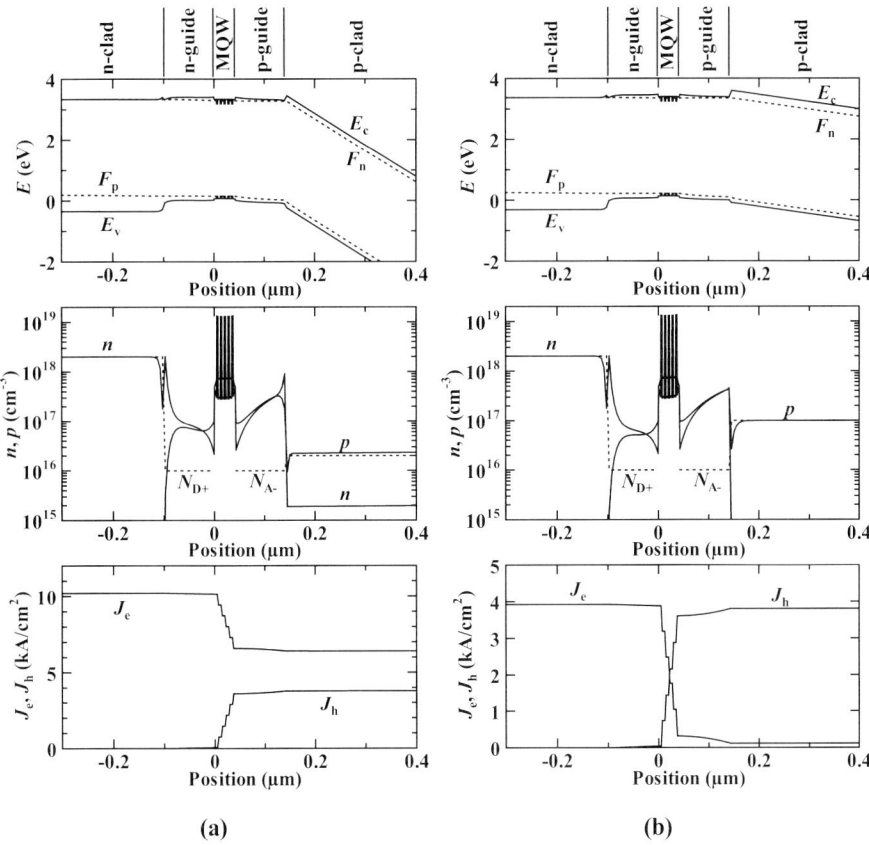

Fig. 16.6 Calculated band diagram, carrier density distributions and current density distributions. (a) and (b) represent the results for the structures (A) and (B), respectively, in Table 16.2.

of the electron barrier layer and the operation principle is similar to that of the conventional barrier structure reported in GaAlAs lasers [26]. The effect of the electron barrier layer can be seen by comparing the structures (A) and (C). As shown in Fig. 16.6(a) and 16.7(a), the overflow electron current is effectively reduced even for the structure with low acceptor concentration in the p-cladding layer. Note that the operation voltage for the structure (C) is still high due to high resistivity of the p-cladding layer, which can be seen as a large inclination of the band diagram in the p-cladding layer region. Both a high acceptor concentration for the p-cladding layer and the electron barrier layer are introduced in the structure (D). In this case, the reduction of the threshold current density is almost the same as for the structure (C), as shown in Fig. 16.5. However, the operation voltage is also remarkably reduced due to the low resistivity of the p-cladding layer, as shown in Fig. 16.7(b). In addition

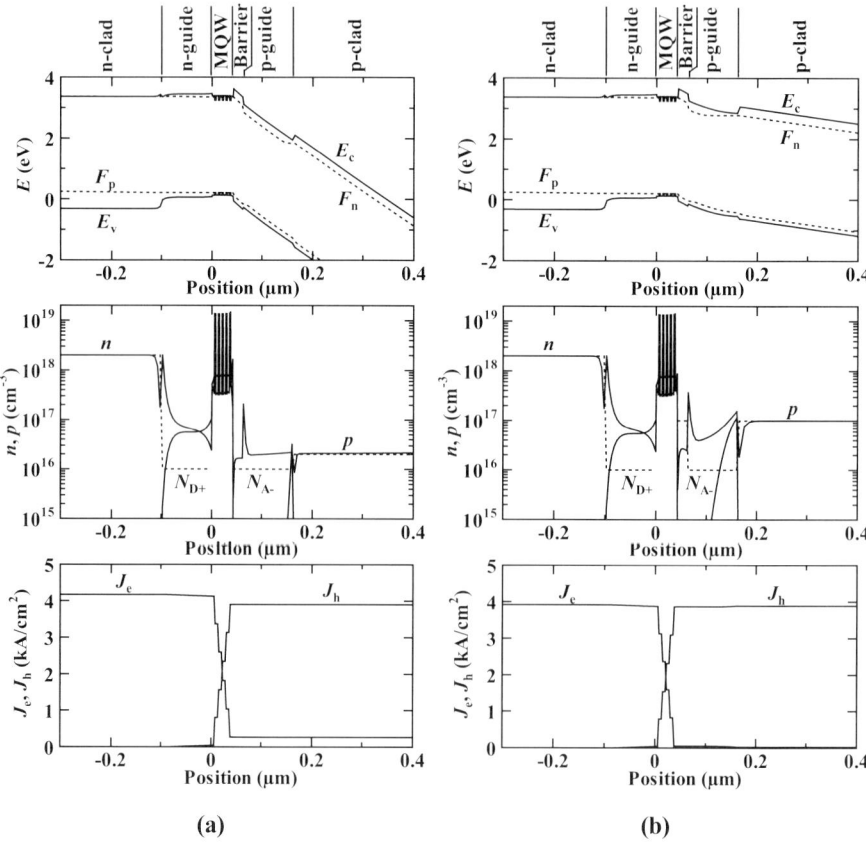

Fig. 16.7 Calculated band diagram, carrier density distributions and current density distributions. (a) and (b) represent the results for the structures (C) and (D), respectively, in Table 16.2.

to this, it can be seen that the slope efficiency of $L-J$ characteristics is higher than in the case of (C), and also higher than in the case of (B). This originates in the electron overflow reduction in the high injected-current region.

Figure 16.8 summarizes the ratio of the overflow current as a function of the total injected current for the structures (A)–(D). It is obvious that, in the structure (D), overflow current is reduced even in the high-injection region.

The structure (E) in Table 16.1 has the same layer structure and the same acceptor and donor concentrations as for (D). However, the threshold current density of (E) is higher than that of (D), as shown in Fig. 16.5. The difference between (D) and (E) is the thickness of the cladding layers. In the structure (E), the cladding layer thickness is smaller than that of (D) and is insufficient for confining guided mode. Figure 16.9 shows a comparison of the guided mode intensity and the far-field pattern of the output light. As shown in the upper figure, a large amount of guided mode penetrates the GaN contact layers, in the structure (E), resulting in a reduction of the optical confinement factor. This causes the increase in the threshold current density and also causes an anti-guide mode behavior [16, 27] as shown in the lower figure of Fig. 16.9(b). The anti-guide mode and related beam quality are described in the following section.

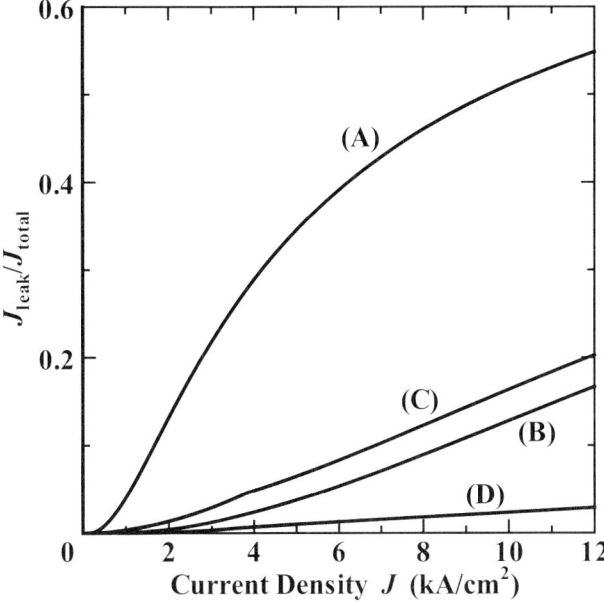

Fig. 16.8 Ratio of overflow current density as a function of total injected current density for the structures (A)–(D). The definition of overflow current is shown in Fig. 16.3.

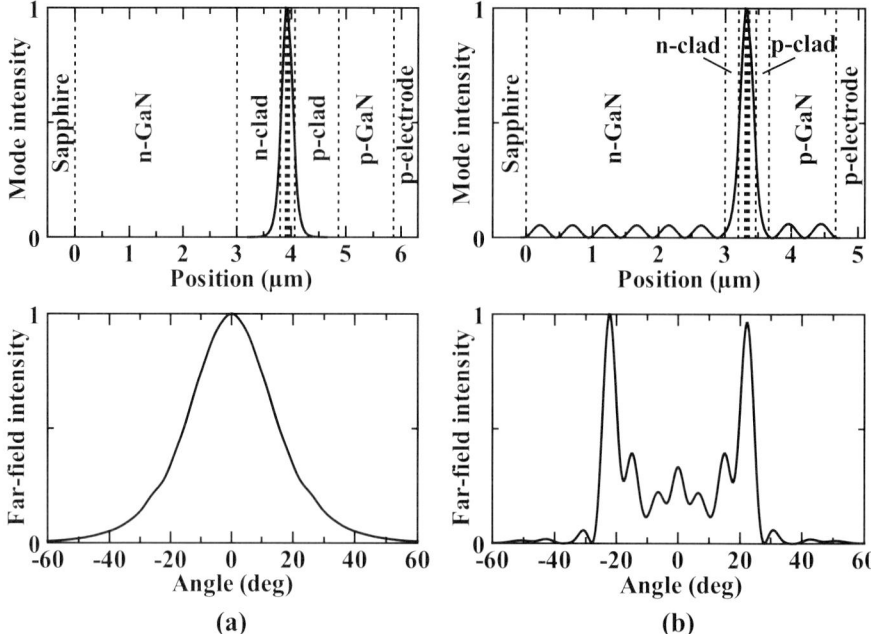

Fig. 16.9 Comparison of guided mode intensity and far-field pattern for the structures (D) and (E). For the calculation of the guided mode, the sapphire substrate and p-electrode were taken into account as outermost layers.

16.4
Perpendicular Transverse Mode and Beam Quality Analysis

The transverse mode is one of the most important characteristics of semiconductor lasers in optical disk applications. In conventional lasers, transverse modes in the perpendicular direction are not a source of concern because the optical confinement is automatically realized by carrier-confinement structures. The optical mode in the perpendicular direction is confined, for example, by low index outer layers in InP material systems or by high-absorbing outer lasers in GaAs material systems. On the other hand, contact layers in GaN lasers have a higher refractive index than neighboring cladding layers and are also almost transparent for oscillating wavelengths. This is a serious problem because the optical mode penetrates into the high-index contact layers and the oscillating mode becomes a high-order mode [16, 28]. Since the effective refractive index of the oscillating mode is lower than the refractive index value of the GaN layer, such a mode behaves like an anti-guide mode [4,16,27]. The optical mode intensity profile shown in Fig. 16.9(b) in the previous section is an example of such an undesired mode. In this case, the far-field pattern has two peaks and is unsuitable for optical disk use.

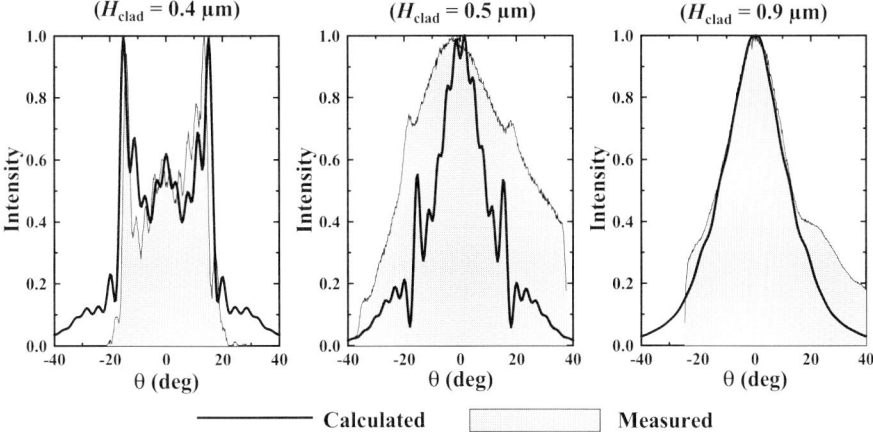

Fig. 16.10 Comparison of measured far-field profiles with the calculated results. It is shown that the anti-guide-like behavior is suppressed by increasing cladding layer thickness. (G. Hatakoshi et al., "Optical, electrical and thermal analysis for GaN semiconductor lasers", Int. J. Numer. Model. 14, pp. 303–323, **2001**, Copyright John Wiley & Sons Limited. Reproduced with permission.)

Figure 16.10 shows a comparison of the measured far-field profiles with the calculated results [4, 27]. It is seen that anti-guide-like behavior can be suppressed by increasing the cladding-layer thickness. It should be noted, however, that the formed waveguide mode for the thick cladding layer case is still a high-order mode. Such a mode order is not necessarily essential for the quality of the output beam, because the near-field pattern of a high-order mode for a structure with sufficient cladding layer thickness will almost coincide with that of the fundamental mode for a structure with infinite cladding thickness. An important point is that undesirable mode amplitude in the GaN contact layers can be suppressed by the optimum design for the layer structures. It is shown, from Fig. 16.10, that the calculated results agree well with the measured profiles and that the perpendicular mode analysis is useful for the device design.

The quality of the output beam can be characterized by a beam quality factor M^2 (M-squared factor), which is defined as the product of the beam waist width and the beam divergence angle [29, 30]. In this case, the beam widths and divergence angles are defined by using the standard deviation; i.e., the square-root of the second-order moment of the beam profile. A characteristic feature of this definition for the beam widths is that the propagation characteristics of an arbitrary beam are represented as those of a Gaussian beam. The aforementioned M^2 factor represents a degree of beam quality deviation from the ideal Gaussian beam which has the smallest M^2 value of unity. In

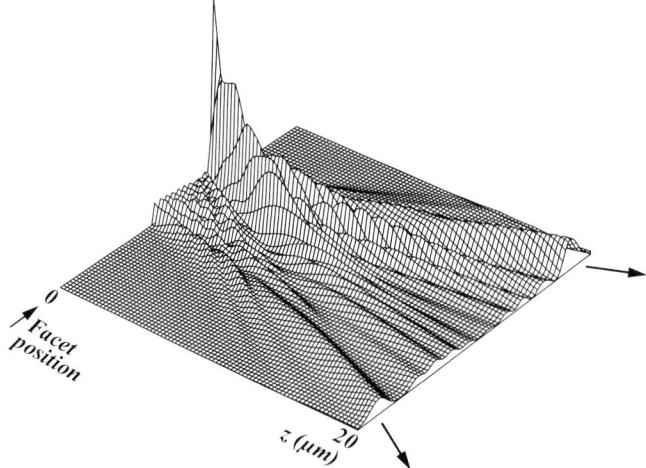

Fig. 16.11 Calculated beam profile near the output facet for an anti-guide mode in a GaN laser. A multi-mode profile at the facet ($z = 0$) is transformed by propagating along the z-axis, into a double-peaked far-field pattern which is a specific feature of the anti-guide mode. (G. Hatakoshi, "Analysis of beam quality factor for semiconductor lasers", Opt. Review 10, pp. 307–314, **2003**, Copyright Optical Society of Japan. Reproduced with permission.)

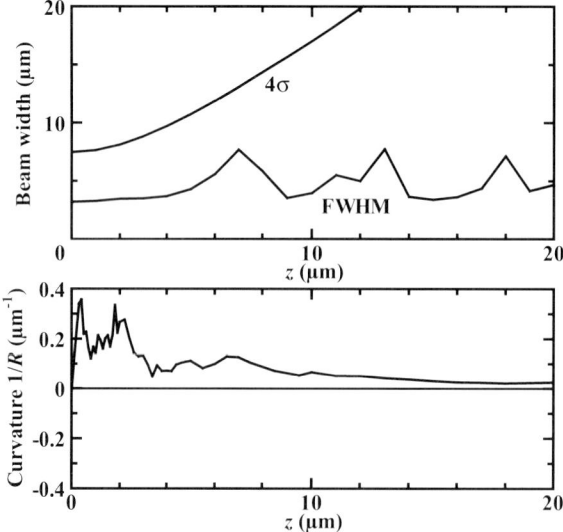

Fig. 16.12 Calculated beam width and wavefront curvature for the beam shown in Fig. 16.11. The beam width (4σ) defined by the standard deviation shows a Gaussian-like propagation profile. The beam waist location, where 4σ takes a minimum value, coincides with the position where the wavefront curvature is zero.

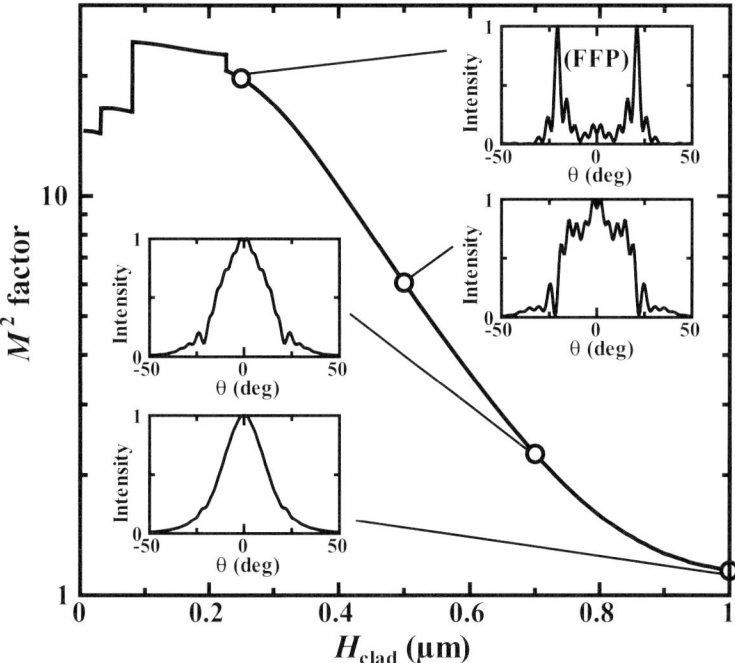

Fig. 16.13 Calculated M^2 factor as a function of cladding layer thickness. Insets show far-field patterns at points indicated by circles. (G. Hatakoshi, "Analysis of beam quality factor for semiconductor lasers", Opt. Review 10, pp. 307–314, **2003**, Copyright Optical Society of Japan. Reproduced with permission.)

general, output beam profiles of semiconductor lasers are not Gaussian, and thus they take various M^2 values greater than unity, depending on their waveguide structure.

Figure 16.11 shows a calculated beam profile [31] near the output facet for an anti-guide mode in the GaN laser. It is shown that a multi-mode profile at the facet is transformed, by propagating along the z-axis, into a double-peaked far-field pattern, which is a specific feature of the anti-guide mode. The beam width defined by the standard deviation shows a Gaussian-like propagation profile, even for such an anti-guide mode with complicated intensity profile, as shown in Fig. 16.12 [31]. The calculated M^2 factor for this beam had a very large value of 20; this means that the beam quality is not good. Such low beam quality originates in the penetration of the guided mode into the contact layer.

It is expected that the effect of the contact layers and the outer region can be suppressed by using thick cladding layers. Figure 16.13 shows the calculated M^2 factor as a function of the cladding layer thickness [31]. Insets in the figure show far-field patterns at several points indicated by circles. An increase in the cladding-layer thickness has a remarkable effect on the improvement in the beam quality due to the increase in the optical confinement factor. It also has an effect on the reduction in the guided-mode loss [16] and, as a result, the reduction in the threshold current density, as described in Section 16.3. A thick cladding layer is desirable because the effect of the outer contact layer can be reduced. However, there is a restriction from the crystal growth in the actual fabrication and a sufficient thickness is not necessarily achievable. Therefore, the device design for the layer structure is very important for obtaining a low-threshold InGaN MQW laser with a high-quality output beam.

16.5
Thermal Analysis

An improvement in the temperature characteristics of GaN laser diodes is important for realizing reliable devices which can operate at high temperatures [4, 18]. In the case of a GaN laser on a sapphire substrate, thermal characteristics are restricted by the thermal conductivity of sapphire, which is as low as about half of that for GaAs, and becomes small at high temperatures. Also, it is difficult for GaN lasers with sapphire substrates to form a junction-down mounting, which is favorable for a reduction in the thermal resistance of the device. Therefore, in designing device structures, it is very important to consider the temperature characteristics.

Figure 16.14 shows an example of a device structure for numerical analysis [18]. Heat sources considered in the calculation, which give the factor S in

Tab. 16.4 Parameters for the calculation of heat generation.

Parameter	Value	Unit	Parameter	Value	Unit
ρ_c	5×10^{-4}	Ω cm^2	h_{p1}	0.5	µm
μ_h	10	cm^2 V^{-1}s^{-1}	h_{p2}	0.8	µm
μ_e	200	cm^2 V^{-1}s^{-1}	h_{n2}	0.8	µm
p_1	3×10^{17}	cm^{-3}	h_{n1}	4	µm
p_2	1×10^{17}	cm^{-3}	W_0	300	µm
n_2	1×10^{18}	cm^{-3}	W	5	µm
n_1	1×10^{18}	cm^{-3}	L	500	µm
J	4	kA cm^{-2}	l	30	µm
V_c	2.0	V	Q_c	200	mW
V_{p1}	0.416	V	Q_{p1}	41.6	mW
V_{p2}	2.0	V	Q_{p2}	200	mW
V_j	3.0	V	Q_j	300	mW
V_{n2}	0.001	V	Q_{n2}	1.0	mW
V_{n1}	0.468	V	Q_{n1}	46.8	mW

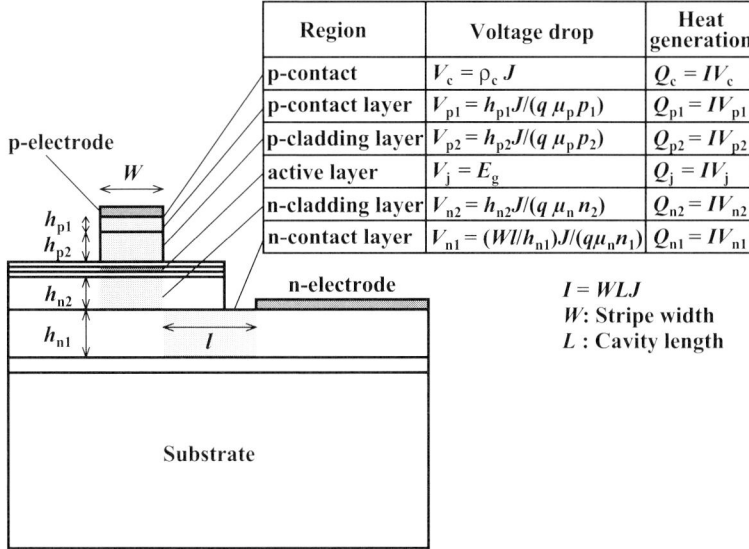

Fig. 16.14 Example of device structure and heat sources for thermal analysis. (G. Hatakoshi et al., "Thermal analysis for GaN laser diodes", Jpn. J. Appl. Phys., 38, pp. 2764–2768, **1999**, Copyright Institute of Pure and Applied Physics. Reproduced with permission.)

Eq. (16.11), are listed in the figure. The relation between S and the heat power Q generated in a volume V_Q is given by

$$S = Q/V_Q \tag{16.15}$$

Table 16.4 shows the parameters for the calculation of heat generation. The parameter W_0 in the table represents the entire width of the device. The other parameters are shown in Fig. 16.14. Calculations were carried out for the threshold current density, where the entire injected power was assumed to be transformed into heat. Material parameters for the thermal analysis are listed in Table 16.5. Parameters for the MQW layers were assumed to be identical to

Tab. 16.5 Material parameters for thermal simulation.

Material	k (W m^{-1} K^{-1})	ρ (kg m^{-3})	C_p (J kg^{-1} K^{-1})
GaN	130	6100	487
GaAlN clad	134	5930	502
Sapphire	25	3980	753
p-electrode	72	21450	130
Heat sink (Cu)	398	8930	380
Heat spreader (Au)	315	19300	120

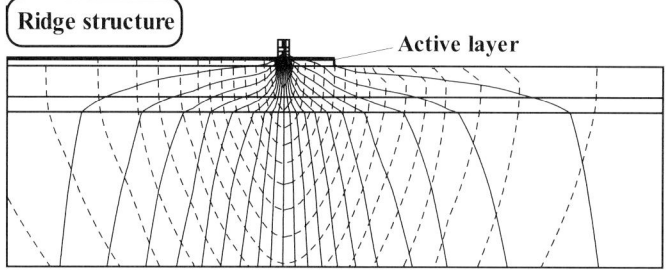

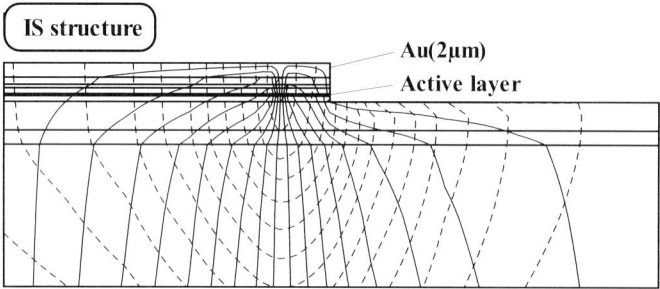

Fig. 16.15 Heat flow distributions for GaN lasers with ridge structure (upper figure) and inner stripe structure (lower figure).

those for GaN. Note that the specific heat C_p and the density ρ are used only for time-dependent analysis as shown by Eq. (16.11).

Figure 16.15 shows heat flow distributions for two kinds of GaN laser structures. The inner stripe (IS) structure shown in Fig. 16.15(b) is basically identical to that shown in Fig. 16.1. In this example, thick metal (Au) is put on the p-electrode, as a heat-spreading layer [32]. The IS structure has the advantage of lower thermal resistance due to the heat flow by the direct path to the heat sink as well as by a roundabout path. As shown in Fig. 16.15(b), heat can flow through the side region and upper region of the stripe. The thick electrode metal also has a remarkable effect in reducing the thermal resistance.

The thermal resistance of the device is given by the following definition.

$$R_{th} = \Delta T_{act}/(IV) \qquad (16.16)$$

where ΔT_{act} is the temperature rise in the active layer and the product IV represents the total injected power. In the simulation, R_{th} was calculated by the following equation.

$$R_{th} = \Delta T_{act}/Q_{total} \qquad (16.17)$$

where, Q_{total} is the total heat power generated in the device. Equations (16.16) and (16.17) are identical to each other under the condition where $IV = Q_{total}$, and this condition is reasonable for the threshold current, as mentioned above. As shown in Fig. 16.14, the total voltage V and total heat power Q_{total} are given as follows.

$$V = V_c + V_{p1} + V_{p2} + V_j + V_{n2} + V_{n1} \tag{16.18}$$

$$Q_{\text{total}} = Q_c + Q_{p1} + Q_{p2} + Q_j + Q_{n2} + Q_{n1} \tag{16.19}$$

It should be noted that R_{th} calculated by Eq. (16.16) is a lumped value and will differ, depending on the spatial distribution of the heat sources. For example, when there are large amounts of heat generation in a region apart from the active layer, such as the p-contact region and the cladding layer, the apparent thermal resistance, defined by Eq. (16.16) or Eq. (16.17), is smaller than the case where heat is generated only in the active layer. Similarly, if the thermal resistance is calculated by Eq. (16.16) for the condition of high output power, the resulting R_{th} value is smaller than that for the condition at the threshold current. Note that Eq. (16.17) gives constant R_{th} independent of the light output power.

Figure 16.16 shows the calculated temperature rise in the active layer and thermal resistance as a function of stripe width W for various mounting configurations [18]. Junction-up (p-side up) and junction-down (p-side down) configurations were calculated for the ridge and IS structures. The difference between the ridge and IS structure is the heat flow into the side region as shown in Fig. 16.15. A result for the IS structure with heat spreading layer (Au 2 µm) is also shown. The thickness of the Cu heat sink was set as 300 µm. In the p-down configuration, the thermal contact area was assumed to be the entire surface ($W_0 \times L$) for the IS structure and a restricted area ($W \times L$) for the ridge structure.

Under the assumption that the threshold current density is constant and the threshold current is proportional to the stripe width W, the total heat power generated in the device is proportional to W. On the other hand, the thermal resistance should be proportional to the inverse of size of the cross-section perpendicular to the heat flow direction. Therefore, the temperature rise ΔT_{act} should be constant, independent of W. However, the calculated results for ΔT_{act} depend significantly on the stripe width, as shown in Fig. 16.16(a). This is due to the difference in the heat spreading effect. In the case of small stripe width, the heat can flow from the active layer to the heat sink, not only through the direct path but also in a roundabout way through the side region. On the other hand, if the stripe width is large, the heat spreading is not as remarkable as that for the narrow stripe case. This causes a stripe-width dependence of the temperature rise as shown in Fig. 16.16. Also, it is understood that the thermal resistance is not necessarily proportional to $1/W$. As shown in Fig. 16.16(b), the dependence of the thermal resistance on the stripe width is not large except for case (e). This indicates that a narrow stripe structure has an advantage in that the temperature rise in the active layer can be reduced, and thus improved temperature characteristics enabling high temperature operation are expected.

Thermal characteristics are closely connected with the electrical and optical characteristics of the device. To be exact, electrical, optical and thermal equations should be solved simultaneously [33, 34]. Self-consistent simulation for the device model of nitride lasers and design optimization for high-power operation have been reported [33]. In the following, a rather simple model for the estimation of the maximum operation temperature is presented.

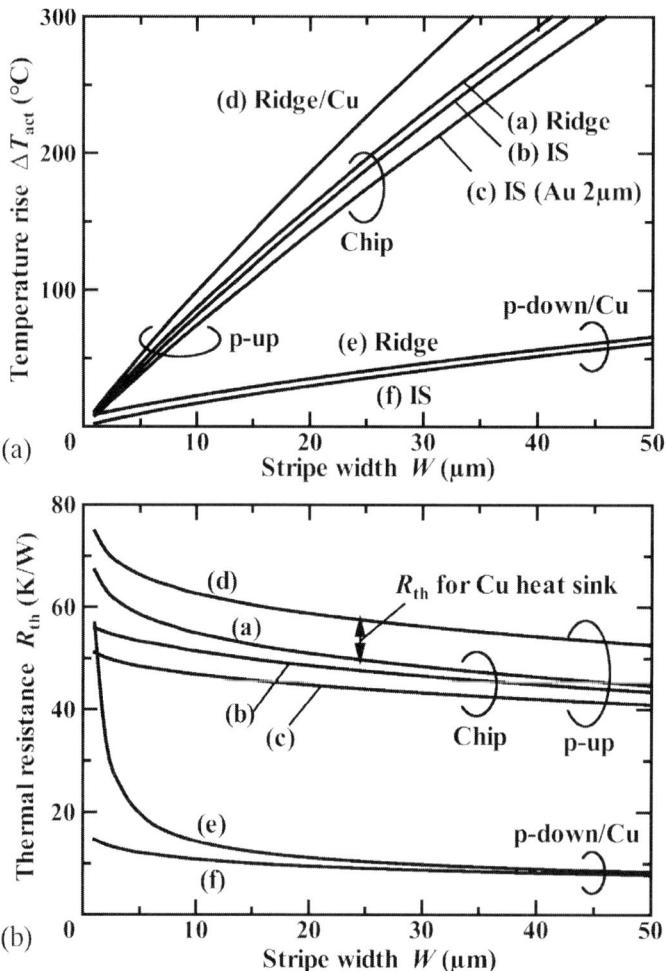

Fig. 16.16 Calculated temperature rise in the active layer (a) and thermal resistance (b) as a function of the stripe width W for various mounting configurations.

The maximum operation temperature, T_{max}, of a semiconductor laser can be estimated from the value of the thermal resistance and the temperature dependence of the threshold current [18,35,36]. In general, the relations between light output power P, injected current I, threshold current I_{th} and active layer temperature T_j are given by the following equations.

$$P = a\eta_d V_j \{I - I_{th}(T_j)\} \tag{16.20}$$

$$I_{th}(T_j) = I_{th0}(T_f) \exp([T_j - T_f]/T_0) \tag{16.21}$$

$$T_j = T_f + R_{th}(IV_j + I^2 R_s - P/a) \tag{16.22}$$

where, η_d is the external differential quantum efficiency, a is the ratio of light output for the output facet, V_j is the junction voltage, T_f is the environment temperature, T_0 is the characteristic temperature of the device, R_{th} is the thermal resistance and R_s is the series electrical resistance. The maximum output power P_{max} is given from Eqs (16.20)–(16.22) by setting $dP/dI = 0$, and the maximum operating temperature T_{max} is given by setting $P_{max} = 0$. Using this procedure, T_{max} is derived as follows.

$$T_{max} = T_{jmax} - R_{th}(I_{max} V_j + I_{max}^2 R_s) \tag{16.23}$$

$$T_{jmax} = T_f + T_0 \ln[I_{max}/I_{th0}(T_f)] \tag{16.24}$$

$$I_{max} = \frac{\sqrt{V_j^2 + 8(R_s/R_{th})T_0} - V_j}{4R_s} \tag{16.25}$$

Figure 16.17 shows T_{max} values estimated by using Eqs (16.23)–(16.25) [18]. In the calculation, the following relations were assumed.

$$I_{th0}(T_f) = WL J_{th0} \tag{16.26}$$

$$R_{th} = R_{th0}/L \tag{16.27}$$

$$R_s = R_{s0}/(WL) \tag{16.28}$$

where W is the stripe width, L is the cavity length and J_{th0} is the threshold current density at room temperature ($T_f = 25\,°C$) in pulsed operation. Note that the thermal resistance was assumed to depend only on L as represented by Eq. (16.27), whereas the electrical series resistance was assumed to depend on W and L. This assumption is reasonable, because the stripe-width dependence of R_{th} is small due to the heat spreading effect as shown in Fig. 16.16, and thus R_{th} can be set independently of W. On the other hand, current spreading is not so remarkable for the structure considered, and thus R_s was assumed to be proportional to $1/W$. The value for R_{th0} was assumed to be 2.5 KW^{-1}cm, which corresponds to the case of $R_{th} = 50$ KW^{-1} for $L = 500$ μm. Reported T_0 values for InGaN MQW lasers are around 80–170 K. In the calculation,

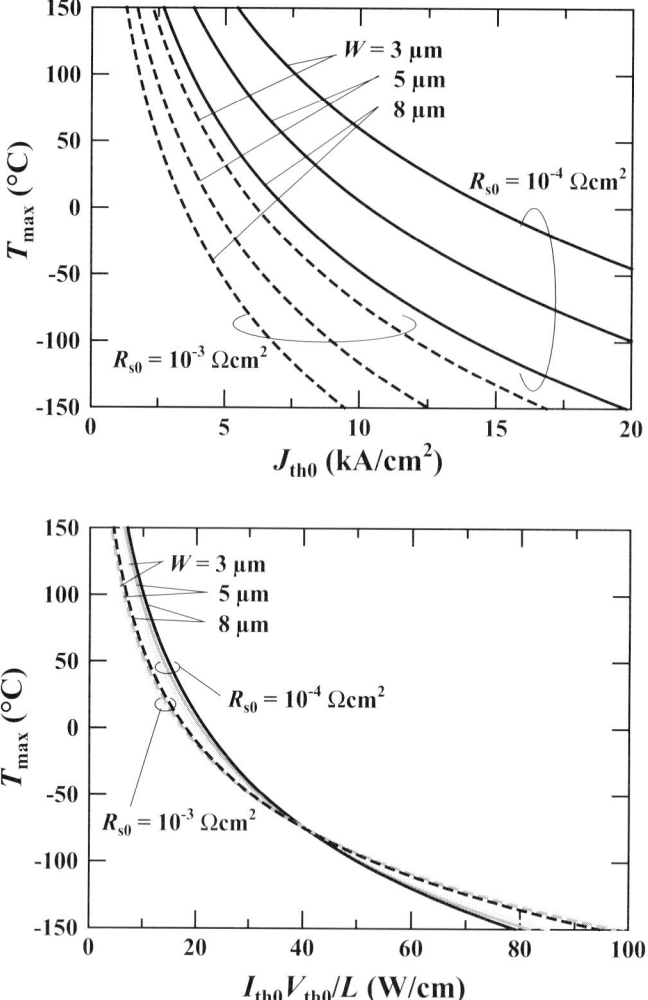

Fig. 16.17 Calculated results for maximum operation temperature. The lower figure indicates that T_{max} can be estimated almost uniquely as a function of required input power per unit cavity length. (G. Hatakoshi et al., "Thermal analysis for GaN laser diodes", Jpn. J. Appl. Phys., 38, pp. 2764–2768, **1999**, Copyright Institute of Pure and Applied Physics. Reproduced with permission.)

$T_0 = 150$ K was assumed. The junction voltage was assumed to be 3 V, which corresponds to the case of $\lambda = 413$ nm.

In Fig. 16.17(a), T_{max} is represented as a function of J_{th0}. It should be noted that a higher T_{max} value is obtained for small W. This arises from the assumption that R_{th} is independent of W due to the heat spreading effect. As shown in Fig. 16.17(b), the dependence of T_{max} curves on W and R_s values can be

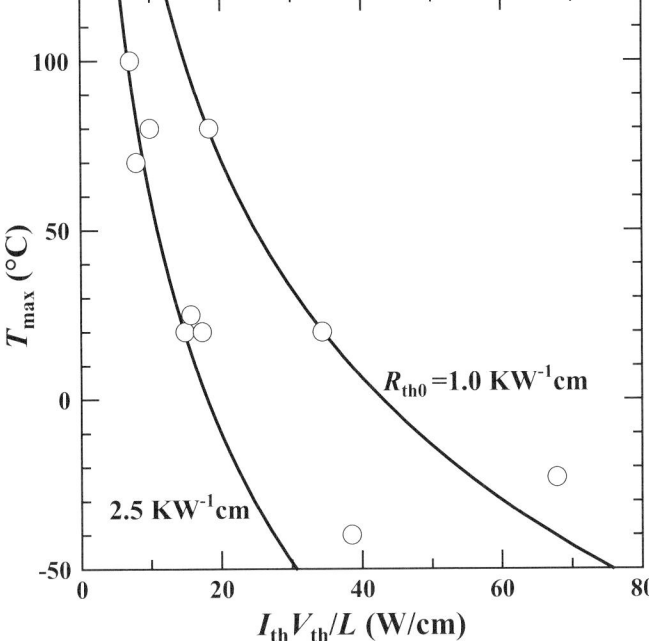

Fig. 16.18 Comparison of reported maximum operation temperatures and calculated results. (G. Hatakoshi et al., "Optical, electrical and thermal analysis for GaN semiconductor lasers", Int. J. Numer. Model. 14, pp. 303–323, **2001**, Copyright John Wiley & Sons Limited. Reproduced with permission.)

remarkably reduced by representing T_{max} as a function of $I_{th0}V_{th0}/L$, where $V_{th0} = V_j + R_s I_{th0}$. This means that T_{max} can be estimated almost uniquely as a function of the required input power per unit cavity length for pulsed operation, if the values of R_{th} and T_0 are given.

Figure 16.18 [18] shows the calculated T_{max} as a function of $I_{th}V_{th}/L$ for the case of $R_{th0} = 2.5$ KW^{-1} cm (junction-up configuration) and $R_{th0} = 1.0$ KW^{-1} cm (junction-down configuration or high-thermal-conductivity substrate). Reported T_{max} values (see references in Ref. [4]) are also plotted in the figure. It can be seen that the reported values agree well with the theoretical curve. The results indicate that the reduction in the required input power $I_{th}V_{th}/L$ per unit cavity length is very important for increasing the T_{max} value. The narrow stripe structure, as well as the reduction in J_{th} and V_{th}, is effective for reducing $I_{th}V_{th}/L$.

16.6
Conclusions

Device simulations for the electrical, optical and thermal characteristics of GaN-based lasers have been described. There are several restrictions in the practical fabrication process of GaN lasers. For example, it is difficult to obtain p-cladding layers with sufficient thickness, high Al composition and high acceptor concentration. This causes anti-guide-like characteristics of the guided mode, and an electron overflow from the active region to the p-cladding layers. Both the anti-guide mode and electron overflow have significant effects on the increase in the threshold current density. Therefore, the device design for the layer structure is very important in order to realize low-threshold devices. The use of sapphire substrate causes high thermal resistance of the device. Thermal characteristics are also very important for GaN-based lasers. The results of thermal analysis indicate that the reduction in the stripe width as well as the reduction in the threshold current density and operation voltage is very important for improving the temperature characteristics of GaN semiconductor lasers.

References

1 M. Kurata, *Numerical Analysis for Semiconductor Devices*, D.C. Heath & Company, Lexington Massachusetts, **1982**

2 D. P. Wilt and A. Yariv, IEEE J. Quantum Electron. QE-17 (**1981**) pp. 1941–1949

3 G. Hatakoshi, M. Kurata, E. Iwasawa and N. Motegi, Trans. IEICE Jpn. E71 (**1988**) pp. 923–925

4 G. Hatakoshi, M. Onomura and M. Ishikawa, Int. J. Numer. Model. 14 (**2001**) pp. 303–323

5 G. Hatakoshi, Proc. IEEE/LEOS 3rd Int. Conf. Numerical Simulation of Semiconductor Optoelectronic Devices, Tokyo (**2003**) paper MC1

6 J.H. Edgar ed., *Properties of Group III Nitrides*, IEE INSPEC, **1994**

7 S. Adachi, *Properties of Group-IV, III-V and II-VI Semiconductors*, John Wiley & Sons Ltd, **2005**

8 J.I. Pankove, S. Bloom and G. Harbeke, RCA Rev. 36 (**1975**) pp. 163–176

9 S. Sakai, Y. Ueta and Y. Terauchi, Jpn. J. Appl. Phys. 32 (**1993**) pp. 4413–4417

10 M. Suzuki, T. Uenoyama and A. Yanase, Phys. Rev. B 52 (**1995**) pp. 8132–8139

11 T. Yang, S. Nakajima and S. Sakai, Jpn. J. Appl. Phys. 34 (**1995**) pp. 5912–5921

12 W. Fang and S.L. Chuang, Appl. Phys. Lett. 67 (**1995**) pp. 751–753

13 S. Kamiyama, K. Ohnaka, M. Suzuki and T. Uenoyama, Jpn. J. Appl. Phys. 34 (**1995**) pp.L821-L823

14 M. Suzuki and T. Uenoyama, Jpn. J. Appl. Phys. 35 (**1996**) pp. 1420–1423

15 M. Asada and Y. Suematsu, IEEE J. Quantum Electron. QE-21 (**1985**) pp. 434–442

16 G. Hatakoshi, M. Onomura, S. Saito, K. Sasanuma and K. Itaya, Jpn. J. Appl. Phys., 38, 3B (**1999**) pp. 1780–1785

17 S. Nunoue, M. Yamamoto, M. Suzuki, C. Nozaki, J. Nishio, L. Sugiura, M. Onomura, K. Itaya and M. Ishikawa, Jpn. J. Appl. Phys. 37 (**1998**) pp. 1470–1473

18 G. Hatakoshi, M. Onomura, M. Yamamoto, S. Nunoue, K. Itaya and M. Ishikawa, Jpn. J. Appl. Phys., 38, 5A (**1999**) pp. 2764–2768

19 Y. Yokono, M. Ishizuka and T, Katoh, Natl. Conv. Rec. IEICE Jpn. (**1987**) paper 1–129

20. C. Kittel and H. Kroemer: *Thermal Physics, Second Edition*, W.H. Freeman and Company, **1980**
21. G. Hatakoshi, M. Suzuki, N. Motegi, M. Ishikawa and Y. Uematsu, Trans. IEICE Jpn. E71 (**1988**) pp. 315–317
22. K. Sasanuma and G. Hatakoshi, Tech. Report of IEICE Jpn. LQE97–152 (**1998**) pp. 49–55
23. K. Domen, R. Soejima, A. Kuramata and T. Tanahashi, MRS Internet J. Nitride Semiconductor Research (http://nsr.mij.mrs.org/), 3 (**1998**) Article 2
24. I. Akasaki and H. Amano, Jpn. J. Appl. Phys. 36 (**1997**) pp. 5393–5408
25. S. Nakamura and G. Fasol, *The Blue Laser Diode*, Springer, **1997**
26. W.T. Tsang, Appl. Phys. Lett. 38 (**1981**) pp. 835–837
27. M. Onomura, S. Saito, K. Sasanuma, G. Hatakoshi, M. Nakasuji, J. Rennie, L. Sugiura, S. Nunoue, J. Nishio and K. Itaya, IEEE J. Sel. Topics Quantum Electron. 5 (**1999**) pp. 765–770
28. D. Hofstetter, D.P. Bour, R.L. Thornton and N.M. Johnson, Appl. Phys. Lett. 70 (**1997**) pp. 1650–1652
29. A.E. Siegman, SPIE Proc. 1224 (**1990**) pp. 2–14
30. International Standard ISO 11146, *Lasers and laser-related equipment - Test methods for laser beam widths, divergence angles and beam propagation ratios*, **2006**
31. G. Hatakoshi, Opt. Review 10, No.4 (**2003**) pp. 307–314
32. O. J. F. Martin, G. -L. Bona and P. Wolf, IEEE J. Quantum Electron. 28 (**1992**) pp. 2582–2588
33. J. Piprek and S. Nakamura, IEE Proc. Optoelectron. 149 (**2002**) pp. 145–151
34. G. Hatakoshi and M. Ishikawa, Proc. IEEE/LEOS 3rd Int. Conf. Numerical Simulation of Semiconductor Optoelectronic Devices, Tokyo (**2003**) paper MC3
35. H. Okuda, M. Ishikawa, H. Shiozawa, Y. Watanabe, K. Itaya, K. Nitta, G. Hatakoshi, Y. Kokubun and Y. Uematsu, IEEE J. Quantum Electron. 25 (**1989**) pp. 1477–1482
36. G. Hatakoshi, K. Nitta, Y. Nishikawa, K. Itaya and M. Okajima, SPIE Proc. 1850 (**1993**) pp. 388–396

17
Resonant Internal Transverse-Mode Coupling in InGaN/GaN/AlGaN Lasers

Gennady A. Smolyakov and Marek Osiński

17.1
Introduction

Group-III-nitride-based diode lasers [1, 2] are of great importance as light sources for a variety of short-wavelength laser applications, such as high-capacity compact disks, UV irradiation, space communications, etc. Mode stability and control of mode shape has been the subject of quite extensive research [3–16] in InGaN-based lasers that very often show multi-lobed near-field and far-field emission patterns [17–21]. The typical design of the lasers is a single- or multiple-quantum-well (MQW) separate-confinement heterostructure that incorporates a multiple optical waveguide structure in the direction perpendicular to the junction plane, with refractive indices of several layers (GaN buffer/substrate, active region, GaN cap layer) being higher than those of adjacent layers (AlGaN cladding layers). The significant misfit stress associated with the lattice mismatch between AlGaN and GaN makes it difficult to grow thick AlGaN cladding layers due to the generation of cracks. The low Al-content and thin cladding layers of AlGaN, on the other hand, do not provide perfect optical isolation between the active region and the passive waveguides of the diode chip, which could lead to higher-order-mode generation and, thus, to additional optical losses, higher threshold current, and lower efficiency of the device.

A particularly strong effect is expected in the case of *phase synchronism*, when the phase velocities of eigenmodes of individual waveguides, treated as uncoupled, would be the same. Under those conditions, resonant mode coupling would occur, with significant reduction of the optical confinement factor and suppression of the modal gain. Therefore, a thin-cladding structure should be carefully optimized when designing an InGaN-based diode laser.

The phenomenon of resonant optical leakage from the active region into the passive waveguides in InGaN-based lasers has been given theoretical treatment in a number of publications [6, 22–29]. Here, we apply the approach of

Nitride Semiconductor Devices: Principles and Simulation. Joachim Piprek (Ed.)
Copyright © 2007 WILEY-VCH Verlag GmbH & Co. KGaA, Weinheim
ISBN: 978-3-527-40667-8

normal modes of coupled waveguides to analyze a typical group-III-nitride-based laser structure, paying attention to various characteristics of laser emission under the resonant conditions of internal mode coupling.

17.2
Internal Mode Coupling and the Concept of "Ghost Modes"

The lasing mode in cladding layers exists as an evanescent wave, whose intensity decays exponentially into the cladding layers. Therefore, the optical losses associated with the penetration of the laser emission via the claddings into the passive waveguides are negligible when the cladding thickness is much larger than the penetration depth, with the latter being dependent on the refractive index step at the GaN waveguide layer/AlGaN cladding layer interfaces. There are, however, certain reasons for using relatively thin cladding layers, which is often the case in the nitride-based diode lasers. Thin cladding layers are of technological advantage because they allow us (i) to avoid an increase in the series resistance of the diode, (ii) to avoid generation of cracks and other defects when there is a significant misfit stress associated with the lattice mismatch between AlGaN and GaN layers, and (iii) to reduce the growth time. In such laser structures, a significant part of laser emission can penetrate via the cladding layers and be accumulated within the passive waveguide layers. As a result, lasing occurs in a high-order mode, with the corresponding modifications in the modal spatial profile, optical confinement factor, propagation constant and internal losses.

Complex behavior of modes in multilayer systems can be understood using the approach of normal modes of coupled waveguides, often approximated by supermodes which are either in-phase or out-of-phase superpositions of the modes of individual uncoupled waveguides (see, for example, [30, 31]). Modes of the passive waveguides (passive modes) can interact with an active-layer mode (active mode), giving rise to two kinds of normal modes or supermodes of a laser structure. Away from resonance, one of the normal modes is localized predominantly in the active region (lasing mode), while the other modes are located mostly in passive waveguides (ghost modes). The lasing mode is the mode at which laser generation occurs. The lossy ghost modes are parasitic modes of a laser structure that can consume energy from the active region. The approximation of supermodes is valid only far away from resonance, when the modes of individual uncoupled waveguides have different phase velocities. At resonance, only the exact solutions (normal modes) of the wave equation for a multilayer waveguide system should be considered.

Consider first the off-resonance situation. In this case, it is easy to identify the modes of individual uncoupled waveguides as the components of which a supermode is built up. For simplicity, consider a system of two waveguides and assume that one of the interacting modes is that of the active layer (active mode, for example, a mode of the active InGaN/GaN/AlGaN waveguide), while the other is a mode of a passive waveguide (passive mode, for example, a mode of the *p*-GaN cap layer). The in-phase and out-of-phase superpositions of these modes give us the supermodes or normal modes that actually exist in this two-waveguide system. One of these modes is localized predominantly in the active region (lasing supermode), while the other is located mostly in the passive waveguide (ghost supermode). In the case of the lasing supermode, the passive mode component can be observed as a local maximum of the near-field intensity profile at the cap layer position. The highly doped cap-layer waveguide has a high internal loss due to both free-carrier absorption and absorption in the contact metal. Therefore, coupling of the active mode to the cap-layer passive modes introduces some additional modal losses, thus reducing the modal gain of the lasing supermode [32]. The modal gain suppression is proportional to the coupling coefficient, with the latter decreasing exponentially, as the thickness of the optical barrier (AlGaN cladding layer) increases. The lossy ghost supermode is a parasitic mode of the laser structure that can consume energy from the active region. In this case, the active mode component can be observed as a local maximum of the near-field intensity profile that the ghost supermode acquires at the active layer position.

Consider now the case when the phase velocities of the modes of individual uncoupled waveguides are very close to each other. In this case the language of supermodes may not be appropriate and we will instead consider only the exact solutions (normal modes) of the wave equation for a multilayer waveguide system. The two normal modes still resemble the in-phase and out-of-phase combinations of the modes of individual uncoupled waveguides. However, contributions of the active mode and the passive mode into each of the two normal modes are now comparable. This implies a much weaker optical confinement for the lasing mode, which now experiences much higher optical losses in the passive waveguide. Consequently, the modal gain of the lasing mode decreases, and this may result in no lasing action at all. On the contrary, the ghost mode "benefits" under such close-to-resonance conditions, since it receives more optical power from the active region. Its modal gain increases, but not necessarily sufficiently to reach the lasing threshold. Obviously, it is possible to have an unlucky laser design (i.e., with resonant mode coupling), where the laser would not operate even though its active region is of perfect quality and properly supplied with high material gain.

17.3
Device Structure and Material Parameters

The InGaN/AlGaN/GaN laser structure under consideration is based on the typical design reported in [33]. We consider an active region composed of four $In_{0.15}Ga_{0.85}N$ quantum-well layers separated by three $In_{0.02}Ga_{0.98}N$ barrier layers. The MQW active region together with the n-GaN and p-GaN waveguide layers form a 235 nm thick active waveguide. The $Al_{0.14}Ga_{0.86}N$ cladding layers separate the active waveguide from the p^+-GaN cap layer and the n-GaN buffer-substrate layer grown on sapphire substrate. Both GaN layers form "parasitic" passive waveguides where the photons of laser emission can be accumulated. With the exception of the contact layers, all the layers outside the active region are quite transparent to the laser emission. Penetration of the optical field from the cap layer into metal is limited to a very short distance. For example, at a wavelength of 400 nm the penetration depth into the gold electrode is ~120 nm. The lasing mode, then, could be sensitive to the optical parameters of a very thin intermediate layer formed at the nitride–metal interface. Since those parameters are not known, we use, instead, the parameters of pure Au often employed as contact material. The values of the optical parameters used in our calculations are listed in Table 17.1 for all the layers in the laser structure.

Tab. 17.1 Optical parameters of the materials comprising the InGaN/AlGaN/GaN MQW laser structure (n: refractive index, k: extinction coefficient, $dn/d\lambda$: material dispersion, d: layer thickness, $\lambda = 400$ nm)

Material	n	k	$dn/d\lambda$ [m^{-1}]	d [nm]	Comments
Au	1.5	1.7	0	200	Electrode material
p^+-GaN	2.55	0.000032	-3.14×10^6	100–2000	p^+-cap layer
p-$Al_{0.14}Ga_{0.86}N$	2.50	0.000032	-2.17×10^6	300–1000	p-cladding
p-GaN	2.55	0.000032	-3.14×10^6	100	p-waveguide layer
$In_{0.15}Ga_{0.85}N$(4) +	2.75	−0.0064–0	-1.133×10^7	3.5	QWs
$In_{0.02}Ga_{0.98}N$(3)	2.63	0.000032	-3.6×10^6	7.0	Barriers
n-GaN	2.55	0.000032	-3.14×10^6	100	n-waveguide layer
n-$Al_{0.14}Ga_{0.86}N$	2.50	0.000032	-2.17×10^6	300–1000	n-cladding layer
n-GaN	2.55	0.000032	-3.14×10^6	200–4000	n-buffer-substrate
α-Al_2O_3	1.77	0.000032	0	∞	Sapphire substrate
SiC	2.738	2.865×10^{-4}	-1.1×10^6	100000	SiC substrate

The refractive indices for AlGaN and InGaN materials were obtained from interpolation of the known data [34–38]. The material gain inside the quantum wells was assumed to be 1000 cm^{-1} for a fixed wavelength of 400 nm, unless otherwise indicated. Note that the extinction coefficient $k = 3.2 \times 10^{-5}$ in the passive layers corresponds to the absorption coefficient of 10 cm^{-1} calculated as $(4\pi/\lambda)k$ at a wavelength $\lambda = 400$ nm. The refractive index profile of the laser structure assumed in our calculations is shown in Fig. 17.1.

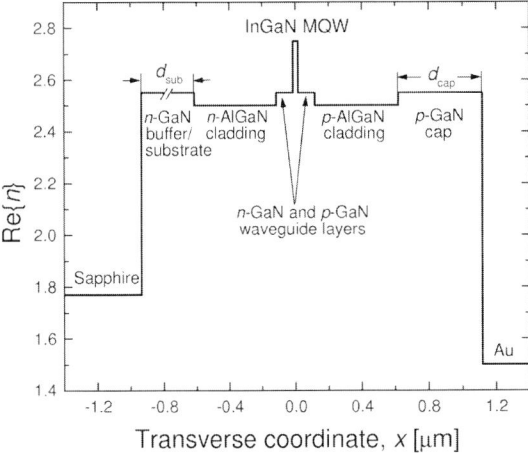

Fig. 17.1 Refractive index profile for a typical InGaN/AlGaN/GaN/sapphire diode laser.

We consider also, at some point, an alternative design of InGaN/AlGaN/GaN laser grown on a SiC substrate. The data for the refractive index of SiC and its dispersion were taken from [39]. The extinction coefficient of 2.865×10^{-4} in SiC corresponds to the absorption coefficient of 90 cm^{-1} at a wavelength of 400 nm.

17.4
Calculation Technique

We consider a unidirectional propagation of modes in a planar multi-waveguide system, with one active waveguide having material gain in the core, and the remaining waveguides being passive. We assume that the field of the waveguide mode varies as $\exp[i(\beta z - \omega t)]$, where t is the time, z is the propagation (longitudinal) direction, ω is the angular optical frequency, and β is the propagation constant in the z direction. Defining further the QW plane of the laser to be yz plane, and assuming only a transverse (vertical) spatial dependence of the electrical field amplitude $E_y(x)$, we can obtain the electric field $E_y(x, z, t) = E_y(x) \exp[i(\beta z - \omega t)]$ of TE modes, known to be predominant in nitride-based MQW lasers, as a solution to the scalar wave equation

$$\frac{\partial^2 E_y}{\partial x^2} + \left[k_0^2 n^2(x) - \beta^2\right] E_y = 0 \tag{17.1}$$

where k_0 is the free-space wave vector, and $n(x)$ is the complex refractive index profile of the laser structure. Using a combination of the complex Newton method and the transfer matrix method, we find the guided wave solutions of the wave equation. The calculated results for normal modes are presented in terms of the real part of the modal effective index $n_{\text{eff}} = \beta/k_0$, net modal gain/loss related to the imaginary part of the propagation constant as $G_{\text{mod}} = -2\,\text{Im}\,\beta$, and normalized near-field and far-field intensity distributions. The far-field pattern perpendicular to the junction plane is calculated according to:

$$I(\theta) = \cos^2(\theta) \left| \int E_y(x) e^{i k_0 \sin\theta \, x} dx \right|^2 \tag{17.2}$$

Note that, although the concept of supermodes (understood as superpositions of individual waveguide modes) is useful in understanding the nature of solutions, the normal modes of the laser structure are found here, without invoking the coupled-mode theory.

17.5
Results of Calculations

17.5.1
Resonant Conditions

Here, we present the results on the resonant behavior of normal modes in the "active waveguide – cap layer" and "active waveguide – GaN buffer/substrate layer" two-waveguide systems reported also in [25, 29].

For the laser structure with four quantum wells, the calculated effective index of the lasing mode is about 2.539, whereas that of the ghost modes varies from ~2.5 to ~2.55, depending on the thickness of the corresponding passive waveguide to which those ghost modes belong. However, the effective index curves for normal modes of coupled waveguides do not cross. Instead, we can specify the points of *anti-crossing* (resonance) when two curves come very close to each other.

17.5.1.1 Resonance in the "active waveguide – cap layer" system

The effective index and modal gain for the first three lowest-order normal modes in the two-waveguide system of the active waveguide and the p^+-GaN cap layer versus the thickness d_{cap} of the p^+-GaN cap layer are shown in Fig. 17.2 for p-AlGaN cladding layer thickness of 500 nm and Au taken as the p-side contact material.

Evolution of the electrical field distribution $E_y(x)$ with increasing cap layer thickness d_{cap} is illustrated in Fig. 17.3 for the first two normal modes. The

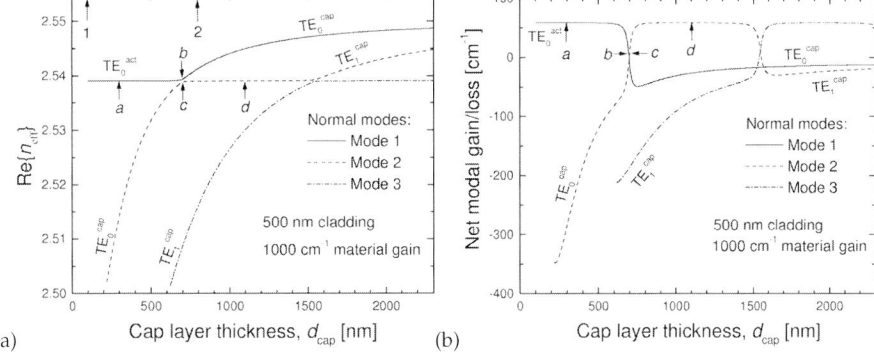

Fig. 17.2 Calculated (a) effective index and (b) net modal gain/loss for the first three lowest-order normal modes in the two-waveguide system of the active waveguide and the p^+-GaN cap layer as a function of the GaN cap layer thickness d_{cap}. The notation TE_N^{act}, TE_N^{cap} ($N = 0, 1, 2\ldots$) identifies the TE_N modes of the corresponding individual uncoupled waveguides (active waveguide, cap layer) to which the normal modes of the structure are very close for a particular range of values of d_{cap}. The arrows labeled with numbers indicate the minimum and maximum values being used for the GaN cap layer thickness in nitride-based lasers as extracted from the literature data (1 – [40], 2 – [41]). The arrows labeled with letters indicate the points at which near- and far-field patterns are calculated (cf. Section 17.5.2).

normal mode 1 resembles the TE_0 mode of the isolated active region waveguide (TE_0^{act}) up to the point when the first anti-crossing occurs. Up to that point, the normal mode 1 is the lasing mode of the structure or, in the language of coupled-mode theory, an in-phase supermode with the main maximum of intensity in the active region and a small intensity peak in the cap layer (cf. Fig. 17.3(a), curve A). After the first anti-crossing point, the normal mode 1 changes its character to become a ghost mode that can be viewed as an in-phase supermode with the main maximum of intensity in the cap layer and a small intensity peak in the active region. Except for this insignificant side peak, it resembles closely the TE_0 mode of the isolated cap layer (TE_0^{cap}) (cf. Fig. 17.3(a), curves B and C).

The normal mode 2 appears first as a ghost mode and resembles the TE_0^{cap} mode up to the point when the first anti-crossing occurs. It corresponds to an out-of-phase supermode with the main maximum of intensity in the cap layer and a small intensity peak in the active region (cf. Fig. 17.3(b), curve A). After the first anti-crossing point, it changes its character to become the lasing mode of the structure. It can be thought of then as an out-of-phase supermode with the main maximum of intensity in the active region and a small intensity peak in the cap layer (cf. Fig. 17.3(b), curve B). It resembles closely the TE_0^{act} mode and has almost constant effective index all the way up to the point of the second anti-crossing, where the normal mode 2 changes its character once again to become a ghost mode. This time, however, it has two

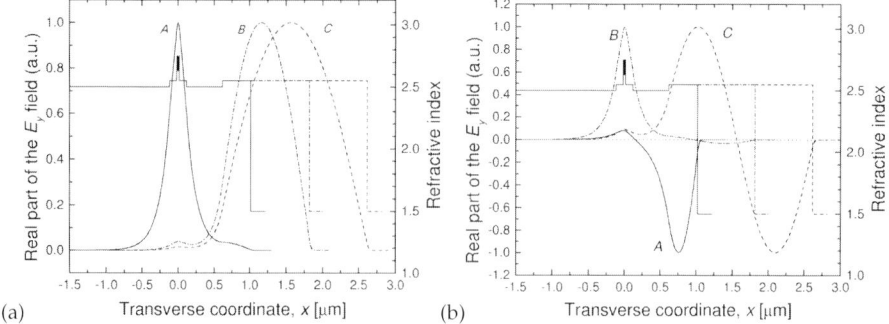

Fig. 17.3 Distribution of electrical field $E_y(x)$ for the two lowest-order normal modes of the two-waveguide system of the active waveguide and p-GaN cap layer, corresponding to conditions below the first resonance (d_{cap} = 400 nm, curves A), between the first two resonances (d_{cap} = 1.2 μm, curves B), and between the second and third resonances (d_{cap} = 2 μm, curves C). (a) Normal mode 1. (b) Normal mode 2.

main intensity peaks in the cap layer and a small one in the active region. The latter being insignificant, the mode is close to the TE_1 mode of the isolated cap layer (TE_1^{cap}) (cf. Fig. 17.3(b), curve C).

The behavior of the normal mode 3 is very similar to that of the normal mode 2 with the only exception that the number of intensity peaks in the cap layer (whether primary or secondary) in each of the three regimes of behavior increases by one compared to mode 2. The same pattern of behavior also applies to all the subsequent normal modes. As the thickness d_{cap} gradually increases, all normal modes, except for the first one, experience resonance twice, first starting as a ghost mode of a certain order (i.e., with a certain number of nodes in the cap layer), then turning into the lasing mode, and ending up as a ghost mode once again but of a higher order (i.e., with the number of nodes in the cap layer increased by one). Note that the number of undesirable small intensity peaks acquired by the lasing mode in the cap layer increases by one after each successive resonance, which accounts for the slow reduction in its modal gain with increasing d_{cap}. Note also that the number of parasitic ghost modes in the structure increases with an increase in d_{cap}, which may have a negative effect on the overall efficiency of the device.

The calculated value of d_{cap} for the first resonance in the two-waveguide Au-clad structure is about 696 nm. The second resonant value of d_{cap} is ~1545 nm, with the periodicity of the resonances being ~850 nm.

Figure 17.4 shows in more detail how the modal gain of the first two normal modes depends on d_{cap} for three different values of the optical barrier thickness d_{clad}. It is evident from this figure that a dip in the modal gain is characteristic of the lasing mode under resonant conditions. Increasing the optical barrier thickness makes the resonance narrower. It is still present even when the optical barrier is very thick, but cannot be fully resolved numerically because of limited computational accuracy. In Fig. 17.4, this is the case when $d_{clad} = 600$ nm, where the narrow resonance region is marked with a dotted line. The maximum value of ~ 59 cm^{-1} for the modal gain of the lasing mode is reached only when the corresponding normal modes have a dominant maximum of intensity within the active waveguide. This occurs only away from resonance, and the modal gain in this case is almost the same as it would be for the uncoupled TE$_0^{act}$ mode. Close to resonance, the modal gain drops dramatically, in spite of the same high value of the material gain (1000 cm^{-1}) in the quantum wells. The minimum of the modal gain (the point of intersection of the modal gain curves at resonance) is almost independent of the thickness of the optical barrier. In the particular case of the 500 nm thick AlGaN optical barrier, the calculated minimum modal gain is ~ 2 cm^{-1}. Obviously, these levels of modal gain are not sufficient to reach the lasing threshold. Therefore, a laser with a resonant value of d_{cap} would not be able to operate. The fact that the resonance becomes broader as the optical barrier gets thinner implies that in order to avoid an inadvertent resonance, it is important to keep the optical barriers reasonably thick.

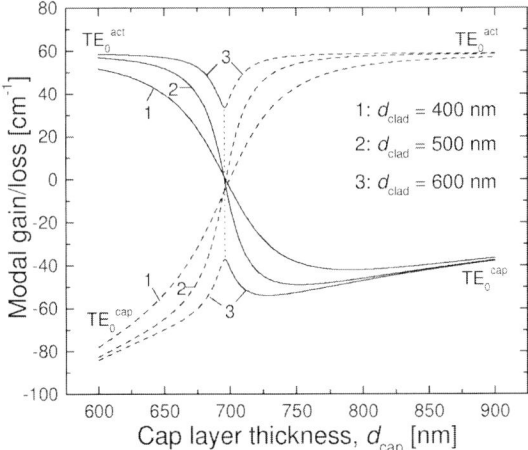

Fig. 17.4 Modal gain for the first two normal modes of TE polarization in the two-waveguide system of the active waveguide and Au-clad p-GaN cap layer as a function of the GaN cap layer thickness d_{cap}, calculated for three different values of the optical barrier thickness d_{clad}.

To see how the spatial characteristics of the laser evolve with increasing d_{cap}, we indicate four characteristic points around the first resonance in Fig. 17.2, at which we calculated the near-field and far-field intensity distributions (see Section 17.5.2). Although they belong to different normal modes, all four of them describe the lasing mode of the structure either far away from resonance (points *a* and *d*) or close to resonance (points *b* and *c*).

17.5.1.2 Resonance in the "active waveguide – GaN buffer/substrate" system

In this section, we consider the possible effects that another passive waveguide in the InGaN-based diode laser chip, namely the n-GaN buffer/substrate, can have on laser performance. The multiple resonances in the two-waveguide system of the active waveguide and the n-GaN buffer/substrate are shown in Fig. 17.5 as a function of the GaN buffer/substrate thickness d_{sub}. Qualitatively, in the case of the "active waveguide – GaN buffer/substrate" interaction, the behavior of the normal modes is quite similar to that of the "active waveguide – cap layer" interaction described in the precedent section, and the periodicity of the resonances is also ∼850 nm. The dips in the modal gain at the points of resonance are not, however, as deep as in the previous case, since the buffer/substrate waveguide is not nearly as lossy as the metal-clad cap layer (cf. Table 17.1). The minimum modal gain is now about 25 cm^{-1}, while the maximum value reached away from resonance remains as ∼59 cm^{-1}. Nevertheless, diode lasers with resonant values of d_{sub} will have a much higher threshold than well-designed devices.

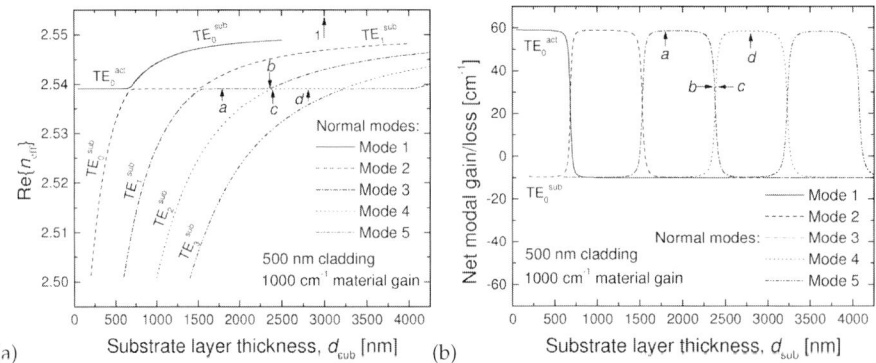

Fig. 17.5 Calculated (a) effective index and (b) net modal gain/loss for the first five lowest-order normal modes of TE polarization in the two-waveguide system of the active waveguide and the n-GaN buffer/substrate layer as a function of the GaN buffer/substrate layer thickness d_{sub}. The notation TE$_N^{\text{act}}$ and TE$_N^{\text{sub}}$, ($N = 0, 1, 2\ldots$) identifies the TE$_N$ modes of the corresponding individual uncoupled waveguides (active waveguide and buffer/substrate layer) to which the normal modes of the structure are very close for a particular range of values of d_{sub}. The arrow labeled "1" indicates the minimum value of d_{sub} used in nitride-based lasers, as extracted from the literature data [40].

We take points $a - d$ for calculating near and far-field patterns in this case around the third anti-crossing point. It should be noted that the exact positions of the resonances depend on the effective index of the lasing mode which can be controlled, for example, by changing the number of quantum wells, the thickness and/or composition of quantum wells and barriers, or the thickness of the GaN waveguide layers.

The important rules applied to the normal modes in a multiple-waveguide system when the thickness d of one of parasitic waveguides changes are as follows:

1. The calculated effective index of the normal modes $n_{\text{eff}}(d)$ closely follows the corresponding curves for the modes of individual uncoupled waveguides everywhere except for the regions where anti-crossing (resonance) occurs. Since no crossing really occurs, there is no spatial mode degeneracy even if the modes of individual uncoupled waveguides might have the same phase velocity.

2. According to this anti-crossing behavior, the lasing mode (located primarily in the active waveguide) is not represented by a single normal mode, but by a sequence of normal modes, with the mode order increasing by one at each subsequent resonance. The normal modes change their character (in terms of the location of the dominant peak in their intensity profile) at each anti-crossing point.

3. Away from anti-crossing points, the modal profile for each normal mode is close to that of the corresponding mode of an individual uncoupled waveguide, but contains additional contributions from the allowed modes of other waveguides.

4. When phase velocities of the modes of individual uncoupled waveguides coincide, the resonant coupling takes place. In such case, the interacting modes share the optical flux almost equally, and the modal gain in both normal modes can be substantially suppressed.

The system of normal modes and their evolution could easily be of even greater complexity should we decide to treat the full system of three waveguides in the laser chip: the p-GaN cap layer, the active waveguide and the n-GaN buffer/substrate. In this case, the increased complexity is manifested by a sequence of additional resonances in the "cap layer – GaN buffer/substrate" system [25].

17.5.2
Spatial Characteristics of Laser Emission under the Resonant Internal Mode Coupling

Normal modes in a system of two coupled waveguide layers are characterized by local field maxima in both waveguide layers. We restrict ourselves to considering a single-mode active waveguide, while the passive waveguides, treated as uncoupled from the active waveguide, can support higher-order modes. This means that the normal modes in our case can have only one field maximum localized within the active waveguide, and a different number of field maxima in a passive layer, depending on its thickness. Recalling that normal modes of a coupled waveguide system can be effectively represented as in-phase or out-of-phase superpositions of the modes of individual waveguides, we can expect the far-field pattern produced by a normal mode to be a rather complex result of interference from those local fields, strongly depending on their relative intensities and mutual phase relations. Far away from resonance, small intensity peaks acquired by the lasing mode in a passive waveguide layer can only slightly perturb the far-field intensity pattern dominated by the main intensity peak in the active waveguide. A particularly strong effect is expected in a close-to-resonance situation, when the lasing mode has peaks of comparable intensity both in the active waveguide and in a passive waveguide. Figure 17.6 illustrates this point, showing the near-field and far-field intensity distributions for the lasing mode calculated at, and away from, the first resonance in a two-waveguide system of the active waveguide and the p^+-GaN cap layer (cf. Fig. 17.2). Since only fundamental modes of the corresponding waveguides are involved in the internal mode coupling, we find this very simple case to be the most illustrative of the mechanism for far-field pattern formation in coupled waveguide systems.

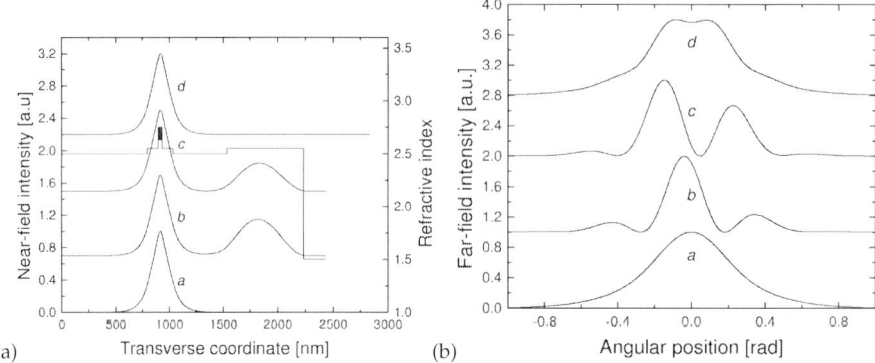

Fig. 17.6 Calculated (a) near-field and (b) far-field intensity distributions at (b, c) and away (a, d) from the first resonance in the "active layer – cap layer" system of Fig. 17.2. The refractive index profile in Fig. 17.6(a) corresponds to $d_{cap} = 700$ nm (point c in Fig. 17.2).

The ideal situation of a laser emitting at the fundamental mode of the active waveguide is represented in Fig. 17.6 by curves a. Since the p^+-GaN cap layer is very thin in this case (300 nm), the effective refractive index of its fundamental mode differs too much from that of the active waveguide mode to give any noticeable contribution to the lasing mode. Both curves b and c were calculated for close-to-resonance waveguide configurations. Although the near-field intensity distributions look very much alike, the phase relations for the fields of individual waveguides are quite different, which changes the far-field pattern dramatically. While the in-phase superposition of the modes of individual waveguides (curves b) gives a strong central (very slightly off-axis) intensity peak and two smaller side peaks, the out-of-phase superposition of the modes of individual waveguides (curves c) gives two characteristic, very pronounced off-axis peaks in the far-field intensity and much smaller side peaks. In both cases the peaks of intensity are much narrower than that obtained for the ideal case a, due to the near-field penetration into the passive waveguide. Curves d in Fig. 17.6 were calculated for $d_{cap} = 1.1$ µm, above the first resonance in the system and sufficiently far away from it. The intensity of the additional peak which the lasing mode acquires in the passive waveguide becomes negligibly small, so that it cannot be seen in Fig. 17.6(a), which causes re-broadening of the far-field intensity distribution. The far-field pattern is, however, very sensitive to the additional tiny peak of intensity in the passive waveguide, as the slight local minimum of the intensity appears at its axis according to the "out-of-phase" character of the normal lasing mode after the first resonance.

Qualitatively, the same features can be observed in the far-field intensity distributions of Fig. 17.7 calculated at some points around the third resonance in the two-waveguide system of the active waveguide and n-GaN buffer/substrate layer (cf. Fig. 17.5). Apart from the obvious increased complexity of the far-field patterns in terms of the number of intensity maxima, the far-field pattern away from the resonance (curves a, d) is still a single, relatively wide, and slightly rippled maximum of intensity. Close to the resonance (curves b, c), it becomes a series of much narrower intensity peaks. The "in-phase" normal lasing mode b produces peaks of comparable intensity, whereas the "out-of-phase" normal lasing mode c produces again two distinct off-axis intensity peaks with several smaller peaks.

The strength of the resonant internal mode coupling has been shown to depend on the cladding layer thickness, and increasing the cladding layer thickness makes the resonance narrower. Here, we again show that it is important to keep the cladding layers reasonably thick in order to avoid an inadvertent resonance. In Fig. 17.8, we start with a laser structure impaired by the resonant coupling of the active waveguide mode with the second-order mode of the GaN buffer/substrate. As the thickness of the cladding layer is increased, the system comes out of resonance, with accompanying significant improvement in the far-field pattern. It should be noted that the results presented in this section are in good qualitative agreement with the experimental results reported in [3–5, 8, 17].

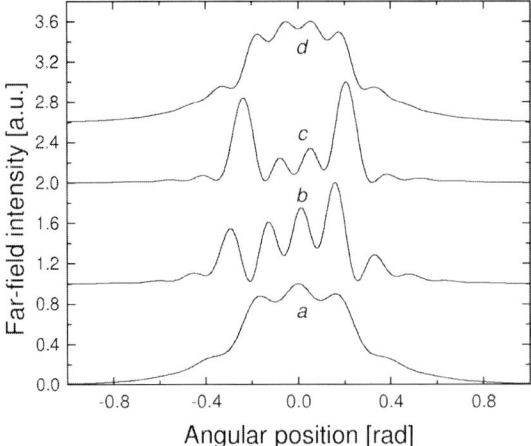

Fig. 17.7 Calculated far-field intensity distributions at (b, c) and away (a, d) from the third resonance in the "active waveguide – GaN buffer/substrate" system of Fig. 17.5.

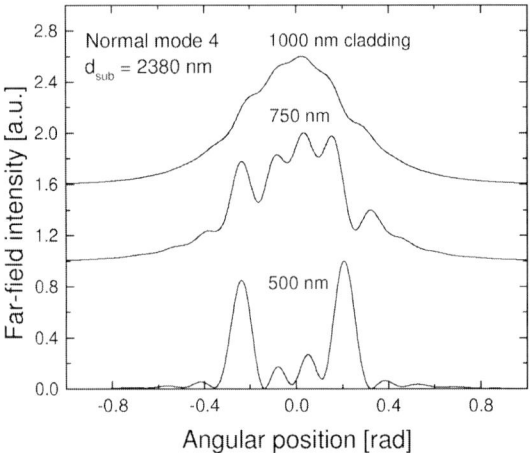

Fig. 17.8 Calculated far-field intensity distributions in the vicinity of the third resonance in the "active waveguide – GaN buffer/substrate" system of Fig. 17.5. The labels indicate the AlGaN cladding layer thickness d_{clad}.

17.5.3
Spectral Effects of the Resonant Internal Mode Coupling

We considered above how the modal gain of the normal modes depends on the thickness of the passive waveguides and optical barriers. Qualitatively, a similar behavior can be expected for the normal modes when the geometry of the device is fixed but the wavelength is varied. The only difference will be due to refractive index dispersion that should be properly included into consideration. Figure 17.9 shows the spectral dependence of the modal gain for the lasing mode calculated for three different values of the cladding layer thickness.

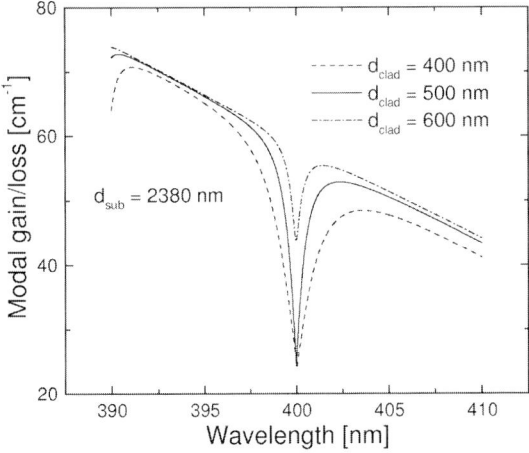

Fig. 17.9 Spectral dependence of the modal gain calculated at the resonant value $d_{sub} = 2380$ nm for three different values of the cladding layer thickness.

The main spectral effect of the internal resonant mode coupling is a periodic modulation of the amplified spontaneous and stimulated emission spectra first reported in InGaAs-based lasers [32, 42–45] and understood in terms of resonant leakage of the lasing mode into a thick GaAs substrate. It should therefore be expected that a similar effect in InGaN-based lasers could arise from the presence of substrate ghost modes. The ghost modes of the GaN buffer/substrate can produce only quite large (over 10 nm in the particular case under consideration, see Fig. 17.9) spectral periodicity due to the relatively small thickness of the n-GaN buffer/substrate. This kind of periodicity can be observed in very broad amplified spontaneous emission (ASE) spectra of InGaN-based lasers grown on sapphire substrates. Distinct peaks in the ASE spectrum with separations from 11 nm to 16 nm were observed in [46] for a laser sample with 2 μm thick GaN buffer layer. The spectral modulation period of 3.8 nm in spontaneous emission spectrum was also reported in [21] for an InGaN laser diode structure incorporating a 4 μm thick GaN buffer layer. The much shorter periodicity of 0.25 nm observed in longitudinal mode spectra in InGaN-based lasers grown on sapphire substrates was reported in [2,33]. To account for that kind of spectral periodicity, one would have to consider a composite sapphire/n-GaN substrate. Obviously, the effect of resonant coupling to ghost modes of the sapphire substrate on the modal gain is bound to be much weaker, since the photon exchange with the sapphire substrate is significantly reduced at the GaN/sapphire interface. For the structure shown in Fig. 17.1, the waveguide calculations predict extremely low penetration of the laser light into the sapphire substrate and no noticeable periodic spectral modulation of the modal gain within the accuracy of the numerical calculations. However, the waveguide theory does not take into account the effect of light scattering, which may also be responsible for the mode coupling. It is reasonable to expect an enhanced level of scattering at the interface of sapphire and the highly defective low-temperature GaN buffer layer.

In order to demonstrate, in the waveguide calculations, the spectral effect of resonant internal mode coupling for very thick substrates, we consider the alternative design of InGaN-based lasers grown on SiC substrates. We give the same numerical treatment to the laser structure of Fig. 17.1 where the composite sapphire/n-GaN substrate is substituted for a 100 µm thick SiC substrate. The spectral periodicity of ∼0.6 nm (∼4.6 meV) in the calculated modal gain (Fig. 17.10) is of the same order of magnitude as that (∼11 meV) observed in the gain spectra of an (In/Al)GaN laser grown on a SiC substrate that was reported to be thinned to approximately 100 µm [47, 48].

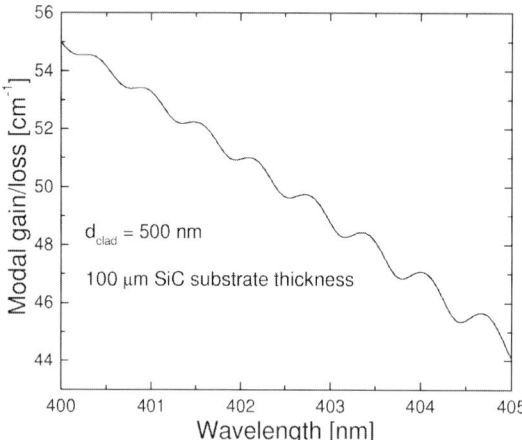

Fig. 17.10 Calculated spectral dependence of the modal gain in the InGaN-based laser on a 100 µm thick SiC substrate.

17.5.4
Carrier-Induced Resonant Internal Mode Coupling

In Sections 17.5.1 and 17.5.2, we drove a system of coupled waveguides into resonance by varying the thickness of a passive waveguide. By doing this, we gradually changed the effective refractive index of a particular mode of the passive waveguide, making it closer to the effective refractive index of the active waveguide mode, while the latter was kept unchanged. Obviously, the thickness of a waveguide is not a parameter that can be easily controlled in a real experiment and, although very interesting, all the theoretical curves shown above (especially those of Figs 17.2 and 17.5) have very little chance of being systematically reproduced experimentally. It is still possible, however, to induce the resonant internal mode coupling in a single device with fixed waveguide geometry, if we employ the effect of carrier-induced variation of the refractive index in the active region, this time to gradually change the effective refractive index of the active waveguide mode. We consider here the same two-waveguide system of the active waveguide and the p^+-GaN cap layer which, according to Fig. 17.2, would be at resonance for $d_{cap} = 696$ nm and material gain in the active region $g = 1000$ cm^{-1}. The system is then definitely not at resonance (although it may already be relatively close to it) at the transparency point, i.e, when $g = 0$ cm^{-1}. We consider this situation to be possible below threshold in the laser structure under consideration, and take it as an initial condition for our numerical experiment. By increasing material gain in the active region, we model the system behavior under increasing injection current. Due to the negative contribution of free carriers into the refractive index of the active region, the effective refractive index of the active waveguide mode would gradually decrease, thus driving the active waveguide mode into resonance with the fundamental mode of the cap layer. The changes in the refractive index are related to changes in the material gain through $\delta \operatorname{Im}\{n\} = -\delta g \lambda/4\pi$ and the linewidth broadening parameter $\alpha = \delta \operatorname{Re}\{n\}/\delta \operatorname{Im}\{n\}$. A whole range of experimentally measured values from $\alpha = 2.1$ to $\alpha = 6.3$ has been reported [47, 49, 50] for the linewidth broadening parameter α in InGaN quantum well lasers depending on the carrier density. In our calculations we used the value $\alpha = 4$ reported in [47] for current densities near the threshold.

The calculated net modal gain as a function of the material gain is shown in Fig. 17.11 for the first two lowest-order normal modes of this coupled waveguide system. The lasing mode in Fig. 17.11 is sequentially represented by two normal modes that go through the resonance and change their roles at $g = 1000$ cm^{-1}. The result is a strongly nonlinear dependence of the modal gain on the injection current, showing a region of *negative differential modal gain*. With the material gain monotonically increasing, this phenomenon is purely of waveguiding origin. On the one hand, as the system approaches resonance, the optical confinement of the lasing mode starts weakening too rapidly for the increase in material gain to be able to compensate for that effect. On the other hand, as the system comes out of resonance, the optical confinement of the lasing mode starts improving equally rapidly, giving rise to the region of superlinear growth of the modal gain. The dotted-dashed line *a* is given for comparison, as it demonstrates the predictable linear growth of the modal gain in a structure properly designed to be far away from the resonance with $d_{cap} = 500$ nm and all other parameters being the same.

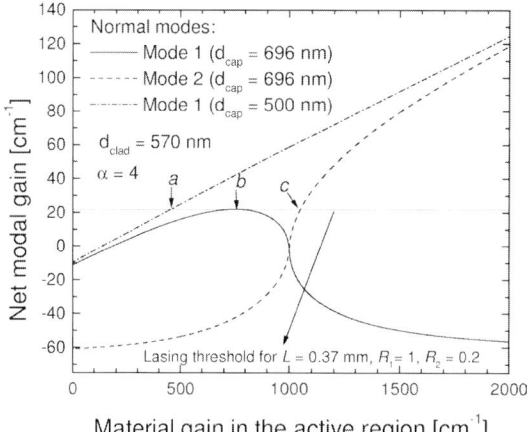

Fig. 17.11 Calculated net modal gain/loss for the first two lowest-order normal modes in the two-waveguide system of the active waveguide and the p^+-GaN cap layer as a function of material gain in the active region.

We indicate three points in Fig. 17.11 corresponding to the same level of net modal gain. Although relatively low ($\sim$22 cm^{-1}), it is sufficient for lasing action in a laser with the reflectance of one of the facets made very high. In the presence of the region of negative differential modal gain, the laser performance can be significantly modified by various kinds of oscillation instabilities (self-sustained pulsations, kinks and hysteresis in light-power characteristics [51–58]). Figure 17.12 shows what kind of far-field pattern, impaired by resonant internal mode coupling, should be expected from such a device.

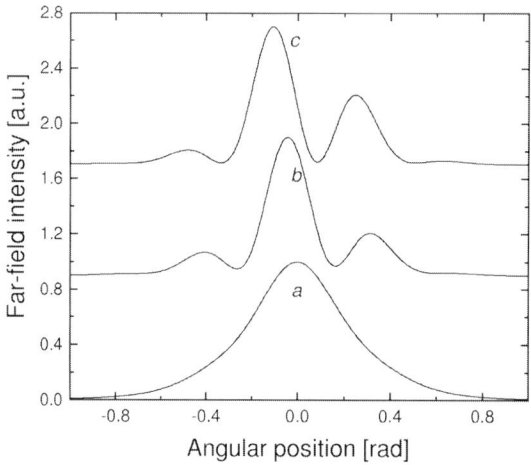

Fig. 17.12 Far-field intensity distributions calculated at points a, b, and c of Fig. 17.11.

It should be noted that, in a real device, the refractive index in the active region will also change due to the increasing internal temperature as the injection current increases, which would to some extent compensate the effect of free carriers described above. An accurate assessment of the effects of temperature increase would require self-consistent electrical-thermal calculations in addition to the optical analysis presented here.

17.6
Discussion and Conclusions

Resonant internal mode coupling is a general phenomenon that may occur in diode lasers whose internal structure incorporates multiple optical waveguides. This effect is particularly prominent in group-III nitride laser structures with thin cladding layers, where it can significantly suppress the modal gain and even stop the lasing action altogether [24–26].

Numerical analysis of a system of multiple coupled waveguides clearly shows the strong effect that ghost modes, located primarily at passive waveguides, can have on the modal gain of the lasing mode. The typical thickness for the optical barrier (AlGaN cladding layer) between the active waveguide and the p-GaN cap layer and between the active waveguide and the n-GaN buffer/substrate is about 500 nm, which is not sufficient for strong optical isolation. Consequently, internal mode coupling may occur and, at some specific values of the passive waveguide thickness, it can become resonantly enhanced.

The modal gain suppression at resonance is the result of a strong decrease in the optical confinement factor Γ. Close to resonance, the volume of the lasing mode is increased due to the substantial optical flux within the passive waveguide. Correspondingly, the lasing mode experiences additional optical losses in the passive waveguide. If the passive waveguide is the cap layer, the losses in the metal cladding are of importance.

In the case of resonant coupling with the Au-clad cap layer, we found the critical thickness d^* of ∼696 nm for the first resonance to occur. The resonance with a cap-layer ghost mode can be easily avoided if a cap layer thickness of less than d^* is used. However, it is also important to note that the first resonant thickness d^* depends on the effective index of the lasing mode. The calculated dependence is shown in Fig. 17.13. Thus, suitable design of the active waveguide is critical, as it determines the effective index of the lasing mode. For example, a greater number of quantum wells in the MQW structure will increase the effective index, making the resonant thickness of the cap layer larger. Also, an increase in the GaN-waveguide-layer thickness would produce an increase in the effective index.

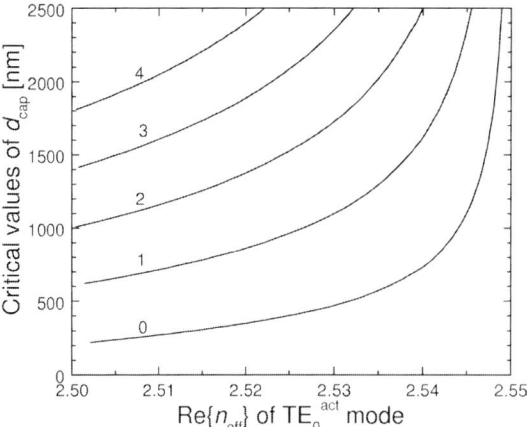

Fig. 17.13 Calculated dependence for the resonant thickness of the cap layer on the effective index of the fundamental mode of the uncoupled active waveguide. The numbers labeling the curves indicate the order of the corresponding mode of the uncoupled cap-layer waveguide.

In the case of internal mode coupling with a GaN buffer/substrate, resonance occurs with the same periodicity, 850 nm, as the buffer/substrate thickness d_{sub} is increased. The critical values for the buffer/substrate thickness are also sensitive to the effective index of the lasing mode, with the dependence being quite identical to that of Fig. 17.13. At resonance with the buffer/substrate ghost modes, the calculated minimum modal gain is not as low as it is at resonance with cap-layer ghost modes. This is due to the substantial contribution of the absorption in metal cladding, in the latter case. Since the waveguide approach does not take into account the light-scattering effects, our analysis neglects possible additional optical losses due to light scattering in the highly defective low-temperature GaN buffer layer adjacent to sapphire substrate. In a real laser structure, such scattering loss can lead to even lower modal gain at the points of resonance in the "active waveguide – GaN buffer/substrate" system.

The typical thickness of n-GaN buffer/substrate is 2–4 µm, therefore it can coincide occasionally with some of the critical values for resonant mode coupling. When such coincidence does occur, the modal gain of the lasing mode can be dramatically suppressed even though the active layers are supplied with high material gain. In diode laser manufacturing, the quality of a wafer with such critical thickness of the GaN buffer/substrate could be questioned, while the true origin of no-lasing (or high lasing threshold) would be the internal mode coupling. Therefore, considerations of resonant mode coupling should be part of the laser design process to select the optimal thickness for the GaN buffer/substrate layer. In order to avoid the effect of resonant suppression of the modal gain by ghost modes, all parasitic waveguide thicknesses should be away from their expected resonant values. It is also possible to reduce the possibility of an unintentional resonance by making the optical barriers stronger and by means of increasing the thickness and/or Al content of cladding layers.

The spatial characteristics of laser emission have been shown to change dramatically under the resonant conditions of the internal mode coupling. We identified the typical features which the far-field intensity pattern acquires under such conditions. We find it to be a great advantage to be able to single out the devices impaired by the resonant mode coupling simply by observation of their far-field patterns.

Furthermore, we have found that the carrier-induced variation of the propagation constant for the lasing mode can significantly modify the performance of laser structures vulnerable to resonant mode coupling. We have shown that the resonant mode coupling can be controlled by carrier-induced variation of the refractive index in the active region, and can result in a strongly nonlinear dependence of the modal gain on injection current, with possible region of negative differential modal gain.

While undesirable in most situations, the resonant mode coupling, on the other hand, could be used in a controllable manner to modify the laser properties (mode tailoring, mode selection). Lasers with negative differential modal gain are expected to display strong nonlinear dynamic behavior, including self-pulsations, bistability, and multistability, which are potentially attractive for a number of applications.

In summary, we have numerically analyzed the optical modes of a system of coupled waveguides, characteristic of InGaN/AlGaN/GaN diode lasers. The main conclusions are as follows.

1. Modes of parasitic waveguides (p^+-GaN cap layer, n-GaN buffer / substrate) in an InGaN/AlGaN/GaN diode laser chip can interact resonantly with modes of the active waveguide.

2. Under resonant conditions, the internal mode coupling can lead to severe modal gain suppression and distortions of both near- and far-field characteristics. Multiple narrow off-axis peaks of intensity in the far-field pattern indicate the resonant internal mode coupling, while a single relatively wide, slightly rippled, on-axis maximum of intensity in the laser far field indicates a properly designed device not susceptible to resonant mode coupling.

3. Due to carrier-induced variation of the refractive index in the active region, the resonant internal mode coupling can result in a strongly nonlinear dependence of the modal gain on the injection current, with a possible region of negative differential modal gain.

References

1 S. Nakamura, M. Senoh, S. Nagahama, N. Iwasa, T. Yamada, T. Matsushita, H. Kiyoku, and Y. Sugimoto, *Jpn. J. Appl. Phys., Pt. 2 (Lett.)*, vol. 35, pp. L74-L76, 1996.

2 S. Nakamura and G. Fasol, *The Blue Laser Diode: GaN based light emitters and lasers*, Springer-Verlag, Berlin, 1997.

3 M. Onomura, S. Saito, K. Sasanuma, G. Hatakoshi, M. Nakasuji, J. Rennie, L. Sugiura, S. Nunoue, M. Suzuki, J. Nishio, and K. Itaya, *Conf. Digest, 16th IEEE Int. Semicond. Laser Conf.*, Nara, Japan, 4–8 Oct. 1998, pp. 7–8.

4 M. Onomura, S. Saito, K. Sasanuma, G. Hatakoshi, M. Nakasuji, J. Rennie, L. Sugiura, S. Nunoue, J. Nishio, and K. Itaya, *IEEE J. Select. Topics Quantum Electron.*, vol. 5, pp. 765–770, 1999.

5 T. Takeuchi, T. Detchprohm, M. Iwaya, N. Hayashi, K. Isomura, K. Kimura, M. Yamaguchi, H. Amano, I. Akasaki, Y. Kaneko, and Yamada N., *Appl. Phys. Lett.*, vol. 75, pp. 2960–2962, 1999.

6 G. Hatakoshi, M. Onomura, S. Saito, K. Sasanuma, K. Itaya, *Jpn. J. Appl. Phys. Part 1 – Regul. Pap. Short Notes Rev. Pap.*, vol. 38, pp. 1780–1785, 1999.

7 T. Sato, M. Iwaya, K. Isomura, T. Ukai, S. Kamiyama, H. Amano, I. Akasaki, *IEICE Trans. Electron.*, vol. E83C, pp. 573–578, 2000.

8 G. Hatakoshi, M. Onomura, and M. Ishikawa, *Int. J. Numerical Modelling – Electronic Networks, Devices, & Fields*, vol. 14, pp. 303–323, 2001.

9 M. Rowe, P. Michler, J. Gutowski, S. Bader, G. Bruderl, V. Kummler, A. Weimar, A. Lell, V. Harle, *Phys. Status Solidi A-Appl. Res.*, vol. 188, pp. 65–68, 2001.

10 M. Rowe, P. Michler, J. Gutowski, S. Bader, G. Bruderl, V. Kummler, S. Miller, A. Weimar, A. Lell, V. Harle, *Phys. Status Solidi A-Appl. Res.*, vol. 194, pp. 414–418, 2002.

11 T. Asano, M. Takeya, T. Tojyo, T. Mizuno, S. Ikeda, K. Shibuya, T. Hino, S. Uchida, M. Ikeda, *Appl. Phys. Lett.*, vol. 80, pp. 3497–3499, 2002.

12 M. Ikeda, S. Uchida, *Phys. Status Solidi A-Appl. Res.*, vol. 194, pp. 407–413, 2002.

13 S. Ito, Y. Yamasaki, S. Omi, K. Takatani, T. Kawakami, T. Ohno, M. Ishida, Y. Ueta, T. Yuasa, M. Taneya, *Phys. Status Solidi A-Appl. Res.*, vol. 200, pp. 131–134, 2003.

14 U. T. Schwarz, M. Pindl, E. Sturm, M. Furitsch, A. Leber, S. Miller, A. Lell, V. Harle, *Phys. Status Solidi A-Appl. Res.*, vol. 202, pp. 261–270, 2005.

15 U. T. Schwarz, M. Pindl, W. Wegscheider, C. Eichler, F. Scholz, M. Furitsch, A. Leber, S. Miller, A. Lell, V. Harle, *Appl. Phys. Lett.*, vol. 86, pp. 161112, 2005.

16 J. Martin, M. Sanchez, *Phys. Status Solidi B-Basic Solid State Phys.*, vol. 242, pp. 1846–1849, 2005.

17 D. Hofstetter, D. P. Bour, R. L. Thornton, and N. M. Johnson, *Appl. Phys. Lett.*, vol. 70, pp. 1650–1652, 1997.

18 A. Kuramata, K. Domen, R. Soejima, K. Horino, S. Kubota, and T. Tanahashi, *Jpn. J. Appl. Phys. Part 2 - Lett.*, vol. 36, pp. L1130-L1132, 1997.

19 S. Nakamura, M. Senoh, S. Nagahama, N. Iwasa, T. Yamada, T. Matsushita, H. Kiyoku, Y. Sugimoto, T. Kozaki, H. Umemoto, M. Sano, and K. Chocho, *Appl. Phys. Lett.*, vol. 72, pp. 2014–2016, 1998.

20 D. K. Young, M. P. Mack, A. C. Abare, M. Hansen, L. A. Coldren, S. P. DenBaars, E. L. Hu, and D. D. Awschalom, *Appl. Phys. Lett.*, vol. 74, pp. 2349–2351, 1999.

21 H. D. Summers, P. M. Smowton, P. Blood, M Dineen, R. M. Perks, D. P. Bour, and M. Kneissl, *J. Cryst. Growth*, vol. 230, pp. 517–521, 2001.

22 V. E. Bougrov and A. S. Zubrilov, *J. Appl. Phys.*, vol. 81, pp. 2952–2956, 1997.

23 M. J. Bergmann and H. C. Casey, Jr., *J. Appl. Phys.*, vol. 84, pp. 1196–1203, 1998.

24 P. G. Eliseev, G. A. Smolyakov, and M. Osiński, *Proc. 2nd Int. Symp. Blue Laser and Light Emitting Diodes (2nd ISBLLED)*, Kisarazu, Japan, 29 Sept. – 2 Oct., 1998, Paper Th-10, pp. 413–416.

25 P. G. Eliseev, G. A. Smolyakov, and M. Osiński, *IEEE J. Select. Topics Quantum Electron.*, vol. 5, pp. 771–779, 1999.

26 G. A. Smolyakov, P. G. Eliseev, and M. Osiński, *Techn. Digest, 19th Annual Conf. Lasers & Electro-Optics CLEO '99*, Baltimore, MD, 23–28 May 1999, Paper CtuU5, p. 203.

27 P. G. Eliseev, *Laser Optics 2000: Semiconductor Lasers and Optical Communication* (S. A. Gurevich and N. N. Rosanov, Eds.), St. Petersburg, Russia, 26–30 June 2000, *Proc. SPIE*, Vol. 4354, pp. 12–23.

28 M. Buda, C. Jagadish, G. A. Acket, and J. H. Wolter, *COMMAD 2000 Proceedings*, pp. 438–442, 2000.

29 G. A. Smolyakov, P. G. Eliseev, M. Osiński, *IEEE J. Quantum Electron.*, vol. 41, pp. 517–524, 2005.

30 H. Haus, *Waves and Fields in Optoelectronics*, Prentice-Hall, Englewood Cliffs, NJ, 1984, Section 7.6, pp. 217–220.

31 W. Streifer, M. Osiński, and A. Hardy, *J. Lightwave Technol.*, vol. LT-5, pp. 1–4, 1987.

32 P. G. Eliseev and A. E. Drakin, *Laser Phys.*, vol. 4, pp. 485–492, 1994.

33 S. Nakamura, M. Senoh, S. Nagahama, N. Iwasa, T. Yamada, T. Matsushita, Y. Sugimoto, and H. Kiyoku, *Appl. Phys. Lett.*, vol. 70, pp. 2753–2755, 1997.

34 M. E. Lin, B. N. Sverdlov, S. Strite, H. Morkoç, and A. E. Drakin, *Electron. Lett.*, vol. 29, pp. 1759–1761, 1993.

35 H. Amano, N. Watanabe, N. Koide, and I. Akasaki, *Jpn. J. Appl. Phys., Pt. 2 (Lett.)*, vol. 32, pp. L1000-L1002, 1993.

36 O Ambacher, M Arzberger, D Brunner, H Angerer, F Freudenberg, N Esser, T Wethkamp, K Wilmers, W Richter, M Stutzmann, *Mrs. Internet J. Nitride Semicond. Res.*, vol. 2, Article 22, 1997.

37 H. Morkoç, *Nitride Semiconductors and Devices*, Springer-Verlag, New York 1999.

38 G. M. Laws, E. C. Larkins, I. Harrison, C. Molloy, and D. Somerford, *J. Appl. Phys.*, vol. 89, pp. 1108–1115, 2001.

39 E. D. Palik (Ed.), *Handbook of Optical Constants of Solids*, II, Academic Press, Orlando 1985.

40 S. Nakamura, M. Senoh, S. Nagahama, N. Iwasa, T. Matsushita, and T. Mukai, *Appl. Phys. Lett.*, vol. 76, pp. 22–24, 2000.

41 I. Akasaki, S. Sota, H. Sakai, T. Tanaka, M. Koike, and H. Amano, *Electron. Lett.*, vol. 32, pp. 1105–1106, 1996.

42 E. V. Arzhanov, A. P. Bogatov, V. P. Konyaev, O. M. Nikitina, and V. I. Shveikin, *Quantum Electron.*, vol. 24, pp. 581–587, 1994.

43 E. P. O'Reilly, A. I. Onischenko, E. A. Avrutin, D. Bhattacharyya, and J. H. Marsh, *Electron. Lett.*, vol. 34, pp. 2035–2037, 1998.

44 D. Bhattacharyya, E. A. Avrutin, A. C. Bryce, J. H. Marsh, D. Bimberg, F. Heinrichsdorff, V. M. Ustinov, S. V. Zaitsev, N. N. Ledentsov, P. S. Kop'ev, Zh. I. Alferov, A. I. Onischenko, and E. P. O'Reilly, *IEEE J. Sel. Top. Quantum Electron.*, vol. 5, pp. 648–657, 1999.

45 J. H. Marsh, D. Bhattacharyya, A. S. Helmy, E. A. Avrutin, A. C. Bryce, *Physica E*, vol. 8, pp. 154–163, 2000.

46 S. Heppel, J. Off, F. Scholz, and A. Hangleiter, *Conf. Digest, 16th IEEE Int. Semicond. Laser Conf.*, Nara, Japan, 4–8 Oct. 1998, pp. 11–12.

47 U. T. Schwarz, E. Sturm, W. Wegscheider, V. Kummler, A. Lell, V. Harle, *Appl. Phys. Lett.*, vol. 83, pp. 4095–4097, 2003.

48 U. T. Schwarz, E. Sturm, W. Wegscheider, V. Kummler, A. Lell, V. Harle, *Phys. Status Solidi A-Appl. Res.*, vol. 200, pp. 143–146, 2003.

49 M. Rowe, P. Michler, J. Gutowski, V. Kummler, A. Lell, V. Harle, *Phys. Status Solidi A-Appl. Res.*, vol. 200, pp. 135–138, 2003.

50 K. G. Gan, J. E. Bowers, *IEEE Photonics Technol. Lett.*, vol. 16, pp. 1256–1258, 2004.

51 H. Kawaguchi, *Jpn. J. Appl. Phys.*, vol. 21, pp. 371–376, 1982.

52 J. Chen, P. E. Barnsley, and H. J. Wickes, *Electron. Lett.*, vol. 27, pp. 1745–1747, 1991.

53 T. Odagawa, T. Machida, T. Sanada, K. Nakai, K. Wakao, and S. Yamakoshi, *IEE Proc.– J. Optoelectron.*, vol. 138, pp. 75–78, 1991.

54 M. Yamada, *IEEE J. Quantum.Electron.*, vol. 29, pp. 1330–1336, 1993.

55 P. G. Eliseev and A. E. Drakin, *Physics and Simulation of Optoelectronic Devices III* (M. Osiński and W. W. Chow, Eds.), San Jose, CA, 6–9 Feb. 1995, *Proc. SPIE*, vol. 2399, pp. 302–306.

56 P. G. Eliseev, G. Beister, A. E. Drakin, I. V. Akimova, G. Erbert, J. Maege, and J. Sebastian, *Kvantovaya Elektronika*, vol. 22, pp. 309–320, 1995.

57 G. Pham and G. H. Duan, *IEEE J. Quant. Electron.*, vol. 34, pp. 1000–1008, 1998.

58 S. K. Hwang and J. M. Liu, *Opt. Commun.*, vol. 83, pp. 195–205, 2000.

18
Optical Properties of Edge-Emitting Lasers: Measurement and Simulation
Ulrich T. Schwarz and Bernd Witzigmann

18.1
Introduction

During the development process of a particular type of laser diode (LD), one of the central problems is to minimize losses. This includes optical losses associated with scattering at the etched waveguide side walls and mirrors, absorption in the waveguide, cladding, and metal contacts. Other critical issues are the stability of the optical mode, high optical output power, and minimization of threshold current density. For the latter it is important to maximize the confinement factor and minimize inhomogeneous broadening of gain spectra. In this chapter we demonstrate how to combine experiments and simulations to understand these topics and then use the knowledge thus gained to optimize the laser structure.

In the following section we start with the basic problems of mode confinement, optical losses, and waveguide mode stability which can be treated to a certain approximation with one-dimensional (1D) and two-dimensional (2D) complex mode solvers. We use a 2D plane-wave expansion method which includes gains and losses. The code can be implemented and tested within days in a script language like Mathematica (Wolfram Research). For a quantitative evaluation of waveguide losses a much more powerful tool is needed. In Section 18.3, we describe how a 2D solver of the complex vector Helmholtz equation can be used to calculate the mode field and wavelength-dependent losses in an (Al,In)GaN laser diode on a GaN bulk substrate. This problem is all but trivial because of the leakage of the waveguide mode to the substrate. Therefore it is necessary for the code to include a rather large device region – in this example $40 \times 100\,\mu m^2$ – with a spectrum of closely spaced modes. In Section 18.4 we present a comparison of experimental and simulated gain spectra. The goal is to estimate the gain reduction due to material fluctuations in the quantum-well active region. For (Al,In)GaN LDs homogeneous and inhomogeneous broadening of the gain spectra are of the same order of magnitude. From the experiment alone it is impossible to distinguish both contributions. The solution is to use a microscopic simulation including many-body effects

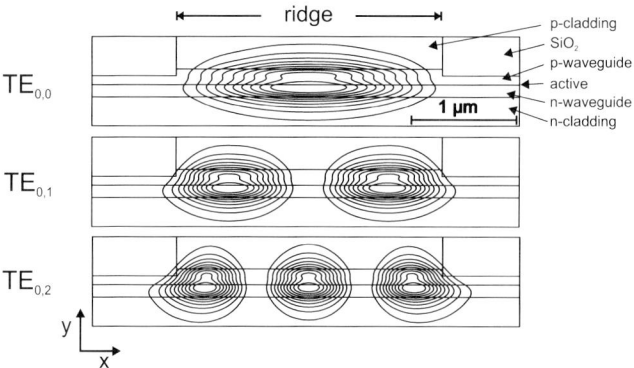

Fig. 18.1 Simulated intensity distribution for a deep etched ridge laser (ridge width $b = 2.5\,\mu m$). The side lobes of the higher lateral modes are shifted towards the n-waveguide and n-cladding. Thus their overlap with the lossy p-waveguide decreases, resulting in an increased gain.

which predicts the homogeneous broadening parameter-free. This is another example where a very complex simulation tool is essential in order to obtain quantitative results from experimental data.

18.2
Waveguide Mode Stability

For certain situations that are not overly complex we use a method for 2D simulations of the optical mode which is relatively straightforward to implement and still capable of treating gain and loss. The method is based on the fact that the wave equation in the paraxial approximation is formally equal to the Schrödinger equation. Thus the same methods used to find solutions of the quantum mechanical wavefunction in a 2D potential can be used to calculate the optical mode in a 2D waveguide. The method is limited to a scalar field, i.e., both TE and TM can be found, but not solutions of the full vector field, which causes limitations for the simulation of small structures or particular geometries. Because the method involves the solution of an eigenvector problem, it is also limited to simple mode structures. We employ a sample Mathematica worksheet on the internet to evaluate this code for an (Al,In)GaN LD waveguide test structure [1].

We start with the scalar wave equation

$$\nabla^2 \Psi + k_0^2 n^2(x,y)\Psi = 0 \qquad (18.1)$$

where Ψ represents the scalar field. In particular, for the TE mode Ψ is the y component of the electric field, pointing in the direction of the GaN c-axis. The

Tab. 18.1 Layer structure with index of refraction n and absorption coefficient α as used in the simulations. The material gain of $813\,\text{cm}^{-1}$ is a typical value. It is averaged over the whole active layer. The gain of the QWs is $2439\,\text{cm}^{-1}$.

Layer	Material	Width [nm]	n	α [cm^{-1}]
gold	Au	100	1.636	$6 \cdot 10^5$
contact	Mg:GaN	155	2.51	80
p-cladding	Mg:AlGaN	439	2.46	80
p-waveguide	Mg:GaN	91	2.525	80
electron barrier	Mg:AlGaN	30	2.37	80
active layer	InGaN/GaN	25	2.5266	−813
n-waveguide	Si:GaN	96	2.525	0
n-cladding	Si:AlGaN	564	2.46	0
buffer layer	Si:AlGaN	400	2.41	0
substrate	SiC	500	2.75	100

complex 2D refractive index distribution $n(x,y)$ describes the geometry of the waveguide, and k_0 corresponds to the frequency ω or vacuum wavelength λ_0 of the mode through $k_0 = \omega/c = 2\pi/\lambda_0$. For the mode propagating in the z direction we use the plane wave ansatz

$$\Psi(x,y,z) = \psi(x,y)e^{-i\beta z} \tag{18.2}$$

with the propagation constant β. Finding solutions for the waveguide mode corresponds to finding solutions of the eigenvalue equation

$$\hat{A}\psi = \beta^2 \psi \tag{18.3}$$

where the operator $\hat{A}$ is defined as

$$\hat{A} = \frac{\partial^2}{\partial x^2} + \frac{\partial^2}{\partial y^2} + k_0^2 \, n(x,y)^2 \tag{18.4}$$

Now the scalar field is expanded in a orthogonal basis $\phi_n(x,y)$,

$$\psi(x,y) \approx \sum_{n=1}^{N} c_n \phi_n(x,y) \tag{18.5}$$

with expansion coefficients c_n. Plane waves are the simplest way to construct a basis, yet the method works with an arbitrary basis which might be chosen to better fit the mode profile and thus allow a more efficient use of a limited number N of basis functions.

Equations (18.3) and (18.5) correspond to a quantum mechanics textbook problem where we finally have to solve the matrix

$$\sum_{n=1}^{N} c_n \left(\hat{A}_{m,n} - \beta^2 \delta_{m,n} \right) = 0 \tag{18.6}$$

Fig. 18.2 Measured (a to c) and simulated (d to f) near-field intensity distribution at the laser facet. (a) Ridge etched to 600 nm above QWs, i.e., with chiefly gain-guided lateral mode. (b) and (c) ridge etched to 200–300 nm above QWs, i.e., with chiefly index-guided lateral mode. The fundamental (b) and first-order (c) mode was observed for two individual LDs under nominal identical conditions.

with the matrix elements

$$\hat{A}_{m,n} = \iint dx\,dy\,\phi_m^* \hat{A} \phi_n \qquad (18.7)$$

Finding solutions takes two steps. First one has to construct the matrix $\hat{A}_{m,n}$ for a given 2D refractive index profile and vacuum wavelength λ_0. Next the matrix is diagonalized by a standard algorithm to search for the eigenvalues and eigenvectors. The eigenvectors together with the basis then constitute the waveguide mode while the corresponding eigenvectors are the complex propagation constant, related to the effective refractive index of the mode by $n_{\text{eff}} = \beta/k_0$. One advantage of this algorithm is that this method returns all modes of the waveguide in one step. This implies that the dimension of the basis must be large enough.

Results of a waveguide-mode simulation for a (Al,In)GaN laser diode is shown in Fig. 18.1. The ridge width was 2.5 µm and the ridge was etched halfway through the p-cladding layer. The vertical structure is listed in Table 18.1. The sides of the ridge are covered with SiO_2. A plane wave basis with 60 vectors in the transverse (y) direction and 15 vectors in the lateral (x) direction was used. To accelerate the simulations, even and odd lateral modes were calculated separately using a reduced set of base vectors of the respective symmetry.

The results of these simulations can be directly compared to the shape of the mode measured by near-field microscopy on the LD facet. Figure 18.2 shows the fundamental mode for a gain-guided structure and both the fundamental and first-order lateral mode for an index-guided LD. The gain-guided LD is realized by etching a 5 µm wide stripe through a SiO_2 layer sputtered onto the laser structure. These oxide stripe LDs are used as test structures, because they are easily processed and produce more comparable results in parameter series studies. For low threshold continuous wave (cw) LDs the tighter confinement of the index-guided or ridge LD is needed. As the 2D mode simulation is based on a complex refractive index, it is capable to predict the optical mode both for gain-guided (stripe) LDs and index-guided (ridge waveguide) LDs.

Measured and calculated mode profiles in Fig. 18.2 are in good quantitative agreement. The fact that the measured modes are a little wider in the y direction is due to the finite spatial resolution of approximately 100 nm of the near-field microscope.

In Fig. 18.3 the threshold current density for an index-guided LD as a function of ridge width is compared to that of a gain-guided LD for corresponding stripe widths. For the index-guided LD both optical mode and current are tightly confined within the ridge waveguide. Therefore the threshold current scales linearly with the ridge width, resulting in the observed constant threshold current density. For the gain-guided LD the threshold current density increases for narrow stripes because the optical mode becomes wider than the stripe.

To explain this observed dependency of threshold current density on the stripe width we simulate the optical mode, taking only gain-guiding into account. The gain region was confined to an area of the same width as the stripe, corresponding to zero lateral current spreading. So, in the lateral direction the imaginary part of the refractive index was varied while the real part was kept constant. To calculate the threshold current density we calculated the material gain necessary to achieve a modal gain $g = 15\,\text{cm}^{-1}$ to compensate the mirror losses

$$\alpha_{\text{mirr}} = \frac{1}{2L} \ln\left(\frac{1}{R_1 R_2}\right) \tag{18.8}$$

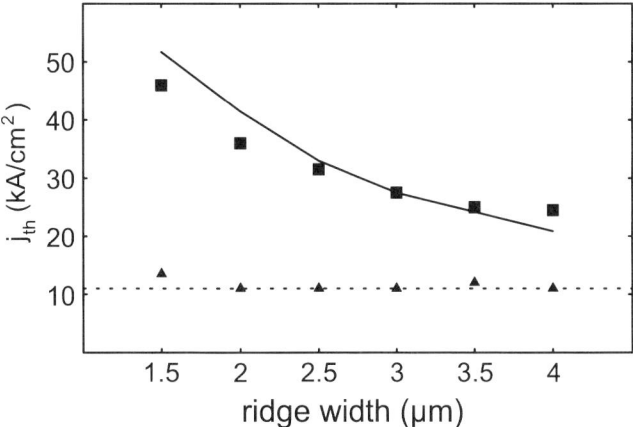

Fig. 18.3 Threshold current density j_{th} as a function of ridge width b. Triangles correspond to LDs with ridges etched through the QWs into the n-cladding (strong index guiding) and squares to hardly etched LDs (gain guiding). The solid curve was calculated for gain-guided LDs without current spreading. The dotted line marks the threshold current $j_{th} = 11\,\text{kA cm}^{-2}$ in the limit of strong index guiding.

with $R_{1,2} = 0.18$ being the facet power reflectivity of the uncoated facets, and $L = 600\,\mu\text{m}$ being the cavity length. The internal losses of approximately $\alpha_i = 40\,\text{cm}^{-1}$ are accounted for by the complex refractive indices of the vertical layer structure as listed in Table 18.1. From the material gain necessary to reach this modal gain we calculate the current density assuming a linear relation between the current density and material gain below the threshold. The proportionality constant is derived from simulations of index-guided LDs (triangles in Fig. 18.3) where the mode is confined to the ridge waveguide. As expected for gain-guided LDs, the optical mode near threshold is considerably wider than the stripe width, and the ratio of the mode width to the stripe width increases with decreasing stripe width. The good agreement of the solid line in Fig. 18.3 with the measurements for the index-guided LD demonstrates that the increasing threshold current density for small stripes can be explained by the gain-guiding mechanism with zero current spreading in the p-cladding layers.

In near-field measurements of the waveguide mode we frequently observed that nominally identical LDs operated either in the fundamental or first-order lateral mode. This near degeneracy of the gain of modes of different orders was confirmed by Hakki–Paoli gain measurements [2]. Here, again, the 2D simulations helped to find the origin of this degeneracy and to optimize the waveguide geometry to suppress higher order lasing, as shown in the following.

Figure 18.4 shows the modal gain and confinement factor for different ridge geometries. One parameter is the ridge width, the other is the etch depth. The etch depth is with respect to the active layer, and positive values correspond to a ridge etched through the active region. In this case, other issues like electrical shortening of the n- and p-layers become a critical issue, so that in general a deep etching is not desirable. Large negative numbers or shallow etching correspond, basically, to gain-guided LDs. The vertical lines at $\pm 120\,\text{nm}$ in Fig. 18.4 mark the boundaries of the p- and n-waveguide layers. For this simulation the material gain was kept at a constant value to achieve a modal gain $g = 15\,\text{cm}^{-1}$ in the limit of a deep-etched ridge. The confinement factor in Fig. 18.4 acts as expected, i.e., it continuously rises with increasing etch depth, corresponding to a tighter confinement of the mode to the ridge width. For the 1.5 μm wide ridge only the two lowest order lateral modes are bound, while for the wider ridges higher order modes are bound. Only the lowest three modes are plotted.

A surprising result is the shape of the modal gain curves. For an etch depth corresponding to the interface between the p-cladding and p-waveguide, the modal gain for all modes is nearly degenerate. For slightly deeper etched ridges, the higher order modes have a higher gain and thus a lower threshold. This behavior is caused by the higher optical losses in the p-layers of the

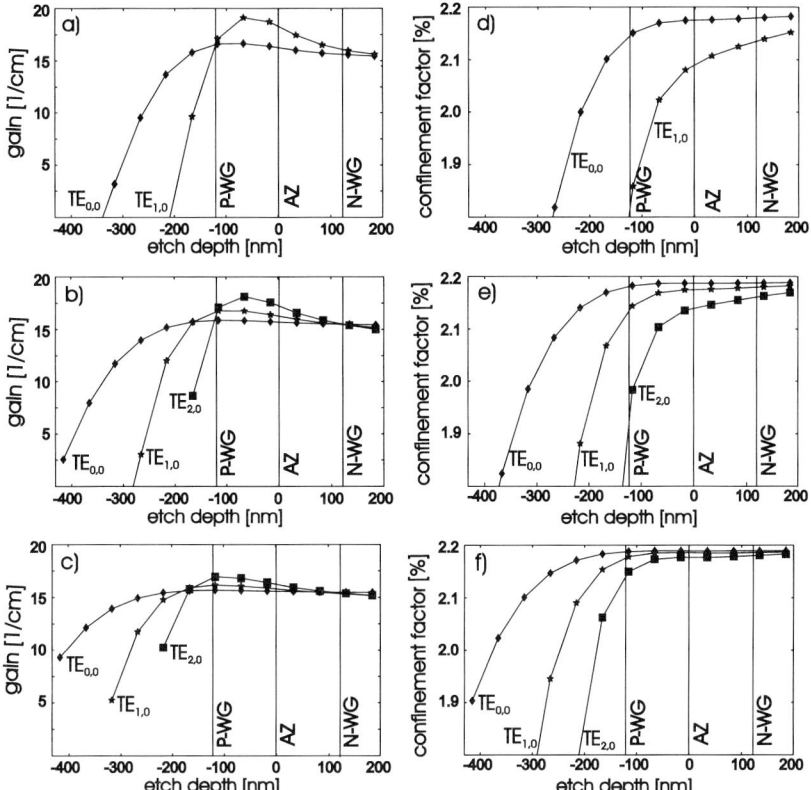

Fig. 18.4 Gain (a to c) and confinement factor (e to f) for different ridge widths and etch depths. (a) and (d) show the gain and confinement factor for a 1.5 μm LD. LDs with a 2.5 μm ridge (b and e) show a stronger lateral confinement of the mode. In 3.5 μm ridge LDs (c and f), a strong lateral confinement can be achieved even in slightly etched LDs.

structure. In the 2D simulations we include losses of $80\,\mathrm{cm}^{-1}$ in the p-layers, assuming no losses on the n-side. This value can be justified from literature data showing that the losses caused by doping are considerably lower in Si doped n-GaN than those of Mg doped p-GaN [3]. The loss averaged over the mode is then $\alpha_i = 40\,\mathrm{cm}^{-1}$ in agreement with the value of α_i measured by the Hakki–Paoli method for our LDs [4].

The higher gain for higher order modes observed in the simulations (see Fig. 18.4) and experiment (see Fig. 18.2 and [2]) stems from a different overlap of the modes with the lossy p-layers. The higher order modes have more degrees of freedom to optimize overlap with the active region and at the same time minimize overlap with the p-layers. For etch depths close to the p-cladding/p-waveguide interface, the ridge edge pushes the outer lobes of

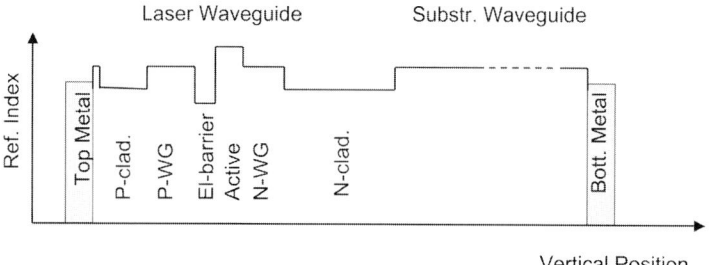

Fig. 18.5 Schematic refractive index profile of vertical laser structure. Both the substrate as well as the active region (plus surrounding layers) form optical waveguides. WG stands for waveguide, a detailed structural description can be found in Table 18.1.

higher order lateral modes into the n-waveguide region. This can be seen nicely in Fig. 18.1.

18.3
Optical Waveguide Loss

Optical losses in an edge-emitting laser cavity come from several mechanisms. First, a fraction of the optical power is not reflected back into the cavity at the facet, and escapes as usable output power. In addition, internal losses originate from absorption by the materials or optical diffraction and scattering. These effects occur in different regions of the cavity. In this section, the goal is to analyze the optical loss budget of a GaN-based edge-emitting laser cavity with the aid of measurement and simulation and to clarify the role of substrate modes. Minimizing internal optical losses is crucial for achieving low threshold current, but in particular high output power. The external efficiency of an edge-emitting laser is directly proportional to the ratio of the mirror loss and the total loss, $\alpha_{mirr}/\alpha_{tot}$ [5], and hence, high efficiency requires low parasitic losses.

The total losses can be spatially divided into three parts:

$$\alpha_{tot} = \alpha_{mirr} + \alpha_{las} + \alpha_{subst} \tag{18.9}$$

The mirror loss is given in Eq. (18.8), and we will focus in the following on the remaining two loss parts (α_{las} and α_{subst}), while neglecting any optical losses due to interface scattering.

An illustration of the vertical geometry is shown in Fig. 18.5. It can be divided into the laser waveguide (upper part) and the substrate waveguide. The internal loss of the laser waveguide $\alpha_{las} = \int_{las} \alpha(r) |\Phi|^2 dV$ is the accumulated loss integrated over the epitaxial structure around the active region plus passivation and top contact. The substrate losses $\alpha_{subst} = \int_{subst} \alpha(r) |\Phi|^2 dV$ are located in the substrate and bottom contact. The normalized optical intensity distribution $|\Phi|^2 = |\vec{E}\vec{E}^*|/\int |\vec{E}\vec{E}^*|dA$ is used as a weighting factor. It is obtained as a solution of the waveguide equation, in our case the complex vectorial Helmholtz equation:

$$\left(\nabla^2 + \frac{\omega^2}{c^2}(n^2(x,y) - n_{\text{eff}}^2)\right)\vec{E}(x,y) = 0 \tag{18.10}$$

In contrast to Eq. (18.1), the electric field $\vec{E}$ is a vectorial quantity, which satisfies transitions at material boundaries in an exact manner. The refractive index $n(x,y)$ is a complex number, representing the local index as well as the local absorption of the material. The effective waveguide index n_{eff} is complex as well, and its imaginary part represents the total waveguide loss:

$$\alpha_{\text{tot}} = \frac{2\omega}{c} Im(n_{\text{eff}}) \tag{18.11}$$

In an optimal edge-emitting laser design, the refractive index profile consists of a high-index region that confines the light around the active region, surrounded by low-index regions for which Eq. (18.10) results in decaying, so-called evanescent solutions. Besides guided modes, there also can be radiating leaky quasi-modes in the waveguide structure. Ideally, the operation frequency is below cutoff of both the higher-order modes and radiating quasi-modes, so that there are no loss mechanisms due to coupling of the main mode with the unwanted parasitic modes. In semiconductor lasers, material configurations are frequently used where the substrate possesses a higher refractive index than the effective index of the laser waveguide.

This can be the case for the GaAs/AlGaAs system in infrared lasers as well as in the GaN/AlGaN or Sapphire/(Al)GaN system for lasers in the UV range. The substrate can be seen as a separate, lossy waveguide, which supports modes in the operating regime of the laser waveguide [6]. The evanescent tail of the mode in the active laser waveguide can excite a resonant mode in the substrate, depending on its amplitude at the cladding–substrate interface and frequency. Therefore, part of the optical power generated in the active laser waveguide is transferred to the substrate, and lost by absorption in the bottom contact or by the substrate itself. The propagation vector of the mode is then tilted towards the substrate.

Now substrate modes in a GaN-based laser are examined having been grown epitaxially on a GaN substrate. AlGaN is used as cladding material around GaN waveguide layers, and the active region consists of InGaN/GaN wells/barriers. The detailed structural data and parameters can be found in Table 18.1, except that the n-cladding layer has a thickness of 800 nm, and the interface layer and SiC substrate are replaced by a GaN substrate with a thickness of 100 µm.

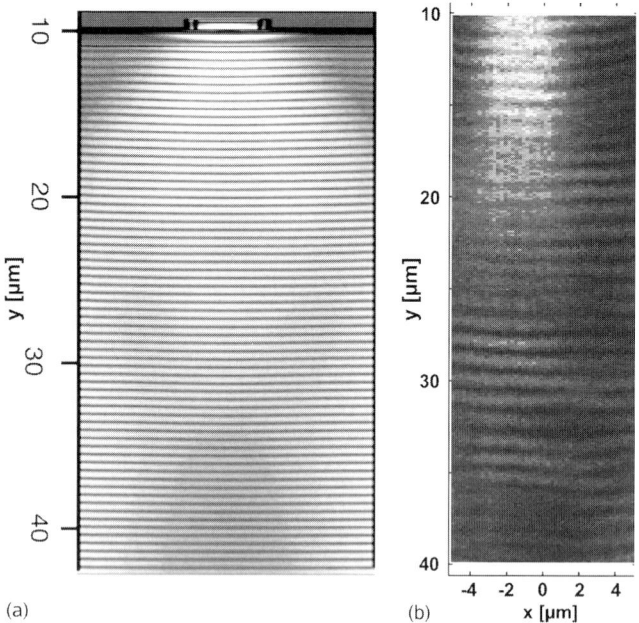

Fig. 18.6 (a) Optical intensity of a longitudinal cross-section of a ridge waveguide laser including the GaN substrate, simulated. The plot area has been truncated for illustration purposes, the substrate thickness is 100 µm. (b) SNOM measurement of the optical intensity on the cleaved substrate facet just below the LD waveguide, taken during lasing operation.

A direct experimental demonstration of the substrate mode is shown in Fig. 18.6, together with the simulation. The experiment has been carried out with a scanning near-field microscope (SNOM), measuring the intensity distribution on the cleaved laser diode facet and substrate below the facet. The laser diode has been driven in a pulsed mode, and details of the measurement can be found in [7,8]. The substrate mode is clearly visible with decreasing intensity in direction from the laser waveguide towards the bottom contact. Additionally, a slight curvature of the mode below the laser waveguide can be observed, which indicates lateral diffraction of the substrate mode. The simulation is the solution of Eq. (18.10), and the simulation domain has a width of 40 µm and a height of 100 µm from the top contact to the bottom contact, including the entire substrate. The left and right sides of the simulation domain are truncated with perfectly matched layer-absorbing boundary conditions, so that lateral diffraction losses are treated directly.

Another signature of substrate modes can be seen in the optical loss spectra as periodic oscillations and increased internal losses [9]. The symbols in Fig. 18.7 show the spectral loss $\alpha_{tot} - \alpha_{mirr}$ measured by the Hakki–Paoli

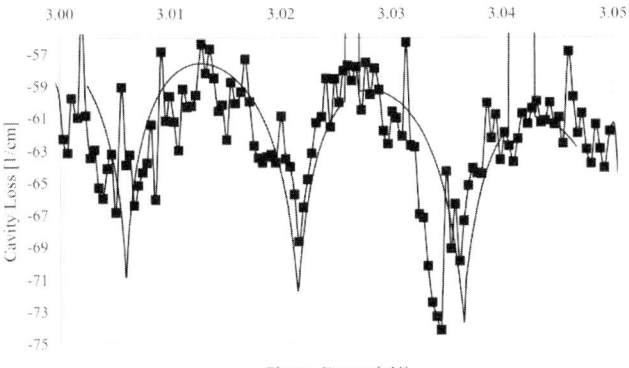

Fig. 18.7 Simulated (small symbols) and measured (large symbols) cavity loss spectrum for a blue laser with GaN substrate ($\alpha_{las} + \alpha_{subst}$). The oscillations are due to a resonant coupling of the laser waveguide modes to substrate modes.

method. The spectral range has been chosen below the bandgap energy of the active region, and the laser has been pumped so that the active region is below transparency. In the figure, the mirror loss $\alpha_{mirr} = 15\,\text{cm}^{-1}$ has been subtracted already, so that the internal optical losses are visible exclusively. The minimum loss level is at $59\,\text{cm}^{-1}$, showing oscillations with a period of 14 meV and a depth of $10\,\text{cm}^{-1}$. The line in the figure is the simulated imaginary part of the eigenvalue, obtained as a solution of Eq. (18.10). The oscillation amplitude and period, as well as the absolute loss value, show a high sensitivity with respect to the parameters chosen. These parameters are, in particular, the refractive index as well as the optical losses of the materials, and their spectral dispersion. The refractive index is listed in Table 18.1, whereas the optical losses have been chosen to be $45\,\text{cm}^{-1}$ for the p-doped material, $3\,\text{cm}^{-1}$ for the n-doped material, and $9\,\text{cm}^{-1}$ for the substrate. As the laser diode used for this measurement has been grown on GaN, compared to the one on SiC in the table, the losses are slightly different. The effective index of the optical mode is $n_{eff} = 2.480$, whereas the substrate material index is $n_{subst} = 2.515$. The oscillation period of the loss spectrum depends crucially on the material dispersion. In the simulation, the dispersion has been set to $-1.30 \times 10^{-3}\,\text{nm}^{-1}$ for InGaN and GaN, and $-1.13 \times 10^{-3}\,\text{nm}^{-1}$ for AlGaN. These numbers are in the range of the values reported in the literature [10,11].

Figure 18.8 shows the optical intensity for a vertical cut in the center of the waveguide for two different operation frequencies ω with a maximal and minimal coupling of the laser waveguide and the substrate. The maximal coupling corresponds to the situation at a maximum loss in Fig. 18.7, and the minimum coupling to the upper minimum loss. If the frequency matches a resonance in the substrate resonator, a maximum amount of energy is transferred

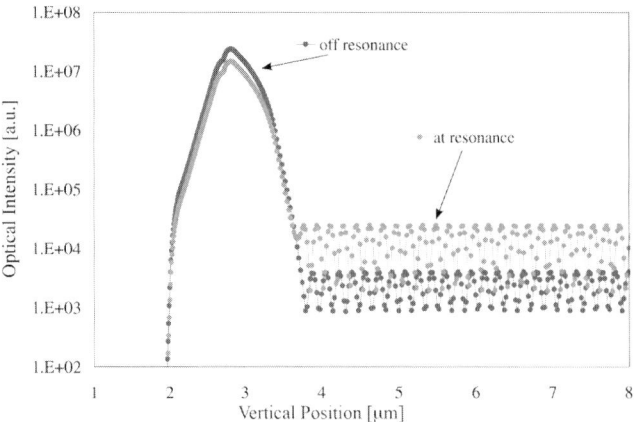

Fig. 18.8 Vertical cut of simulated optical intensity for two slightly different wavelengths. At the resonance condition, the maximum power is coupled into the substrate, which corresponds to a maximum loss in Fig. 18.7. Off-resonance corresponds to a minimum in the loss oscillations.

to the substrate, which results in an increased loss. Due to the open boundaries at the side of the substrate, the quality factor of the substrate resonator is rather low, so that the resonances are wide. A one-dimensional simulation performed on a vertical cut would result in much sharper resonances.

The calibrated simulation of the loss spectrum allows us to extract the loss components α_{las} and α_{subst} for the specific design. This is shown in Fig. 18.9, where the loss components are plotted for the off and on resonance frequency. Between 57% (off-resonance) and 72% of the internal parasitic losses in the

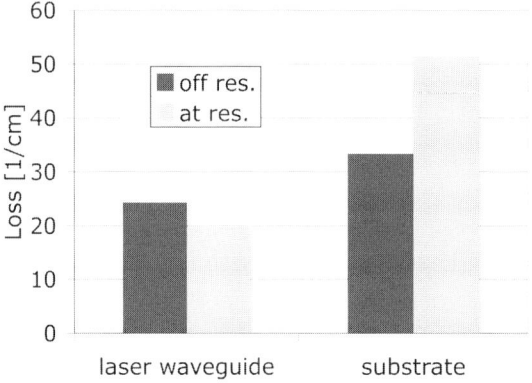

Fig. 18.9 Simulated loss contributions of the laser waveguide (top contact, p- and n-doped regions) and the substrate resonator (substrate, and bottom contact) for resonant and off-resonant coupling of the two resonators.

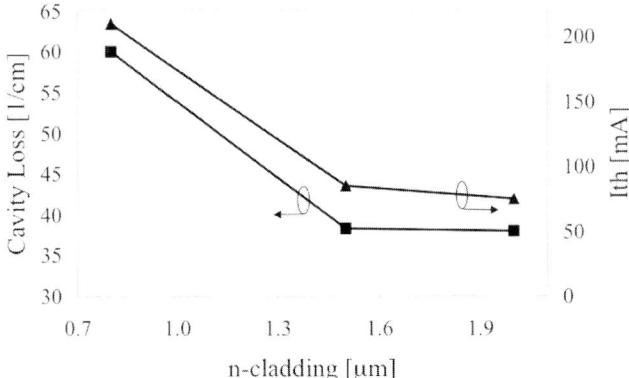

Fig. 18.10 Measured threshold current I_{th} (right axis) and simulated cavity losses (left axis) for different n-cladding layer thicknesses. The n-cladding layer separates the substrate from the laser waveguide (see Table 18.1). Both experiment and simulation show a saturation above 1.5 µm due to an efficient suppression of the substrate mode.

cavity are caused by absorption and diffraction in the substrate and bottom contact. In order to reduce the mode coupling into the substrate, devices with increased n-cladding thickness have been manufactured. The thickness has been increased from initially 0.8 µm up to 2.0 µm. Figure 18.10 plots the simulated cavity losses on the left axis and the measured threshold current on the right axis with varying n-cladding thickness. The simulation shows a reduction of the cavity loss from approx. $59\,\text{cm}^{-1}$ to $37\,\text{cm}^{-1}$, with a saturation at a thickness of approx. 1.5 µm. The threshold current is reduced from 210 mA to 75 mA, and exhibits a similar saturation behavior.

18.4
Mode Gain Analysis

In Section 18.3, the optical losses in the waveguide have been analyzed in detail. The performance of the laser depends on the optical losses as well as on the gain in the active region: the lasing condition requires the mode gain to reach the total optical loss, and the current for this condition is the threshold current. The gain is supplied by the material gain of the InGaN quantum wells. In GaN-based lasers, one peculiarity is the occurrence of compositional fluctuations in the InGaN active region. In this section, the aim is to quantify the gain reduction due to the compositional fluctuations. Compared to a structure with an ideal topology, this can be viewed as an additional loss factor for the performance of the device, and hence it is discussed in the following.

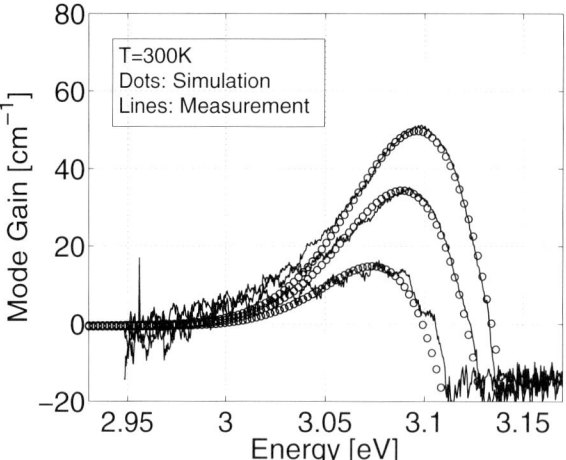

Fig. 18.11 Comparison of measured (lines) and simulated (dots) mode gain for a GaN based edge-emitting laser at 405 nm. The currents for the measurements are $[60, 100, 130]$ mA and the carrier densities for the simulation $n = [2.35, 2.67, 2.90] \times 10^{19}$ cm^{-3}, respectively. The inhomogeneous broadening parameter has been set to $\Delta E = 30$ meV. The corresponding full width at half maximum value is $\Delta E_{\text{FWHM}} = 71$ meV.

Figure 18.11 shows a comparison of the measured and simulated optical mode gain of an edge-emitting laser. The active region consists of three 2 nm In$_{0.1}$Ga$_{0.9}$N multiple quantum wells (MQW) with 6 nm GaN barrier layers, and is described in more detail in [12]. The measurement has been performed using the Hakki–Paoli method [13]. The gain simulation is done in a two-step procedure: first, the crystal band structure is computed using a 6-band $k \cdot p$ method [14]. It includes band mixing for the heavy-hole, light-hole and the crystal-field split hole bands. In a second step, the spectral optical gain is obtained from the following semi-classical relation [15]:

$$G_{hom} = -\frac{2\omega}{\epsilon_0 n c E_o V} \text{Im} \left(e^{i\omega t} \sum_{ji} \sum_{\mathbf{k}} (\mu_{\mathbf{k}}^{ji})^* p_{\mathbf{k}}^{ji} \right) \tag{18.12}$$

Here, ω is the laser frequency, ϵ_0 the vacuum permittivity, n the optical refractive index, c the speed of light in vacuum, E_o the slowly varying electric field amplitude, V the active region volume, $\mathbf{k} = (k_x, k_y)$ the quantum well 2D-wave vector, $\mu_{\mathbf{k}}^{ji}$ the matrix element between the conduction subband i and the valence subband j, and $p_{\mathbf{k}}^{ji} = \langle b_{-\mathbf{k}}^j a_{\mathbf{k}}^i \rangle$ the microscopic polarization induced between subbands i and j. The microscopic polarization is obtained from the semiconductor Bloch equations with an inclusion of the carrier correlations on a quantum kinetic level. The polarization dephasing rates are calculated within the second Born approximation [15]. The equation of motion is solved in the steady state, and iterated with the Poisson equation, in order to include the screening effect from the piezo-charges and quantum-well charges in the electrostatic potential [16]. The microscopic theory contains the correct treatment of plasma and excitonic effects, which leads to a quantitative determination of homogeneous broadening and bandgap renormalization without any phenomenological parameters. In addition to homogeneous broadening, inhomogeneous broadening results from local well width or composition fluctuations. In the simulation, a convolution of the homogeneously broadened gain spectrum with a Gaussian distribution $\sigma(\omega, \Delta E)$ of given spectral width is performed [14]:

$$G_{inh}(\omega) = \int d\omega' \, G_{hom}(\omega - \omega') \sigma(\omega', \Delta E) \tag{18.13}$$

The broadening parameter ΔE is obtained by a comparison of experimental and simulated spectra. For the simulations, standard parameters from the literature are used. The crystal potentials and effective masses for the bandstructure calculation are taken from [17] for GaN and InN, using Vegard's law for the ternary alloy. The bandgap relation is taken from [18], and the spontaneous and piezo-induced polarization lead to interface charges of 4.5×10^{12} cm^{-2}, which is smaller than the theoretical value [19]. This reduction can be well explained by interface impurities, compositional inhomogeneities or partial relaxation. The optical confinement factor which converts mode gain to material gain is used from the waveguide analysis in the previous chapter, and is set to $\Gamma = 0.017$.

Using Fig. 18.11 as a reference point, the characteristic shape of the measured gain spectra at different drive currents can be represented by a rigorous calculation of a homogeneously broadened spectrum in combination with phenomenological inhomogeneous gain broadening. Assuming that deviations from the ideal homogeneous semiconductor layer are the cause of inhomogeneous broadening, the simulation allows us to gradually decrease the inhomogeneous broadening and approach this ideal situation. Although difficult to achieve experimentally, it is quite instructional in order to compare the InGaN quantum film luminescence to the one in GaAs or InP films, which have been shown to exhibit almost no inhomogeneous broadening effects. Starting from the value of $\Delta E = 30$ meV (or $\Delta E_{FWHM} = 71$ meV) extracted from the experimental data, Fig. 18.12 shows the mode gain for a gradual decrease of the inhomogeneous broadening down to 0 meV. For this number, the peak mode gain is increased to 86 cm^{-1}, compared to 52 cm^{-1} for the experimental value. As shown in Fig. 18.13 this decrease is almost constant in the range above inversion up to 100 cm^{-1} of the mode gain (or material gain of 5800 cm^{-1}). A direct consequence is a substantially increased threshold current density in GaN-based lasers. Assuming a nonradiative recombina-

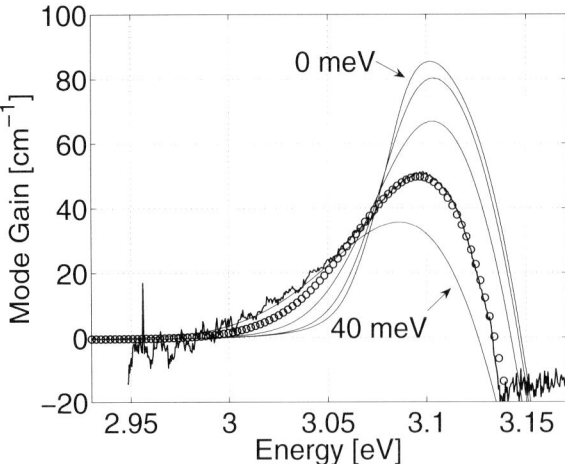

Fig. 18.12 Simulated mode gain at $n = 2.90 \times 10^{19}$ cm^{-3} for different values of inhomogeneous broadening. It varies between 0 meV and 40 meV in steps of 10 meV. Measured mode gain is also plotted for comparison.

tion time of $\tau = 1$ ns for the active region carriers [12] and a cavity loss of 60 cm^{-1}, the threshold current density can be estimated. In order to do so, the integrated spontaneous emission spectrum gives the radiative recombination current from the simulation, and the resulting threshold current densities are $j_{th} = 2.75$ kA cm^{-2} and $j_{th} = 3.50$ kA cm^{-2} for zero and 30 meV inhomogeneous broadening, respectively.

18.5
Conclusion

In this contribution, the optical mode stability and loss mechanisms in group-III nitride edge-emitting lasers have been analyzed. We demonstrate how a two-dimensional plane-wave expansion method, which includes optical loss and gain, can be used for a qualitative interpretation of the experimental results. As examples we discuss the impact of gain-guiding on the threshold current density in gain-guided and index-guided laser diodes as well as mode competition in ridge waveguide laser diodes as a function of the etch depth. The latter was explained by a varying overlap of different order lateral modes with lossy p-regions. Other problems which can be addressed with this method are, e.g., the impact of the ridge shape or the insertion of lossy layers on mode stability and the suppression of kinks in the current–optical power (IP) curve. The code used for the above simulations is relatively simple and was briefly described.

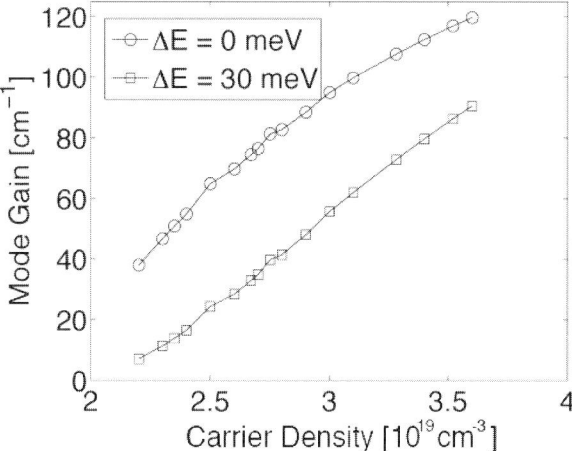

Fig. 18.13 Peak Mode Gain versus carrier density in the case of no inhomogeneous broadening and 30 meV broadening. Differential gain remains almost unchanged, however, mode gain drops by approximately $25\,\text{cm}^{-1}$.

For a quantitative analysis and for more complex problems we use a sophisticated two-dimensional simulation of the optical mode in the laser waveguide including the full substrate. Substantial losses spatially located in the substrate have been identified. Optical near-field measurements confirmed the presence of substrate modes, and improved designs have been presented. Finally, spectral gain measurements have been analyzed with a quantum-kinetic gain theory. There is a strong indication that compositional fluctuations in the active quantum films deteriorate the peak gain by almost 50 %, and therefore the laser performance.

Acknowledgments

We would like to thank Mathieu Luisier and Valerio Laino at ETH Zurich, for their contributions in code development and calibration; Markus Pindl and Christoph Lauterbach from the University of Regensburg for the near-field measurements and development of the matrix method code; and Alfred Lell and Volker Härle at Osram Opto Semiconductors GmbH for their continuing support.

References

1. 2D-Waveguide-MM for download from NUSOD software directory (http://www.nusod.org).
2. U. T. Schwarz, M. Pindl, E. Sturm, M. Furitsch, A. Leber, S. Miller, A. Lell, and V. Härle, phys. stat. sol. (a) 202, 261 (2005).
3. M. Kuramoto, C. Sasaoka, F. Futagawa, M. Nido, and A. A. Yamaguchi, phys. stat. sol. (a) **192**, 329 (2002).
4. U. T. Schwarz, E. Sturm, W. Wegscheider, V. Kümmler, A. Lell, and V. Härle, phys. stat. sol. (a) 200, 143 (2003).
5. L.A. Coldren and S.W. Corzine, Diode Lasers and Photonic Integrated Circuits, Wiley, 1995.
6. P. G. Eliseev, G.A. Smolyakov, M. Osiński, IEEE Journal Sel. Top. Quant. El., vol. 5, No. 3, pp. 771, May 1999.
7. U.T. Schwarz, M. Pindl, W. Wegscheider, C. Eichler, F. Scholz, M. Furitsch, A. Leber, S. Miller, A. Lell, V. Härle, Appl. Phys. Lett., vol. 86, pp. 16112, Sep. 2005.
8. U.T. Schwarz, C. Lauterbach, M.O. Schillgalies, C. Rumbolz, M. Furitsch, A. Lell, V. Härle, Proc. SPIE, vol. 6184, 2006.
9. G.A. Smolyakov, P. G. Eliseev, M. Osiński, IEEE Journal Quant. El., vol. 41, No. 4, 2005.
10. C.X. Lian, X.Y. Li, J. Liu, Semic. Sci. Technology, vol. 19, No. 3, pp. 417, 2004.
11. H.Y. Zhang, X.H. He, Y.H. Shih, M. Schurman, Z.C. Feng, R. A. Stall, Optics Letters, vol. 21, No. 19, 1996.
12. B. Witzigmann, V. Laino, M. Luisier, U. T. Schwarz, G. Feicht, W. Wegscheider, K. Engl, M. Furitsch, A. Leber, A. Lell, and V. Härle, Appl. Phys. Lett., vol. 88, pp. 021104, Jan. 2006.
13. B. W. Hakki and T. L. Paoli, J. Appl. Phys., vol. 44, pp. 4113–4119, Sep. 1973.
14. W. W. Chow, A. Girndt and S. W. Koch, Optics Express, vol. 2, pp. 119–124, Feb. 1998.
15. W. W. Chow and S. W. Koch, Semiconductor-Laser Fundamentals, Springer, (1999), Berlin.
16. B. Witzigmann, V. Laino, M. Luisier, F. Roemer, G. Feicht, U. T. Schwarz, Proc. SPIE Photonics Europe 06, vol. 6184, 2006.
17. J. Piprek, Semiconductor Optoelectronic Devices: Introduction to Physics and Simulation, Academic Press, (2003), San Diego.
18. I. Vurgaftman and J. R. Meyer, Journal Appl. Phys., vol. 94, pp. 3675–3696, 2003.
19. F. Bernardini and V. Fiorentini, Phys. Rev. B, vol. 64, pp. 085207, Aug. 2001.

19
Electronic Properties of InGaN/GaN Vertical-Cavity Lasers

Joachim Piprek, Zhan-Ming Li, Robert Farrell, Steven P. DenBaars, and Shuji Nakamura

19.1
Introduction to Vertical-Cavity Lasers

In vertical-cavity surface-emitting lasers (VCSELs), the optical cavity is formed by mirrors above and below the gain region (Fig. 19.1(a)). The laser light propagates in a vertical direction and typically exhibits a circular beam shape. Internally, the photons pass the gain region in a perpendicular direction, i.e., optical gain is provided over a short propagation distance only and the amplification per photon round trip is small. Therefore, the mirrors need to be highly reflective so that photons make many round trips before they are emitted. To achieve high reflectivity, distributed Bragg reflectors (DBRs) are used with two alternating layers of high refractive index contrast. With quarter-wavelength layer thickness, the reflected waves from all DBR interfaces add up constructively, allowing for DBR reflectivities above 99% [1].

Within the active layers (quantum wells), optical gain arises from the stimulated recombination of electrons and holes, which may be generated by optical absorption of pump light or by current injection. The latter is more difficult to accomplish and it is strongly affected by the electrical DBR properties. Semiconductor DBRs allow for vertical carrier injection through the DBR directly into the active gain region (Fig. 19.1(b)) which gives good overlap of the lateral carrier and photon profile. However, suitable semiconductor materials often exhibit small index contrast and the many hetero-interfaces tend to generate a high electrical DBR resistance. An alternative choice is dielectric DBRs which typically provide a large refractive index contrast so that a few layer pairs are often sufficient for high mirror reflectivity. However, dielectric DBRs are electrically insulating and the injection current needs to be funneled into the active region from the side, typically by using ring contacts around the DBR (Fig. 19.1(c)). Some type of electrical confinement structure is required that forces the carriers to move into the small center region where the optical mode is located [2]. Last, but not least, the small lateral extension of the active region causes a potentially high thermal resistance so that good thermal

Nitride Semiconductor Devices: Principles and Simulation. Joachim Piprek (Ed.)
Copyright © 2007 WILEY-VCH Verlag GmbH & Co. KGaA, Weinheim
ISBN: 978-3-527-40667-8

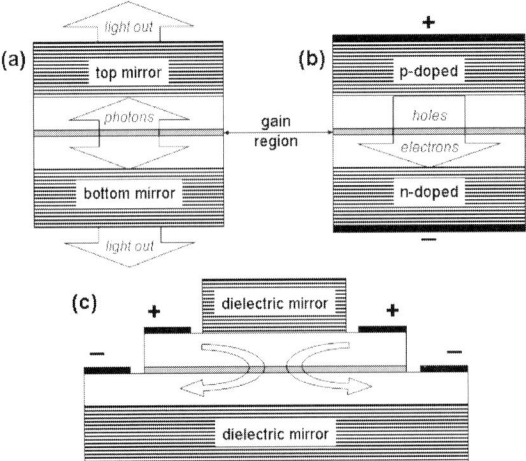

Fig. 19.1 Illustration of vertical-cavity laser principles: (a) vertical photon flux; (b) vertical current flow with semiconductor mirrors; (c) current flow with dielectric mirrors.

conductivity is another essential requirement of VCSEL design [3]. A more detailed review on VCSELs is given in [4].

VCSELs exhibit several advantages over in-plane lasers, including lower manufacturing costs, circular output beams, and longer lifetime [2]. However, the complex interaction of electrical, thermal, and optical processes in VCSELs often requires advanced computer simulation for device analysis and design optimization [5]. Using recently fabricated GaN-VCSELs as an example [6], this chapter demonstrates how advanced simulations are used to analyze internal physics and reveal performance limitations of real devices. Section 19.2 explains the device structure. Section 19.3 describes the VCSEL model and the material parameters used. Simulation results are employed in Section 19.4 to gain deeper insight into internal physical processes.

19.2
GaN-based VCSEL Structure

In contrast to the success of GaAs-based VCSELs in recent years, the demonstration of GaN-based VCSELs still faces significant challenges [7]. Current-injected GaN-VCSELs often suffer from high optical loss due to insufficient reflectance of semiconductor DBRs, low conductivity of p-GaN, carrier leakage, or the resistance of GaN to conventional wet etching. Some of these problems have been addressed by the device design presented in [6] (Fig. 19.2, Table 19.1). Eleven-period dielectric DBRs are used on both sides which ex-

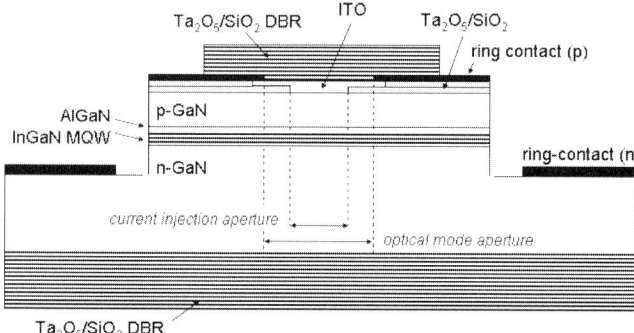

Fig. 19.2 Schematic structure of the vertical-cavity laser investigated.

hibit a high reflectance above 99%. The multi-quantum well (MQW) active region consists of five 4 nm thick $In_{0.1}Ga_{0.9}N$ quantum wells and 8 nm wide $In_{0.035}Ga_{0.965}N$ barriers. It is covered by a 20 nm p-doped $Al_{0.18}Ga_{0.82}N$ electron stopper layer to reduce electron leakage into the p-GaN spacer layer. Indium tin oxide (ITO) is employed for lateral leveling of the injection current. However, due to common difficulties with thinning GaN in a well-controlled fashion, the cavity length is as large as 5.5 µm. The measured longitudinal mode spacing is correspondingly small (4–5 nm) which eliminates the usual VCSEL design challenge of aligning gain peak and cavity mode [8]. The dense mode spectrum promises a reduced temperature sensitivity of the threshold current.

19.3
Theoretical Models and Material Parameters

Due to the complexity of VCSEL physics, realistic device simulations often need to include models for electronic, optical, and thermal processes [9]. In this chapter we focus on electronic processes at room temperature considering pulsed laser operation without significant self-heating. We employ the simulation software PICS3D [10] which self-consistently combines the computation of carrier transport, electron band structure, optical gain, and optical mode. Each model is described below. A similar approach was previously used to study InGaN/GaN in-plane lasers, resulting in excellent agreement between measurement and simulation [11].

An important issue in any device simulation is the selection of appropriate values for the various material parameters. Published parameter values for GaN-based compounds often vary substantially (see, e.g., the review of band structure parameters in [12]). We mainly employ the parameters used successfully in our previous investigations, most of which are listed in Table 19.1 and Table 19.3. Crucial parameters are discussed below.

Tab. 19.1 VCSEL layer structure and material parameters (t - layer thickness, N - carrier concentration from doping, n_r - refractive index).

Parameter	t	N	n_r
Unit	nm	cm^{-3}	
SiO$_2$ (11× in top DBR)	70	—	1.47
Ta$_2$O$_5$ (11× in top DBR)	47	—	2.2
ITO (contact)	40	$10. \times 10^{18}$	2.1
p-GaN (spacer)	540	0.4×10^{18}	2.55
p-Al$_{0.18}$Ga$_{0.82}$N (stopper)	20	10^{16}–10^{18}	2.27
i-In$_{0.035}$Ga$_{0.965}$N (barrier)	8	—	2.68
i-In$_{0.1}$Ga$_{0.9}$N (quantum well)	4	—	3.0
n-In$_{0.035}$Ga$_{0.965}$N (barrier)	8	1.0×10^{18}	2.68
i-In$_{0.1}$Ga$_{0.9}$N (quantum well)	4	—	3.0
n-In$_{0.035}$Ga$_{0.965}$N (barrier)	8	1.0×10^{18}	2.68
i-In$_{0.1}$Ga$_{0.9}$N (quantum well)	4	—	3.0
n-In$_{0.035}$Ga$_{0.965}$N (barrier)	8	1.0×10^{18}	2.68
i-In$_{0.1}$Ga$_{0.9}$N (quantum well)	4	—	3.0
n-In$_{0.035}$Ga$_{0.965}$N (barrier)	8	1.0×10^{18}	2.68
i-In$_{0.1}$Ga$_{0.9}$N (quantum well)	4	—	3.0
n-In$_{0.035}$Ga$_{0.965}$N (barrier)	8	1.0×10^{18}	2.68
n-GaN (spacer)	5300	2.5×10^{18}	2.55
SiO$_2$ (11× in bottom DBR)	70	—	1.47
Ta$_2$O$_5$ (11× in bottom DBR)	47	—	2.2

19.3.1
Carrier Transport

PICS3D employs the traditional drift-diffusion model for semiconductors. The current density of electrons $\vec{j}_n$ and holes $\vec{j}_p$ is caused by the electrostatic field $\vec{F}$ (drift) and by the concentration gradient of electrons and holes, ∇n and ∇p, respectively,

$$\vec{j}_n = q\mu_n n\vec{F} + qD_n \nabla n \tag{19.1}$$

$$\vec{j}_p = q\mu_p p\vec{F} - qD_p \nabla p \tag{19.2}$$

with the elementary charge q, the carrier densities n and p, and their mobilities μ_n and μ_p, respectively. The diffusion constants D_n and D_p are replaced by mobilities using the Einstein relation $D = \mu k_B T/q$ with the Boltzmann constant k_B and the temperature T. The electric field is affected by the charge distribution, which includes electrons n and holes p, dopant ions (p_D, n_A), and other fixed charges N_f that are of special importance in GaN-based devices in order to account for built-in polarization. This relationship is described by the

Poisson equation

$$\nabla \cdot (\varepsilon\varepsilon_0 \vec{F}) = q(p - n + p_D - n_A \pm N_f) \tag{19.3}$$

with the free-space permittivity ε_0. Changes in the local carrier concentration are connected to a spatial change in current flow $\nabla \vec{j}$ or to the local recombination rate R of electron–hole pairs, as described by the continuity equations

$$q\frac{\partial n}{\partial t} = \nabla \cdot \vec{j}_n - qR \tag{19.4}$$

$$q\frac{\partial p}{\partial t} = -\nabla \cdot \vec{j}_p - qR \tag{19.5}$$

The relevant carrier recombination mechanisms in GaN-based VCSELs are stimulated photon emission, spontaneous photon emission, and nonradiative Shockley–Read–Hall (SRH) recombination. The stimulated emission of photons is the key physical mechanism in lasers and it is described in Section 19.3.4 together with the spontaneous photon emission in quantum wells. Within bulk layers, the local spontaneous emission rate is approximated by

$$R_{sp} = B(np - n_i^2) \tag{19.6}$$

using the bimolecular recombination coefficient $B = 5 \times 10^{-11} \text{cm}^3 \text{s}^{-1}$ [13] (n_i - intrinsic density). The defect-related SRH recombination rate is given by

$$R_{SRH} = \frac{np - n_i^2}{\tau_p^{SRH}\left(n + N_c \exp\left[\frac{E_t - E_c}{k_B T}\right]\right) + \tau_n^{SRH}\left(p + N_v \exp\left[\frac{E_v - E_t}{k_B T}\right]\right)} \tag{19.7}$$

and it is governed by the SRH lifetimes τ_n^{SRH} and τ_p^{SRH} ($N_{c,v}$ - density of states of conduction, valence band; E_t - mid gap defect energy). SRH lifetimes are different for electrons and holes but the SRH recombination rate is usually dominated by the minority carrier lifetime. We assume $\tau_{nr} = \tau_n^{SRH} = \tau_p^{SRH}$ in the following. The nonradiative carrier lifetime τ_{nr} is a crucial material parameter for GaN-based devices. The defect density and nonradiative lifetime depend on the substrate used and on the growth quality. Since SRH lifetimes are hard to predict, we assume a common value of $\tau_{nr} = 1$ ns in our simulations. The SRH lifetime in quantum wells is of particular importance and it is sometimes employed as a fit parameter to find agreement with experimental characteristics [14].

Our model includes Fermi statistics and thermionic emission of carriers at hetero-interfaces [15]. The conduction band offset ΔE_c of hetero-interfaces is one of the critical parameters in our simulation as it controls electron confinement to the active region [16]. An offset ratio $\Delta E_c/\Delta E_g$ of 0.7 is assumed for our device which represents an average of reported values [17–19].

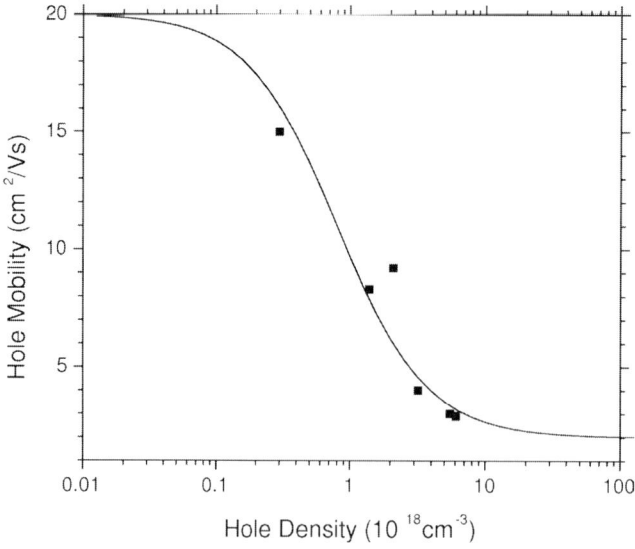

Fig. 19.3 GaN hole mobility vs hole density. The solid line gives our fit of Eq. (19.8) to measurements (dots) [22] using the parameters in Table 19.2.

The doping data in Table 19.1 give the actual densities of free carriers. While the Si donor exhibits a low ionization energy, the Mg acceptor has a high activation energy of about 170 meV [20]. Large Mg densities near $10^{20} cm^{-3}$ are required to obtain free hole densities above $10^{18} cm^{-3}$ [21]. The high defect density contributes to a very low hole mobility, which is another obstacle to GaN device operation. The hole mobility measured on Mg-doped MOCVD-grown GaN layers is less than 20 cm^{2}V^{-1}s^{-1} and it decreases with higher doping density [22]. We employ the Caughey–Thomas approximation for the mobility as function of carrier density [23]

$$\mu(N) = \mu_{min} + \frac{\mu_{max} - \mu_{min}}{1 + (N/N_{ref})^\alpha} \tag{19.8}$$

Tab. 19.2 GaN mobility parameters for Eq. (19.8).

Parameter Unit	μ_{max} cm^{2}V^{-1}s^{-1}	μ_{min} cm^{2}V^{-1}s^{-1}	N_{ref} cm^{-3}	α —
Electrons [24]	1405	80	0.771×10^{17}	0.71
Holes	20	2	8×10^{17}	1.3

which can be fitted to mobility measurements using the parameters in Table 19.2. Figure 19.3 plots Eq. (19.8) for holes in comparison to mobility measurements on Mg-doped GaN layers grown by MOCVD. We here exclude

Tab. 19.3 Parameters for nitride wurtzite semiconductors at room temperature.

Parameter	Symbol	Unit	InN	GaN	AlN
Electron eff. mass (c-axis)	m_c^z	m_0	0.11	0.20	0.33
Electron eff. mass (transversal)	m_c^t	m_0	0.11	0.18	0.25
Hole eff. mass parameter	A_1	—	−9.24	−7.24	−3.95
Hole eff. mass parameter	A_2	—	−0.60	−0.51	−0.27
Hole eff. mass parameter	A_3	—	8.68	6.73	3.68
Hole eff. mass parameter	A_4	—	−4.34	−3.36	−1.84
Hole eff. mass parameter	A_5	—	−4.32	−3.35	−1.92
Hole eff. mass parameter	A_6	—	−6.08	−4.72	−2.91
Valence band reference level	E_v	eV	−1.59	−2.64	−3.44
Direct band gap	E_g	eV	0.8	3.42	6.28
Spin–orbit split energy	Δ_{so}	eV	0.001	0.014	0.019
Crystal-field split energy	Δ_{cr}	eV	0.041	0.019	−0.164
Lattice constant	a_0	Å	3.548	3.189	3.112
Elastic constant	C_{33}	GPa	200	392	382
Elastic constant	C_{13}	GPa	94	100	127
Deform. potential (E_c)	a_c	eV		−4.08	
Deform. potential	D_1	eV		−0.89	
Deform. potential	D_2	eV		4.27	
Deform. potential	D_3	eV		5.18	
Deform. potential	D_4	eV		−2.59	
Dielectric constant	ε	—	15.0	9.5	8.5

Note. $\Delta_{cr} = \Delta_1$, $\Delta_{so} = 3\Delta_2 = 3\Delta_3$, $a_c = a/2$

measurements of C-doped or MBE-grown GaN layers [25] which lead to inappropriate fit parameters as in [26]. In ternary and MQW layers, the mobilities are further reduced by additional scattering mechanisms, that are related to alloy disorder, interface roughness, or compositional fluctuations. These effects are hard to predict and we assume an electron mobility of 100 cm^{2}V^{-1}s^{-1} and a hole mobility of 5 cm^{2}V^{-1}s^{-1} for all ternary layers. In addition to the low hole conductivity, a high contact resistance may significantly increase the device bias of GaN-based light emitters.

19.3.2
Electron Band Structure

PICS3D includes the 6 × 6 $\vec{k} \cdot \vec{p}$ model for the valence-band structure of wurtzite semiconductors as developed by Chuang and Chang [27, 28], which is summarized in this section with an emphasis on strain effects. The three valence bands are referred to as heavy-hole (hh), light-hole (lh), and crystal-field split-hole (ch) band. Material parameters are listed in Table 19.3.

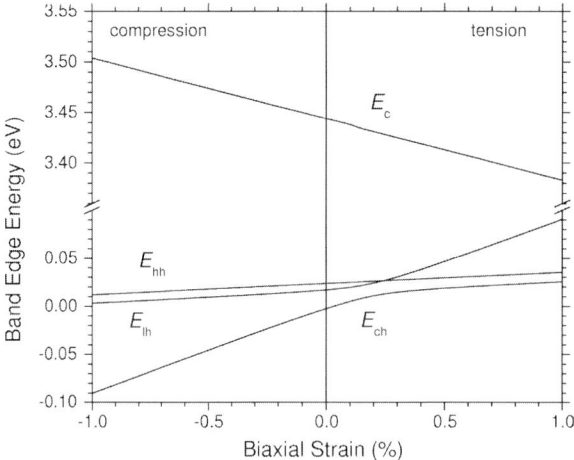

Fig. 19.4 GaN band-edge shift with strain.

The epitaxial growth of nitride devices is typically along the c-axis of the wurtzite crystal, which is parallel to the z-axis in our coordinate system. The natural $In_xGa_{1-x}N$ lattice constant $a_0(x)$ is reduced to that of the GaN substrate, a_s, imposing biaxial compressive strain in the transverse plane

$$e_t = \frac{a_s - a_0}{a_0} \tag{19.9}$$

and tensile strain in the growth direction

$$e_z = -2\frac{C_{13}}{C_{33}}e_t \tag{19.10}$$

The nondiagonal elements of the strain tensor are zero. The valence band edge energies are

$$E_{hh} = E_v + \Delta_1 + \Delta_2 + \theta_e + \lambda_e \tag{19.11}$$

$$E_{lh} = E_v + \frac{\Delta_1 - \Delta_2 + \theta_e}{2} + \lambda_e + \sqrt{\left(\frac{\Delta_1 - \Delta_2 + \theta_e}{2}\right)^2 + 2\Delta_3^2} \tag{19.12}$$

$$E_{ch} = E_v + \frac{\Delta_1 - \Delta_2 + \theta_e}{2} + \lambda_e - \sqrt{\left(\frac{\Delta_1 - \Delta_2 + \theta_e}{2}\right)^2 + 2\Delta_3^2} \tag{19.13}$$

with the average valence band edge E_v and

$$\theta_e = D_3 e_z + 2D_4 e_t \tag{19.14}$$

$$\lambda_e = D_1 e_z + 2D_2 e_t \tag{19.15}$$

Without strain, the spin–orbit interaction leads to only slight separations between the three valence band edges. Figure 19.4 shows the influence of strain.

In contrast to traditional zinc blende III-V compounds like GaAs, the GaN light hole and heavy hole bands hardly separate under compressive strain. This contributes to the high quantum well carrier densities and threshold current densities observed with GaN-based lasers [22], which are much higher than with GaAs-based lasers.

The conduction band-edge is calculated from

$$E_c = E_v + E_g + P_{ce} \tag{19.16}$$

with the energy bandgap E_g and the hydrostatic energy shift

$$P_{ce} = a_{cz}e_z + 2a_{ct}e_t \tag{19.17}$$

The hydrostatic deformation potential is anisotropic (a_z, a_t) and half of the deformation is assumed to affect the conduction band (a_{cz}, a_{ct}). The bandgaps of $In_xGa_{1-x}N$ and $Al_xGa_{1-x}N$ are known to deviate from the linear Vegard law, and they are approximated by

$$E_g^0(x) = xE_g^0(\text{AlN or InN}) + (1-x)E_g^0(\text{GaN}) - x(1-x)C_g \tag{19.18}$$

using the bowing parameter C_g. A wide range of bowing parameters has been reported [12]. For unstrained layers with a low mole fraction of the alloy element we adopt $C_g = 2.6$ eV for InGaN [29] and $C_g = 1.3$ eV for AlGaN [30].

The dispersion $E_c(\vec{k})$ of the conduction band can be characterized by a parabolic band model with electron effective masses m_c^t and m_c^z perpendicular and parallel to the c-growth direction, respectively. The three valence bands are nonparabolic. Near the Γ point, the hole effective masses can be approximated as

$$m_{hh}^z = -m_0(A_1 + A_3)^{-1} \tag{19.19}$$

$$m_{hh}^t = -m_0(A_2 + A_4)^{-1} \tag{19.20}$$

$$m_{lh}^z = -m_0\left[A_1 + \left(\frac{E_{lh} - \lambda_e}{E_{lh} - E_{ch}}\right)A_3\right]^{-1} \tag{19.21}$$

$$m_{lh}^t = -m_0\left[A_2 + \left(\frac{E_{lh} - \lambda_e}{E_{lh} - E_{ch}}\right)A_4\right]^{-1} \tag{19.22}$$

$$m_{ch}^z = -m_0\left[A_1 + \left(\frac{E_{ch} - \lambda_e}{E_{ch} - E_{lh}}\right)A_3\right]^{-1} \tag{19.23}$$

$$m_{ch}^t = -m_0\left[A_2 + \left(\frac{E_{ch} - \lambda_e}{E_{ch} - E_{lh}}\right)A_4\right]^{-1} \tag{19.24}$$

using the hole effective mass parameters A_i given in Table 19.3 (m_0 – free electron mass). More details on the calculation procedure for quantum well valence bands are given in [28]. Figure 19.5 shows the electron band structure calculated for our $In_{0.1}Ga_{0.9}N$ quantum well.

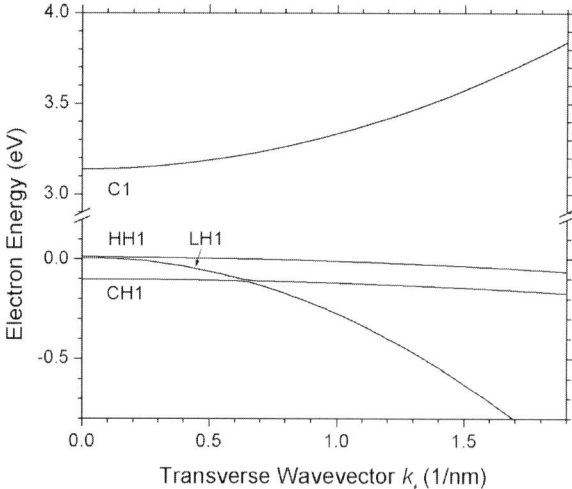

Fig. 19.5 Quantum well band structure within the transverse plane (lowest levels).

19.3.3
Built-In Polarization

Spontaneous and piezoelectric polarization of nitride compounds is larger than in other III-V semiconductors. It depends on the compound's composition. Net polarization charges remain at each hetero-interface. We here use the nonlinear model described in [31]. Accordingly, the spontaneous polarization P_{sp} [C m^{-2}] is calculated as

$$P_{sp}(Al_xGa_{1-x}N) = -0.090x - 0.034(1-x) + 0.019x(1-x) \qquad (19.25)$$

$$P_{sp}(In_xGa_{1-x}N) = -0.042x - 0.034(1-x) + 0.038x(1-x) \qquad (19.26)$$

For binary compounds, the piezoelectric polarization P_{pz} [C m^{-2}] is given as nonlinear functions of the transverse strain e_t by

$$P_{pz}(GaN) = -0.918e_t + 9.541e_t^2 \qquad (19.27)$$

$$P_{pz}(InN) = -1.373e_t + 7.559e_t^2 \qquad (19.28)$$

$$P_{pz}(AlN) = -1.808e_t - 7.888e_t^2 \quad (e_t > 0) \qquad (19.29)$$

$$P_{pz}(AlN) = -1.808e_t + 5.624e_t^2 \quad (e_t < 0) \qquad (19.30)$$

and it is linearly interpolated for ternary compounds. Spontaneous and piezoelectric polarization add up and result in a strong built-in field (Fig. 19.6). In PICS3D, the built-in polarization is represented by a fixed surface charge density. At hetero-interfaces, the difference of these surface charge densities gives

Tab. 19.4 Built-in polarization charges at VCSEL interfaces.

Interface	Net density
GaN/In$_{0.035}$Ga$_{0.965}$N	-3.2×10^{12} cm^{-2}
In$_{0.035}$Ga$_{0.965}$N/In$_{0.1}$Ga$_{0.9}$N	-6.5×10^{12} cm^{-2}
In$_{0.1}$Ga$_{0.9}$N/In$_{0.035}$Ga$_{0.965}$N	$+6.5 \times 10^{12}$ cm^{-2}
In$_{0.035}$Ga$_{0.965}$N/Al$_{0.18}$Ga$_{0.82}$N	$+11.0 \times 10^{12}$ cm^{-2}
Al$_{0.18}$Ga$_{0.82}$N/GaN	-7.8×10^{12} cm^{-2}

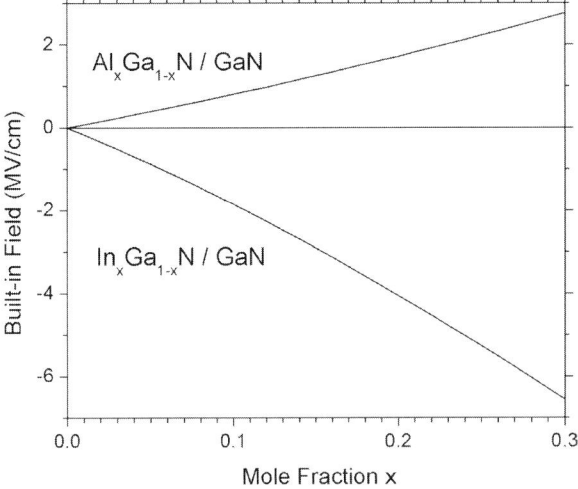

Fig. 19.6 Net built-in field in thin ternary layers grown on GaN.

the net polarization charge density, which is listed in Table 19.4 for the interfaces in our device.

Built-in polarization fields are expected to strongly affect the VCSEL performance [32]. Within the quantum wells, the polarization field separates electrons and holes, thereby reducing stimulated and spontaneous emission. However, experimental investigations of InGaN quantum wells often result in weaker built-in fields than predicted, ranging from 20% [33] to 80% [34] of the theoretical value, with typical results near 50% [35]. This broad variation has been attributed to partial compensation of the polarization field by fixed defect and interface charges [36] or to inappropriate analysis of measured data [37]. On the other hand, the theoretical polarization formulas may deviate from reality, especially for InGaN, as only AlGaN measurements have been used for validation [31]. Since the actual magnitude of the built-in polarization in our device is unknown, we are using the polarization charges as a variable in some of our simulations.

19.3.4
Photon Generation in the Quantum Wells

The carrier recombination mechanisms in InGaN quantum wells are not yet fully understood. The small impact of the high defect density still puzzles many researchers. Indium segregation, defect self-screening, or a short hole diffusion length may play an important role. Essential quantum well parameters like nonradiative carrier lifetime and net polarization field are not exactly known. All these uncertainties make it difficult to verify models for the photon generation rates. We here use the same free-carrier model which previously resulted in good agreement with measurements on InGaN/GaN in-plane lasers [11]. More sophisticated many-body models are presented in [38].

The net optical gain $g(h\nu)$ of stimulated carrier recombination (photon emission) and carrier generation (photon absorption) is a function of the photon energy $h\nu$. For transitions between parabolic subbands i and j it is given by

$$g(h\nu) = \left(\frac{q^2 h}{2m_0^2 \varepsilon_0 n_r c}\right)\left(\frac{1}{h\nu}\right) \sum_{i,j} \int |M|^2 D_r(f_c - f_v) L \, dE \qquad (19.31)$$

with Planck's constant h, the refractive index n_r, the photon velocity c, and the transition energy E. The Fermi functions f_c and f_v for the electron population in the conduction and valence subband, respectively, determine whether more photons are generated ($f_c > f_v$) or absorbed ($f_c < f_v$). D_r is the reduced density of states between the two subbands. Carrier scattering within each band causes an energy broadening which is considered by the Lorentzian lineshape function

$$L(h\nu - E) = \frac{1}{\pi} \frac{\Gamma_s}{(h\nu - E)^2 + \Gamma_s^2} \qquad (19.32)$$

with the half-width $\Gamma_s = 6.6$ meV (scattering time $\tau_s = 0.1$ ps) [39]. $|M|^2$ is the momentum matrix element and it gives the transition strength. Its computation is based on the $\vec{k} \cdot \vec{p}$ electron band structure model outlined above. The matrix element depends on the photon polarization which has two principal directions: parallel (TE) and perpendicular (TM) to the quantum well plane. For a typical VCSEL, only the TE polarization contributes to lasing. However, the TM polarization needs to be included in the computation of spontaneous photon emission. For a quantum well grown in the hexagonal c direction, the energy dependent matrix elements for transitions involving heavy (hh), light (lh), and crystal-field holes (ch), respectively, can be written as

$$|M_{hh}^{TE}|^2 = \frac{3}{2} O_{ij}(M_b^{TE})^2 \quad (19.33)$$

$$|M_{lh}^{TE}|^2 = \frac{3}{2} \cos^2(\theta_k) O_{ij}(M_b^{TE})^2 \quad (19.34)$$

$$|M_{ch}^{TE}|^2 = 0 \quad (19.35)$$

$$|M_{hh}^{TM}|^2 = 0 \quad (19.36)$$

$$|M_{lh}^{TM}|^2 = 3 \sin^2(\theta_k) O_{ij}(M_b^{TM})^2 \quad (19.37)$$

$$|M_{ch}^{TM}|^2 = 3 O_{ij}(M_b^{TM})^2 \quad (19.38)$$

The overlap integral O_{ij} of the electron and hole wavefunctions can assume values between 0 and 1. At the Γ point, O_{ij} is nonzero only for subbands with the same quantum number. Away from the Γ point, O_{ij} may be nonzero for any transition. The angle θ_k of the electron wave vector $\vec{k}$ to the k_z direction introduces an additional energy dependence to the matrix element, with $\cos(\theta_k) = 1$ at the Γ point. The bulk momentum matrix elements are given by [39]

$$(M_b^{TE})^2 = \frac{m_0}{6}\left(\frac{m_0}{m_c^t} - 1\right) \frac{E_g[(E_g + \Delta_1 + \Delta_2)(E_g + 2\Delta_2) - 2\Delta_3^2]}{(E_g + \Delta_1 + \Delta_2)(E_g + \Delta_2) - \Delta_3^2} \quad (19.39)$$

$$(M_b^{TM})^2 = \frac{m_0}{6}\left(\frac{m_0}{m_c^z} - 1\right) \frac{(E_g + \Delta_1 + \Delta_2)(E_g + 2\Delta_2) - 2\Delta_3^2}{E_g + 2\Delta_2} \quad (19.40)$$

Note that the bulk electron mass is different in the transversal (m_c^t) and parallel directions (m_c^z) relative to the hexagonal c-axis. The material parameters are given in Table 19.3.

Figure 19.7 plots the calculated gain spectra for our quantum well at different carrier densities. Built-in polarization is not considered here. However, relatively large densities are required to obtain positive gain. A second gain peak from transitions between higher quantum levels emerges with carrier densities above 5×10^{19} cm^{-3}. The gain peaks' red-shift is caused by band gap renormalization which is included as a function of the quantum well 2D carrier density N_{2D} using the simple formula

$$\Delta E_g = -\zeta N_{2D}^{1/3} \quad (19.41)$$

with $\zeta = 6 \times 10^{-6}$ eV cm$^{2/3}$ [40].

The spontaneous photon emission rate is calculated as

$$r_{sp}(h\nu) = \left(\frac{q^2 h}{2m_0^2 \varepsilon \varepsilon_0}\right)\left(\frac{1}{h\nu}\right) \sum_{i,j} \int |\overline{M}|^2 D_{opt} D_r f_c(1-f_v) L \, dE \quad (19.42)$$

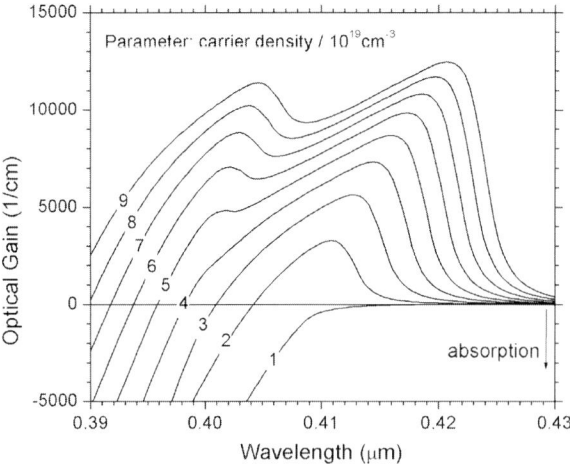

Fig. 19.7 Gain spectra at different carrier densities.

with the density of photon states D_{opt} and the polarization-averaged matrix element $|\overline{M}|^2$. The Fermi factor $f_c(1-f_v)$ is different from that in the gain formula (19.31), i.e., the spontaneous emission spectrum peaks at slightly higher photon energies than the gain spectrum.

19.3.5
Optical Mode

Calculation of the internal optical field is one of the most challenging tasks of VCSEL simulations. A precise optical analysis requires sophisticated solutions to Maxwell's equations for an open resonator. Various numerical models have been developed, most of which are reviewed in [41]. As this chapter focuses on electronic effects, we use a simplified optical model and restrict our simulation to the fundamental lasing mode. Higher-order VCSEL modes are investigated in the next chapter.

Our model is based on the effective index method [42] and it decouples the optical fields in the vertical (z) and the transverse (r) direction. The fundamental mode profile is approximated by a zeroth-order Bessel function which is a solution to the reduced scalar Helmholtz equation for the lateral optical field $\Phi(r)$ in cylindrical waveguides

$$\frac{d^2\Phi}{dr^2} + \frac{1}{r}\frac{d\Phi}{dr} + (k^2 - k_z^2)\Phi = 0 \tag{19.43}$$

with the optical wave number

$$k = n_r \frac{2\pi}{\lambda_0} \tag{19.44}$$

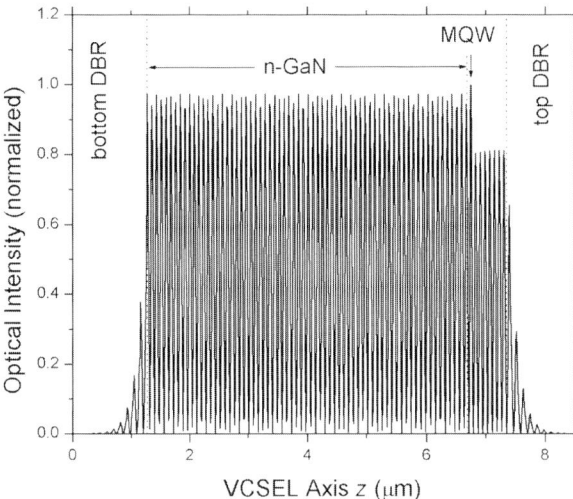

Fig. 19.8 Vertical standing wave.

(n_r – refractive index, λ_0 – free space wavelength). The mode diameter is given by the ring contact aperture of 12 μm (see Fig. 19.2). The refractive index is calculated using the model in [43] which shows good agreement with measurement. The resulting index data at 400 nm wavelength are listed in Table 19.1. A modal optical loss of $\alpha_i = 20\,\text{cm}^{-1}$ is assumed.

In the vertical direction, the transmission matrix method [9] is utilized to obtain the IDX[optical!standing wave]standing optical wave for our large VCSEL cavity (Fig. 19.8). Due to the quarter-wavelength thickness of the DBR layers, null or peak of the optical field are located at the DBR interface. For maximum modal gain, the MQW region is placed at a peak of the standing wave. Due to the thick n-GaN layer, many vertical modes are allowed in our VCSEL with a narrow mode spacing of a few nanometers (Fig. 19.9). This is unusual for VCSELs, but it eliminates a crucial VCSEL design issue, namely the alignment of mode wavelength and optical gain spectrum [8] (cf. Fig. 19.7). In our VCSEL, the optical mode nearest to the gain peak is expected to lase so that the shape of the gain spectrum is less important. The peak gain can reasonably be approximated by a free-carrier model as described above.

19.4
Simulation Results and Device Analysis

In the following, we utilize our VCSEL simulation to analyze various internal device processes, including current confinement, polarization effects, and electron leakage. Some of those processes limit the device performance and

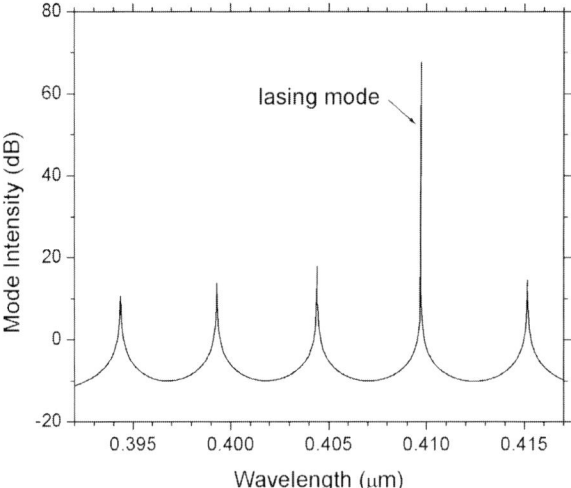

Fig. 19.9 Mode spectrum.

contribute to the fact that no lasing was observed in the experimental investigation of our example device [6].

19.4.1
Current Confinement

The optical mode size in our device is laterally confined by the top ring contact with about 12 μm aperture (cf. Fig. 19.2). Hole injection is provided by the top ITO layer with manufactured apertures between 2 μm and 10 μm. Due to the poor hole mobility, the lateral hole spreading is very small and the carrier profile within the active layers is well confined to the injection aperture. Figure 19.10 illustrates this situation for the device with 2 μm ITO aperture. Due to the narrowly confined carrier density, positive optical gain is only provided for $r < 1.5$ μm. Beyond that radius, the optical mode experiences strong quantum well absorption leading to a net modal gain below zero. Thus, the 2 μm VCSEL never reaches the lasing threshold in our simulation. In the following, we therefore focus our investigation on VCSELs with 10 μm injection aperture, which seem to have a better chance to lase.

19.4.2
Polarization Effects

The built-in polarization strongly deforms the energy band diagram of our MQW active region. Figure 19.11 compares the MQW band diagrams as calculated at forward bias with and without the built-in polarization charges given in Table 19.4. Without polarization, the quantum wells are almost rectangu-

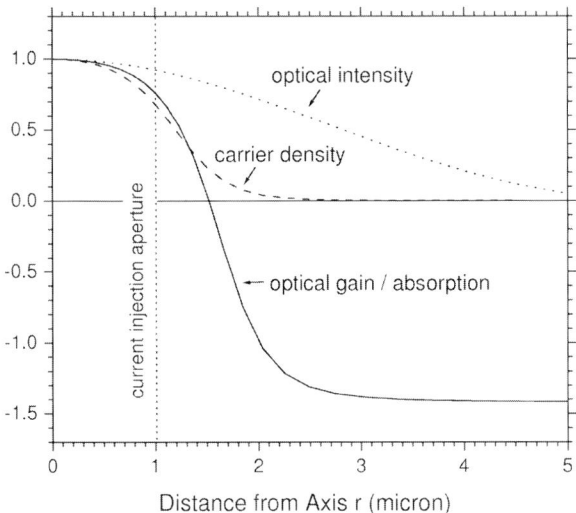

Fig. 19.10 Normalized lateral profiles within the quantum well for 2 μm current injection aperture.

lar and the AlGaN layer imposes a considerable energy barrier of 250 meV on electrons trying to leak out of the MQW active region. With full polarization, the energy band diagram is significantly deformed. This deformation is even more remarkable considering the high current density of $j = 50\,\text{kA}\,\text{cm}^{-2}$ used in this calculation, which is 16 times higher than the threshold current density of similar edge-emitting lasers [11]. Surprisingly, even with strong carrier injection, the built-in polarization field is not completely screened as often assumed for laser operation.

The corresponding electrostatic field profile is plotted in Fig. 19.12. The polarization charge densities at the MQW interfaces translate into a built-in quantum well field of $1.8\,\text{MV}\,\text{cm}^{-1}$. The actual electrostatic field within the quantum wells is about $0.5\,\text{MV}\,\text{cm}^{-1}$ due to partial screening by electrons and holes. Figure 19.13 gives the carrier density profile. The built-in polarization clearly leads to a separation of electrons and holes within the quantum wells. But even with a current density of $50\,\text{kA}\,\text{cm}^{-2}$, the injected quantum well carrier density is not large enough to completely screen the built-in field. This can be easily understood by converting the interface charge densities given in Table 19.4 into a uniform quantum well carrier density of $2.4 \times 10^{19}\,\text{cm}^{-3}$ needed for full screening.

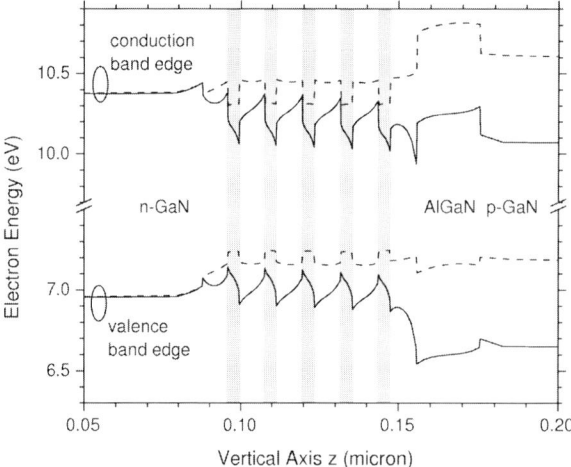

Fig. 19.11 Forward-bias MQW energy band diagram at the VCSEL axis with (solid) and without (dashed) the built-in polarization charges given in Table 19.4 (grey: quantum wells, $j = 50$ kA cm^{-2}, $p_{AlGaN} = 10^{16}$ cm^{-3}).

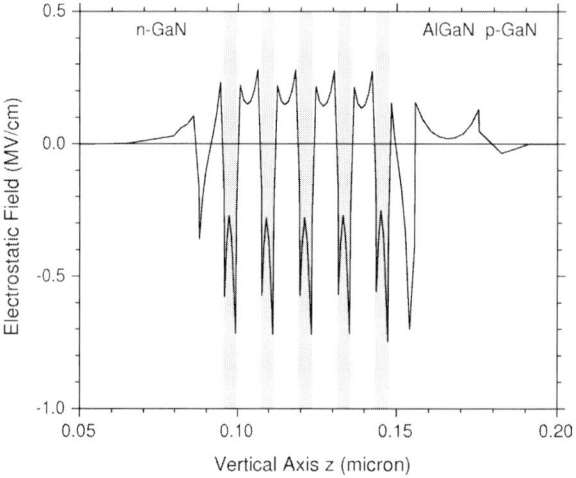

Fig. 19.12 Electrostatic field at the VCSEL axis with full polarization (grey: quantum wells, $j = 50$ kA cm^{-2}, $p_{AlGaN} = 10^{16}$ cm^{-3}).

19.4.3
Threshold Current

The separation of electrons and holes within the quantum well reduces the radiative recombination rate and leads to an increased threshold current of the laser. Figure 19.14 shows the calculated VCSEL threshold current as a function of polarization strength with 100% corresponding to the theoretically

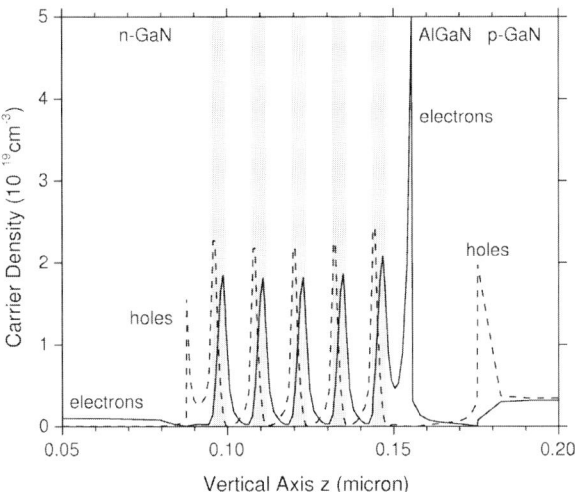

Fig. 19.13 Carrier densities at the VCSEL axis with full polarization (solid: electrons, dashed: holes, grey: quantum wells, $j=50$ kA cm^{-2}, $p_{AlGaN}=10^{16}$ cm^{-3}).

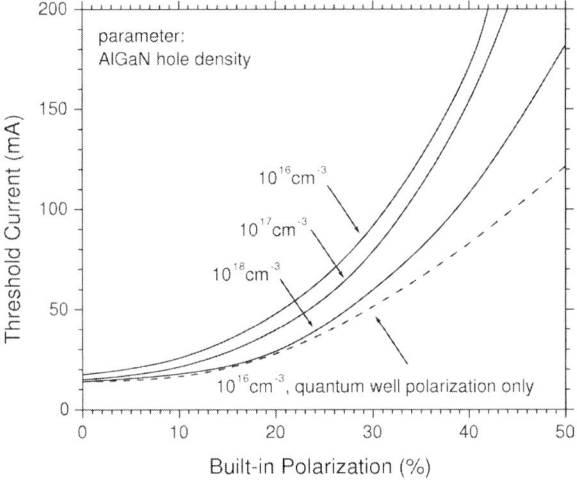

Fig. 19.14 VCSEL threshold current as function of built-in polarization strength (100% = Table 19.4).

predicted values of Table 19.4. Without polarization, the threshold current is 17 mA. Experimentally, these VCSELs suffered from catastrophic short-circuiting at about half that current, probably due to threading dislocations [6].

Considering only the polarization charges at the quantum well interfaces, 50% polarization enlarges the threshold current by a factor of 7 (dashed line in Fig. 19.14). This is mainly due to the reduced overlap of electron and hole wavefunctions. However, with polarization charges at all hetero-interfaces,

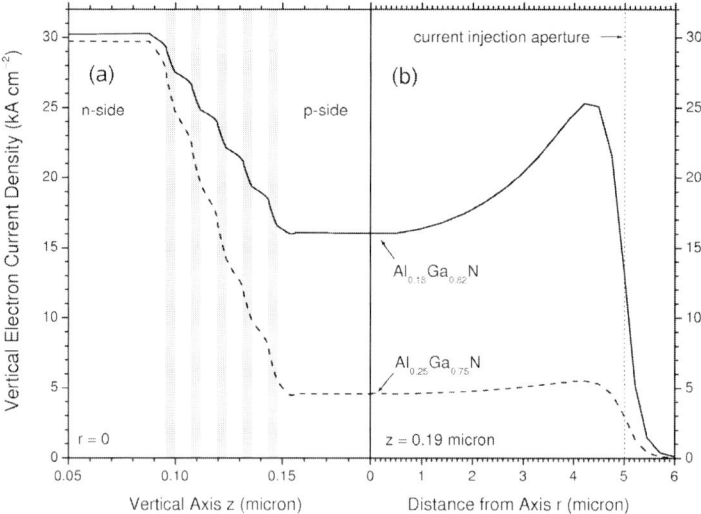

Fig. 19.15 Vertical electron current density $j_{n,z}$ with the default $Al_{0.18}Ga_{0.82}N$ stopper layer (solid) and with an $Al_{0.25}Ga_{0.75}N$ stopper layer (dashed). (a) Vertical profile $j_{n,z}(z)$ at the VCSEL axis; (b) Radial profile $j_{n,z}(r)$ above the stopper layer; (grey: quantum wells, $p_{AlGaN} = 10^{18} cm^{-3}$).

the threshold current rises even more strongly (solid lines in Fig. 19.14). The only difference in both cases is the polarization introduced by the AlGaN stopper layer. As shown in Figs 19.11 and 19.13, the polarization charges at the InGaN/AlGaN interface attract a large density of electrons which lead to significant band bending. Consequently, the AlGaN energy barrier is dramatically reduced allowing for strong electron leakage from the MQW. Such leakage effects are also indicated by the measured photoluminescence spectrum of this VCSEL [6] as well as by investigations on similar devices [16].

19.4.4
AlGaN Doping

Electron leakage is a common problem with laser diodes and stronger p-doping of the stopper layer is a typical countermeasure [44]. Figure 19.14 demonstrates the effect of AlGaN doping in our case. The free hole density is increased from $10^{16} cm^{-3}$, as used in the previous section, to $10^{18} cm^{-3}$, which is close to the maximum value achieved to date. With higher p-doping, the AlGaN band bending increases and so does the effective electron energy barrier imposed by the stopper layer. As our simulation results show in Fig. 19.14, raising the hole density from $10^{16} cm^{-3}$ to $10^{18} cm^{-3}$ results in a significant threshold current reduction.

19.4.5
AlGaN Composition

The electron leakage can also be reduced by increasing the Al mole fraction of the AlGaN stopper layer. This will enlarge the AlGaN bandgap but it will also lead to a higher density of polarization charges at the AlGaN interfaces. The combined effect of both changes on the electron current is illustrated in Fig. 19.15. The vertical current profile across the MQW region is shown on the left-hand side. Electrons enter the MQW from the n-side and partially recombine with holes within the quantum wells. Electrons leaving the MQW and entering the p-side constitute the leakage current. With 25% Al mole fraction (dashed), the leakage current is significantly smaller than with the original $Al_{0.18}Ga_{0.82}N$ stopper layer (solid).

The right-hand side of Fig. 19.15 plots the lateral profile of the leakage current above the stopper layer. Its shape is similar to that of the hole current. The current peaks near the injection aperture given by the ITO contact diameter of 10 μm. Such current crowding reduces the quantum well carrier density in the center of the device and the modal gain, leading to an increased threshold current. Obviously, the higher $Al_{0.25}Ga_{0.75}N$ barrier causes a more uniform current distribution (dashed line) compared to the original stopper layer (solid line).

19.5
Summary

We have demonstrated how advanced simulation of GaN VCSELs can be used to gain deeper insight into internal device physics that is not available through measurements. Our simulation results help to understand performance limiting mechanisms and to improve the design of future devices.

References

1 T. E. Sale. *Vertical Cavity Surface Emitting Lasers*. Wiley, New York, 1995.

2 C. Wilmsen, H. Temkin, and L. A. Coldren, editors. *Vertical-Cavity Surface-Emitting Lasers*. Cambridge Univ. Press, Cambridge, UK, 1999.

3 J. Piprek. *phys. stat. sol. (a)*, 188:905–912, 2001.

4 H. Li and K. Iga, editors. *Vertical-Cavity Surface-Emitting Laser Devices*. Springer, Berlin, 2003.

5 J. Piprek, editor. *Optoelectronic Devices: Advanced Simulation and Analysis*. Springer, New York, 2005.

6 T. Margalith. *Development of Growth and Fabrication Technology for Gallium Nitride-Based Vertical-Cavity Surface-Emitting Lasers*. PhD thesis, University of California at Santa Barbara, 2002.

7 A. V. Nurmikko and J. Han. Progress in blue and near-ultraviolet vertical-cavity emitters: A status report. In: H. Li and K. Iga, editors, *Vertical-Cavity Surface-*

Emitting Laser Devices. Springer, Berlin, 2003.

8. J. Piprek, Y. A. Akulova, D. I. Babic, L. A. Coldren, and J. E. Bowers. *Appl. Phys. Lett.*, 72(15):1814–1816, 1998.

9. S. F. Yu. *Analysis and Design of Vertical Cavity Surface Emitting Lasers*. Wiley, Hoboken, 2003.

10. PICS3D by Crosslight Software Inc., Burnaby, Canada, 2005.

11. J. Piprek and S. Nakamura. *IEE Proc., Part J: Optoelectron.*, 149:145–151, 2002.

12. I. Vurgaftman and J. R. Meyer. *J. Appl. Phys.*, 94:3675–3691, 2003.

13. A. V. Dmitriev and A. L. Oruzheinikov. *J. Appl. Phys.*, 86:3241–3246, 1999.

14. J. Piprek, T. Katona, S. P. DenBaars, and S. Li. 3D simulation and analysis of UV AlGaN/GaN LEDs. In: *Light-Emitting Diodes: Research, Manufacturing and Applications VII*, volume 5366, SPIE, Bellingham, 2004.

15. J. Piprek. *Semiconductor Optoelectronic Devices: Introduction to Physics and Simulation*. Academic Press, San Diego, 2003.

16. J. Piprek and S. Li. GaN-based light-emitting diode. In: J. Piprek, editor, *Optoelectronic Devices: Advanced Simulation and Analysis*. Springer, New York, 2005.

17. S.-H. Wei and A. Zunger. *Appl. Phys. Lett.*, 69:2719–2721, 1996.

18. A. C. Abare. *Growth and Fabrication of Nitride-Based Distributed Feedback Laser Diodes*. PhD thesis, University of California at Santa Barbara, 2000.

19. C. G. Van de Walle and J. Neugebauer. *Nature*, 423:626–628, 2003.

20. W. Götz, N. M. Johnson, J. Walker, D. P. Bour, and R. A. Street. *Appl. Phys. Lett.*, 68:667–669, 1996.

21. K. H. Ploog and O. Brandt. *J. Vac. Sci. Technol.*, A16:1609–1614, 1998.

22. S. Nakamura, S. Pearton, and G. Fasol. *The Blue Laser Diode*. Springer-Verlag, Berlin, 2000.

23. D. M. Caughey and R. E. Thomas. *Proc. IEEE*, 55:2192–2193, 1967.

24. F. Schwierz. *Solid State Electron.*, 49:889–895, 2005.

25. D. K. Gaskill, L. B. Rowland, and K. Doverspike. Electrical transport properties of AlN, GaN, and AlGaN. In: J. H. Edgar, editor, *Properties of Group-III Nitrides*, pages 101–116. IEE/INSPEC, London, 1994.

26. T. T. Mnatsakanov, M. E. Levinshtein, L. I. Pomortseva, S. N. Yurkov, G. S. Simin, and M. A. Khan. *Solid State Electron.*, 47:111–115, 2003.

27. S. L. Chuang and C. S. Chang. *Phys. Rev. B*, 54:2491–2504, 1996.

28. S. L. Chuang and C. S. Chang. *Semicond. Sci. Technol.*, 12:252–263, 1997.

29. M. D. McCluskey, C. G. Van de Walle, L. T. Romano, B. S. Krusor, and N. M. Johnson. *J. Appl. Phys.*, 93:4340–4342, 2003.

30. C. Wetzel, T. Takeuchi, S. Yamaguchi, H. Katoh, H. Amano, and I. Akasaki. *Appl. Phys. Lett.*, 73:1994–1996, 1998.

31. V. Fiorentini, F. Bernardini, and O. Ambacher. *Appl. Phys. Lett.*, 80:1204–1206, 2002.

32. J. Piprek, R. Farrell, S. DenBaars, and S. Nakamura. *IEEE Photon. Technol. Lett.*, 18:7–9, 2006.

33. S. F. Chichibu, A. C. Abare, M. S. Minsky, S. Keller, S. B. Fleischer, J. E. Bowers, E. Hu, U. K. Mishra, L. A. Coldren, and S. P. DenBaars. *Appl. Phys. Lett.*, 73:2006–2008, 1998.

34. F. Renner, P. Kiesel, G. H. Döhler, M. Kneissl, C. G. Van de Walle, and N. M. Johnson. *Appl. Phys. Lett.*, 81:490–492, 2002.

35. H. Zhang, E. J. Miller, E. T. Yu, C. Poblenz, and J. S. Speck. *Appl. Phys. Lett.*, 84:4644–4646, 2004.

36. J. P. Ibbetson, P. T. Fini, K. D. Ness, S. P. DenBaars, J. S. Speck, and U. K. Mishra. *Appl. Phys. Lett.*, 77:250–252, 2000.

37. I. H. Brown, I. A. Pope, P. M. Smowton, P. Blood, J. D. Thomson, W. W. Chow, D. P. Bour, and M. Kneissl. *Appl. Phys. Lett.*, 86:131108, 2005.

38. W. W. Chow and S. W. Koch. *Semiconductor-Laser Fundamentals*. Springer-Verlag, Berlin, 1999.

39. S. L. Chuang. *IEEE J. Quantum Electron.*, 32(10):1791–1799, 1996.

40 S. H. Park and S. L. Chuang. *Appl. Phys. Lett.*, 72:287–289, 1998.

41 P. Bienstman, R. R. Baets, J. Vukusic, A. Larsson, M.J. Noble, M. Brunner, K. Gulden, P. Debernardi, L. Fratta, G.P. Bava, H. Wenzel, B. Klein, O. Conradi, R. Pregla, S.A. Riyopoulos, J.-F.P. Seurin, and S. L. Chuang. *IEEE J. Quantum Electron.*, 37:1618 –1631, 2001.

42 G.R. Hadley, K.L. Lear, M.E. Warren, K.D. Choquette, J.W. Scott, and S.W. Corzine. *IEEE J. Quantum Electron.*, 32:607–616, 1996.

43 G. M. Laws, E. C. Larkins, I. Harrison, C. Molloy, and D. Somerford. *J. Appl. Phys.*, 89:1108–1115, 2001.

44 R. F. Kazarinov and M. R. Pinto. *IEEE J. Quantum Electron.*, 30:49–53, 1994.

20
Optical Design of Vertical-Cavity Lasers

Włodzimierz Nakwaski[1], Tomasz Czyszanowski, and Robert P. Sarzała

20.1
Introduction

As in all diode lasers, the correct operation of vertical-cavity surface-emitting diode lasers (VCSELs) depends mostly on efficient confinement of both a radiation field and recombining carriers to the area of their interaction known as the active region [1]. In standard GaAs-based VCSELs, the above is usually accomplished with the aid of oxide apertures. Unfortunately, a simple technology similar to the AlAs radial oxidation used to create the above apertures is not known in nitride devices [2]. Therefore, in possible GaN-based VCSELs, only the radial gain-guiding effect may be used, much less, however, efficient in optical waveguiding than the radial built-in index guiding achieved with the aid of the oxide apertures. Together with a strongly nonuniform radial current injection into the active region and an intense Joule heat generation within high-electrical-resistivity p-type layers, the above is probably one of the main reasons for problems in achieving room-temperature (RT) operation of GaN-based VCSELs. In the light of the above, a possible application of the real built-in radial waveguide effect induced by photonic crystals in cavities of GaN-based VCSELs may be a very tempting solution [3].

Optical properties of all optoelectronic devices are described by the Maxwell equations. Usually they may be reduced to the wave equation for the vector of the electric field E or the vector of the magnetic field H; the electromagnetic wave vectors. When the lateral sizes of a device optical cavity and its active region are much larger than the wavelength of the propagated radiation, the plane-wave approximation may be applied, leading to a scalar form of the above equation. Then relatively simple scalar optical approaches may be used. However, in modern micro-cavity devices, a distinct curvature of the spatial mode profiles may be observed, which can no longer be

1) Corresponding author

treated as plane waves. The above is followed by the indispensability of the application of fully vectorial optical models.

In the case of cylindrically symmetric VCSEL structures, an application of standard PC-class computers is sufficient for scalar optical models, whereas vectorial optical models usually require much more extensive and time-consuming calculations, carried out with the aid of more powerful computers. However, simplified scalar optical approaches lead sometimes to quite exact results even beyond the limits of their confirmed validity. Therefore it is to be desired to determine for GaN-based VCSELs the real area of a proper operation of simple scalar models.

For a homogeneous lossless medium, the Maxwell equations may be combined to give the vector wave equation [1]:

$$\left(\nabla^2 + k_0^2 n_R^2\right) \boldsymbol{\Psi} = 0 \tag{20.1}$$

where n_R is the real index of refraction, $k_0 = \omega/c$ is the wave number, where ω stands for the circular frequency of an electromagnetic wave and c is the speed of light in vacuum. Operator ∇^2 is the vector Laplacian and $\boldsymbol{\Psi}$ may be either the $\boldsymbol{E}$ or the $\boldsymbol{H}$ vectors. The above equation is approximately satisfied also in inhomogeneous media, provided that, along a distance equal to the wavelength $\lambda = 2\pi c/\omega$, the dielectric constant ε changes by much less than unity [4]. Using the expansion of the vector Laplacian, the above vector equation (20.1) may be, for each Φ_u component of $\boldsymbol{E}$ and $\boldsymbol{H}$ vectors, reduced to the following scalar wave equations (for $u = r, z, \phi$ or x, y, z):

$$\left(\Delta + k_0^2 n_R^2\right) \Phi_u = 0 \tag{20.2}$$

where Δ is the scalar Laplacian.

In lossy (or gain) medium, instead of the real index of refraction n_R in Eqs (20.1), (20.2), the complex index of refraction N_R should be used:

$$N_R = n_R - ik_e \tag{20.3}$$

where $k_e = \alpha/(2k_0)$ is the extinction coefficient, directly related to the absorption α (or gain) coefficient. Both (real and imaginary) parts of N_R are influenced by three-dimensional (3D) profiles of the carrier concentration n and temperature T, which is described by the following expression:

$$N_R = n_{R0} + \frac{dn_R}{dT}(T - T_R) + \frac{dn_R}{dn}n - i\frac{\alpha - g}{2k_0} \tag{20.4}$$

where n_{R0} symbolizes a 3D distribution of the real index of refraction without carriers and at the reference temperature T_R (equal usually to that of the ambient) whereas n stands for the 3D free-carrier concentration (electrons or holes or both); g is the temperature- and carrier-concentration-dependent optical material gain within the active region, and α stands for all the optical losses, including both the free-carrier and the intervalence absorption, scattering and diffraction losses and so on, but not the band-to-band absorption within the active region, which has already been taken into account in the gain determination.

The above simple relation (20.4) carries information about the various interrelations between physical phenomena including the index-guiding (IG) thermal focusing (second term) and the IG self-focusing (third term) as well as the gain-guiding (GG) mechanism (last term). As one can see, both the current-spreading and the heat-flux-spreading phenomena strongly influence VCSEL waveguiding properties. Therefore, comprehensive VCSEL modeling should include an important interaction part, comprising a real network of many, usually strongly nonlinear and mutual interrelations between individual physical phenomena, crucial for a VCSEL operation. In more advanced VCSEL designs with built-in waveguiding effects, the 3D n_R and α profiles may contain designed fixed distributions of an index of refraction and/or absorption, respectively, intentionally introduced to stabilize the VCSEL waveguiding properties. Then temperature- and carrier-concentration-dependent changes of both N_R parts have a nearly insignificant impact on the radiation field which is mostly confined by a stable built-in waveguide.

20.2
The GaN VCSEL Structure

Room-temperature (RT) lasing operation of nitride VCSELs has not been achieved until now, although analogous RT operation of nitride edge-emitting (EE) diode lasers had already been reported in 1996 [5] and these devices are currently commercially available. There are many possible reasons for such a situation. It is obvious that nitride VCSELs need stronger excitation (they exhibit higher lasing thresholds) than nitride EE lasers. Also, nitride p-type materials exhibit very high electrical resistivity. Consequently, in modern VCSEL structures with both n-side and p-side annular contacts, current injection into a centrally located VCSEL active region is extremely nonuniform: the nearly insignificant current density reaches the active-region center, whereas a dramatic current-crowding effect may be observed close to its edge. Therefore, an overlapping of the radial profile of an optical gain with the analogous profile of an intensity of the usually lowest-threshold fundamental transverse mode is extremely low, which is followed by a very high lasing threshold. The above situation may be drastically improved by an enhancement in the uniformity of current injection into a VCSEL active region.

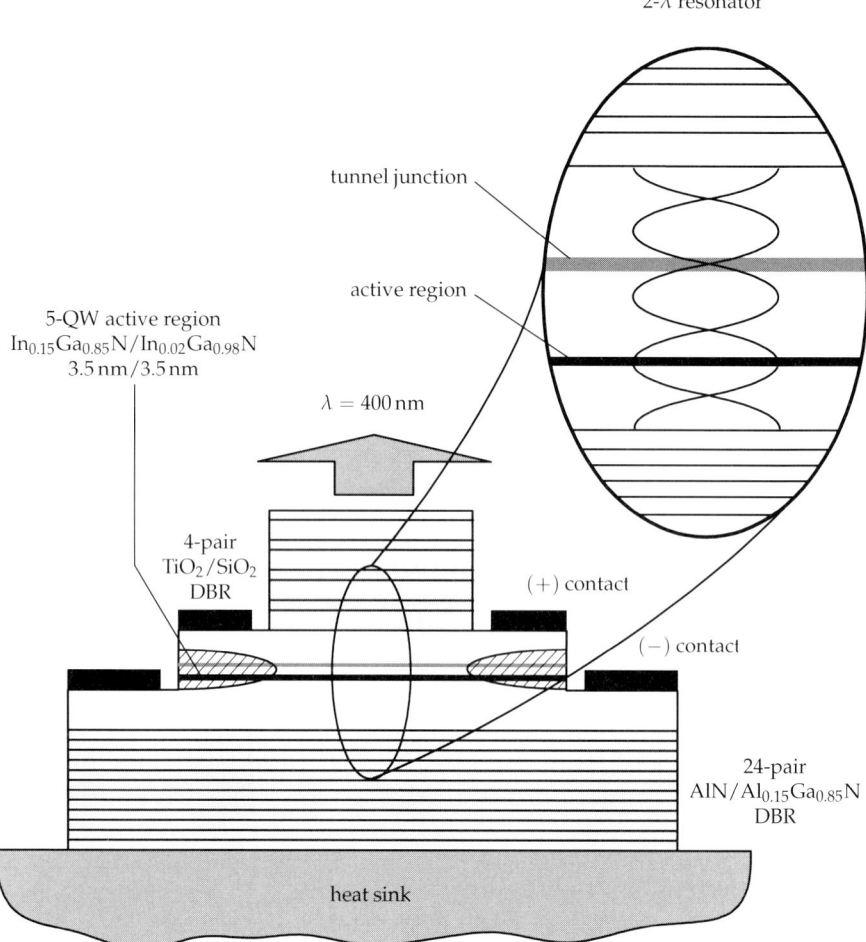

Fig. 20.1 The structure of the modeled GaN-based VCSEL with the active region comprising five quantum wells (5-QW) and the tunnel junction. Dashed regions corresponds to the proton implantation.

As has happened many times in the past, structures, which have confirmed to be the best for earlier devices, have been found not to be optimal for new ones. Because of the special features of nitride materials and nitride devices, the physics of nitride VCSELs is often quite different from the case of earlier arsenide and phosphide ones. For example, lattice constants of available for nitride device substrate materials differ substantially from those of nitride materials. As a result of this lattice mismatch, misfit dislocations or, in the case of plastic deformation, stress fields, are generated. Nitride materials exhibit very strong piezoelectric properties. Therefore, strains induced in nitride structures by the lattice mismatch generate an electric field perpendicular to

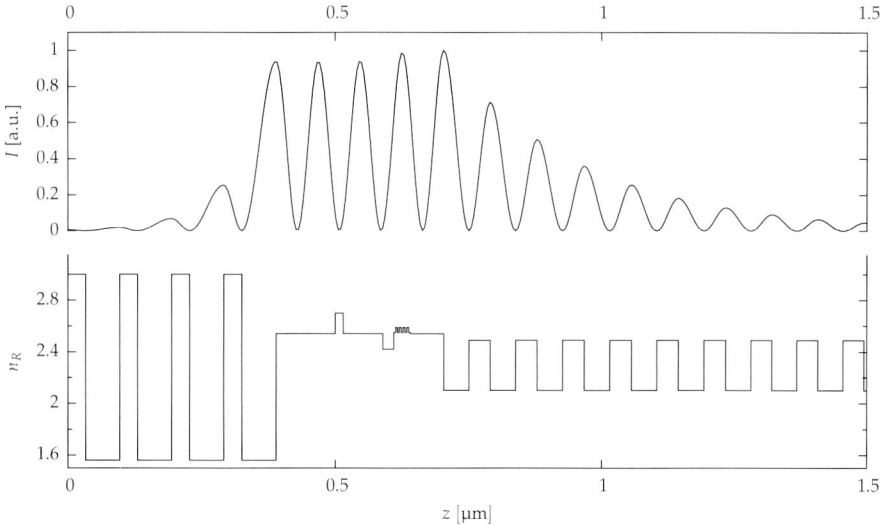

Fig. 20.2 VCSEL cavity design, i.e., the transverse-fundamental-mode intensity profile $I(z)$ of the standing optical wave, and a profile of an index of refraction $n_R(z)$ within the cavity in the z propagation direction. As one can see, in the nitride design under consideration, the QW material exhibits a lower index of refraction than that of the barriers.

the quantum-well edges, pulling electrons and holes in opposite directions [6] and influencing their recombination. Fortunately, these piezoelectric-related phenomena in diode lasers are effectively screened by high carrier concentrations [7], although very recently the efficiency of this behavior has been called into question [8]. A possible unexpectedly strong band-to-band absorption in InGaN and AlGaN layers as a result of their sometimes very nonuniform composition, is another feature of nitrides. Also, a temperature increase within the VCSEL volume, usually not profitable for the efficiency of arsenide and phosphide VCSELs, may sometimes play a quite advantageous role in nitride ones, improving the uniformity of current injection into the active region [9] or introducing a real radial waveguiding with the aid of the thermal focusing effect. Because of all the above special features of nitride materials, earlier VCSEL structures regarded to be optimal are often not suited for nitride ones. Optimal nitride VCSELs may differ substantially from earlier ones.

A traditional VCSEL design with a classical double lateral injection into its active region has been found to create a very nonuniform carrier-concentration profile in nitride lasers. Then the carrier density (and an optical gain roughly proportional to it), although practically insignificant within the broad central circular part of the VCSEL active region, increases dramatically when approaching its perimeter. Because of this very nonuniform radial optical gain profile and also due to problems with the effective confinement of an optical field in a radial direction, nitride VCSELs of such standard designs are expected to exhibit extremely high lasing thresholds, so RT operation of electrically pumped nitride VCSELs has not been at all reported until now.

In many VCSEL applications, the single fundamental mode operation is required. To reduce its lasing threshold, its bell-like radial intensity profile should be as similar to the analogous optical gain profile as possible (to obtain higher modal gain for the fundamental mode than those for higher-order ones, see Eq. (20.37)), which is not achievable in the above standard VCSEL designs. It is, however, well known that the radial gain profile is quite similar to the above bell-like one for a uniform current injection into the active region. Therefore, to enhance the desired radial uniformity of the above injection, we would like to propose the application of a tunnel junction over the active region, which additionally enables replacement of the extremely high-electrical-resistivity p-type GaN in the radial-current-spreading layer by the n-type one with resistivity lower by more than two orders of magnitude. It has been confirmed by our comprehensive simulation that both the above-mentioned structure modifications effectively enhance the uniformity of current injection into the active region and the radial gain profiles obtained usually resemble the bell-like intensity profile of the fundamental mode, which ensures its lowest lasing threshold.

Let us consider a possible structure of the GaN-based VCSELs (Fig. 20.1) emitting 400 nm radiation. The structure is similar to, but technologically considerably simpler, than that reported in [10]. The active region is assumed to be composed of five 3.5 nm $In_{0.15}Ga_{0.85}N$ quantum wells (QWs) separated by the 3.5 nm $In_{0.02}Ga_{0.98}N$ barriers. The 20 nm blocking p-$Al_{0.2}Ga_{0.8}N$ layer is additionally grown on the p-side of the active region. The active region is sandwiched by GaN spacers, the n-type 62.82 nm one and the p-type 75.16 nm one. The standard double intra-cavity-contacted VCSEL structure (Fig. 20.1) is proposed with two n- and p-side ring contacts. Radial uniformity of the carrier injection into the active region has been found to be crucial in order to reach the expected RT operation [11]. Therefore, as compared with standard GaAs-based VCSELs, we suggest additionally employing the tunnel junction [12, 13] (the 30 nm n^{++}-GaN layer and the 15 nm p^{++}-$In_{0.2}Ga_{0.8}N$ layer). Between the p-side contact and the tunnel junction, the radial-current-spreading 80.14 nm n-GaN layer is grown. Additionally, the proton implantation [14] is assumed to be used to form high-resistivity lateral areas to funnel the current flow from annular contacts towards the centrally located active region. The 2λ cavity is terminated on both sides by distributed Bragg reflector (DBR) mirrors [15]: the dielectric 4-pair SiO_2 (33.33 nm)/TiO_2 (64.10 nm) p-side DBR and the 28-pair AlN (47.62 nm)/$Al_{0.15}Ga_{0.85}N$ (40.16 nm) n-side DBR. To minimize the lasing threshold, the QW active region is placed at the

Tab. 20.1 VCSEL parameters assumed in the calculations: d is the layer thickness, n_R the real refractive index, α –the absorption coefficient, κ the thermal conductivity at 300 K, σ the electrical conductivity. Dielectric, top DBR mirrors consist of 4 periods, while the bottom ones of 28 periods. 15% of Al mol fraction in the bottom AlGaN DBR layer has been used. The active region consists of 5 $In_{0.15}Ga_{0.85}N$ quantum wells and 4 $In_{0.02}Ga_{0.98}N$ barriers. The proton implanted areas are marked by 'p.i.'

Material	d [nm]	n_R (300 K)	α [cm^{-1}] (300 K)	$\frac{dn_R}{dT}$ [K^{-1}]	$\frac{dn_R}{dn}$ [cm^3]	κ [W mK^{-1}]	σ [Ωcm^{-1}]
TiO$_2$ DBR	33.33	3.0	10	-5×10^{-6}	0	8.9	—
SiO$_2$ DBR	64.10	1.56	10	-5×10^{-6}	0	1.38	—
n GaN	80.14	2.54	10	2.88×10^{-4}	10^{-22}	195	270.27
n^{++} GaN	30.00	2.54	150	2.88×10^{-4}	10^{-22}	195	—
n^{++} GaN p.i.			1000		0		0
p^{++} In$_{0.2}$Ga$_{0.8}$N	15.00	2.7	150	2.58×10^{-4}	10^{-22}	100	—
p^{++} In$_{0.2}$Ga$_{0.8}$N p.i.			1000		0		0
p GaN	75.16	2.54	10	2.88×10^{-4}	10^{-22}	100	3.82
p Al$_{0.2}$Ga$_{0.8}$N	20.00	2.42	10	5.7×10^{-5}	10^{-22}	60	0.95
MQW barrier	3.50	2.59	—	2.58×10^{-4}	10^{-22}	100	—
MQW well	3.50	2.55	—	2.35×10^{-3}	10^{-22}	100	—
n GaN	62.82	2.54	10	2.88×10^{-4}	10^{-22}	195	270.27
n AlGaN DBR	40.16	2.49	10	5.7×10^{-5}	10^{-22}	65	1.41
n AlN DBR	47.62	2.1	10	6×10^{-5}	10^{-22}	285	10^{-11}
Cu heat sink	—	—	—	—	—	398	588200

anti-node position of an optical standing wave within the VCSEL cavity and the tunnel junction, in its node positions (Fig. 20.2). The values of the material parameters used in our simulation are listed in Table 20.1.

20.3
The Scalar Optical Approach

In our scalar approach, the cylindrical (r, z, ϕ) co-ordinate system is used with the z-axis directed along the VCSEL symmetry axis, r perpendicular to it, and ϕ as the azimuthal angle. We are using the most advanced scalar optical VCSEL approach, i.e., the Effective Frequency Method, proposed by Wenzel and Wünsche [16]. Then the optical field $E(r, z, \phi)$ is assumed to be of the following form:

$$E(r, z, \phi) = f(r, z) \Phi_L(r) \exp(iL\phi) \tag{20.5}$$

where $L = 0, 1, 2, \ldots$ is the azimuthal mode number. The above enables the reduction of the wave equation within the VCSEL cavity to the following two mutually inter-related nearly-one-dimensional wave equations along both the axial and the radial VCSEL directions:

$$\left[\frac{d^2}{dz^2} + k_0^2 N_P^2(r,z)\right] f(r,z) = v_{\text{eff}} k_0^2 N_P(r,z) N_G(r,z) f(r,z) \tag{20.6}$$

$$\left[\frac{d^2}{dr^2} + \frac{1}{r}\frac{d}{dr} - \frac{L^2}{r^2} v_{\text{eff}} k_0^2 \langle N_P N_G \rangle_r \right] \Phi_L(r) = v k_0^2 \langle N_P N_G \rangle_r \Phi_L(r) \tag{20.7}$$

where v_{eff} stands for the effective frequency, $k_0 = \omega_0/c$ is the vacuum wave number and ω_0 is the real-valued nominal angular frequency corresponding to the designed periodicity of DBR mirrors. N_P and N_G are the complex phase and group indices, respectively, evaluated at the nominal frequency ω_0. The dimensionless complex parameter $v = 2(1 - \omega/\omega_0)$ plays a role of an eigenvalue. As boundary conditions, the outgoing plane waves are assumed at both the bottom and the top cavity surfaces, whereas the cylindrical outgoing wave is presumed for a sufficiently large radial distance. As a consequence of the first above boundary condition, diffraction losses are neglected in this scalar approach. The lasing threshold for each cavity mode is determined from the condition of its real propagation constant, i.e., the threshold is reached when the propagation-constant imaginary part is reduced to zero. Then, for the mode under consideration, its modal gain (see Eq. (20.37)) and modal losses (Eq. (20.38)) are equal to each other.

20.4
The Vectorial Optical Approach

In the vectorial approach, the Cartesian (x, y, z) coordinate system is applied with the xy-plane parallel to the layer boundaries and z in the direction of the light propagation. The new optical Plane Wave Admittance Method (PWAM) proposed by Dems, Kotyński and Panajotov [17] will be used. The PWAM combines two approaches successfully used in photonic simulations: the Plane Wave Method and the Methods of Lines (MoL). In a similar way to MoL, the 3D structure is divided into uniform layers along the z-axis: the solution in each of the layers is therefore given as a sum of forward and backward travelling waves. The original MoL solves for the lateral distribution of the field by carrying out Finite Difference Discretization. The expansion of the field in the basis of orthonormal functions can be significantly more computationally efficient. The set of exponential functions provides a very convenient tool to expand both the field and the spatial distribution of the refractive index. Furthermore, due to the special properties of the Fourier transform, the refractive index is expanded within an infinite basis, which allows for precise representation not only of smooth lateral variations of the refractive index (as in the case of thermal lensing in proton-implanted VCSELs) but also of abrupt ones (as in the case of confined oxide). The preciseness of the PWAM method has been successfully confirmed by comparing its results [18] with those determined by other, more time-consuming, vectorial methods [19], both given for oxide confined VCSELs. For the planar VCSEL structure composed of a number of uniform layers, the E and H vectors of the cavity standing wave are governed by the following Maxwell equations:

$$\nabla \times E(x,y,z,t) = -\mu\mu_0 \frac{\partial H}{\partial t}(x,y,z,t) \tag{20.8}$$

$$\nabla \times H(x,y,z,t) = \varepsilon\varepsilon_0 \frac{\partial E}{\partial t}(x,y,z,t) \tag{20.9}$$

where μ and μ_0 as well as ε and ε_0 are the diagonal tensors of the magnetic permittivity and the dielectric constant, respectively, for the material and for the vacuum. Then, after some mathematical manipulation, the following set of two coupled equations is achieved:

$$\partial_z^2 \begin{bmatrix} E_x \\ E_y \end{bmatrix} = \frac{1}{\omega^2 \varepsilon_0 \mu_0} \begin{bmatrix} -\partial_x \frac{1}{\varepsilon_z} \partial_x - \omega^2 \varepsilon_0 \mu_y \mu_0 & \partial_x \frac{1}{\varepsilon_z} \partial_y \\ \partial_y \frac{1}{\varepsilon_z} \partial_x & \partial_y \frac{1}{\varepsilon_z} \partial_y + \omega^2 \varepsilon_0 \mu_x \mu_0 \end{bmatrix} \times \begin{bmatrix} -\partial_y \frac{1}{\mu_z} \partial_y - \omega^2 \mu_0 \varepsilon_x \varepsilon_0 & \partial_y \frac{1}{\mu_z} \partial_x \\ -\partial_x \frac{1}{\mu_z} \partial_y & \partial_x \frac{1}{\mu_z} \partial_x + \omega^2 \mu_0 \varepsilon_y \varepsilon_0 \end{bmatrix} \begin{bmatrix} E_x \\ E_y \end{bmatrix} \tag{20.10}$$

To describe the electromagnetic field, for $u = x, y, z$, we use the orthonormal complete basis of the exponential functions:

$$\Phi_u = \sum_{n,m=1}^{2N+1} \tilde{\Phi}_u^{n,m} \varphi_{n,m} \tag{20.11}$$

$$\eta_u = \sum_{n,m=1}^{\infty} \tilde{\eta}_u^{n,m} \varphi_{n,m} \tag{20.12}$$

where, $\tilde{\Phi}$ and $\tilde{\eta}$ are the expansion coefficients of the field and the material parameters, respectively; N stands for a number of plane waves used to represent the field, and, as previously, Φ_u are arbitrary components of E and H, whereas η_u are the corresponding components of the dielectric constant or magnetic permittivity, or their reverse values (see Eq. (20.10)). The basis functions have been defined as the product of two exponential functions (still satisfying orthonormality and completeness of the basis) of the following form:

$$\varphi_{n,m} = \exp\left[i\left(\frac{2\pi n}{L_x} + k_x\right)x + i\left(\frac{2\pi m}{L_y} + k_y\right)y\right] \equiv \exp\left[i(G^{n,m} + k) \cdot r\right] \tag{20.13}$$

where L_x and L_y are dimensions of the calculating window for the x and y-axis, whereas k_x and k_y are components of the wavevector in the xy-plane.

Then Eq. (20.10) may be expressed in the following form:

$$\partial_z^2 \begin{bmatrix} \tilde{E}_y^{nm} \\ \tilde{E}_x^{nm} \end{bmatrix} = -\frac{1}{k_0^2} \begin{bmatrix} \chi_\kappa^{i,j,n,m} - k_0^2 \tilde{\mu}_x^{i-n,j-m} & -\chi_\kappa^{i,j,n,m} \\ \chi_\kappa^{i,j,n,m} & k_0^2 \tilde{\mu}_x^{i-n,j-m} - \chi_\kappa^{i,j,n,m} \end{bmatrix} \times$$
$$\begin{bmatrix} \chi_\gamma^{i,j,n,m} - k_0^2 \tilde{\varepsilon}_y^{i-n,j-m} & -\chi_\gamma^{i,j,n,m} \\ \chi_\gamma^{i,j,n,m} & k_0^2 \tilde{\varepsilon}_x^{i-n,j-m} - \chi_\gamma^{i,j,n,m} \end{bmatrix} \begin{bmatrix} \tilde{E}_y^{ij} \\ \tilde{E}_x^{ij} \end{bmatrix} \equiv \mathbf{R}_H \mathbf{R}_E \begin{bmatrix} \tilde{E}_y^{ij} \\ \tilde{E}_x^{ij} \end{bmatrix} \quad (20.14)$$

where

$$\chi_\kappa^{i,j,n,m} \equiv (G^{nm} + k) \cdot (G^{ij} + k) \tilde{\kappa}_z^{i-n,j-m} \qquad \kappa_z \equiv 1/\varepsilon_z \quad (20.15)$$
$$\chi_\gamma^{i,j,n,m} \equiv (G^{nm} + k) \cdot (G^{ij} + k) \tilde{\gamma}_z^{i-n,j-m} \qquad \gamma_z \equiv 1/\mu_z \quad (20.16)$$

The boundary conditions are assumed to be in the form of the absorbing Perfectly Matched Layers [20, 21].

The above Eq. (20.14) and the analogous equation for the magnetic field may be expressed as:

$$\partial_z^2 \mathbf{E} = \mathbf{R}_H \mathbf{R}_E \mathbf{E} \equiv \mathbf{Q}_E \mathbf{E} \quad (20.17)$$
$$\partial_z^2 \mathbf{H} = \mathbf{R}_E \mathbf{R}_H \mathbf{H} \equiv \mathbf{Q}_H \mathbf{H} \quad (20.18)$$

where

$$\mathbf{E} = \begin{bmatrix} \tilde{E}_y^{nm} \\ \tilde{E}_x^{nm} \end{bmatrix} \qquad \mathbf{H} = \begin{bmatrix} \tilde{H}_x^{nm} \\ \tilde{H}_y^{nm} \end{bmatrix} \quad (20.19)$$

Let us use the basis which simplifies the above $\mathbf{Q}_E$ and $\mathbf{Q}_H$ matrices to their diagonal forms $\mathbf{\Gamma}_E$ and $\mathbf{\Gamma}_H$, respectively. Then Eqs (20.17), (20.18) are given as:

$$\partial_z^2 \hat{\mathbf{E}} + \mathbf{\Gamma}_E^2 \hat{\mathbf{E}} = 0 \quad (20.20)$$
$$\partial_z^2 \hat{\mathbf{H}} + \mathbf{\Gamma}_H^2 \hat{\mathbf{H}} = 0 \quad (20.21)$$

where $\hat{\mathbf{E}}$ and $\hat{\mathbf{H}}$ stand for the electric and magnetic fields in the new base and can be defined as:

$$\hat{\mathbf{E}} = \mathbf{T}_E^{-1} \mathbf{E} \quad (20.22)$$
$$\hat{\mathbf{H}} = \mathbf{T}_H^{-1} \mathbf{H} \quad (20.23)$$

where the $\mathbf{T}_E$ and $\mathbf{T}_H$ matrices diagonalize $\mathbf{Q}_E$ and $\mathbf{Q}_H$, respectively:

$$\mathbf{T}_E^{-1} \mathbf{Q}_E \mathbf{T}_E = \mathbf{\Gamma}_E^2 \quad (20.24)$$
$$\mathbf{T}_H^{-1} \mathbf{Q}_H \mathbf{T}_H = \mathbf{\Gamma}_H^2 \quad (20.25)$$

Solutions of Eqs (20.20), (20.21) have the well-known form of standing waves:

$$\widehat{E}(z) = A_E \cosh(i\Gamma_E z) + B_E \sinh(i\Gamma_E z) \tag{20.26}$$
$$\widehat{H}(z) = A_H \cosh(i\Gamma_H z) + B_H \sinh(i\Gamma_H z) \tag{20.27}$$

with the A_E, A_H, B_E and B_H parameters, which should be determined from the continuity conditions for the field and its z-derivative at the *i*th layer boundaries in a multi-layer structure and which leads, for each *i*th layer, to the following relations between the fields:

$$\begin{bmatrix} \widehat{H}_0^{(i)} \\ -\widehat{H}_d^{(i)} \end{bmatrix} = \begin{bmatrix} y_1^{(i)} & y_2^{(i)} \\ y_2^{(i)} & y_1^{(i)} \end{bmatrix} \begin{bmatrix} \widehat{E}_0^{(i)} \\ \widehat{E}_d^{(i)} \end{bmatrix} \tag{20.28}$$

where the subscripts 0 and *d* correspond to the bottom and the top layer boundaries, respectively, and where $\mathbf{y}_1$ and $\mathbf{y}_2$ read as follows:

$$\mathbf{y}_1^{(i)} = \mathbf{a}_H^{(i)} \tanh\left(i\Gamma_E^{(i)} d^{(i)}\right) \tag{20.29}$$
$$\mathbf{y}_2^{(i)} = -\mathbf{a}_H^{(i)} \sinh\left(i\Gamma_E^{(i)} d^{(i)}\right) \tag{20.30}$$

with $d^{(i)}$ the *i*th layer thickness and

$$\mathbf{a}_H^{(i)} = \left(\mathbf{T}_E^{(i)}{}^{-1} \mathbf{R}_H^{(i)} \mathbf{T}_H^{(i)}\right)^{-1} \mathbf{\Gamma}_E^{(i)} \tag{20.31}$$

Transformation of the fields from the *i*th layer to the $(i+1)$th one is given in the following forms:

$$\widehat{E}_0^{(i+1)} = \mathbf{T}_E^{(i+1)}{}^{-1} \mathbf{T}_E^{(i)} \widehat{E}_d^{(i+1)} \equiv \mathbf{t}_E^{(i)} \widehat{E}_d^{(i+1)} \tag{20.32}$$
$$\widehat{H}_0^{(i+1)} = \mathbf{T}_H^{(i+1)}{}^{-1} \mathbf{T}_H^{(i)} \widehat{H}_d^{(i+1)} \equiv \mathbf{t}_H^{(i)} \widehat{H}_d^{(i+1)} \tag{20.33}$$

As the boundary conditions, the decay of the electric field at the boundaries of the simulating domain is assumed, whereas, for the magnetic field, the matching interface placed at its maximum, between the *l* and *m* layers is chosen, to reduce the numerical errors. Then the above relations, enabling field transformation between and within the layers, lead to the following equation:

$$\mathbf{T}_H^{(m)} \mathbf{Y}_{up}^{(m)} \widehat{E}_d^{(m)} = \mathbf{T}_H^{(l)} \mathbf{Y}_{down}^{(l)} \widehat{E}_d^{(l)} \tag{20.34}$$

where the layers are numbered from 0 at the bottom to *l* and from 0 at the top to *m*, and where

$$\mathbf{Y}^{(i)} = -\left[\mathbf{y}_2^{(i)} \left(\mathbf{t}_H^{(i)} \mathbf{Y}^{(i-1)} \mathbf{t}_E^{(i)}{}^{-1} - \mathbf{y}_1^{(i)}\right)^{-1} \mathbf{y}_2^{(i)} - \mathbf{y}_1^{(i)}\right] \tag{20.35}$$

Y matrices are determined with the aid of the recurrence relation starting from the bottom $\mathbf{Y}_{up}^{(0)}$ and from the top $\mathbf{Y}_{down}^{(0)}$ of the VCSEL. Then Eq. (20.34) will be of the following form:

$$\left(\mathbf{T}_H^{(m)} \mathbf{Y}_{up}^{(m)} \mathbf{T}_E^{(m)^{-1}} - \mathbf{T}_H^{(l)} \mathbf{Y}_{down}^{(l)} \mathbf{T}_E^{(m)^{-1}} \right) \mathbf{E} = \mathbf{Y}\mathbf{E} = 0 \qquad (20.36)$$

The above equation is used to find the complex modal wavelength of the emitted radiation, the imaginary part of which allows the determination of an excess of the modal gain (Eq. (20.37)) over modal losses (Eq. (20.38)). The threshold condition for the mode under consideration is determined from the condition of its modal gain which is equal to the modal losses, so the complex modal wavelength becomes real.

Contrary to most other rigorous optical vectorial methods, the new Plane Wave Admittance Method [17] has proved to be surprisingly rapidly convergent in the case of the GaN-based VCSELs.

20.5
The Self-consistent Calculation Algorithm

The comprehensive VCSEL model used here to simulate an operation of GaN-based VCSELs consists not only of the above scalar or vectorial optical part, but equally important electrical, thermal and recombination parts. Also, all the most important interactions between individual physical phenomena [1] are taken into account with the aid of the self-consistent iteration algorithm (Fig. 20.3) including temperature and carrier-concentration dependences of both refractive indices and absorption, as well as temperature- and doping-dependent thermal conductivities, electrical resistivities and optical material gain. In all successive iteration self-consistency loops, new 3D profiles of all model parameters are recalculated taking into account current 3D profiles of temperature, current density, carrier concentration and radiation intensity. The above multilateral dependences create a real network of mutual and, usually strongly, nonlinear inter-relations between individual physical processes and have been found to be more important in the case of GaN-based lasers [9, 11] than in earlier arsenide and phosphide ones. Therefore, a suitable design for nitride VCSELs should take into consideration not only simulation of their optical characteristics, but, as has been done in this work, equally important electrical, thermal and recombination phenomena, which should also be included along with the very important interactions between them all.

Guidelines for simulating the operation of VCSEL devices have been given in [1], the VCSEL model used here is described in more detail in [22] and its application for the GaN-based VCSELs is in [23]. The model will be used to simulate an RT operation of the possible GaN-based VCSEL design (Fig. 20.3)

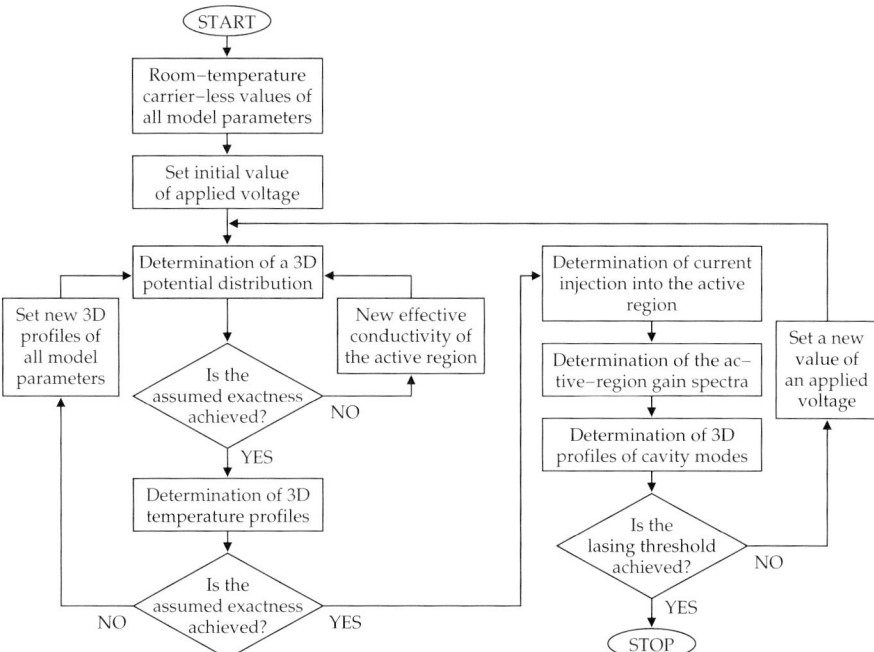

Fig. 20.3 A flow chart of the self-consistent calculation algorithm.

emitting 400 nm radiation and also to determine the areas of an approximately accurate performance of simplified scalar optical approaches.

Various modes of different order k have different intensity profiles $I_k(x,y,z)$. Therefore they may take advantage of a current optical-gain distribution $g(x,y,z)$ depending on the overlapping of this distribution within the active region and the mode intensity profile. It is described by a modal gain G_k defined as [24]:

$$G_k = \frac{\int_A g(x,y,z) I_k(x,y,z)\,dxdydz}{\int_V I_k(x,y,z)\,dxdydz} \tag{20.37}$$

where A and V symbolize the active-region volume and the laser volume, respectively. In an analogous way, the mode suffers from different optical losses depending on its local intensity at the place of those losses. It is described by the modal losses A_k written as:

$$A_k = \frac{\int_V \alpha(x,y,z) I_k(x,y,z)\,dxdydz}{\int_V I_k(x,y,z)\,dxdydz} \tag{20.38}$$

where $\alpha(x,y,z)$ symbolizes the distribution of all optical losses within the laser cavity. For all cavity modes, their lasing thresholds are determined from the condition of their modal gains equal to their modal losses.

20.6
Simulation Results

Continuous-wave (CW) operation of GaN-based VCSELs is much more dependent on their thermal properties than it has been in the case of earlier arsenide ones [9,11]. In addition, possible standard structures of those devices are not equipped with any built-in radial waveguiding mechanism. Therefore, the first simulation is devoted to an examination of the influence on VCSEL lasing thresholds of thermal focusing, introducing this type of weak radial waveguiding. At first, for the RT ambient temperature of 300 K and two VCSEL designs (Fig. 20.1) with the active region radius r_A equal to 2 μm and 4 μm, the model is used to determine, in a fully self-consistent way, RT CW lasing thresholds of two of the lowest-order scalar and vectorial transverse modes (Fig. 20.4). They are the fundamental LP_{01} scalar mode and the fundamental vectorial HE_{11} mode as well as the first-order LP_{11} scalar mode and the three vectorial HE_{21}, TE_{01} and TM_{01} first-order modes corresponding to the LP_{11} mode [11]. All three of these last vectorial modes degenerate to the LP_{11} mode in the scalar limit, but, as the difference between the three modes is rather small, we have restricted our simulation to only the HE_{21} one. The vectorial approach is definitely more exact than the scalar one [18], therefore any difference between the results predicted by both approaches is regarded as a measure of the inaccuracy of the scalar one.

In both the above VCSEL structures under consideration, the fundamental transverse modes have been found usually to exhibit the lowest RT CW lasing thresholds because of their advantageous bell-like radial profiles of the threshold carrier concentration $n_{th}(r)$ within the active region (Fig. 20.5(a)). The above is a result of an impact on a radial current spreading of both the tunnel junction introduced over the active region into the current path and a replacement of the usually used in GaN-based VCSELs p-type radial-current-spreading layer with the n-type one (Fig. 20.1). The latter is followed by a very important decrease in the resistivity of this layer by nearly two orders of magnitude (cf. Table 20.1). Both the above-mentioned structure modifications effectively enhance uniformity of current injection into the active region. Then the resultant bell-like radial optical gain profile (roughly proportional to the active-region carrier concentration (Fig. 20.5(a)) is very similar to the analogous profile of an intensity of the fundamental mode (Fig. 20.4(a)) which ensures its highest modal gain (see Eq. (20.37)) as compared with other modes and, as its result, the lowest fundamental-mode lasing threshold. As expected, values of a threshold gain determined for VCSELs with relatively small ($r_A = 2$ μm) active regions are distinctly higher than those found for VCSELs with larger active regions ($r_A = 4$ μm).

An increase in temperature is followed in VCSELs by many various physical phenomena: the QW active-region gain spectrum is somewhat decreased

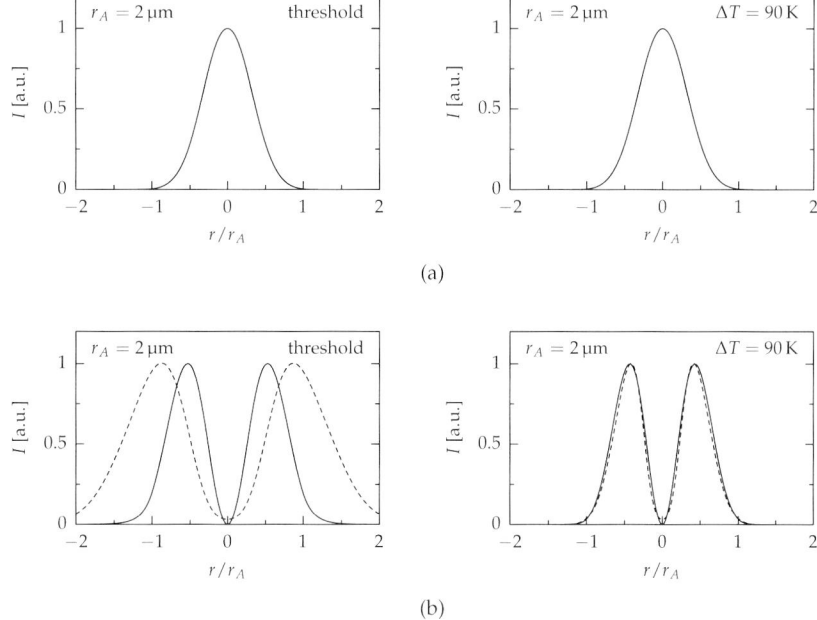

Fig. 20.4 Radial intensity profiles (within the relatively small active region $r_A = 2\,\mu m$) of two of the lowest-threshold transverse modes: (a) the fundamental LP_{01} scalar mode (solid line) and the fundamental HE_{11} vectorial mode (dashed line); (b) the first-order LP_{11} scalar mode (solid line) and the first-order HE_{21} vectorial mode (dashed line), determined for the RT lasing threshold (left-hand diagrams) and the artificially induced temperature profile of the maximal temperature increase of 90 K (right-hand diagrams). For the fundamental mode, both curves plotted for both considered cases are practically indistinguishable, whereas, for the first-order mode, significant differences are seen for the threshold condition which again disappear when an additional thermal focusing is added.

and considerably shifted on the wavelength axis; cavity sizes and refractive indices are increased, which is followed by a slower shifting of wavelengths corresponding to cavity modes; thermal conductivities are decreased; electrical resistivities are increased (with the important exception of the p-type nitrides); absorption coefficients are increased, and so on. Therefore, to investigate exclusively the impact of thermal focusing (cf. Eq. (20.4)) on lasing thresholds of nitride VCSELs, all the above phenomena are neglected here and only the top temperature value at the active-region center is gradually artificially increased and decreased (with respect to the above values determined with the aid of the fully self-consistent approach for the RT ambient temperature – see Fig. 20.5(b)) for successive cases, while its relative radial distribution is kept constant. Then, for each case, new radial profiles of the carrier concentration necessary to reach the lasing threshold, and new modal lasing thresholds, are found as a function of the assumed active-region temperature profile.

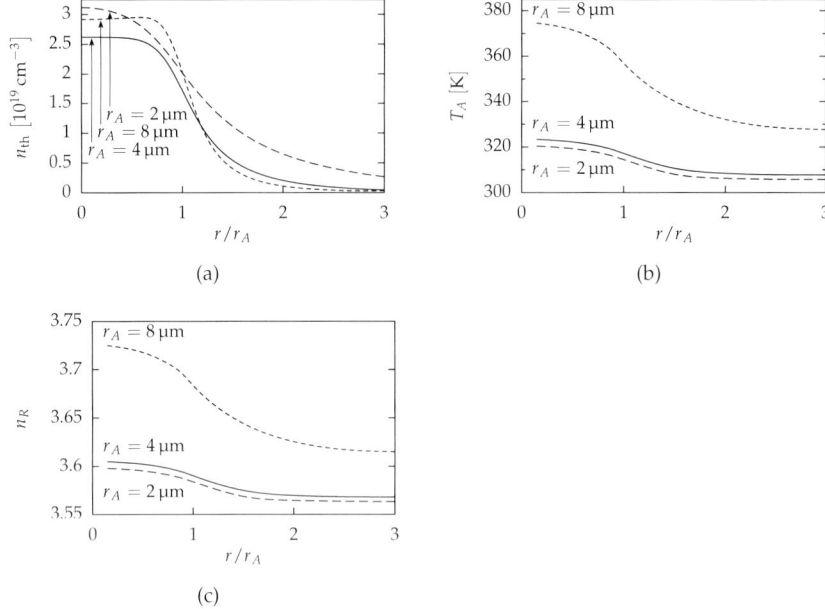

Fig. 20.5 Radial profiles of: a) the carrier concentration n_{th} b) the active-region temperature $T_A(r)$ and c) the active-region refractive index $n_R(r)$ determined for the RT CW threshold operation of VCSELs within their active regions of various radii r_A.

For two of the lowest-order scalar and vectorial transverse modes, Fig. 20.6 presents the maximal values (for $r = 0$) of the optical gain g_{th} necessary to reach their lasing thresholds, as a function of the hypothetical maximal active-region temperature increase ΔT over the ambient RT = 300 K temperature for two active region radii: (a) 4 µm and (b) 2 µm. It should, however, be stressed that values of the above threshold gain have been only determined as the necessary optical gain in order to reach the lasing threshold and may happen to be beyond the reach of the considered active region. In fact, because of some saturation effects, the maximal achievable (at 360 K) optical gain in the nitride QW active regions is estimated to reach only about the 9000 cm^{-1} value for the extremely high carrier concentration of 10^{20} cm^{-3}. As one can see, with an increase in the above maximal temperature, which means that, with an increase in the thermal focusing followed by a better radial mode confinement within the active region (see, e.g., Fig. 20.4(b)), the lasing thresholds are usually dramatically reduced. This effect is especially important for higher-order modes (cf. Fig. 20.4), whereas, for example, for the fundamental one and the smaller active region of a radius equal to 2 µm, the additional thermal focusing effect does not seem to be required in order to confine the mode effectively. Also, for this case, intensity profiles determined using the scalar approach are practically identical with those of the vectorial one. Nevertheless, the beneficial role of the thermal focusing, being the only radial waveguiding mechanism (see radial profiles of the active-region refractive indices in Fig. 20.5(c)) supporting here radial-gain guiding, has been confirmed. It can also be seen, how a considerable threshold decrease may be expected in nitride VCSELs if an efficient built-in radial-waveguiding mechanism (introduced, for example, with the aid of photonic crystals) is applied in their cavities.

As expected, for both VCSEL designs considered here and for all considered temperatures, the fundamental transverse mode remains the lowest-threshold mode (Fig. 20.6). In addition, threshold LP_{01} values, determined with the aid

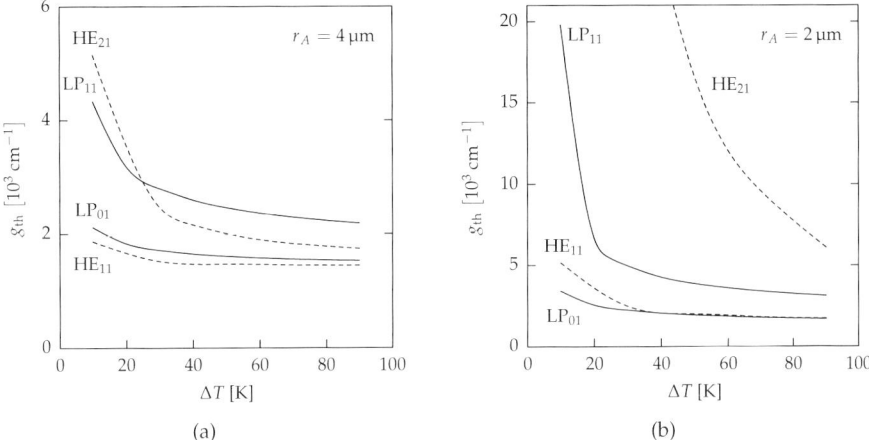

Fig. 20.6 Maximal threshold optical gain g_{th} (for the active-region center) necessary to reach the threshold lasing condition as a function of the hypothetical temperature increase ΔT over the ambient RT = 300 K temperature determined for two the lowest-order scalar and vectorial transverse modes and two active region radii r_A: (a) 4 μm (b) 2 μm

of a simplified scalar optical approach, have been found to be usually slightly higher than those found for the fundamental HE_{11} mode using the more rigorous vectorial method. The above is definitely connected with the earlier observed fact [25], that *scalar* modes are somewhat less confined by relatively weak real waveguiding than are *vectorial* ones (Fig. 20.4). In the considered case of nitride VCSELs, there is no built-in radial waveguiding mechanism as, for example, oxide apertures in arsenide ones, so the radiation field is confined only by the gain-guiding and the weak, real-temperature focusing (cf. Fig. 20.5(c)). Therefore, for the same active-region carrier concentration, the better confined HE_{11} *vectorial* mode exhibits a higher modal gain, which is followed by its lower lasing threshold. For smaller active regions ($r_A = 2\,\mu m$) and lower temperature increases within the active region (Fig. 20.6(b)), the thermal focusing effect is too small to effectively confine the radiation field in a radial direction, so practically only gain-guiding waveguiding influences the radial distribution of the field. As one can see, this time the scalar fundamental mode exhibits a lower lasing threshold, which is direct evidence that it has better radial confinement than that of the vectorial one. The total lasing threshold is also influenced by diffraction losses, not included in scalar optical approaches and more important for higher-order modes and smaller devices. Therefore, the first-order *scalar* LP_{11} mode (Fig. 20.4(b)) usually exhibits (especially for smaller devices – Fig. 20.6(b)) distinctly lower lasing thresholds than the first-order vectorial HE_{21} one.

Any difference between the results predicted by both of the above optical approaches is regarded as a measure of inaccuracy of the scalar one. It is evident from the above results that, for the considered GaN-based VCSEL design working at the RT ambient temperature, the best agreement between both the approaches, which means the most exact performance of the scalar optical approach, is reached for the fundamental mode for medium active regions, i.e., for r_A values between about $2\,\mu m$ and $6.5\,\mu m$. However, for the first-order transverse mode, the lower limit should be shifted to $4.5\,\mu m$.

20.7
Discussion and Conclusions

The usability of two optical approaches for the modeling of an operation of GaN-based VCSELs is compared: the rigorous fully vectorial Plane Wave Admittance Method [17] and the approximated scalar Effective Frequency Method [16]. For our modified VCSEL design equipped with a tunnel junction and the n-type radial-current-spreading layer (instead of the p-type one), both of the above optical methods have been found to give usually quite close results for the fundamental transverse mode, which confirms the general opinion, that the scalar Effective Frequency Method gives approximately good results even beyond the area of its confirmed validity. Larger discrepancies between the results of both approaches have been found for higher-order transverse modes and for smaller active regions. For the same carrier concentration, the modal gain of the fundamental mode is usually somewhat higher (and the lasing threshold slightly lower), when it is determined using the vectorial approach, than it is in the case of the scalar one. The above is partly compensated by diffraction losses, not included in scalar approaches, which are more important for higher-order modes and smaller active regions. Therefore, contrary to the fundamental mode, the modal gain determined for the same carrier concentration is higher (and the lasing threshold lower) for the first-order mode using the scalar optical approach than it is in the case of the vectorial one, especially in smaller devices. To summarize this comparison: simplified scalar optical approaches have been found to give approximately good results for fundamental transverse modes in standard GaN-based VCSEL designs, but their accuracy is considerably worse for higher-order modes and especially for modern micro-resonator devices.

The performance of GaN-based VCSELs might be drastically improved if an efficient built-in radial waveguiding mechanism could be applied in their cavities. To examine its possible impact on a VCSEL performance, the influence of an artificially introduced radial waveguiding mechanism has been investigated with the aid of our simulation model. A possible reduction in the VCSEL lasing threshold has been found even to exceed 50%. Therefore, the possibility of the application of such a mechanism introduced, for example, by photonic crystals, is recommended.

Acknowledgments

The authors would like to acknowledge support from the Polish Ministry of Science and Information Society Technologies (MNiI), grant No 3-T11B-073-29. Special thanks go to Dr. Michał Wasiak for his help in the preparation of the final form of the manuscript.

References

1 Osiński, M., Nakwaski, W., Chapter 5 in: *Vertical-Cavity Surface-Emitting Laser Devices*, Springer Verlag, Berlin Heidelberg, **2003**, p. 135

2 Maćkowiak, P., Sarzała, R.P., Wasiak, M., Nakwaski, W., *J. Phys. D.: Appl. Phys.* 36 (**2003**), p. 2041

3 Yokouchi, N., Danner, A.J., Choquette, K.D., *IEEE J. Select. Topics Quantum Electron.* 9 (**2003**), p. 1439

4 Marcuse, D., *Light Transmission Optics*, Oxford University Press, Oxford, **1992**, Chapter 1.3

5 Nakamura, S., Senoh, M., Nagahama, S., Iwasa, N., Yamada, T., Matsushita, T., Sugimoto, Y., Kiyoku, H., *Appl. Phys. Lett.* 70 (**1997**), p. 868

6 Takeuchi, T., Wetzel, Ch., Yamaguchi, S., Sakai, H., Amano, H., Akasaki, I., Kaneko, Y., Nakagawa, S., Yamaoka, Y., Yamada, N., *Appl. Phys. Lett.* 73 (**1998**), p. 1691

7 Domen, K., Kuramata, A., Soejima, R., Horino, K., Kubota, S., Tanahashi, T., *IEEE J. Select. Topics Quantum Electron.* 4 (**1998**), p. 490

8 Piprek, J., Farrell, R., DenBaars, S., Nakamura, S., *IEEE Photon. Techn. Lett.* 18 (**2006**), p. 7

9 Sarzała, R.P., Maćkowiak, P., Nakwaski, W., *Semicond. Sci. Technol.* 17 (**2002**), p. 255

10 Maćkowiak, P., Sarzała, R.P., Wasiak, M., Nakwaski, W., *IEEE Photon. Techn. Lett.* 15 (**2003**), p. 495

11 Nakwaski, W., Maćkowiak, P., *Opt. Quantum Electron.* 35 (**2003**), p. 1037

12 Jeon, S.R., Song, Y.H., Jang, H.J., Yang, G.M., Hwang, S.W., Son, S.J., *Appl. Phys. Lett.* 78 (**2001**), p. 3265

13 Takeuchi, T., Hasnain, G., Corzine, S., Huechen, M., Schneider, Jr., R.P., Kocot, Ch., Blomqvist, M., Chang, Y., Lefforge, D., Krames, M.R., Cook, L.W., Stockman, S.A., *Jpn. J. Appl. Phys.*, 40 (**2001**), p. L861

14 Pearton, S.J., Wilson, R.G., Zavada, J.M., Han, J., Shul, R.J., *Appl. Phys. Lett.*, 73 (**1998**), p. 1877

15 Maćkowiak, P., Wasiak, M., Czyszanowski, T., Sarzała, R.P., Nakwaski, W., *Optica Applicata* 32 (**2002**), p. 493

16 Wenzel, H., Wünsche, H.-J., *IEEE J. Quantum Electron.*, 33 (**1997**), p. 1156

17 Dems, M., Kotynski, R., Panajotov, K., *Optics Express*, 13 (**2005**), p. 3196

18 Czyszanowski, T., Dems, M., Thienpont, H., Panajotov, K., *Photonics Europe 2006*, 3–7 **April 2006**, Strasbourg, France, talk no. 6185 36

19 Bienstman, P., Baets, R., Vukusic, J., Larsson, A., Noble, M.J., Brunner, M., Gulden, K., Debernardi, P., Fratta, L., Bava, G.P., Wenzel, H., Klein, B., Conradi, O., Pregla, R., Riyopoulos, S.A., Seurin, J.-F.P., Chuang, S.L., *IEEE J. Quantum Electron.*, 37 (**2001**), p. 1618

20 Berenger, J.P., *J. Comput. Phys.*, 114 (**1994**), p. 185

21 Derudder, H., Olyslager, F., De Zuter, D., Van den Berghe, S., *IEEE Trans. Antennas and Propagation* 49 (**2001**), p. 185

22 Sarzała, R.P., Nakwaski, W., *J. Phys.: Condens. Matter* 16 (**2004**), p. S3121

23 Maćkowiak, P., Sarzała, R.P., Wasiak, M., Nakwaski, W., *Appl. Phys. A*, 78 (**2004**), p. 315

24 Mroziewicz, B., Bugajski, M., Nakwaski, W., *Physics of Semiconductor Lasers*, North Holland, **1991**, Chapter 5.1

25 Czyszanowski, T., Nakwaski, W., *J. Phys. D: Appl. Phys.*, 39 (**2006**), p. 30

21
GaN Nanowire Lasers

Alexey V. Maslov and Cun-Zheng Ning

21.1
Introduction

Semiconductor nanowires are a relatively new member of the increasingly large family of nanostructures. They differ from the more conventional one-dimensional structures – quantum wires – which are not discussed here. Quantum wires are typically made using planar growth techniques similar to quantum wells and then lithographically defined or processed. They provide quasi one-dimensional electronic confinement, but with very poor or absent transverse optical confinement because of the similar materials used for the wires and the surrounding matrix. The quantum wire structures mostly studied are the T-type wires [1]. In contrast, nanowires are single-crystal semiconductor wires that are grown in vacuum by the tip-led growth mechanism in one direction. The growth techniques are discussed in recent review articles [2–4]. The typical width of such wires is tens of nanometers; hence they are referred to as nanowires. The length can be from a few to tens of micrometers. A typical image of GaN nanowires is shown in Fig. 21.1.

Since fabricated nanowires are surrounded by air, they form natural optical waveguides with strong optical confinement provided by the large contrast of the refractive indexes. Most of the nanowires studied so far do not have electronic confinement: their electronic structure is similar to that in bulk with the possible existence of surface states. Lasers demonstrated using individual nanowires represent the smallest lasers of any kind. An array of such nanowires may provide an ideal two-dimensional array of light sources similar to vertical-cavity surface-emitting lasers (VCSELs). Such tiny nanowires or arrays are potentially important as light sources, waveguides, and detectors for very large-scale integrated photonic systems, parallel optical interconnects, integrated bio or chemical sensing and detection.

Nanowires have several advantages over such heterostructures as quantum wells, wires, and dots. One of the most appealing advantages is insensitivity to substrates. Nanowires can be grown either with or without an epitaxial connection to a substrate. For wires that are grown epitaxially from a sub-

Nitride Semiconductor Devices: Principles and Simulation. Joachim Piprek (Ed.)
Copyright © 2007 WILEY-VCH Verlag GmbH & Co. KGaA, Weinheim
ISBN: 978-3-527-40667-8

Fig. 21.1 Scanning electron microscope image of GaN nanowires. Courtesy of M. Sunkara (Univ. of Louisville).

strate, the crystal orientation of the substrate is epitaxially transferred to the wire and determines the wire growth direction and orientation. Even in this case, the requirement of lattice matching is substantially relaxed compared with the planar growth, mainly due to the very small wire cross-section. A much larger strain can therefore be tolerated. Nanopillars and nanorods of GaN have been grown on Si substrate using a hot-wall epitaxy technique [5] or molecular beam epitaxy (MBE) [6,7]. There is also experimental evidence of strain relaxation in GaN nanowires grown either by MBE or prepared by etching GaN layers [8]. Single-crystal nanowires can also be synthesized without being epitaxially connected to a substrate. In this mode of growth, any substrate can be used for wire growth from the lattice matching point of view. Such substrate insensitivity opens a wide array of material choices that are not available for planar nanostructures. Synthesis of almost any III-V semiconductors on silicon epitaxially or nonepitaxially is definitely an exciting opportunity for silicon photonics and for integrated detection and sensing applications.

This chapter is organized as follows. We start by reviewing the progress in nanowire growth, characterization and device fabrication. We then describe some theoretical aspects of nanowire-based lasers that distinguish them from other existing lasers, such as planar heterostructure lasers and VCSELs. Among these aspects are the strong anisotropy of material gain in GaN, the ability of nanowires to support well-guided modes, and the possibility of obtaining large modal gain and spontaneous emission factors. While the anisotropy is mostly relevant to GaN nanowires, the other properties are common for all semiconductor nanowires. The final part is our conclusion.

21.2
Nanowire Growth and Characterization

Nitride nanowires have been one of the most studied nanowires because of the interest in ultraviolet (UV) emission. The first GaN nanowires were grown less than ten years ago using carbon nanotubes as templates [9,10]. The wire size was in the range of 4–50 nm in diameter and up to 25 microns in length. Since then nitride nanowire structures have been grown using various methods [2–4]. Similar to other nanowires, most of GaN nanowires are synthesized using a vapor-liquid solid (VLS) method with gold as the metal catalyst, but other metals such as indium, iron, nickel, and cobalt were also used to synthesize GaN [11,12]. A self-catalyzed nitridation process was demonstrated in [13]. In addition to GaN nanowires, the growth of single-crystal GaN nanotubes was achieved [14] as well as of radial structures consisting of several layers of GaN-based alloys [15,16].

One of the most remarkable developments in the area of nanowire applications was the demonstration of lasing in a single GaN nanowire under optical pumping [17]. Both the waveguiding and cavity effects were observed in individual nanowires of 100 nm in diameter and tens of microns in length with a reasonably high cavity quality factor of up to 1500. The lasing nanowire showed a clear threshold behavior in terms of the linewidth narrowing and change in differential efficiency. Recently, a lower threshold was achieved in GaN nanowires [18] as well as lasing in a nanowire that was fashioned into a ring structure [19].

Achieving *p-n* junctions with high and controllable dopings in nitride nanowires is the key issue for making electrical pumped light-emitting diodes (LEDs) and nanolasers. The first nanowire GaN LED report was based on a segmental *p-n* junction along the wire axis [20,21]. Such simple *p-n* junctions, however, provide only a small volume where electron–hole recombination occurs. An interesting structure is the radial *p-i-n* heterostructure [16], where an intrinsic nanowire shell is sandwiched between a Si-doped core and a magnesium doped outshell. This structure can provide a large intrinsic (active) volume for electron–hole recombination and could potentially lead to more efficient LEDs and lasers. In addition, using alloys $In_xGa_{1-x}N$ instead of pure GaN sandwiched between an *n*-type core and a *p*-type shell offers wavelength tunability [16,22]. One possible shortcoming of the radial heterostructure is that In composition could be potentially limited by the lattice-matching requirement between the core and the outshell. In this sense, the *p-i-n* heterostructure along the wire axis could be advantageous for allowing much larger mismatch, and thus a larger In composition.

One of the important advances made was the control of growth direction along different crystal directions using temperature [23], single-crystal substrate orientation [24], or Ga-flux [13]. For example, in the self-catalyzed growth scheme [13], a high flux of Ga vapor leads to growth along the c-direction, while low Ga flux leads to a-direction growth. GaN nanowires grown along different crystal directions also show different photoluminescence emission peak positions [13, 24]. While the polarization field was speculated as a reason for such a difference, a clear physical understanding is still lacking.

21.3
Nanowire Laser Principles

The general principles of nanowire laser operation are the same as for the standard semiconductor lasers – the nanowire acts as an active waveguide and the feedback is provided by the reflection of the guided modes from the nanowire facets. The transverse size of the nanowire is chosen to guide at least one mode. The high refractive index contrast between the semiconductor and air allows for efficient guiding even for a nanowire transverse size comparable to or smaller than the wavelength. The gain is created by injecting carriers into the nanowire. The injection can be either optical or electrical. So far, lasing has been demonstrated only under optical injection in GaN nanowires [17, 18]. For this reason, we consider only a free-standing nanowire without electrical contacts. The realization of the electrical injection in GaN nanowires and its optimization will most likely take place in the very near future. Figure 21.2(a) shows schematically the geometry of a nanowire standing on a substrate. Figure 21.2(b) shows a typical spatial distribution of a wavepacket (one component of the field) that bounces between the two nanowire facets and is being amplified inside the nanowire. The emitted field in the air is very divergent and decays rapidly with distance from the facets. Lasing starts when the injected carriers provide sufficient gain in the structure to overcome all losses. Keeping only the losses of the mirrors and focusing on one mode, the threshold condition is

$$GL = -\ln(|r_1 r_2|) \qquad (21.1)$$

where G is the modal gain per unit length, L is the length of the nanowire, and $r_{1,2}$ are the reflection coefficients for the given mode from the top and bottom facets. The gain G depends not only on the carrier concentration (material parameter) but also on the specific mode (waveguide parameter). The material gain in GaN depends strongly on the orientation of the electric field with respect to the c-axis of the crystal. The presence of several components in the guided mode, in addition to the gain anisotropy of GaN, complicates the estimation of the modal gain. The reflectivities that enter into the expression for the threshold gain (21.1) also cannot be calculated analytically as one encounters a typical diffraction problem of guided wave scattering upon dis-

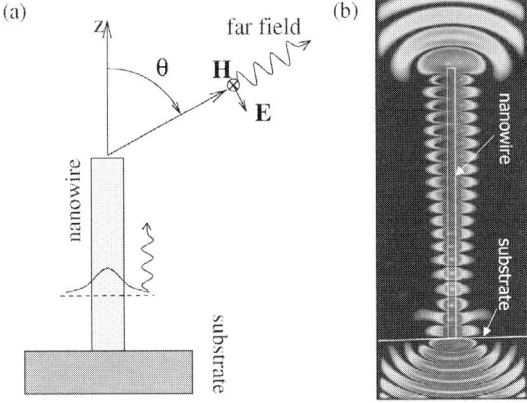

Fig. 21.2 (a) Schematic of a free-standing nanowire on a substrate. A guided mode (indicated by its profile) bounces between the facets and forms a far-field distribution upon its diffraction at the facets (only one θ-polarized component is shown). (b) Snapshot of field distribution from finite-difference time-domain simulation of a nanowire laser.

continuity. For the same reason, the behavior of the far-field pattern is best approached by numerical means.

Due to their ideal crystalline structure, nanowires form perfect optical waveguides with sizes comparable to the wavelength. The strong waveguiding in nanowires leads to very large values of the modal gain, that can exceed the bulk gain. This results in lasing which is possible even in rather short ($\sim$ 10 μm) nanowires with no specifically fabricated mirrors. The presence of only a few guided modes in nanowires allows an efficient coupling of spontaneous emission into the guided modes. This reduces the threshold current and suppresses relaxation oscillations. We will discuss all these issues in the subsequent sections.

21.4
Anisotropy of Material Gain

In this section we describe the main features of the bandstructure of bulk GaN and the resultant optical gain. We consider only the bulk gain using the free-carrier model. The bulk model is directly applicable to GaN nanowires with radius $R \gtrsim 20$ nm. Our presentation closely follows Ref. [25]. We will later comment on the bandstructure of very thin nanowires where quantization becomes important.

Gain calculations for GaN were reported previously in the literature (see, for example, [26–28]). In [27], the transverse gain was calculated using approximate dipole matrix elements for each band. The work in [26] focused on many-body effects and only two parabolic bands were considered. We use exact diagonalization of the 6 × 6 valence band Hamiltonian for GaN [29] to calculate the optical gain using the free-carrier model. In particular, we emphasize the presence of strong anisotropy for gain near the band-edge. Similar anisotropy was also emphasized in [28].

We will treat wurtzite GaN. Crystals with a wurtzite lattice are uniaxial. However, the difference in the background dielectric constants is rather small ($\lesssim 10\%$) [30] and thus, we will neglect it and assume that the material is characterized by a real isotropic dielectric constant $\varepsilon(\omega)$. In contrast, the absorption or gain in GaN depends strongly on the polarization of the light with respect to the crystal c-axis. The absorption (or gain) in semiconductors can be described by putting a macroscopic current density into Maxwell's equations. The current density is related to the creation or destruction of electron–hole pairs in the presence of an optical field which induces the interband transitions. The optical conductivity tensor σ can be defined as the proportionality coefficient between the current density and the electric field:

$$J_i = \sigma_{ij} E_j \quad i = x, y, z \tag{21.2}$$

Using (21.2) and solving Maxwell's equations we obtain that a plane wave polarized along a direction specified by some index i experiences gain (the so-called material gain) given by [1]

$$G_i^0 = -\frac{4\pi \sigma_{ii}}{c\sqrt{\varepsilon}} \quad i = x, y, z \tag{21.3}$$

where c is the speed of light in a vacuum. Under the proper choice of the coordinate axis, the conductivity tensor of uniaxial crystals has only two different components $\sigma_{xx} = \sigma_{yy}$ and σ_{zz}.

The components of the conductivity tensor are calculated by taking the expectation value of the current operator evaluated for all possible transitions that take place in the semiconductor at a given energy. Thus, one needs to calculate the bandstructure, wavefunctions and expectation values of the momentum operator. We can explicitly treat the situation when the optical transitions involve one pair of bands – one conduction and one valence band; subsequently, we must add the contributions from all existing bands. Let $\Psi_k^c(\mathbf{r})$ and $\Psi_k^v(\mathbf{r})$ be the wavefunctions (with quasi-momentum $\mathbf{k}$) of the two bands. Then we define the dimensionless momentum matrix elements $\tilde{p}_k$ between the states as

$$P\frac{m_0}{\hbar}\tilde{p}_k = p_k = <\Psi_k^v|\hat{p}|\Psi_k^c> \tag{21.4}$$

where the Kane parameter is $P = \sqrt{\hbar^2 E_P/(2m_0)}$ with E_p the Kane energy and m_0 the free electron mass. After solving the Schrödinger equation for the

---
1) We use cgs units throughout the paper

two bands (with the light-matter interaction included in the dipole approximation) and summing over all possible k states, we arrive at the following conductivity components at an arbitrary optical energy $E = \hbar\omega$:

$$\sigma_{ij}(E) = -\frac{p^2 e^2}{8\pi^2 \hbar E_g} \int_{S(E)} dS \, \frac{1 - f_e(k) - f_h(k)}{|\nabla_k E|} \Sigma_{ij}^k \qquad (21.5)$$

where $\Sigma_{ij}^k = \tilde{p}_{k,i}^* \tilde{p}_{k,j}$ and $i, j = x, y, z$; e is the electron charge and E_g is the bandgap energy. The integral in Eq. (21.5) is taken over the constant energy surface $S(E)$ in the k-space. Expression (21.5) for the conductivity does not contain any approximations regarding the symmetry of the crystal. In general, the tensor Σ_{ij}^k has both diagonal (if $i = j$) and off-diagonal (if $i \neq j$) components. However, in uniaxial crystals (with z the optical axis) the total conductivity σ_{ij} (which is obtained after integrating in the k-space (see Eq. (21.5)) and summing contributions from all existing bands) has only diagonal components. Moreover, in zinc blende materials all diagonal elements will be equal. The population of electrons f_e and holes f_h in Eq. (21.5) are the usual Fermi–Dirac distribution functions which are calculated from the density of states and the total number of carriers.

The line-broadening factor omitted in Eq. (21.5) can be included by convoluting the conductivity (21.5) and a line-broadening function. The convolution smooths out the sharp features (on an energy scale smaller than the linewidth) of the conductivity.

To perform the surface integration (21.5), we divide the k-space into a cubic mesh and approximate the constant-energy surfaces inside every small cube by a set of parallel planes [31]. Using the density of state and the Fermi–Dirac distribution function (with some fixed chemical potentials), we can calculate the carrier density by integrating their product over energy and summing the contributions from all bands. This yields the chemical potentials for electrons and holes at different carrier densities.

The conductivity (21.5) and density of states depend on the bandstructure, i.e., the dispersion and momentum matrix elements between the states. For wurtzite GaN, we assume a parabolic and isotropic conduction band, while for the valence band we numerically diagonalize the 6×6 Hamiltonian [29]. The parameters for GaN used in the calculations are listed in Table 21.1 [25]. The valence band consists of three double-degenerate branches. These branches are labeled (starting from the top of the valence band) as heavy-holes (HH), light-holes (LH) and crystal-field split-off holes (CH). The branches are strongly nonparabolic and anisotropic. This gives rise to a noticeable deviation in the valence-band density of states from the $\sqrt{E}$-dependence, which is common for isotropic and parabolic bands. The density of states in the conduction band is significantly smaller than in the valence band.

Tab. 21.1 Material parameters for GaN used in bandstructure calculations.

E_g	E_P	Δ_{cr}	Δ_{so}	A_1	A_2	A_5	m_e/m_0
3.507 eV	14 eV	19 meV	14 meV	−6.56	−0.91	−3.13	0.20

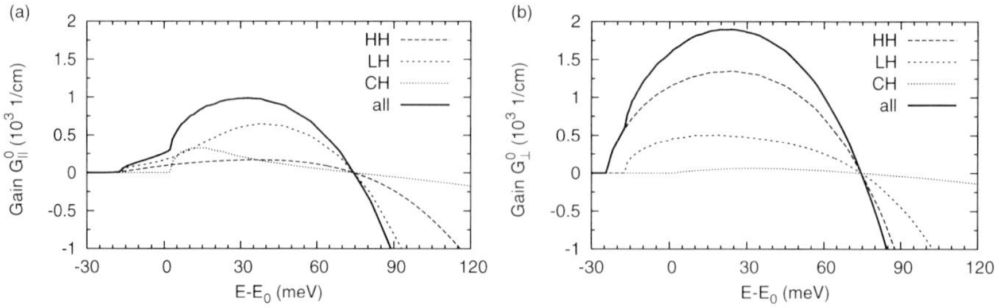

Fig. 21.3 Contributions to the gain spectrum from the HH-, LH-, and CH-to-conduction band transitions and the total gain for $N = 2 \times 10^{19}$ cm^{-3} and $T = 300$ K.

The optical gain is obtained by substituting the numerically calculated conductivity (21.5) into Eq. (21.3). Figure 21.3 shows the gain for carrier density $N = 2 \times 10^{19}$ cm^{-3} and temperature $T = 300$ K. We can clearly see a large difference between the longitudinal and transverse components of the gain. The difference is the most pronounced below the energy of LH-C (light-hole-to-conduction) transitions where the longitudinal gain is very small. For the given parameters, the longitudinal gain is determined mostly by the LH and CH states. The transverse gain is determined mostly by the HH states. At the energy of the maximum gain, around $E = E_0 + 25$ meV ($E_0 = E_g + \Delta_{cr} + \Delta_{so}/3$), the longitudinal gain is about one-half of the transverse gain. A smaller value of anisotropy (the transverse gain is 1.6 times larger than the longitudinal) was obtained in Ref. [28]. At a lower temperature, the anisotropy of gain becomes even larger and reaches, for example, a factor of three at $T = 77$ K [25].

There are experimental reports on the anisotropy of optical transitions in GaN. The first measurements date back to early 1970 [32]. A direct comparison between the experimental measurements and theoretical predictions was reported much later [33] where it was found that the intensity of light with polarization perpendicular to the c-axis dominates. A high degree of polarization (~ 0.9 at $T = 10$ K) of luminescence was obtained for M-plane GaN quantum wells. It was suggested that not-complete-polarization can be explained by a small admixture of other bands due to quantization effects [34]. Later, luminescence with a smaller degree of polarization (~ 0.8) in GaN epilayers was reported [35]. Despite some general agreement of the theoretical and experimental results on luminescent properties, the anisotropy of optical gain and its dependence on temperature and carried density appears to be less studied. In particular, near-band-gap emission and optical gain in *a*-plane GaN layers, did not show any polarization dependence [36].

Let us briefly discuss how the bandstructure of very thin GaN nanowires ($R \lesssim 10$ nm) changes as compared to that in bulk. Although such thin nanowires do not support well-guided modes and cannot form nanolasers by themselves, there are several interesting effects related to quantization that we would like to mention. Quite generally, size quantization in semiconductors can give rise to the appearance of anisotropy of the optical transitions even if the bulk semiconductor is isotropic [37, 38]. It is natural to expect that the presence of bulk anisotropy in GaN will make the quantization effects quite complicated. We describe here only the simplest case – when the nanowire axis is along the *c*-axis of GaN. The standard $\mathbf{k} \cdot \mathbf{p}$ theory predicts several new features in such nanowires. First, the optical transitions from the top valence state can be either allowed or forbidden, depending on the radius of the nanowire [39, 40]. Second, in the very narrow GaN nanowires ($R \lesssim 3$ nm) the top state in the valence band is found to contain mostly $|z\rangle$ components [39]. This is in contrast to bulk GaN in which the top state has only $|x\rangle$ and $|y\rangle$ components. Thus, the quantization in GaN can completely reverse the polarization anisotropy as compared to bulk material. This polarization anisotropy that arises from quantization should be distinguished from the one that comes from purely electromagnetic effects, i.e., screening of the optical field inside the nanowire by polarization charges [41–43].

To conclude this section, we want to emphasize that the material gain in bulk GaN depends strongly on the polarization of the electric field. It is greater when the electric field is perpendicular to the *c*-axis. The degree of anisotropy is a factor of two but also depends on the temperature and carrier concentration. Theoretical models based on the $\mathbf{k} \cdot \mathbf{p}$ theory applied to extremely narrow GaN nanowires grown along the *c*-axis predict a complete reversal of anisotropy, due to size quantization.

21.5
Guided Modes

In this section we will use a simple waveguiding structure – a dielectric cylinder – to describe the basic properties of nanowire lasers and LEDs. We assume that the nanowire has a circular cross-section with radius R. The dielectric constant of the material ε is assumed to be frequency-independent. We will take $\varepsilon = 6$, which roughly corresponds to GaN close to the band-edge [30], in all numerical estimates. The nanowire is surrounded by air with unit refractive index. We note that most nanowires have shapes that differ from circular, for example, hexagonal if grown along the [0001] or triangular if grown

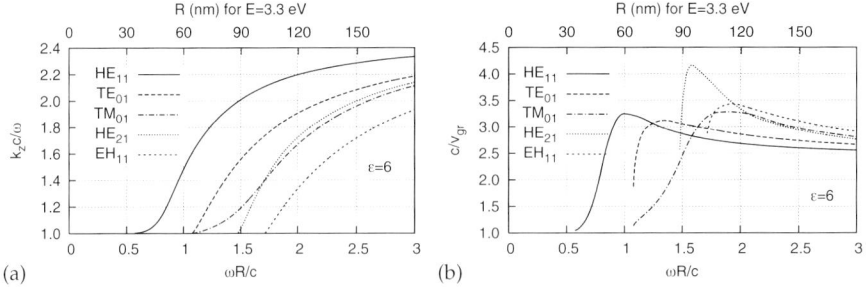

Fig. 21.4 Phase index (a) and group index (b) as functions of dimensionless $\omega R/c$ for the lowest modes supported by a dielectric cylinder with $\varepsilon = 6$ in air. The top axis shows the values of R when the photon energy $E = \hbar\omega = 3.3$ eV.

along $\langle 11\bar{2}0 \rangle$ [18, 24]. Nanowires grown along $\langle 10\bar{1}0 \rangle$ resemble rectangular-shaped nanobelts [13]. The model of a circular nanowire should still provide good insight into the role of the contrast of the nanowire/air dielectric constants on the optical properties. The dispersions and types of guided modes of dielectric cylinders are well known and have been studied in detail in the literature [44, 45]. Here we only emphasize the waveguiding properties that make nanowire lasers different from heterostructure lasers. These properties include large waveguide dispersion, large gain confinement factors, strong frequency and radius dependence of the facet reflectivities and of the far-field pattern. These properties are not unique to GaN but are common to all semiconductor nanowires.

21.5.1
Guided Modes, Dispersions, and Mode Spacing

We start with dispersion properties of the nanowires which were calculated by solving numerically the well-known dispersion equations for dielectric cylinders [44]. The phase and group velocities of several lowest order modes HE_{11}, TE_{01}, TM_{01}, HE_{21} and EH_{11} are shown in Fig. 21.4. The first subscript denotes angular symmetry, the second subscript denotes the radial dependence of the fields.

The fundamental mode HE_{11} and all other hybrid modes have all six field components which behave like $\cos\phi$ or $\sin\phi$ with ϕ the azimuthal angle. The purely transverse modes TE_{01} and TM_{01} have only three field components with no dependence on ϕ. All modes except HE_{11} have low-frequency cut-offs. The cutoff frequency is $\omega_c R/c = x_{11}/\sqrt{\varepsilon-1} \approx 1.71$ for HE_{12} and EH_{11}, $\omega_c R/c = x_{01}/\sqrt{\varepsilon-1} \approx 1.08$ for TE_{01} and TM_{01}, and $\omega_c R/c = x_{02}/\sqrt{\varepsilon-1} \approx 2.47$ for TE_{02} and TM_{02}, where x_{kl} is the lth zero of the Bessel function $J_k(x)$, $k=0,1$. The cutoff for the HE_{21} mode is $\omega_c R/c \approx 1.47$ [45].

The large refractive index contrast leads to a strong confinement of the lowest optical mode inside the nanowire even for small radii $\omega R/c \sim 1$. In this range the nanowires have very strong dispersions (see Fig. 21.4) that affect many key properties, in particular the mode spacing in a finite-length nanowire. Both the group and phase refractive indices start from unity at the cut-offs and asymptotically reach the refractive index of the waveguide ($\sqrt{\varepsilon} = 2.45$) far away from the cut-off. However, while the phase refractive index is always less than the asymptotic limit, the group refractive index can exceed it.

Taking L as the length, we obtain the mode spacing between the cavity modes formed by the two ends as

$$\Delta\omega = \frac{2\pi}{2L\frac{dk_z}{d\omega} + \frac{d\varphi_1}{d\omega} + \frac{d\varphi_2}{d\omega}} \tag{21.6}$$

where $\varphi_{1,2}$ are the phases of the reflection coefficients for the given guided mode at the two ends. Although the reflectivities depend very strongly on the frequency and radius, Eq. (21.6) suggests that the contribution of $\varphi_{1,2}$ into the mode spacing can be neglected for long nanowires.

Neglecting $\varphi_{1,2}$, we have $\Delta\omega = \pi v_{gr}/L$. The group velocity $v_{gr} = d\omega/dk_z$ should include the waveguide dispersion as well as that of the material, i.e., $\varepsilon = \varepsilon(\omega)$. We can separate these two contributions by representing the dispersion dependence as $k_z = k_z(\omega, \varepsilon(\omega))$ where the first argument describes the waveguiding dispersion for constant ε and the second argument describes the material dispersion. This gives

$$\frac{dk_z}{d\omega} = \frac{\partial k_z}{\partial \omega} + \frac{\partial k_z}{\partial \varepsilon}\frac{\partial \varepsilon}{\partial \omega} = \frac{\partial k_z}{\partial \omega} + \frac{\Gamma}{2}\frac{\omega}{c\sqrt{\varepsilon}}\frac{\partial \varepsilon}{\partial \omega} \tag{21.7}$$

The first term gives the group velocity for a constant $\varepsilon = 6$ and its behavior is shown in Fig. 21.4. The second term accounts for the material dispersion $\varepsilon(\omega)$. The material dispersion is also quite large. For example, at $\hbar\omega = 3.3$ eV we estimate $(\omega/\sqrt{\varepsilon})(\partial \varepsilon/\partial \omega) \approx 2.6$ based on the data of [46]. Thus, the terms in Eq. (21.7) can have comparable values. Despite the fact that we neglected the material dispersion in obtaining the waveguide dispersion, Eq. (21.7) allows us to account correctly for it, as long as our operating frequency of the laser is where $\varepsilon = 6$.

The material dispersion enters into the group velocity (21.7) with a weight factor

$$\Gamma = 2\frac{c\sqrt{\varepsilon}}{\omega}\frac{\partial k_z}{\partial \varepsilon} = \frac{\partial n_{ph}}{\partial n} \tag{21.8}$$

where n_{ph} is the phase index and $n = \sqrt{\varepsilon}$. This factor shows how sensitive the dispersion equation is to the change in the dielectric constant. This factor is also known as the gain confinement factor, which we will discuss in

Section 21.5.4. We note that a factor of 2 in Eqs (21.7, 21.8) was introduced to conform to the commonly accepted definition of the confinement factor as the rate of power growth with distance, not the amplitude of the mode. The group velocity (21.7) is explicitly written for the case when only one material (the nanowire itself) is dispersive. For the case when the waveguide consists of several dispersive materials one needs to add factors similar to the second term in Eq. (21.7).

The knowledge of dispersion equations and modal fields is necessary for the calculation of the gain confinement factors and then modal gains. If modes are degenerate, the presence of anisotropy of gain can lift the degeneracy and lead to the preferred excitation of one of the modes. A reasonable approximation is to assume that the gain does not change the real part of the propagation wavevectors but only changes the modal gain. In this case, we need to choose the superposition of the modes that satisfies the symmetry of the material gain. There are two convenient ways to represent the fields of the degenerate HE_{11} mode. In the first one we have, for the E_z component,

$$E_z = E_z(\rho) \left\{ \begin{array}{c} \cos\phi \\ \sin\phi \end{array} \right\} \sin(\omega t - k_z z) \tag{21.9}$$

where the upper and lower lines give the two orthogonal modes; ρ is the radial coordinate. The other components are related to E_z through Maxwell's equations. These modes represent fields with linear polarizations either along the x or y-axis. Alternatively, one can use modes in which the polarization is circular:

$$E_z = E_z(\rho) \left\{ \begin{array}{c} \cos(\omega t - k_z z - \phi) \\ \sin(\omega t - k_z z - \phi) \end{array} \right\} \tag{21.10}$$

In the case when the c-axis is perpendicular to the nanowire axis, the modes with fixed polarizations (see Eq. (21.9)) will experience different modal gains. Figure 21.5 shows the electric field in the x–y plane for the case when the guided mode is x-polarized. The y-component of the mode is much smaller than the x-component. Consequently, if the c-axis is along y then the x-component of the gain will be larger than the y-component. As a result, the x-polarized mode will experience the larger gain.

In finishing this section we want to emphasize that semiconductor nanowires, unlike the usual optical fibers, provide a large refractive index contrast with surrounding air. These give rise to well-confined modes even for wires where $\omega R/c \sim 1$. For such small waveguides, however, one needs to account for strong waveguide dispersion.

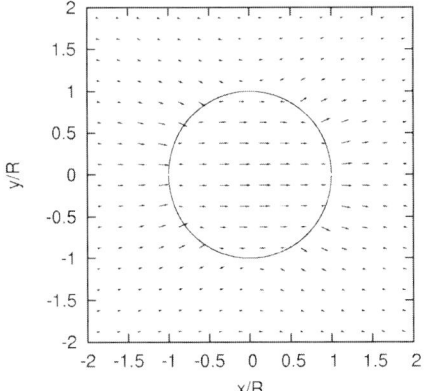

Fig. 21.5 Electric field in the x–y plane for a linearly x-polarized HE_{11} mode when $\varepsilon = 6$ and $\omega R/c = 1$.

21.5.2 Reflection from Facets

According to Eq. (21.1), the reflection coefficient for the guided modes from the nanowire ends determine the threshold gain, i.e., the gain which is required to compensate all losses. In heterostructure lasers, the reflection coefficient of a mode from the facet is given by the Fresnel formula for normal incidence because of the large cross-section of the mode. For a VCSEL, one can use a transfer-matrix approach for plane waves to calculate the transmission through the two distributed Bragg reflectors that form the laser cavity. In the case of nanowires that have open ends, the situation is somewhat different since the cross-sectional dimensions of the nanowire and of the guided mode are comparable to the wavelength. Thus, the guided mode diffracts from the facet giving rise to reflected waves as well as to radiation that leaves the nanowire. Furthermore, reflections can occur into more than one guided mode. For example, EH_{11} will also reflect into HE_{11}. However, for quite small nanowires when only three or four lowest modes exist, such intermode coupling does not take place.

The diffraction of guided modes at waveguide discontinuities has been treated in the past using a combination of analytical and numerical approaches, with applications to various optical and microwave devices [47,48]. As for most diffraction problems, there are no analytical solutions available. We applied the brute-force numerical modeling using the finite-difference time-domain (FDTD) technique [49]. For cylindrical nanowires, the symmetry allows one to reduce the problem to two dimensions. The error is typically determined by the grid and we took ten points per radius. In our approach, we initially excite a wavepacket that consists of only one guided mode. The

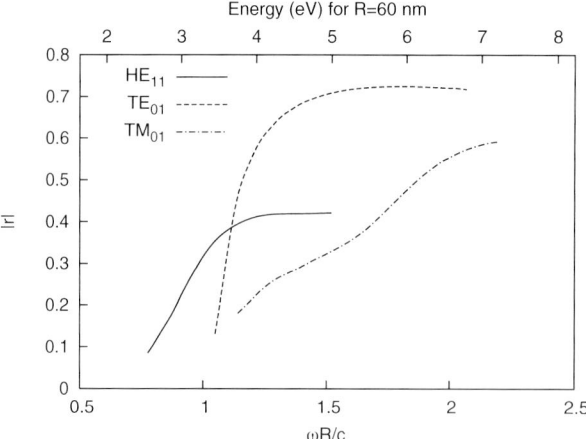

Fig. 21.6 Absolute value of the reflection coefficient for the first three guided modes from the open end of the nanowire.

wavepacket propagates towards the end of the nanowire and is reflected with a reduced amplitude. For sufficiently small nanowires, only one guided mode with specific angular dependence exists, and thus there is only one reflected guided mode. Another way to estimate the reflectivity is to calculate the quality factor of the resonator formed by the two ends and extract the reflectivity from the data [50]. The plane-wave-based transfer-matrix approach can also be applied in order to analyze the reflection properties of isolated nanowires as well as a one-dimensional periodic array of nanowires [51].

Figure 21.6 shows the reflection coefficient for the three lowest guided modes from an open end of the nanowire with dielectric constant $\varepsilon = 6$ [49]. A common feature of the reflection coefficient for the modes is its growth with $\omega R/c$. However, the actual values of the reflectivity depend to a large extent on the mode type. To highlight this even more, we note that the reflection coefficient for normal incidence on the plane semiconductor/air interface would be $r = (\sqrt{\varepsilon} - 1)/(\sqrt{\varepsilon} + 1) = 0.42$ for $\varepsilon = 6$. Only for large $\omega R/c$ does the reflection for the HE_{11} mode reach this value. Interestingly, the reflection for the higher order modes, TE_{01} and TM_{01}, can exceed this value. The analysis of reflectivity at large values of $\omega R/c$ is complicated by the presence of inter-mode coupling and was not performed. Unlike the reflectivity from the open end of the nanowire, the reflectivity from the substrate end of the nanowire is not a monotonic function of $\omega R/c$ but still depends strongly on the mode type [49].

The knowledge of how reflectivities depend on nanowire size allows one to estimate the quality factor of the resonators formed by the nanowire and the threshold gain. Due to the rather small reflection coefficients, the Q-factors are quite small, and long nanowires are needed to compensate for the losses. For example, the length of the free-standing ZnO nanowires in which lasing was demonstrated was around 3 μm. For GaN, lasing was demonstrated only in nanowires ($L \sim$ 30–40 μm) laying on a substrate [17]. One can, however, increase the reflectivity quite dramatically by growing nanowire superlattices at the ends. The superlattices act as distributed Bragg reflectors and are very similar to those used in fibers [52]. One of the limitations of the reflection coefficient appears to be the diffraction of the incident wave into free-space radiation in the air. Despite this, simulations predict relatively high reflectivities (98.5%) of distributed Bragg reflectors based on GaN [53]. Such high reflectivities should lead to lasing in much shorter nanowires.

21.5.3
Far-field Pattern

The angular distribution and polarization of the laser radiation in the far-field zone is among the most important of the laser characteristics. To describe it we consider the situations when the radiation originates from the top end of a free-standing nanowire (see Fig. 21.2). The problem of finding far fields of a free-standing nanowire is identical to that of optical fibers. However, for fibers, one can safely use the scalar (or paraxial wave) approximation, which neglects the longitudinal components of the fields in comparison to the transverse components, see for example, [54]. This approximation is not valid for nanowires and thus, the diffraction problem has to be solved rigorously using Maxwell's equations. We approached this by using the FDTD technique [55].

While the fields near the end can have quite complicated behavior because of diffraction, their distribution in the far-field zone is what we are mostly interested in. In the far-field region, the radiation emitted in the direction specified by the unit vector $\hat{r}$ can be represented as a superposition of two locally plane waves (see Fig. 21.2). If $\hat{\theta}$ and $\hat{\phi}$ are the usual basis vectors orthogonal to $\hat{r}$, one plane wave has its electric field in the $\hat{\theta}$ direction, the other – in the $\hat{\phi}$ direction. Thus, the far-field emission is completely characterized by its polarization and the energy density $S(\hat{r}, \omega)$ emitted in the unit solid angle in the direction $\hat{r}$ per unit time.

The polarization properties of the far fields can be deduced from the symmetry of the modes. We focus on the lowest-order modes (HE_{11}, TM_{01}, and TE_{01}); their dispersion properties are shown in Fig. 21.4 [44]. The transverse modes TM_{01} and TE_{01} produce far fields that are independent of the ϕ-coordinate, i.e., $S(\hat{r}, \omega) = S(\theta, \omega)$. The TM_{01} mode has only E_ρ, E_z, and H_ϕ components. Thus, the far-field radiation produced by the TM_{01} mode will have only an E_θ component; $E_\phi = 0$ everywhere. By the same token, the far fields of the TE_{01} mode will have only E_ϕ component; $E_\theta = 0$ everywhere. The HE_{11} mode has all six components and it is degenerate, i.e., two possible angular dependencies exist. Let us consider one situation when the components E_z, E_ρ, $H_\phi \sim \cos\phi$, while H_z, H_ρ, $E_\phi \sim \sin\phi$. This choice

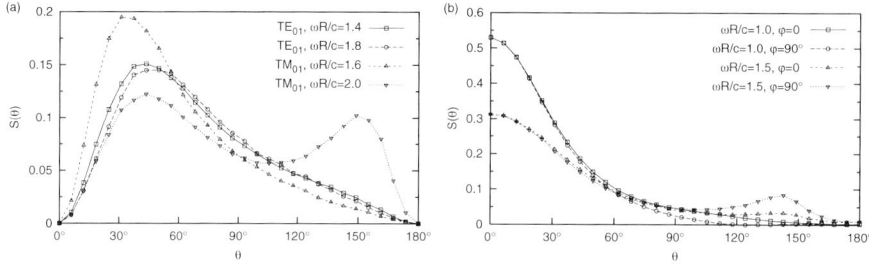

Fig. 21.7 Intensity $S(\hat{r},\omega)$ of the far fields as a function of θ for (a) TE_{01}, TM_{01} and (b) HE_{11} modes.

describes an x-polarized mode (21.9). It follows from the assumed angular dependencies that the far fields behave like $E_\theta \sim \cos\phi$ and $E_\phi \sim \sin\phi$. Thus, unlike the TE_{01} and TM_{01} modes, the far fields of the HE_{11} mode depend not only on θ but also on ϕ. Using the assumed angular dependence for the HE_{11} mode we can write the energy density of the generated far fields as $S(\hat{r},\omega) = \left[S_\theta(\hat{r},\omega)\cos^2\phi + S_\phi(\hat{r},\omega)\sin^2\phi\right]$, where the subscripts θ and ϕ denote polarization of the corresponding plane waves. In a typical experimental situation, the HE_{11} modes with the two possible polarizations are excited, and thus the intensity of the emitted radiation does not have any ϕ-dependence.

Figure 21.7 shows the energy density of the emitted radiation for the TE_{01}, TM_{01}, HE_{11} modes as a function of θ. Focusing on the HE_{11}, the emission at small θ is quite pronounced and is independent of the polarization at $\omega R/c = 1.0$. As $\omega R/c$ increases, the emission in the forward direction decreases while the emission in the backward ($\theta > 90°$) direction increases. Interestingly, the backward scattering occurs also when a guided mode of aluminum coated tapered optical fiber tips interacts with subwavelength circular aperture [56]. The broadening of the emission angle with increase in $\omega R/c$ can be attributed to the decrease of the transverse extent of the mode for larger $\omega R/c$. The difference in the intensities between the θ- and ϕ-polarized components also increases with $\omega R/c$. A strong maximum in the backward direction develops for the θ-polarized component. The most directional emission with a strong maximum at $\theta = 0$ is achieved when $\omega R/c$ is small. This is simply because the guided mode is weakly localized in thin nanowires. However, in this regime the reflection coefficient from the top facet is very small. This leads to high values of the threshold gain. Thicker nanowires can provide lower threshold gain but their emission has a very broad angular distribution.

The emission diagram for the transverse TE_{01} and TM_{01} modes is quite different from that for the HE_{11} mode (21.7). The main difference is the complete absence of emission at $\theta = 0$. This absence follows directly from the symmetry of the transverse modes. Indeed, the far field emitted at $\theta = 0$, if present, must have the electric and magnetic components perpendicular to the nanowire axis. Such components cannot be produced by the transverse modes which are independent of the angle ϕ.

By forming a superlattice at the end of the nanowire [53] and thus increasing the reflectivity, one will also change the far-field pattern for the laser.

21.5.4
Confinement Factors for Anisotropic Nanowires

A given guided mode experiences a gain as it propagates along a waveguide with an active medium. The power carried by the mode $I(z)$ obeys the equation

$$\frac{dI}{dt} = GI \tag{21.11}$$

where G is the modal gain that enters into the threshold condition (21.1). The modal gain is determined by the material gain of Eq. (21.3) and how the guided mode interacts with the active region.

The material properties of the active region are characterized by a conductivity tensor (see Eqs (21.2, 21.5)). We can assume that the tensor has only three components, σ_{xx}, σ_{yy} and σ_{zz}, that describe conductivity for the three orientations of the electric field. We can further write the modal gain as a sum of three components $G = G_x + G_y + G_z$ where each component of the gain depends only on the corresponding component of the conductivity and electric field. It is now convenient to represent each component of gain as a product of two factors, such as

$$G_x = \Gamma_x G_x^0 \tag{21.12}$$

The second factor is defined as the material gain (21.3). The first factor accounts for the geometry of the waveguide and type of the mode. Since both the modal gain G_x and material gain G_x^0 are in principle measurable quantities, Eq. (21.3) simply serves as a definition of the confinement factor Γ_x. Similar to Γ_x, we define the other components Γ_y and Γ_z. The introduction of three confinement factors, $\Gamma_{x,y,z}$, for the waveguide allows us to calculate the modal gain for various orientation of the crystal axis with respect to the guiding direction.

To obtain an analytical formula for the modal gain, we consider a mode that propagates along the waveguide. The expressions that we derive are valid for waveguides of arbitrary cross-section. We write the real electric field at an arbitrary position r as

$$E(r) = A(z)\tilde{E}(\rho)e^{-i\omega t + ik_z z} + \text{c.c.} \tag{21.13}$$

and similar expressions exist for $H(r)$. The wavenumber k_z and ω are related by the dispersion equation for the given mode; c.c. stands for complex conjugate. Calculating the energy transferred from the currents into the mode, we can obtain the modal gain. Dividing it by the material gain we have

$$\Gamma_x = \frac{2}{I_0} \frac{c\sqrt{\varepsilon}}{4\pi} \int_{\text{active}} d\rho \, |\tilde{E}_x|^2 \tag{21.14}$$

where the integral is taken over the active region only. The constant I_0 is the power carried by the mode (21.13) of unit amplitude:

$$I_0 = \frac{c}{4\pi} \int_{\text{mode}} d\rho \, [\tilde{E}(\rho) \times \tilde{H}^*(\rho) + \tilde{E}^*(\rho) \times \tilde{H}(\rho)] \, \hat{z} \tag{21.15}$$

where the integral is taken over the whole extent of the guided mode. The expressions for the other confinement factors, i.e., Γ_y and Γ_z are similar to Eq. (21.14).

It is important to realize that the confinement factor is not the ratio of the power inside the active region to the total power carried by the mode. This was also emphasized earlier [47, 57, 58]. Moreover, the confinement factor (21.3) has no upper limit; in particular, it can exceed unity. This counterintuitive result is a direct consequence of the fact that a large confinement factor describes the situation when the group velocity is very small and therefore, the gain per unit length (modal gain) is large. Similarly, the absorption coefficient in dielectric waveguides can exceed that in the bulk material [59].

The confinement factors for a mode given by (21.14) and (21.15) can also be calculated directly from the dispersion curves without any knowledge of the modal fields. To show this, we assume that the region with gain has a complex dielectric constant with a dominant real part ε and a very small imaginary part ε'' that determines gain or loss. For simplicity, we assume an isotropic material. Let $k_z(\varepsilon)$ be the wavenumber for the given mode at a given frequency calculated for real ε. To account for the presence of ε'' we expand $k_z(\varepsilon)$ into a Taylor series:

$$k_z \approx k_z(\varepsilon) + i\varepsilon'' \frac{\partial k_z}{\partial \varepsilon} \tag{21.16}$$

Thus, the gain experienced by the mode is simply $G = -2\varepsilon''(\partial k_z/\partial \varepsilon)$. Dividing this modal gain by the material gain $G^0 = -(\omega/c)(\varepsilon''/\sqrt{\varepsilon})$, we obtain the confinement factor given by Eq. (21.8). Equations (21.8) and (21.14) give identical results. We note that Eq. (21.8) follows directly from the definition of the confinement factor given by Eq. (21.3) (see also [47]).

Equation (21.8) can be generalized for the case when the gain is anisotropic. This is achieved by taking the derivative with respect to the corresponding

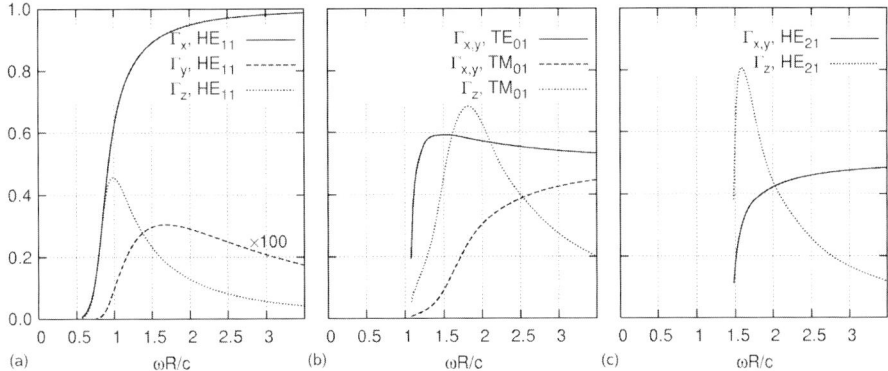

Fig. 21.8 Confinement factors for (a) x-polarized HE_{11}, (b) TE_{01} and TM_{01}, and (c) HE_{21}. The curve for Γ_y in (a) was multiplied by the indicated scale factor.

components of the dielectric constant. The advantage of using (21.8) is that we do not need to know the modal fields; only the change in dispersion with dielectric constant in the gain region is needed. This is particularly useful for simple waveguiding structures with isotropic materials. The main disadvantage is that analytical calculation of the dispersion relations can be quite involved in the case of anisotropic material. However, in many numerical schemes (for example, finite-difference or finite-element schemes [44]), the introduction of an anisotropic dielectric constant does not complicate the problem.

Now let us consider a specific situation of a circular waveguide with dielectric constant ε in air. Our purpose is to describe the gain for modes when the c crystal axis is along z and along x. We first discuss the gain for the two degenerate fundamental modes. If the axis c is along z then the two modes experience the same gain. Further, any mode that is a linear combination of these two will also experience the same gain. This is no longer true when the c-axis is along the x-axis. In this case we must choose an appropriate superposition of the modes that will correspond to the symmetry of the crystal. We will focus on the properties of one mode and choose the x-polarized HE_{11} (see Eq. (21.9)).

The confinement factors for the x-polarized HE_{11} and for TE_{01}, TM_{01} and HE_{21} are shown in Fig. 21.8. The factors were calculated using Eq. (21.14) but the same results follow from the FDTD approach [25]. Focusing on the HE_{11} mode, we can clearly see that the confinement factor Γ_y is practically negligible. The confinement factor Γ_x increases with $\omega R/c$ and reaches unity; Γ_z also initially increases but after reaching a maximum goes to zero for large $\omega R/c$. The transverse mode TE_{01} has only components Γ_x and Γ_y that are equal in magnitude. The transverse mode TM_{01} in addition to equal Γ_x and

Γ_y also has Γ_z. The behavior of the confinement factors for TE_{01}, TM_{01} and HE_{21} are similar. The transverse factors Γ_x and Γ_y increase and eventually reach their asymptotic value, one-half or one. The longitudinal confinement factors Γ_z increase initially but then drop to zero for large $\omega R/c$.

If the material gain is isotropic, the total confinement factor is simply the sum of Γ_x, Γ_y, and Γ_z. Just by looking at the corresponding curves we can clearly see that their sum can easily exceed unity. We also performed calculations of the confinement factors using Eq. (21.8), rather than Eq. (21.14), for the case when the gain is isotropic, and obtained identical results.

To conclude this section, we emphasize that the confinement factors for nanowire lasers can be comparable to unity even for very small nanowires with $\omega R/c \sim 1$.

21.5.5
Spontaneous Emission Factors

One of the advantages of having a large dielectric contrast between the nanowire and surrounding air is the possibility of very efficient coupling of the spontaneous emission into waveguiding modes. This can lead to a reduced threshold current [60] and suppressed relaxation oscillations [61] in lasers.

In general, the spontaneous emission factor, or the β-factor, is the probability that a spontaneously emitted photon will go into the lasing mode. As a result, the β-factor can be represented as a product of two factors [61–63]. The first one (geometrical factor) describes the probability of a photon emitted at a particular frequency being trapped by the lasing mode. The second factor (spectral factor), is the probability of emission at this particular frequency. The second factor is determined by the overlap between the spectrum of spontaneous emission and the cavity spectrum. We will only focus on the geometrical β-factor.

Let us adopt a classical picture of the emission process, i.e., emission is produced by a distribution of dipoles inside a nanowire. The distribution is uniform and each dipole emits independently from the others. Thus, to calculate the emitted powers we add the powers emitted by each dipole. We can explicitly calculate the powers emitted into the guided modes of an infinite nanowire. By doing this we neglect the modification of the spontaneous emission due to formation of the longitudinal cavity modes. This modification will depend on the reflectivity of the end facets of the nanowire that are determined by a specific laser design.

The β-factor depends on the orientation of the optical transitions in the nanowire. To calculate the β-factor for the case when the dipole does not have any preferable direction, we add the ρ-dependent powers emitted into a specific guided mode (P_g) by unit dipoles oriented along the three orthogonal

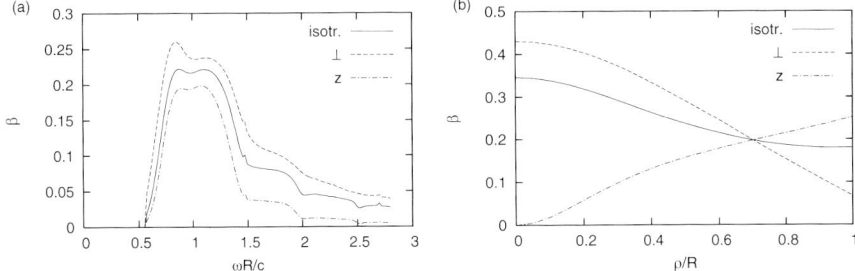

Fig. 21.9 (a) The β-factor for the HE_{11} mode as a function of $\omega R/c$ for uniform distribution of dipoles inside a nanowire with $\varepsilon = 6$. (b) The β-factor for the HE_{11} mode as a function of the dipole position inside the nanowire for $\omega R/c = 1$ and $\varepsilon = 6$.

directions ($\hat{\rho}, \hat{\varphi}, \hat{z}$) and divide by the total emitted power (P_{tot}):

$$\beta_{isotr} = \frac{\int \rho d\rho \left[P_g(\rho, \hat{\rho}) + P_g(\rho, \hat{\varphi}) + P_g(\rho, \hat{z}) \right]}{\int \rho d\rho \left[P_{tot}(\rho, \hat{\rho}) + P_{tot}(\rho, \hat{\varphi}) + P_{tot}(\rho, \hat{z}) \right]} \tag{21.17}$$

where the integration is performed in the interval $0 < \rho < R$. It is important that this β factor is not an average of the spontaneous emission factors over dipole locations and orientations. To describe the dipoles oriented perpendicular to z (denoted by $\perp$), we add the radiation emitted by ρ and φ-oriented dipoles of the same magnitude and multiply by $1/2$. To include the isotropic case, we add the radiation for all three orthogonal orientations ($\hat{\rho}, \hat{\varphi}, \hat{z}$) and multiply by $1/3$. To calculate $\beta_\perp$ we keep only the $\hat{\rho}$ and $\hat{\varphi}$ terms in Eq. (21.17); for β_z we keep only the $\hat{z}$ terms. For the calculations we use the approach presented in [64]. Figure 21.9(a) shows the β-factor for one HE_{11} mode. In this case it makes no difference whether we use the mode of type (21.9) or (21.10). There we added the guided powers emitted both in the $+z$ and $-z$ directions. The same power is emitted into the other degenerate mode.

As one can see from Fig. 21.9(b), the maximum value for the β-factor is about 0.25. The dependence of the β-factor on $\omega R/c$ looks like two peaks merging together. The first peak, at smaller values of $\omega R/c$ is due to the maximum of the guided mode excitation. The second peak is due to a decrease in the total emitted power [64]. Such high values for the β-factor should be compared to the ones for typical semiconductor diode lasers where $\beta \sim 10^{-3}$ [63].

Figure 21.9(b) shows the spontaneous emission factor for a single dipole as a function of its position for $\omega R/c = 1$. The most efficient excitation occurs when the dipole is transverse and in the center of the nanowire. As much as 43% of the energy goes into one of the HE_{11} in this case. This also means that the other guided mode also takes the same fraction of energy. Interestingly, if the dipole transitions have only one polarization, say x, then 86% of the energy will go into the HE$_{11}$ mode with the same polarization (21.9); no energy will go into the orthogonally polarized one. Although the structures in which a GaN wire is grown inside an AlGaN core can be grown [15], the polarization of the optical transitions remains to be studied. If the dipoles move away from the center, the β-factor decreases for the transverse dipoles but increases for the longitudinal one.

The high probabilities of photon emission into lowest-order guided modes in nanowires can be attributed to their strong localization even for rather small values of $\omega R/c$. Despite the increase in mode localization for even larger $\omega R/c$, the appearance of other guided modes reduces the efficiency of dipole emission into the lowest-order modes.

21.6
Modal Gain and Threshold

The knowledge of material gain and confinement factors allows us to estimate the modal gain that can be written as

$$G = \Gamma_x G_x^0 + \Gamma_y G_y^0 + \Gamma_z G_z^0 \tag{21.18}$$

where $\Gamma_{x,y,z}$ are the confinement factors and $G_{x,y,z}^0$ are the material gain coefficients. In addition to this, the knowledge of reflectivities from the facets allows us to find the threshold gain (21.1).

We performed such estimates for a GaN nanowire grown along the c-axis and standing on a sapphire substrate [25]. The solid lines in Fig. 21.10(a) show the threshold gain that was calculated for the HE$_{11}$ and TE$_{01}$ modes using the facet reflectivities. The connected data points show the modal gain for various densities. The modal gain was calculated using the material gain (see Section 21.4) and confinement factors (see Section 21.5.4). The intersections determine the threshold density for given nanowire size. The threshold densities are plotted for the two modes in Fig. 21.10(b). For small radii the fundamental mode HE$_{11}$ will be lasing. However, for nanowires with $R > 70$ nm, the TE$_{01}$ mode will have a smaller threshold, larger modal gain and thus will lase first. The larger modal gain, despite a slightly weaker mode localization, can be attributed to larger transverse confinement factors for the TE$_{01}$ mode. The most straightforward way to determine the actual lasing mode in experiments is to measure the far-field pattern which differs substantially for HE$_{11}$ and TE$_{01}$ modes (see Fig. 21.7).

If the c-axis of the crystal is perpendicular to the nanowire axis, the two HE$_{11}$ modes of type (21.9) will experience different gain. In this case, the mode that is polarized in a direction perpendicular to the c-axis will have the larger gain. The large confinement factors, together with large transverse gain, may

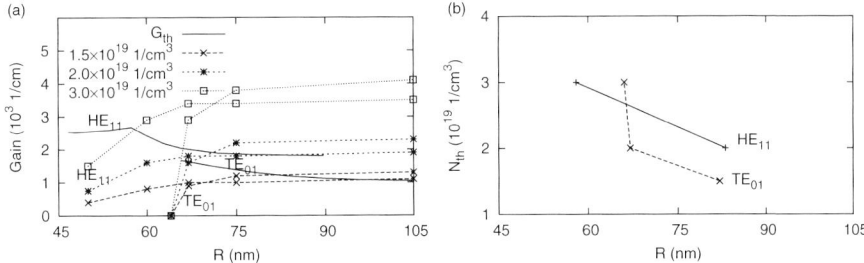

Fig. 21.10 (a) Threshold gain (solid lines) for a $L = 16$ μm long nanowire and maximum values of the modal gain (dashed and dotted lines) for different modes and for different carrier densities. (b) Threshold density for HE_{11} and TE_{01} modes extracted from (a).

explain the lower threshold for nanowires grown along $\langle 11\bar{2}0 \rangle$ direction as compared to those grown along $[0001]$ axis [18].

21.7 Conclusion

We have briefly summarized the experimental progress made in the growth and fabrication of GaN nanowires and have focused on theoretical modeling of GaN nanowire lasers. Our results show that semiconductor nanowires in general, and GaN nanowires in particular, have potential advantages over the standard planar technologies for future optoelectronics.

To conclude, the emergence of GaN nanowires with diameters comparable to, or smaller than, the vacuum UV wavelength opens the possibility of creating LEDs and lasers with unprecedently small sizes. The small number of lattice defects, combined with strong photon confinement and large spontaneous emission factors, are among the most important features that can make nanowires very efficient lasers. Light emission and lasing was demonstrated in single GaN nanowires only recently [15–17, 20] and the potential advantages of nanowires remain to be evaluated carefully. In addition to GaN, virtually any other semiconductor material used in optoelectronics have been demonstrated for wire growth. Because of their small sizes, semiconductor nanowires may eventually serve as elemental building blocks (waveguides, light emitters and detectors) for optoelectronic integrated circuits. Such highly integrated photonic circuits can enable a wide array of applications, such as in computing, communication, and imaging systems.

References

1. H. Akiyama, M. Yoshita, L. N. Pfeiffer, and K. W. West. *J. Phys.: Condens. Matter*, 16:S3549–S3566, 2004.
2. Y. Xia, P. Yang, Y. Sun, Y. Wu, B. Mayers, B. Gates, Y. Yin, F. Kim, and H. Yan. *Adv. Mater.*, 15:353–389, 2003.
3. M. Law, J. Goldberger, and P. Yang. *Annu. Rev. Mater. Res.*, 34:83–122, 2004.
4. Y. Huang and C. M. Lieber. *Pure Appl. Chem.*, 76(12):2051–2068, 2004.
5. Y. Inoue, T. Hoshino, S. Takeda, K. Ishino, A. Ishida, H. Fujiyasu, H. Kominami, H. Mimura, Y. Nakanishi, and S. Sakakibara. *Appl. Phys. Lett.*, 85(12):2340–2342, 2004.
6. Y. S. Park, C. M. Park, D. J. Fu, T. W. Kang, and J. E. Oh. *Appl. Phys. Lett.*, 85:5718–5720, 2004.
7. C. L. Hsiao, L. W. Tu, T. W. Chi, H. W. Seo, Q. Y. Chen, and W. K. Chu. *J. Vac. Sci. Technol. B*, 24:845–851, 2006.
8. N. Thillosen, K. Sebald, H. Hardtdegen, R. Meijers, R. Calarco, S. Montanari, N. Kaluza, J. Gutowski, and H. Lüth. *Nano Lett.*, 6(4):704–708, 2006.
9. W. Han, S. Fan, Q. Li, and Y. Hu. *Science*, 277:1287, 1997.
10. J. Zhu and S. Fan. *J. Mater. Res.*, 14:1175–1177, 1999.
11. C. C. Chen and C. C. Yeh. *Adv. Mater.*, 12:738, 2000.
12. C. C. Chen, C. C. Yeh, C. H. Chen, M. Y. Yu, H. L. Liu, J. J. Wu, K. H. Chen, L. C. Chen, J. Y. Peng, and Y. F. Chen. *J. Am. Chem. Soc.*, 123:2791, 2001.
13. H. Li, A. H. Chin, and M. K. Sunkara. *Adv. Mater.*, 18:216, 2006.
14. J. Goldberger, R. He, Y. Zhang, S. Lee, H. Yan, H. J. Choi, and P. Yang. *Nature*, 422:599–602, 2003.
15. H. J. Choi, J. C. Johnson, R. He, S. K. Lee, F. Kim, P. Pauzauskie, J. Goldberger, R. J. Saykally, and P. Yang. *J. Phys. Chem. B*, 107:8721–8725, 2003.
16. F. Qian, Y. Li, S. Gradečak, D. Wang, C. J. Barrelet, and C. M. Lieber. *Nano Lett.*, 4:1975–1979, 2004.
17. J. C. Johnson, H. J. Choi, R. D. Schaller K. P. Knutsen, P. Yang, and R. J. Saykally. *Nature Mater.*, 1:106–110, 2002.
18. S. Gradečak, F. Qian, Y. Li, H. G. Park, and C. M. Lieber. *Appl. Phys. Lett.*, 87:173111, 2005.
19. P. J. Pauzauskie, D. J. Sirbuly, and P. Yang. *Phys. Rev. Lett.*, 96:143903, 2006.
20. H. M. Kim, T. W. Kang, and K. S. Chung. *Adv. Mater.*, 15:567–569, 2003.
21. Z. Zhong, F. Qian, D. Wang, and C. M. Lieber. *Nano Lett.*, 3:343–346, 2003.
22. H. M. Kim, Y. H. Cho, H. Lee, S. Kim, S. R. Ryu, D. Y. Kim, T. W. Kang, and K. S. Chung. *Nano Lett.*, 4:1059–1062, 2004.
23. H. Y. Peng, N. Wang, X.T. Zhou, Y. F. Zheng, C. S. Lee, and S. T. Lee. *Chem. Phys. Lett.*, 359:241–245, 2002.
24. T. Kuykendall, P. J. Pauzauskie, Y. Zhang, J. Goldberger, D. Sirbuly, J. Denlinger, and P. Yang. *Nature Mater.*, 3:524, 2004.
25. A. V. Maslov and C. Z. Ning. *IEEE J. Quantum Electron.*, 40(10):1389–1397, 2004.
26. W. W. Chow, A. Knorr, and S. W. Koch. *Appl. Phys. Lett.*, 67:754–756, 1995.
27. K. Domen, K. Kondo, A. Kuramata, and T. Tanahashi. *Appl. Phys. Lett.*, 69:94–96, 1996.
28. J. B. Jeon, B. C. Lee, Yu. M. Sirenko, K. W. Kim, and M. A. Littlejohn. *J. Appl. Phys.*, 82:386–391, 1997.
29. S. L. Chuang and C. S. Chang. *Phys. Rev. B*, 54:2491–2504, 1996.
30. O. Madelung, editor. *Semiconductors – Basic Data*. Springer-Verlag, 1996.
31. G. Gilat and L. J. Raubenheimer. *Phys. Rev. B*, 144:390–395, 1966.
32. R. Dingle, D. D. Sell, S. E. Stokowski, and M. Ilegems. *Phys. Rev. B*, 4:1211–1218, 1971.
33. K. Domen, K. Horino, A. Kuramata, and T. Tanahashi. *Appl. Phys. Lett.*, 71:1996–1998, 1997.
34. B. Rau, P. Waltereit, O. Brandt, M. Ramsteiner, K. H. Ploog, J. Puls, and F. Henneberger. *Appl. Phys. Lett.*, 77:3343–3345, 2000.
35. K. B. Nam, J. Li, M. L. Nakarmi, J. Y. Lin, and H. X. Jiang. *Appl. Phys. Lett.*, 84:5264–5266, 2004.

36. E. Kuokstis, C. Q. Chen, J. W. Yang, M. Shatalov, M. E. Gaevski, V. Adivarahan, and M. Asif Khan. *Appl. Phys. Lett.*, 84:2998–3000, 2004.
37. P. C. Sercel and K. J. Vahala. *Phys. Rev. B*, 42:3690–3710, 1990.
38. P. C. Sercel and K. J. Vahala. *Phys. Rev. B*, 44:5681–5691, 1991.
39. A. V. Maslov and C. Z. Ning. *Phys. Rev. B*, 72:125319, 2005.
40. X. W. Zhang and J. B. Xia. *J. Phys.: Condens. Matter*, 18:3107–3115, 2006.
41. J. Wang, M. S. Gudiksen, X. Duan, Y. Cui, and C. M. Lieber. *Science*, 291:1455, 2001.
42. J. Qi, A. M. Belcher, and J. M. White. *Appl. Phys. Lett.*, 82(16):2616, 2003.
43. H. E. Ruda and A. Shik. *Phys. Rev. B*, 72:115308, 2005.
44. R. E. Collin. *Field Theory of Guided Waves*. IEEE Press, New York, NY, 1990.
45. A. Yariv. *Optical Electronics*. CBS College Publishing, New York, 1985.
46. E. Ejder. *Phys. Status Solidi (a)*, 6:K39, 1971.
47. C. Vassallo. *Optical Waveguide Concepts*. Elsevier, 1991.
48. P. Gelin, S. Toutain, P. Kennis, and J. Citerne. *IEEE Trans. Microwave Theory Tech.*, 29(7):712–719, 1981.
49. A. V. Maslov and C. Z. Ning. *Appl. Phys. Lett.*, 83:1237–1239, 2003.
50. M. Q. Wang, Y. Z. Huang, Q. Chen, and Z. P. Cai. *IEEE J. Quantum Electron.*, 42:146–151, 2006.
51. Z. Y. Li and K. M. Ho. *Phys. Rev. B*, 71:045315, 2005.
52. T. Erdogan. Fiber grating spectra. *J. Lightwave Technol.*, 15(8):1277–1294, 1997.
53. L. Chen and E. Towe. *Appl. Phys. Lett.*, 87:103111, 2005.
54. W. Freude and A. Sharma. *J. Lightwave Technol.*, 3(3):628–634, 1985.
55. A. V. Maslov and C. Z. Ning. *Optics Lett.*, 29:572–574, 2004.
56. C. Obermüller and K. Karrai. *Appl. Phys. Lett.*, 67:3408–3410, 1995.
57. T. D. Visser, H. Blok, B. Demeulenaere, and D. Lenstra. *IEEE J. Quantum Electron.*, 33(10):1763–1766, 1997.
58. Y. Z. Huang. In *Proc. IEE Optoelectron.*, volume 148, pages 131–133, June 2001.
59. V. F. Vzyatyshev. *Dielectric Waveguides [in Russian]*. Sov. Radio, 1970.
60. Y. Yamamoto, S. Inoue, G. Björk, H. Heitmann, and F. Matinaga. *Semiconductor Lasers I*, chapter 'Quantum optics effects in semiconductor lasers'. Academic Press, 1999.
61. Y. Suematsu and K. Furuya. *Trans. Inst. Electron. Commun. Eng. Japan*, 60:467–472, 1977.
62. K. Petermann. *IEEE J. Quantum Electron.*, 15:566–570, 1979.
63. T. D. Cassidy. *J. Opt. Soc. Am. B.*, 8:747–752, 1991.
64. A. V. Maslov, M. I. Bakunov, and C. Z. Ning. *J. Appl. Phys.*, 99:024314, 2006.

Index

δ-doping 257, 262
(0001) crystal orientation 169
(10$\bar{1}$0) crystal orientation 170

a

absorption 117, 238, 266
– coefficient 96
– of phosphors 334
– saturation 272
acoustic phonon 74
Adachi's model 290
alloy
– ordering 60
– scattering 8
anisotropy 73, 175
– of material gain 471
anti-guide mode 365
APSYS 280–282, 299
Arora model 78
ARPACK 125
ATLAS 216
Auger recombination 146, 152, 164

b

band diagram 359, 440
band nonparabolicity 101, 240, 245
band offset 35, 292
band structure 14, 70, 146, 418, 429
band-filling 101
BandEng 216, 217
bandgap renormalization 150, 157, 176, 195, 419
basal plane 77
beam quality factor 367
Berry's phase 49
bimolecular recombination coefficient 427
Bloch equations 149, 236, 418
Boltzmann Transport Equation 70
Born charges 53, 57
bound excitonic states 102
bowing 18, 58–60, 63, 107

breakdown voltage 227, 228
Brillouin zone 71
built-in field 198, 235, 242, 243, 258, 432

c

carrier
– leakage 312, 314, 318, 442
– lifetime 152, 163, 165, 427
– overflow 292, 359
carrier–carrier scattering 193
carrier–phonon interaction 200
Caughey–Thomas model 77, 225, 428
chirp parameter 266, 269
chromaticity 330, 331, 334, 350
color coordinates 345, 348
color perception 327, 330, 331
complex refractive index 96, 260, 266, 448
compositional fluctuations 310, 312, 323, 417, 419
conductivity tensor 472, 473
confinement factor 410, 483
Coulomb interaction 150, 157, 164, 193
coupled waveguides 382
critical
– field 76
– point 96
crystal
– field 17, 97, 285, 429
– orientation 169
current
– confinement 438
– continuity equation 354
– spreading 409

d

Debye temperature 6
density functional perturbation theory 57, 58
density functional theory 51
density matrix 237
dephasing time 237, 240, 241
depolarization 121, 267

DFPT, *see* density functional perturbation theory
DFT, *see* density functional theory
dielectric
– constant   54, 56–58, 64, 65
– function   96
– tensor   96
dislocation   73, 243, 247, 248
distributed Bragg reflector   423, 452
drift velocity   76
drift-diffusion model   228, 287, 305, 426
Drude
– expression   264
– model   266

e
edge-emitting laser diode   353, 381, 412
effective medium   265
effective-mass approximation   13, 118, 171
Einstein relation   426
elastic constants   13, 52, 54, 75
electroabsorption modulator   253
electron
– barrier   439
– blocking layer   280, 281, 293, 296–299
– leakage   443
– mobility   78, 224, 225
– velocity   218, 225
electron–electron scattering   150, 240
electron–phonon scattering   74, 150
ellipsometry   99
energy balance model   228
Euler angles   172
exchange correlation   51, 118
exciton binding energy   102, 155, 157
exciton continuum   110
excitonic effect   101, 121
excitonic resonance   150, 157–159

f
far-field pattern   365, 381, 481
FCA, *see* free-carrier absorption
Fermi Golden Rule   72
ferroelectrics   49, 53
fiber-optical communication   253, 255
field plate   228
finite-difference time-domain   235, 244, 249, 479, 481
first-order mode   465
free-carrier absorption   265
Fröhlich formulation   73
fundamenta mode   460
fundamental mode   367, 408, 436, 449, 464, 476

g
gain   145
– guiding   447, 449, 463, 464
– spectra   405
gain guiding   408
gain spectra   436, 474
garnet   332
Gaussian lineshape function   175
graded barriers   138
guided modes   365, 475

h
Hakki–Paoli method   410, 414, 418
HE_{21} first-order mode   460
heat flux equation   288
high electron mobility transistor   214
homogeneous
– broadening   405, 419
– linewidth   238, 240, 246
hybrid LED   321

i
Implicitly Restarted Arnoldi Method   125
index
– guided laser diode   408
– of refraction   96
indium-free LED   320, 321
InGaP   50, 59
inhomogeneous broadening   155, 158, 240, 245, 246, 405, 419
injection efficiency   314, 318
inner-stripe laser   355
internal quantum efficiency   308, 310, 312, 314, 321
intersubband
– absorption   119, 264
– relaxation time   235, 238, 245, 249
– scattering   239, 240
intraband scattering relaxation time   290
ionized acceptor concentration   354
ionized impurity scattering   74, 241

l
Lagrange polynomials   124
lateral mode   410
leaky quasi-modes   413
light-emitting diode   279, 303, 327
longitudinal optical phonon   200, 235, 239–241
Lorentzian lineshape function   290, 434
LP_{11} scalar mode   460
luminescence
– conversion   328, 329
– of phosphors   334
Luttinger parameters   171

m

M^2 factor 367
many-body effects 121, 170, 194, 405
maximum operation temperature 374
Maxwell equations 337, 447, 448, 455
mobility 69, 428
modal effective index 386
modal gain 381, 488
mode spacing 476
Modern Theory of Polarization 49
Monte Carlo methods 71, 227
MTP 49, 51, 54–59
multi-waveguide system 385

n

N-polar LED 318, 319
nanowire 467
near-field microscope 408, 414
negative differential modal gain 397
Newton–Richardson approach 127
non-Markovian gain model 175
nonparabolicity 29, 101, 118, 265
nonperiodicity 138
nonradiative recombination 292, 307, 420
nonuniform
– carrier distribution 296
– radial current injection 442, 447
normal modes 386

o

optical
– anisotropy 96
– loss 405, 417
– momentum matrix elements 174
– phonons 73
– standing wave 339, 453
– switches 235, 249
oscillator strength 98, 257, 264, 267
oxide stripe laser diode 408

p

particle scattering 329, 335, 336
phonon bottleneck 201
phononic bandgap shift 155
phosphor materials for LEDs 327
photoluminescence 151
piezoelectric
– coefficient 50, 53, 61, 66
– components of polarization 61
– constant 173
– effect 235, 242
– moduli 53
– nonlinearity 60
– polarization 50, 52, 55, 60, 61, 66, 148, 164, 282, 286, 287, 432
plane-wave expansion method 405

plasma effect 259
Poisson's equation 262, 263, 282, 287, 354, 427
polarization 49, 218–220, 222, 261, 418
– charge screening 439
– effect on energy bands 439
– effect on threshold current 440
polaron 201
predictor–corrector approach 128
principle of stationary action 125
pump and probe responses 248
pyroelectrics 49, 50

q

QCSE, see quantum confined Stark Effect
quantum confined Stark Effect 157, 253, 316
quantum-kinetic equation 196
quasi-cubic model 97
quasi-saturation 226
quaternary alloys 8, 34, 58, 62, 63, 280

r

radial waveguiding 447, 451, 460, 463, 465
rate equations 236
– for photons 356
ray tracing 290, 344, 345, 348
relaxation time 69, 239
– approximation 194, 265
resonant mode coupling 381
rotation matrix 172

s

saturation velocity 76, 218
scattering
– rate 70, 195, 239
– theory 335
– time 235, 241, 265
Schrödinger equation 120, 263
Schrödinger solver 216
Schrödinger–Poisson problem 123
semiconductor Bloch equations 149
semiconductor luminescence equations 151
Shockley–Read–Hall recombination 282, 306, 427
single-photon detectors 117
Special Quasi-random Structure (SQS) 59
specific heat 6
spectral function 204
spectral modulation 395
spin–orbit energy 16, 97, 124, 285, 309
spontaneous emission 145, 146, 151, 161, 164, 166, 177, 288, 290, 427, 486

spontaneous polarization 49, 50, 53, 55, 58, 61, 148, 164, 174, 235, 242
standing wave 339, 437, 451, 455, 457
step quantum well 257
strain effect on band energy 430
strain tensors 172
stress 17, 130
substrate modes 414
superinjection 310
supermodes 386
surface state 222
surface trap 219, 222, 223
switching energy 238, 248, 249
switching pulse energy 241, 248, 250

t
TE_{01} first-order mode 460
temperature focusing 463
thermal
– analysis 357
– conductivity 6, 273, 280, 288, 291, 357
– focusing 449, 451, 460–464
– resistance 9, 370
TM_{01} first-order mode 460
transfer matrix technique 123
transistor 213

transparency carrier density 355
transverse-fundamental-mode 451
tunnel junction 452, 453, 460, 464
tunnel radiative recombination 322

u
under-relaxation method 127
uniformity of current injection 449

v
Van Hove singularities 96
vector wave equation 448
vertical-cavity surface-emitting laser 423, 447

w
wave equation 356, 385, 406, 436, 448
waveguide 235, 245–248, 250, 381, 405, 410, 436, 470
white light generation in LEDs 327, 331, 332, 346

y
YAG, *see* Yttrium Aluminum Garnet
Yttrium Aluminum Garnet 328

Related Titles

Ng, K. K.

Complete Guide to Semiconductor Devices

764 pages
2002
Hardcover
ISBN-13: 978-0-471-20240-0
ISBN-10: 0-471-20240-1

Ruterana, P., Albrecht, M., Neugebauer, J. (eds.)

Nitride Semiconductors

Handbook on Materials and Devices

686 pages with 387 figures and 37 tables
2003
Hardcover
ISBN-13: 978-3-527-40387-5
ISBN-10: 3-527-40387-6

Deveaud-Pledran, B.

The Physics of Semiconductor Microcavities

From Fundamentals to Nanoscale Devices

approx. 340 pages with 200 figures
2006
Hardcover
ISBN-13: 978-3-527-40561-9
ISBN-10: 3-527-40561-5

Bechstedt, F., Meyer, B. K., Stutzmann, M. (eds.)

Group III-Nitrides and Their Heterostructures: Growth, Characterization and Applications

approx. 390 pages
2003
Hardcover
ISBN-13: 978-3-527-40475-9
ISBN-10: 3-527-40475-9